STUDY GUIDE TO ACCCOMP

Anatomy &Physiology

THIRD EDITION

Seeley

Stephens

Tate

Philip Tate
Phoenix College

James Kennedy
Phoenix College

Rod Seeley
Idaho State University

Original Illustrations by D. Michael Dick

Mosby

St. Louis Baltimore Boston Carlsbad Chicago Naples New York Philadelphia Portland
London Madrid Mexico City Singapore Sydney Tokyo Toronto Wiesbaden

Mosby

Dedicated to Publishing Excellence

A Times Mirror
Company

Editor-in-Chief: James M. Smith
Editor: Robert J. Callanan
Developmental Editor: Jean Babrick
Project Manager: Linda McKinley

Printed in the United States of America

Printing/binding by Plus Communications

Mosby–Year Book, Inc.
11830 Westline Industrial Drive
St. Louis, Missouri 63146

ISBN 0-8151-8823-4
24133

95 96 97 98 99 / 9 8 7 6 5 4 3 2 1

PREFACE

To the Student

This study guide is designed to accompany *Anatomy and Physiology* by Seeley, Stephens, and Tate. Each chapter in the study guide, and the order of topics within the chapter, corresponds to a chapter in the text. This makes it possible for you to study systematically and also makes it easier for you to find or to review information. Read the corresponding material in the text before you use this study guide. It is designed to help you understand and master the subject of anatomy and physiology.

FEATURES

Focus

Each chapter begins with a focus statement, briefly reviewing some of the main points of the text chapter. This is not a chapter summary; you will find that in the text. The chapter summary is useful. Reading it should become a routine part of your study habits. In the study guide, the focus statement sets the stage by reminding you of the major concepts you should have learned by reading the textbook.

Content Learning Activity

This section of the study guide contains a variety of exercises including matching, completion, ordering, and labeling activities, arranged by order of topics in the text. Each part begins with a quotation from the text or a statement that identifies the subject to be covered. Occasionally, you will find a "bulletin" statement describing important information. Just because that information is not a question does not mean it is unimportant. Quite the contrary. The "bulletin" statements are added to the study guide because they will help you to understand the material, so pay attention to them.

The content learning activity is not a test; it is a strategy to help you learn. Don't guess! If you learn something incorrectly it

is difficult to relearn it correctly. Use the textbook or your lecture notes for help whenever you are not sure of an answer. The emphasis here is on learning the content, hence the name of this section. The content questions cover the material in the same sequence as it is presented in the text. Learning the material in this order makes it easier to relate pieces of information to each other, and makes it easier to remember the information.

After completing the exercises check your answers against the answer key. If you missed the correct answer to a question, check the text to make sure you now understand the correct answer. Before going on to the next section of the study guide, review this section to be sure you understand and remember the content. Cover the answers you have written with a piece of paper and mentally answer each exercise once more as you review.

Quick Recall

The quick recall section asks you to list, name, or briefly describe some aspect of the chapter's content. Although this section can be completed rapidly, do not confine yourself to quickly writing down the answers. As you complete each quick recall question, use it to trigger more information in your mind. For example, if the quick recall question asks you to name the two major regions of the body, do that, then think of their definition, what their various sub-parts are, visualize them, and so on. This section should be enjoyable and satisfying because it will demonstrate that you have learned the basic information about the material. Verify your answers against the answer key.

Mastery Learning Activity

The mastery learning activity lets you see what you have learned and if you can use that information. It consists of multiple choice questions that are similar to the questions on the exams you will take for a grade, so it is really a "practice" test and should be taken as a test. However, don't guess. This "practice" test is also a learning tool. If you don't know the answer for sure, admit it and then find out what the correct answer is. Some of the questions require recall of information. Others may state the information somewhat differently than the way it appeared in either the text or study guide. This is entirely fair, because in real life you must be able to recognize the information no matter how it is reworded, and you should even be able to express the information in your own words. Another goal of this section is to make you think about the relationship between different bits of information or concepts, so some of the questions are more complex than those requiring only recall. Finally, some questions in this section ask you to use what you have learned to solve new problems.

After you have answered these questions, check the answer key. In addition to the answers, there is a detailed explanation of why a particular answer was correct. Sometimes an explanation of why a choice is incorrect is also given. These explanations are provided because this section is more difficult than the preceding sections. Make sure you understand why each answer is correct. Check the textbook, ask another student, talk with your instructor, but make sure you know. The mastery learning activity will show you the areas that you need to concentrate on further. Use it to improve your understanding of anatomy and physiology.

The format of this section allows you to write the answers to the questions beside each question. If you cover the answers, you can retake the test. Don't be satisfied until you get at least 90% of the questions correct.

Final Challenges

This section of the study guide corresponds to the concept questions at the end of each chapter in the text. These questions challenge you to apply information to new situations, analyze data and come to conclusions, synthesize solutions, and evaluate problems. Some of the problem solving questions in the mastery learning section have given you practice for the questions in this section. In addition, explanations are provided to help you see

how to go about solving questions of this type. Even though explanations are given, write down your answers to the questions on a separate piece of paper. Writing is a good way to organize your thoughts and most of us can benefit from practice in writing. A good way to determine if you have communicated your thoughts effectively is to have another student read your answers and see if they make sense.

The questions contain useful information, but they are not designed primarily to help you learn specific information. Rather, they emphasize the thought processes necessary to solve problems. If all you do is read the question and quickly look up the answer, you have defeated the purpose of this section. Think about the questions and develop your reasoning skills. Long after you have forgotten a particular bit of information, these skills will be useful, not only for anatomy and physiology related problems, but for many other aspects of your life as well. We hope that you not only see the benefit of possessing problem solving skills, but will come to appreciated that solving problems is fun!

A Final Thought

Good luck with all aspects of the anatomy and physiology course you are about to begin. We hope that the study guide makes things a little easier and a little clearer for you. When you have completed the course we are confident that you will be proud of what you have accomplished. Just remember to enjoy the learning process as you go along.

Philip Tate
James Kennedy
Rodney Seeley

ACKNOWLEDGMENTS

The development and production of this study guide involved much more than the work of the authors and we gratefully acknowledge the assistance of the many other individuals involved. We wish to thank our families for their support, for their encouragement and understanding from the beginning to the end of the project helped to sustain our efforts. Erica Michaels contributed to the development of the instructional design for the study guide. We also wish to acknowledge Robert Callanan and Jean Babrick at Mosby-Year Book College Publishing for their assistance in making the study guide a unique and valuable asset for the student. Our thanks to everybody involved with the artwork taken from the textbook, and to D. Michael Dick for his original study guide illustrations. We would like to thank Minyon Bond for the original poetry. We also wish to recognize the contribution of Sue Pepe to the design and layout of the study guide. To the reviewer listed below, our gratitude for her thorough and thoughtful critique, which resulted in a significant improvement of the study guide. Thank you.

Anita Herl-Peterson
Phoenix College

Contents

1   The Human Organism

FOCUS: The human organism is often examined at seven structural levels: molecule, organelle, cell, tissue, organ, organ system, and the organism. Anatomy examines the structure of the human organism, and physiology investigates its processes. Structure and process interact to maintain homeostasis through negative-feedback mechanisms. Human anatomy can be examined by sectioning the body into three planes: sagittal, transverse, and frontal. The trunk contains three cavities: thoracic, abdominal, and pelvic. These cavities and the organs they contain are lined with serous membranes.

CONTENT LEARNING ACTIVITY

Anatomy and Physiology

❝Anatomy is the scientific discipline that investigates the body's structure.❞

Match these terms with the correct statement or definition:

    Anatomic imaging   Regional anatomy
    Cytology   Surface anatomy
    Histology   Systemic anatomy

_____ 1. Study of the structural features of cells.

_____ 2. Study of tissues.

_____ 3. Study of the body by systems (a group of structures that have one or more common functions).

_____ 4. Study of the body's organization by areas.

_____ 5. Use of external landmarks such as bony projections to locate deeper structures.

_____ 6. Involves the use of x-rays, ultrasound, nuclear magnetic resonance, and other technologies to create pictures of internal structures.

☞   Physiology is the scientific discipline that deals with the vital processes or functions of living things.

Structural and Functional Organization

The body can be considered conceptually at seven structural levels.

A. Match these terms with the correct statement or definition:

Cell
Molecule
Organ
Organelle
Organism
Organ system
Tissue

_____ 1. Structure within a cell that performs one or more specific functions.

_____ 2. Two or more tissue types that perform one or more common functions.

_____ 3. Basic living unit of all plants and animals.

_____ 4. Group of organs classified as a unit because of a common set of functions.

_____ 5. Group of cells with similar structure and function, together with the extracellular substances located between the cells.

B. Match these terms with the correct statement or definition:

Cardiovascular
Digestive
Endocrine
Integumentary
Lymphatic
Muscular
Nervous
Reproductive
Respiratory
Skeletal
Urinary

_____ 1. Organ system that consists of skin, hair, and nails; protects and prevents water loss.

_____ 2. Organ system that consists of the brain, spinal cord, and nerves; detects sensation and controls movements.

_____ 3. Organ system that consists of the lungs; exchanges gases between blood and the air.

_____ 4. Organ system that consists of the kidneys and urinary bladder; removes waste products from the circulatory system.

_____ 5. Organ system that consists of the mouth, pharynx, esophagus, stomach, and intestines; breaks down and absorbs nutrients.

_____ 6. Organ system that consists of bones and cartilage; protects and supports the body, and produces blood cells.

_____ 7. Organ system that consists of the heart, blood vessels, and blood; transports nutrients, wastes, and gases.

_____ 8. Organ system that consists of glands such as the pituitary and the thyroid gland; a major regulatory system.

_____ 9. Organ system that consists of muscles attached to the skeleton; allows body movement, maintains posture, and produces body heat.

The Human Organism

66 *Much of our knowledge about humans has come from studying other organisms.* **99**

Match these terms with the correct statement or definition:

Development	Morphogenesis
Differentiation	Organization
Growth	Reproduction
Metabolism	Responsiveness

_____ 1. Ability to use energy to perform vital functions.

_____ 2. Ability to sense changes in the environment and make adjustments that help to maintain life.

_____ 3. Ability of cells to increase in size or number.

_____ 4. The changes an organism undergoes through time.

_____ 5. Changes in the shape of tissues, organs, and the entire organism.

Homeostasis

66 *Homeostasis is the existence and maintenance of a relatively constant environment* **99** *within the body.*

A. Match these terms with the correct statement or definition:

Negative feedback
Positive feedback

_____ 1. Maintains homeostasis by reducing or resisting any deviation from an ideal normal value.

_____ 2. Medical therapy is often designed to help this type of feedback.

_____ 3. When a deviation from a normal value occurs, the response is to increase the deviation.

_____ 4. Causes heart rate to increase in response to a decrease in blood pressure.

_____ 5. Maintains an elevated blood pressure during exercise.

_____ 6. Causes a decrease in blood pressure as a result of losing blood.

_____ 7. Increases the strength of uterine contractions during delivery.

3

B. Match these terms with the correct statement or definition:

Control center   Set point
Effector   Stimulus
Receptor   Variable
Response

_____ 1. A condition, such as temperature, that can change.

_____ 2. Ideal, normal value maintained by a homeostasis.

_____ 3. Functions to monitor the value of a variable.

_____ 4. Establishes the set point.

_____ 5. Functions to change the value of a variable.

_____ 6. A deviation from a set point.

_____ 7. Returns a variable back toward the set point; produced by an effector.

Directional Terms

" *Directional terms always refer to the body in the anatomical position.* "

Match these terms with the correct statement or definition:

Anterior   Lateral
Caudal   Medial
Cephalic   Posterior
Deep   Proximal
Distal   Superficial
Dorsal   Superior
Inferior   Ventral

_____ 1. Lower than or toward the tail (two terms).

_____ 2. Higher than or toward the head (two terms).

_____ 3. Toward the front or toward the belly (two terms).

_____ 4. Toward the back (of the body) (two terms).

_____ 5. Farther than another structure from the point of attachment to the trunk.

_____ 6. Away from the midline.

_____ 7. Away from the surface.

Planes

"A plane is an imaginary flat surface passing through the body or an organ."

A. Match these terms with the correct statement or definition:

Cross (transverse) section
Frontal (coronal) plane
Longitudinal section

Oblique section
Sagittal plane
Transverse plane

_____ 1. Runs vertically through the body and divides the body into right and left portions.

_____ 2. Runs vertically through the body and divides the body into anterior and posterior portions.

_____ 3. Divides the body into superior and inferior portions and runs parallel to the surface of the ground.

_____ 4. A cut through the long axis of an organ.

_____ 5. A right angle cut across the long axis of an organ.

 A midsagittal section divides the body into equal right and left halves. A parasagittal section divides the body into right and left parts to one side of the midline.

B. Match these terms with the correct planes labeled in Figure 1-1:

Frontal (coronal) plane
Midsagittal plane
Transverse plane

1. _____

2. _____

3. _____

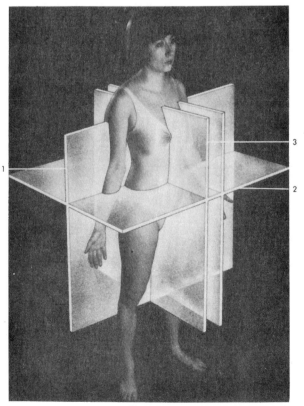

Figure 1-1

C. Match these terms with the correct section in Figure 1-2:

Longitudinal section
Oblique section
Cross (transverse) section

1. _____

2. _____

3. _____

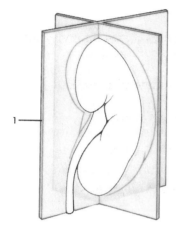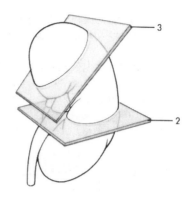

Figure 1-2

Body Regions

❝*The body commonly is divided into several regions.***❞**

Using the terms provided, complete these statements:

Abdomen   Pelvis
Arm   Thigh
Forearm   Thorax
Leg   Upper limb
Lower limb

The _(1)_ consists of the arm, forearm, wrist, and hand.
The _(2)_ extends from the shoulder to the elbow, and the
_(3)_ extends from the elbow to the wrist. The _(4)_ consists
of the thigh, leg, ankle, and wrist. The _(5)_ extends from
the hip to the knee, and the leg _(6)_ extends from the knee
to the ankle. The trunk consists of the _(7)_ , _(8)_ , and _(9)_ .

1. _____

2. _____

3. _____

4. _____

5. _____

6. _____

7. _____

8. _____

9. _____

☞ The abdominal region can be subdivided by imaginary lines into four quadrants or nine regions. The quadrants or regions can be used as reference points for locating underlying organs.

Body Cavities

"_The body contains several large trunk cavities that do not open to the exterior of the body._**"**

A. Match these terms with the
correct statement or definition:

Abdominal cavity
Pelvic cavity
Thoracic cavity

_____ 1. Cavity surrounded by the rib cage, bounded inferiorly by the diaphragm, and divided by the mediastinum.

_____ 2. Cavity bounded primarily by the abdominal muscles.

_____ 3. Cavity containing the stomach and kidneys.

_____ 4. Small space enclosed by the bones of the pelvis.

 There is no physical separation between the abdominal and pelvic cavities, which sometimes are called the abdominopelvic cavity.

B. Match these terms with the
correct statement or definition:

Mesentery   Pleural membrane
Parietal serous membrane   Retroperitoneal
Pericardial membrane   Visceral serous membrane
Peritoneal membrane

_____ 1. Portion of a serous membrane in contact with an organ.

_____ 2. Serous membrane that surrounds the lungs and lines the thoracic cavity.

_____ 3. Serous membrane that surrounds the heart and lines a connective tissue sac.

_____ 4. Serous membrane that lines the abdominal and pelvic cavities and their organs.

_____ 5. Double-layered serous membrane that anchors some abdominal organs to the body wall.

_____ 6. Location of organs covered only by parietal peritoneum.

 A potential space or cavity is located between the visceral and parietal serous membranes. The cavity is filled with serous fluid that reduces friction between the visceral and parietal serous membranes.

C. Match these terms with the correct parts labeled
in Figure 1-3:

Abdominal cavity
Abdominopelvic cavity
Diaphragm
Mediastinum
Pelvic cavity
Pericardial cavity
Pleural cavity
Thoracic cavity

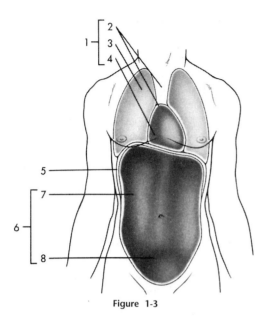

Figure 1-3

1. _____

2. _____

3. _____

4. _____

5. _____

6. _____

7. _____

8. _____

1. From smallest to largest, list the seven structural levels of the human body.

2. List the four primary tissue types.

3. Name the three components of many negative-feedback mechanisms.

4. Describe the anatomical position.

5. List the three major planes used to cut the human body and the three sections used to cut an organ.

6. List the three trunk cavities of the human body.

7. Name the three serous membranes lining the trunk cavities and their organs.

8. List four retroperitoneal organs.

Place the letter corresponding to the correct answer in the space provided.

_____ 1. Physiology
   a. deals with the processes or functions of living things.
   b. is the scientific discipline that investigates the body's structure.
   c. is concerned with organisms and does not deal with different levels of organization such as cells and systems.
   d. recognizes the unchanging (as opposed to dynamic) nature of living things.

_____ 2. An organ is
   a. a specialized structure within a cell that carries out a specific function.
   b. at a lower level of organization than a cell.
   c. two or more tissues that perform a specific function.
   d. a group of cells that perform a specific function.

_____ 3. The systems that are most important in the regulation or control of the other systems of the body are the
   a. circulatory and muscular systems.
   b. circulatory and endocrine systems.
   c. nervous and muscular systems.
   d. nervous and endocrine systems.

_____ 4. Negative-feedback mechanisms
   a. make deviations from normal smaller.
   b. maintain homeostasis.
   c. are responsible for an increased sense of hunger the longer a person goes without eating.
   d. all of the above

_____ 5. The following events are part of a negative-feedback mechanism.
   1. Blood pressure increases.
   2. Control center compares actual blood pressure to the blood pressure set point.
   3. The heart beats faster.
   4. Receptors detect a decrease in blood pressure.

   Choose the arrangement that lists the events in the order they occur.
   a. 1, 2, 3, 4
   b. 1, 3, 2, 4
   c. 4, 2, 3, 1
   d. 4, 3, 2, 1

_____ 6. Which of the following terms mean the same thing when referring to a human in the anatomical position?
   a. superior and dorsal
   b. deep and distal
   c. anterior and ventral
   d. proximal and medial

_____ 7. A term that means nearer the attached end of a limb is
   a. medial.
   b. lateral.
   c. distal.
   d. proximal.
   e. superficial.

_____ 8. Which of the following directional terms are paired most appropriately as opposites?
   a. superficial and deep
   b. medial and proximal
   c. distal and lateral
   d. superior and posterior
   e. anterior and inferior

_____ 9. The chin is _____ to the umbilicus (belly button).
   a. lateral
   b. posterior
   c. distal
   d. superior

10. A plane that divides the body into anterior and posterior portions is a
    a. frontal plane.
    b. sagittal plane.
    c. transverse plane.

11. The pelvic cavity contains the
    a. kidneys.
    b. liver.
    c. stomach.
    d. spleen.
    e. urinary bladder.

12. The thoracic cavity is separated from the abdominal cavity by the
    a. diaphragm.
    b. mediastinum.
    c. mesentery.
    d. sternum.

13. The heart is
    a. part of the mediastinum.
    b. surrounded by the pericardial cavity.
    c. found within the thoracic cavity.
    d. all of the above

14. Given the following characteristics:
    1. reduce friction between organs
    2. line fluid-filled cavities
    3. line trunk cavities that open to the exterior of the body

    Which of the characteristics describe serous membranes?
    a. 1,2
    b. 1,3
    c. 2,3
    d. 1,2,3

15. Given the following cavities:
    1. abdominal
    2. pelvic
    3. oral
    4. pericardial

    Which cavities are lined by serous membranes?
    a. 1,2
    b. 1,2,3
    c. 1,2,4
    d. 2,3,4
    e. 1,2,3,4

16. Given the following organ and cavity combinations:
    1. heart and pericardial cavity
    2. lungs and pleural cavity
    3. stomach and peritoneal cavity
    4. kidney and peritoneal cavity

    Which of the organs is correctly paired with a space that surrounds that organ?
    a. 1,2
    b. 1,2,3
    c. 1,2,4
    d. 2,3,4
    e. 1,2,3,4

17. Given the following body cavity and membrane combinations:
    1. abdominal cavity and peritoneum
    2. thoracic cavity and pleural membrane
    3. pericardial cavity and pericardial membrane
    4. pelvic cavity and peritoneum

    Which of the body cavities are correctly paired with a membrane lining that body cavity?
    a. 1,2
    b. 2,3
    c. 3,4
    d. 1,2,3
    e. 1,2,3,4

18. Which of the following membrane combinations line portions of the diaphragm?
    a. parietal pleura - parietal peritoneum
    b. parietal pleura - visceral peritoneum
    c. visceral pleura - parietal peritoneum
    d. visceral pleura - visceral peritoneum

19. Which of the following organs are retroperitoneal?
    a. kidneys
    b. pancreas
    c. adrenal glands
    d. urinary bladder
    e. all of the above

_____20. Body temperatures were measured during an experiment. On the graph below at point A, the subject moved from a cool room into a hot room. As a result, body temperature increased to point B.

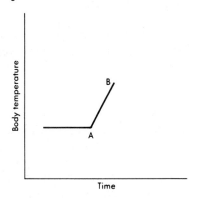

Graphed below are two possible responses to the increase in body temperature.

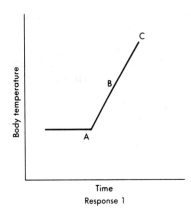

Response 1

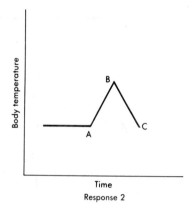

Response 2

Which of the responses graphed above represents a negative-feedback mechanism?

a. Response 1
b. Response 2

Use a separate sheet of paper to complete this section.

1. What sections through the body can be produced by cutting through the body from a superior to inferior direction? From right to left?

2. The esophagus is a tube that connects the throat (pharynx) and the stomach. What planes through the body make a longitudinal section through the esophagus? A cross section?

3. Complete the following statements, using the correct directional term for a human being.
    a. The knee is _____ to the ankle.
    b. The ear is _____ to the nose.
    c. The nose is _____ to the lips.
    d. The lips are _____ to the teeth.
    e. The heart is _____ to the sternum (breastbone).

4. When blood sugar levels decrease, the hunger center in the brain is stimulated. Is this part of a negative-feedback or positive-feedback system? Explain.

5. When food enters the stomach, it stimulates the stomach to secrete acid and enzymes. The secretions convert the food into a partially digested, semiliquid material called chyme, which passes into the small intestine, where the digestive process is completed. Is this an example of a positive-feedback or negative-feedback system? Explain.

6. A man has been shot in the abdomen. The bullet passed through the abdominal wall and the large intestine and lodged in the kidney. Name, in order, the serous membranes through which the bullet passed.

ANSWERS TO CHAPTER 1

Anatomy and Physiology
1. Cytology; 2. Histology; 3. Systemic anatomy;
4. Regional anatomy; 5. Surface anatomy;
6. Anatomic imaging

Structural and Functional Organization
A. 1. Organelle; 2. Organ; 3. Cell; 4. Organ System; 5. Tissue
B. 1. Integumentary; 2. Nervous; 3. Respiratory;
4. Urinary; 5. Digestive; 6. Skeletal;
7. Cardiovascular; 8. Endocrine; 9. Muscular

The Human Organism
1. Metabolism; 2. Responsiveness; 3. Growth;
4. Development; 5. Morphogenesis

Homeostasis
A. 1. Negative feedback; 2. Negative feedback;
3. Positive feedback; 4. Negative feedback;
5. Negative feedback; 6. Positive feedback;
7. Positive feedback
B. 1. Variable; 2. Set point; 3. Receptor;
4. Control center; 5. Effector; 6. Stimulus;
7. Response

Directional Terms
1. Inferior, Caudal; 2. Superior, Cephalic;
3. Anterior, Ventral; 4. Posterior, Dorsal; 5. Distal;
6. Lateral; 7. Deep

Planes
A. 1. Sagittal plane; 2. Frontal (coronal) plane;
3. Transverse plane; Longitudinal section;
5. Cross (transverse) section
B. 1. Frontal (coronal) plane; 2. Transverse plane; 3. Midsagittal plane
C. 1. Longitudinal section; 2. Cross (transverse) section; 3. Oblique section

Body Regions
1. Upper limb; 2. Arm; 3. Forearm; 4. Lower limb;
5. Thigh; 6. Leg; 7. Thorax; 8. Abdomen; 9. Pelvis

Body Cavities
A. 1. Thoracic cavity; 2. Abdominal cavity;
3. Abdominal cavity; 4. Pelvic cavity
B. 1. Visceral serous membrane; 2. Pleural membrane; 3. Pericardial membrane;
4. Peritoneal membrane; 5. Mesentery;
6. Retroperitoneal
C. 1. Thoracic cavity; 2. Mediastinum; 3. Pleural cavity; 4. Pericardial cavity; 5. Diaphragm;
6. Abdominopelvic cavity; 7. Abdominal cavity; 8. Pelvic cavity

1. Molecule, organelle, cell, tissue, organ, organ system, organism
2. Epithelial, connective, muscle, and nervous tissues
3. Receptor, control center, effector
4. A person standing erect with feet facing forward, arms hanging to the sides, and palms of the hands facing forward with the thumbs to the outside
5. Body: sagittal, transverse (horizontal), and frontal (coronal) planes; Organ: longitudinal, cross (transverse), and oblique sections
6. Thoracic, abdominal, and pelvic cavities
7. Pleural, pericardial, and peritoneal serous membranes
8. Kidneys, adrenal glands, pancreas, portions of the intestine, and the urinary bladder

1. A. Physiology deals with the processes or functions of living things. It is divided according to the organisms involved or the levels of organization within a given organism. Physiology emphasizes the dynamic nature of living things.

2. C. An organ is two or more tissues that perform a specific function. An organelle is a specialized structure within a cell. Organelles are at a lower level of organization than a cell, but organs are at a higher level. A tissue is a group of cells that perform a specific function.

3. D. The nervous and endocrine systems are the most important regulatory systems of the body. The circulatory system transports gases, nutrients, and waste products. The muscular system is responsible for movement.

4. D. Negative-feedback mechanisms maintain homeostasis by making deviations from normal smaller. The sense of hunger is part of a negative-feedback mechanism that maintains adequate nutrient levels (homeostasis). The sense of hunger increases as a result of decreased nutrient levels. The sense of hunger stimulates eating, which helps to restore nutrient levels.

5. C. Receptors detect a stimulus, a decrease in blood pressure. The control center in the brain compares the decreased blood pressure to an ideal, normal value, the set point. Because blood pressure is too low, the control center causes an effector, the heart, to produce a response by beating faster. As a result, blood pressure increases back toward the set point and homeostasis is maintained.

6. C. Anterior (toward the front of the body) and ventral (toward the belly) can be used interchangeably.

7. D. Proximal means nearer the attached end, and distal means away from the attached end of a limb. Medial is toward the midline, and lateral is away from the midline (toward the side). Superficial is toward or on the body's surface.

8. A. The correct answer is based on knowing the definitions of the terms listed and making a judgment about which pair of terms is the best set of opposites.

9. D. The chin is superior to (higher than) the umbilicus.

10. A. A frontal plane divides the body into anterior and posterior portions. A sagittal plane divides the body into left and right portions, and a transverse plane divides the body into superior and inferior portions.

11. E. The pelvic cavity contains the urinary bladder, the internal reproductive organs, and the lower part of the digestive tract. The other organs listed are found in the abdominal cavity.

12. A. The diaphragm separates the thoracic cavity from the abdominal cavity. The mediastinum divides the thoracic cavity into left and right parts.

13. D. The mediastinum is a partition containing the heart, trachea, esophagus, and other structures. The heart is surrounded by the pericardial cavity. The mediastinum, pericardial cavity, and heart are found in the thoracic cavity.

14. A. Serous membranes line cavities into which they secrete serous fluid. The serous fluid reduces friction between organs. Serous membranes line cavities that do NOT open to the exterior of the body.

15. C. Serous membranes line the trunk cavities and related organs. This includes peritoneal, pleural, and pericardial membranes. Mucous membranes line the surface of organs and cavities that open to the exterior of the body. The oral cavity is lined by a mucous membrane.

16. B. The kidneys are retroperitoneal; therefore they are NOT surrounded by the peritoneal cavity.

17. E. The abdominal and pelvic cavities are lined by the peritoneum, the thoracic cavity by the pleural membrane, and the pericardial cavity by the pericardial membrane.

18. A. The thoracic cavity, including the superior surface of the diaphragm, is lined by parietal pleura. The abdominal cavity, including the inferior surface of the diaphragm, is lined by the parietal peritoneum. Visceral pleura covers the lungs and visceral peritoneum covers many of the organs in the abdominopelvic cavity.

19. E. Retroperitoneal refers to organs that are located behind the peritoneum, i.e., they are between the body wall and the parietal peritoneum. All of the organs listed are retroperitoneal.

20. B. First, you must be able to interpret the graphs. For response 1, body temperature increased still further from the normal value, and for response 2 body temperature returned to normal. Next, the definitions of positive and negative feedback must be applied to the graphs. Because positive-feedback mechanisms increase the difference between a value and its normal level (homeostasis), response 1 is a positive-feedback mechanism. Negative-feedback mechanisms resist further change or return the values to normal as in response 2.

1. A superior to inferior cut can produce a sagittal or frontal section of the body. A right to left cut can produce a transverse or frontal section.

2. A sagittal or frontal plane through the body also makes a longitudinal section through the esophagus. A transverse plane makes a cross section.

3. a. Proximal (superior)
   a. Lateral
   b. Superior
   c. Anterior
   d. Deep (posterior or dorsal)

4. It is part of a negative-feedback system. Stimulation of the hunger center can result in eating, and the ingested food causes blood sugar levels to increase (return to homeostasis).

5. The homeostatic set point for the stomach is empty. Food entering the stomach is a stimulus that activates negative-feedback mechanisms such as increased secretions. As a result, the stomach empties, stimulation of secretions decreases, and homeostasis is restored.

6. After the bullet passes through the abdominal wall, the parietal peritoneum is the first membrane pierced. In passing through the large intestine, the bullet penetrates the visceral peritoneum, the large intestine, and then the visceral peritoneum. To enter the kidney, which is retroperitoneal, the bullet passes through the parietal peritoneum.

2 The Chemical Basis of Life

FOCUS: Chemistry is the study of the composition and structure of matter and the reactions that matter can undergo. Matter is anything that occupies space. Matter is composed of atoms, which consist of a nucleus (protons and neutrons) surrounded by electrons. Atoms can combine to form molecules. The chemical bonds between atoms of molecules include ionic, covalent, and hydrogen bonds. Chemical reactions are classified as synthesis, decomposition, exchange, or oxidation-reduction reactions. Enzymes catalyze the reactions in the body, greatly increasing the rate at which the reactions occur. Energy is the ability to do work; potential energy (stored energy) and kinetic energy (caused by an object's moving) are two subdivisions of energy. Reactions that release energy are exergonic, whereas endergonic reactions require the input of energy from another source. Organic molecules contain carbon (and usually hydrogen) atoms, and inorganic molecules include all other molecules except for carbon dioxide and carbon monoxide. Water has many important physical and chemical properties that make it an essential compound in living organisms. Acids are defined as proton (H^+) donors, and bases are proton acceptors. The pH scale measures the hydrogen ion concentration of a solution. Carbohydrates are built from monosaccharides, triacylglycerols (the most common type of fat in the body) are built from glycerol and fatty acids, proteins are built from amino acids, and nucleic acids (DNA and RNA) are formed from nucleotides.

CONTENT LEARNING ACTIVITY

Matter and Elements

❝All living and nonliving things are composed of matter.❞

Match these terms with the correct statement or definition:

Mass
Matter
Weight

_____ 1. Anything that occupies space.

_____ 2. Amount of matter in an object.

_____ 3. Gravitational force acting on an object of a given mass.

Elements, Atoms, and Atomic Structure

66The characteristics of living and nonliving matter result from the structure, organization,99 and behavior of its atoms.

A. Match these terms with the correct statement or definition:

Atom   Neutron
Electron   Proton
Element

_____ 1. Matter that cannot be broken down into simpler substances by ordinary chemical reactions.

_____ 2. Smallest piece of an element having the chemical characteristics of that element.

_____ 3. Subatomic particle with no electrical charge.

_____ 4. Subatomic particle with negative charge.

B. Match these terms with the correct statement or definition:

Electron orbital   Nucleus
Electron shell

_____ 1. Central part of an atom, which contains protons and neutrons.

_____ 2. Region in which any given electron is most likely to be found.

_____ 3. Orbitals grouped together; can contain more than one electron.

C. Match these terms with the correct parts labeled in Figure 2-1:

Electron
Nucleus
Neutron
Orbital
Proton

1. _____

2. _____

3. _____

4. _____

5. _____

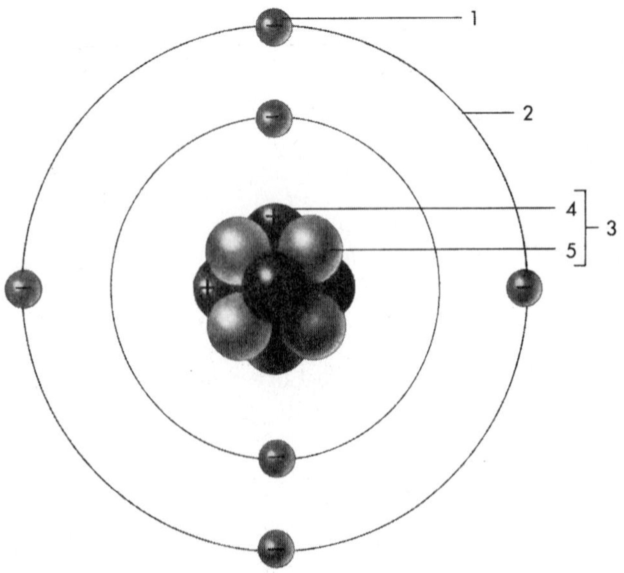

Figure 2-1

Atomic Number, Mass Number, and Atomic Weight

66*The numbers of protons, neutrons, and electrons in an atom determine the chemical***99** *characteristics of each atom.*

Match these terms with the correct statement or definition:

Atomic number   Isotopes
Atomic weight   Mass number
Avogadro's number   Radioactive

_____ 1. Number of protons in an atom.

_____ 2. Number of protons and neutrons in an atom.

_____ 3. Weight in grams of Avogadro's number of atoms of an element.

_____ 4. Number of atoms in one mole of an element.

_____ 5. 6.023×10^{23}.

_____ 6. Two or more forms of the same element that have the same atomic number, but different mass numbers, e.g., hydrogen, deuterium, and tritium.

_____ 7. Isotopes with unstable nuclei that spontaneously change to form more stable nuclei.

Electrons and Chemical Bonds

66*Chemical behavior of an atom is determined by the electrons in its outermost electron shell.***99**

A. Using the terms provided, complete these statements:

Chemical bond   Molecular formula
Compound   Stable
Molecule

A _(1)_ is formed when the outermost electrons are transferred or shared between atoms. The resulting combination of atoms is called a _(2)_. If a molecule has two or more different kinds of atoms, it can be referred to as a _(3)_. A molecule can be represented by a _(4)_, which consists of the symbols of the atoms in the molecule plus a subscript denoting the number of each type of atom. When atoms combine into molecules, the outermost electron shell of the each atom becomes _(5)_, or complete.

1. _____

2. _____

3. _____

4. _____

5. _____

B. Using the terms provided, complete these statements:

Covalent
Dissociate
Electrolytes
Gained
Hydrogen

Ionic
Ions
Lost
Negative
Positive

1. _____

2. _____

3. _____

4. _____

5. _____

6. _____

7. _____

8. _____

9. _____

10. _____

_(1)_ bonds result when one atom loses an electron and another atom accepts that electron. Atoms that have donated or accepted electrons are called _(2)_ and may be positively or negatively charged. A cation is an ion that has _(3)_ one or more electrons, and an anion is an ion that has _(4)_ one or more electrons. Cations and anions are sometimes called _(5)_ because they can conduct electrical current when they are in solution. When ionic substances dissolve in water, the cations and anions separate, or _(6)_ , and water molecules surround and isolate the ions. _(7)_ bonds result when two atoms complete their energy shells by sharing electrons. Hydrogen atoms covalently bound to oxygen or nitrogen have a slight _(8)_ charge. The weak attraction of a hydrogen atom to the small _(9)_ charge associated with the oxygen or nitrogen atoms of other molecules is called a _(10)_ bond.

C. Match these terms with the correct statement or definition:

Double
Nonpolar

Polar
Single

_____

_____

_____

_____

1. Covalent bond in which two atoms share two electron pairs.

2. Covalent bond in which two atoms share electrons equally.

3. Covalent bond in which two atoms share electrons unequally.

4. Molecules with this type of covalent bond readily dissolve in water.

 The substances that enter into a chemical reaction are called the reactants, and the substances that result from the chemical reactions are the products.

Chemical Reactions

66 *A chemical reaction is the process by which atoms, ions, or molecules interact either to form or break chemical bonds.* 99

Using the terms provided, complete these statements:

Anabolism   Hydrolysis
Catabolism   Metabolism
Decomposition   Oxidation
Dehydration   Reduction
Equilibrium   Reversible
Exchange   Synthesis

When two or more atoms, ions, or molecules combine to form a new and larger molecule, the process is called a __(1)__ reaction; all these reactions that occur within the body are collectively referred to as __(2)__. When larger molecules are broken down to form smaller molecules, ions, or atoms, the process is called a __(3)__ reaction. All of these breakdown reactions that occur within the body are called __(4)__. All of the anabolic and catabolic reactions in the body are called __(5)__. __(6)__ reactions occur when part of one molecule is exchanged for part of another molecule. Synthesis reactions in which water is a product are called __(7)__ reactions, whereas decomposition reactions that require water are called __(8)__ reactions. The loss of an electron by an atom is called __(9)__, whereas the gain of an electron is called __(10)__. Reactions that can proceed from reactants to products and from products to reactants are called __(11)__ reactions. At __(12)__ the ratio of products to the reactants tends to remain constant.

1. _____
2. _____
3. _____
4. _____
5. _____
6. _____
7. _____
8. _____
9. _____
10. _____
11. _____
12. _____

Rate of Chemical Reactions

66 *The rate at which a chemical reaction proceeds is influenced by several factors.* 99

Using the terms provided, complete these statements:

Catalyst   Enzymes
Concentration   Increases
Decreases

Chemical reactions are influenced by several factors, including how easily the substances react with one another. If the concentration of the reactants increases, the rate of the chemical reaction __(1)__. When the temperature decreases, the speed of chemical reactions __(2)__. Protein molecules in the body that act as catalysts are called __(3)__. A __(4)__ is a substance that increases the rate at which a chemical reaction proceeds without itself being permanently changed or depleted. Regulation of chemical events in the cell is due primarily to mechanisms that control either the __(5)__ or the activity of enzymes.

1. _____
2. _____
3. _____
4. _____
5. _____

Energy

"Energy is defined as the capacity to do work, that is, to move matter."

A. Match these terms with the correct statement or definition:

Kinetic energy
Potential energy

_____ 1. Stored energy that could do work but is not doing so, e.g., chemical bonds.

_____ 2. Energy caused by the movement of an object; the form of energy that actually does work.

B. Match these terms with the correct statement or definition:

Chemical   Electromagnetic
Electrical   Heat

_____ 1. Form of energy involving movement of ions or electrons, e.g., nerve impulses.

_____ 2. Form of energy that moves in waves, e.g., gamma rays, x-rays, and UV rays.

_____ 3. Form of potential energy caused by the attraction of negatively charged electrons to positively charged protons, e.g., bonds between atoms.

_____ 4. Form of energy resulting from random movement of atoms, ions, or molecules; energy form into which all other forms of energy can be converted.

C. Match these terms with the correct statement or definition:

Endergonic
Exergonic

_____ 1. Reaction that releases energy.

_____ 2. Type of reaction that is typical of catabolism.

_____ 3. Type of reaction that is typical of anabolism.

_____ 4. Type of reaction used by plants to convert the sun's energy into chemical bonds in glucose.

☞ Energy, unlike matter, does not occupy space, and has no mass.

Inorganic Molecules

"*Inorganic molecules are defined as those that do not contain carbon.***"**

A. Match these terms with the correct statement or definition:

Hydrogen bonding   Water
Polar covalent bonding

_____ 1. Molecule that composes 60% to 80% of most cells' volume.

_____ 2. Force that holds an oxygen atom and two hydrogen atoms together in a molecule of water.

_____ 3. Major force that holds water molecules together as a liquid.

B. Using the terms provided, complete these statements:

Dissolved   Mixes
Hydrolysis and dehydration Specific heat
Ions   Transports
Lubricates

Water has many physical and chemical properties that are important for living organisms. Water has high _(1)_; as a result, water tends to resist large temperature fluctuations. Water _(2)_ body parts and protects them from damage caused by friction. The chemical reactions necessary for life do not take place unless reacting molecules are _(3)_ in water. For instance, molecules such as sodium chloride dissociate in water; the resulting _(4)_ can then react with other molecules. Water participates directly in _(5)_ reactions. Water _(6)_ with other substances to form solutions, suspensions, or colloids; as a result, water _(7)_ nutrients, gases, and wastes throughout the body.

1. _____

2. _____

3. _____

4. _____

5. _____

6. _____

7. _____

C. Match these terms with the correct statement or definition:

Colloid   Suspension
Solution

_____ 1. Any liquid that contains dissolved substances, e.g., sweat.

_____ 2. Liquid that contains nondissolved substances that will settle out of the liquid unless continuously shaken, e.g., red blood cells in plasma.

_____ 3. Liquid that contains nondissolved materials that do not settle out of the liquid, e.g., water and proteins inside cells.

Solution Concentrations

"The osmotic concentration of body cells is important because it influences the movement of water **"** into or out of cells.

Match these terms with the
correct statement or definition:

Milliosmole   Solute
Osmole   Solvent
Percent

_____ 1. Liquid portion of a solution.

_____ 2. Substance dissolved in the liquid portion of a solution.

_____ 3. Method of expressing concentration that states the weight of solute in a given volume of solvent (usually water).

_____ 4. One mole of particles in one kilogram of water.

_____ 5. 1/1000 of an osmole.

Acids and Bases

"Many molecules are classified as acids or bases.**"**

A. Match these terms with the
correct statement or definition:

Acids   Salts
Bases   Strong
Buffers   Weak

_____ 1. Substances that are proton (H^+) donors.

_____ 2. Substances that accept hydrogen ions.

_____ 3. Acids or bases that dissociate almost completely in water.

_____ 4. Molecules consisting of a cation other than H^+ and an anion other than OH^-.

_____ 5. Chemicals that resist changes in the pH of a solution when acids or bases are added.

B. Match these terms with the
correct statement or definition:

Acidic solution   Neutral solution
Alkaline (basic) solution

_____ 1. pH of 7, e.g., pure water.

_____ 2. pH less than 7.

_____ 3. Greater concentration of H^+ than OH^-.

☞ A change in the pH of a solution by one pH unit represents a tenfold change in the hydrogen ion concentration.

Oxygen and Carbon Dioxide

"Oxygen and carbon dioxide are two important inorganic molecules.**"**

Match these terms with the
correct statement or definition:

Carbon dioxide
Oxygen

_____ 1. Composes about 21% of the gas in the atmosphere.

_____ 2. Used in the final step of reactions that extract energy from
food particles.

_____ 3. Produced when organic molecules such as glucose are
metabolized.

Carbohydrates

"In most carbohydrates, for each carbon atom there are two hydrogen atoms and one oxygen atom.**"**

A. Match these terms with the
correct statement or definition:

Monosaccharides   Polysaccharides
Disaccharides

_____ 1. Simple sugars (e.g., glucose) that are building blocks for other
carbohydrates.

_____ 2. Sucrose, lactose, maltose, and other double sugars.

_____ 3. Many monosaccharides bound together to form long chains,
e.g., glycogen, starch, and cellulose.

☞ Isomers such as glucose, fructose, and galactose are molecules that have the same number
and type of atoms but differ in their three-dimensional arrangements.

B. Match these terms with the
correct statement or definition:

Cellulose   Glycogen
Glucose   Starch

_____ 1. Carbohydrate that is found in blood and is a major nutrient for
most cells of the body.

_____ 2. Energy storage molecule in muscle and liver.

_____ 3. Structural component of plant cell walls that is not digestible
by humans.

Lipids

"Lipids contain a lower ratio of oxygen to carbon than do carbohydrates, and dissolve in nonpolar organic solvents such as acetone or alcohol, but are insoluble in water."

A. Match these terms with the correct statement or definition:

Carboxyl group   Monounsaturated
Fatty acids   Polyunsaturated
Glycerol   Saturated

_____ 1. These two combine to form triacylglycerols.

_____ 2. Straight chain of carbon atoms with carboxyl group attached.

_____ 3. Oxygen and a hydroxyl group attached to a carbon atom; responsible for the acidic nature of some molecules.

_____ 4. Fatty acid that has only single covalent bonds between carbon atoms.

_____ 5. Fats with one double covalent bond between carbon atoms.

_____ 6. Fats with two or more double covalent bonds between carbon atoms.

B. Match these terms with the correct statement or definition:

Fats   Prostaglandins
Fat-soluble vitamins   Steroids
Leukotrienes   Thromboxanes
Phospholipids

_____ 1. Monoacylglycerols, diacylglycerols, and triacylglycerols.

_____ 2. Components of the cell membrane that have polar and nonpolar ends.

_____ 3. Regulatory molecules; important in response of tissues to injuries.

_____ 4. Cholesterol, bile acids, estrogens, progesterone, and testosterone.

_____ 5. Small nonpolar molecules that aid normal function by aiding vision, promoting calcium uptake, promoting wound healing, and promoting synthesis of blood clotting proteins.

Proteins

"Proteins are large molecules whose function depends upon their shape."

A. Using the terms provided, complete these statements:

Amino acid   Primary
Denaturation   Quaternary
Peptide   Secondary
Polypeptide   Tertiary

The building blocks for proteins are 20 types of __(1)__ molecules. Covalent bonds formed between amino acid molecules during protein synthesis are called __(2)__ bonds. A molecule of many amino acids bonded together by peptide bonds is called a __(3)__ chain. The __(4)__ structure of a protein is determined by the sequence of amino acids bound to each other, whereas the __(5)__ structure of a protein is determined by the hydrogen bonds between amino acids that cause the protein to coil into helices or pleated sheets. If these hydrogen bonds are broken, __(6)__ of the protein occurs, and the protein is nonfunctional. Folding of helices or pleated sheets by the formation of covalent bonds between sulfur atoms of two amino acids and attraction or repulsion of part of the protein molecule to water produces the __(7)__ structure of proteins. The shape produced as two or more proteins combine to produce a functional unit is the __(8)__ structure.

1. _____

2. _____

3. _____

4. _____

5. _____

6. _____

7. _____

8. _____

B. Using the terms provided, complete these statements:

Activation energy   Cofactors
Active site   Induced fit model
Catalysts   Lock-and-key
Chemical reactions   model
Coenzymes   Protease

Enzymes control the rate of most __(1)__, thus they control essentially all cellular activities. Enzymes are protein __(2)__ that increase the rate at which a reaction proceeds without themselves being permanently changed. Enzymes function by lowering the __(3__ required to start a reaction. Enzymes are highly specific because of their three-dimensional shape, which determines the enzyme's __(4)__. According to the __(5)__, reactants must bind to a specific active site on the enzyme. The view of enzymes and reactants as rigid structures has been modified by the __(6)__, in which the enzyme is able to slightly change shape and better fit the reactants. Some enzymes require additional, nonprotein substances called __(7)__ to be functional. These substances can be ions or a complex of organic molecules. Organic cofactors, such as certain vitamins, are called __(8)__. Enzymes are often named by adding -*ase* to the name of the molecules on which they act. For instance, an enzyme that breaks down proteins is called a __(9)__.

1. _____

2. _____

3. _____

4. _____

5. _____

6. _____

7. _____

8. _____

9. _____

27

Nucleic Acids and ATP

66 *Nucleic acids direct activities of the cell and are involved in protein synthesis; ATP provides* 99 *energy for nearly all the endergonic reactions within cells.*

Match these terms with the correct statement or definition:

Adenosine triphosphate (ATP)
Chromatin
Chromosome

Deoxyribonucleic acid (DNA)
Nucleotide
Ribonucleic acid (RNA)

_____ 1. Building block for nucleic acids, consisting of a mono-saccharide, a nitrogenous organic base, and a phosphate group.

_____ 2. Double helix of nucleotides; its organic bases include thymine.

_____ 3. A single strand of nucleotides; its organic bases include uracil.

_____ 4. Structure formed by condensed chromatin during cell division.

_____ 5. Energy currency of cells; synthesized from oxidation of glucose.

👉 The loss of an electron by one atom (oxidation) is accompanied by the gain of that electron by another atom (reduction); these reactions are oxidation-reduction reactions.

Molecular Diagrams

Match these terms with the correct molecular diagram in Figure 2-2:

Amino acid
ATP
Triacylglycerol
Fatty acid
Glycerol

Monosaccharide
Nucleotide
Polypeptide
Polysaccharide

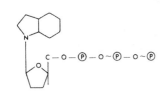

1. _____

2. _____

3. _____

4. _____

5. _____

6. _____

7. _____

8. _____

9. _____

Figure 2-2

1. List three subatomic particles and give their charge.

2. List three types of bonds between atoms.

3. Name four types of chemical reactions.

4. List four factors that affect the rate of chemical reactions.

5. List four types of energy and give an example that applies to the human organism for each.

6. List four properties of water that make it well suited for living organisms.

7. Name the four types of organic molecules found in living things. For each type of organic molecule, list its building blocks.

8. List three kinds of carbohydrates and give an example for each in the human organism.

9. List three kinds of lipids and give an example for each in the human organism.

10. List the organic bases for DNA, RNA, and ATP.

MASTERY LEARNING ACTIVITY

Place the letter corresponding to the correct answer in the space provided.

_____ 1. The smallest portion of an element that still has the chemical characteristics of that element is a (an)
a. electron.
b. molecule.
c. neutron.
d. proton.
e. atom.

_____ 2. The number of electrons in an atom is equal to
a. Avogadro's number.
b. the mass number.
c. the atomic number.
d. the number of neutrons.

_____ 3. Carbon forms chemical bonds with other atoms by achieving a stable or complete outermost energy shell. This is accomplished when carbon
a. loses two electrons.
b. gains two electrons.
c. shares two electrons.
d. loses four electrons.
e. shares four electrons.

_____ 4. A cation is a (an)
a. uncharged atom.
b. positively charged atom.
c. negatively charged atom.
d. atom that has gained an electron.
e. c and d

_____ 5. A polar covalent bond between atoms occurs when
a. one atom attracts electrons more strongly than the other atom.
b. atoms attract electrons equally.
c. an electron from one atom is completely transferred to another.
d. the molecule becomes ionized.
e. a hydrogen atom is shared between two different atoms.

_____ 6. In a decomposition reaction
a. anabolism occurs.
b. proteins are formed from amino acids.
c. large molecules are broken down to form small molecules.
d. all of the above

_____ 7. Reversible reactions
a. are synthesis, decomposition, or exchange reactions.
b. have reactants forming products and products forming reactants.
c. reach equilibrium if the ratio of products to reactants is constant.
d. all of the above

_____ 8. The rate of chemical reactions is influenced by
a. how easily substances react with each other.
b. the concentration of the reactants.
c. temperature.
d. enzymes.
e. all of the above

_____ 9. Potential energy
   a. is energy caused by an object's moving.
   b. is the form of energy that actually does work.
   c. includes energy within chemical bonds.
   d. all of the above

_____ 10. Heat energy
   a. results from the random movement of atoms, ions, and molecules.
   b. can be produced from other forms of energy.
   c. is released during exergonic reactions.
   d. all of the above

_____ 11. Water
   a. is composed of two oxygen atoms and one hydrogen atom.
   b. has a low specific heat.
   c. is composed of polar molecules into which ionic substances dissociate.
   d. is produced in a hydrolysis reaction.

_____ 12. When sugar is dissolved in water, the water is called the
   a. solute.
   b. solvent.
   c. solution.

_____ 13. Which of these would be an example of a suspension?
   a. sweat
   b. water and proteins inside cells
   c. red blood cells in plasma
   d. sugar dissolved in water
   e. none of the above

_____ 14. A 1-osmolal solution of glucose and a 1-osmolal solution of sodium chloride
   a. have the same number of water molecules.
   b. have the same number of solute particles.
   c. have the same number of molecules.
   d. a and b
   e. a and c

_____ 15. A solution with a pH of 5 is a (an) _____ and contains _____ hydrogen ions than a neutral solution.
   a. base, more
   b. base, less
   c. acid, more
   d. acid, less

_____ 16. A buffer
   a. slows down chemical reactions.
   b. speeds up chemical reactions.
   c. increases the pH of a solution.
   d. maintains a relatively constant pH.
   e. b and c

_____ 17. Oxygen
   a. is an inorganic molecule.
   b. is produced when glucose is broken down for energy.
   c. combines with carbon dioxide and is released from the body.
   d. all of the above

_____ 18. Which of these is an example of a carbohydrate?
   a. glycogen
   b. wax
   c. steroid
   d. DNA
   e. none of the above

_____ 19. The basic units or building blocks of triacylglycerols are
   a. simple sugars (monosaccharides).
   b. double sugars (disaccharides).
   c. amino acids.
   d. glycerol and fatty acids.
   e. nucleotides.

_____ 20. A peptide bond joins together
   a. amino acids.
   b. fatty acids.
   c. disaccharides.
   d. nucleotides.

_____ 21. The energy required to start a chemical reaction is called
   a. equilibrium energy.
   b. nonequilibrium energy.
   c. endergonic energy.
   d. exergonic energy.
   e. activation energy.

_____22. According to the lock-and-key model of enzyme action,
a. reactants must first be heated.
b. enzyme shape is not important.
c. reactants must bind to an active site on the enzyme.
d. enzymes are highly specific.
e. c and d

_____23. Which of these chemicals would most likely be an enzyme?
a. ribose
b. lactic acid
c. lipase
d. sodium hydroxide
e. glycerol

_____24. DNA
a. is the genetic material.
b. is a single strand of nucleotides.
c. contains the nucleotide uracil.
d. all of the above

_____25. ATP
a. is formed by the addition of a phosphate group to ADP.
b. is formed with energy released during catabolism reactions.
c. provides the energy for anabolism reactions.
d. contains three phosphate groups.
e. all of the above

FINAL CHALLENGES

Use a separate sheet of paper to complete this section.

1. The atomic number of nitrogen is 7 and the mass number is 14. How many protons, neutrons, and electrons in a nitrogen atom?

2. Nitrogen atoms have five electrons in their outermost shell. How many hydrogen atoms, which have one electron in their outermost shell, will combine with nitrogen to form an uncharged molecule? What kind of chemical bond is formed?

3. Two substances, A and B, can combine to form substance C:

$$A + B \longrightarrow C$$

Substances A and B each dissolve in water to form a colorless solution, whereas substance C forms a red solution. Using this information, explain these experiments:

A. When solutions A and B are combined, no color change takes place. However, when the combined solution is heated, it turns red.

B. When solutions A and B are combined, no color change takes place. However, when substance D is added to the combined solution, it turns red. Later, the exact amount of substance D that was added is recovered from the solution.

4. When placed in a solution, potassium chloride (KCl) dissociates to form potassium ions (K^+) and chloride ions (Cl^-). On the other hand, fructose (a sugar) does not dissociate in water. Suppose you have a 1-osmolal solution of potassium chloride and a 1-osmolal solution of fructose. Which solution has the larger number of molecules? Compared with the solution that has the smaller number of molecules, how many more molecules are in the solution with the larger number of molecules? Give the actual number of molecules.

5. Given that blood is buffered by these reactions:

$$CO_2 + H_2O \longleftrightarrow H_2CO_3 \longleftrightarrow H^+ + HCO_3^-$$

What will happen to blood pH if a person holds his breath?

6. Suppose you have two substances and you know that one is a carbohydrate and one is a lipid. How could you tell which one was the carbohydrate and which one was the lipid, if you had available the materials found in a bathroom (hint: medicine cabinet, make-up kit)?

32

ANSWERS TO CHAPTER 2

Matter and Elements
1. Matter; 2. Mass; 3. Weight

Atomic Structure
A. 1. Element; 2. Atom; 3. Neutron; 4. Electron
B. 1. Nucleus; 2. Electron orbital; 3. Electron shell
C. 1. Electron; 2. Orbital; 3. Nucleus; 4. Proton; 5. Neutron

Atomic Number, Mass Number, and Atomic Weight
1. Atomic number; 2. Mass number; 3. Atomic weight; 4. Avogadro's number; 5. Avogadro's number; 6. Isotopes; 7. Radioactive

Electrons and Chemical Bonds
A. 1. Chemical bond; 2. Molecule; 3. Compound; 4. Molecular formula; 5. Stable
B. 1. Ionic; 2. Ions; 3. Lost; 4. Gained; 5. Electrolytes; 6. Dissociate; 7. Covalent; 8. Positive; 9. Negative; 10. Hydrogen
C. 1. Double; 2. Nonpolar; 3. Polar; 4. Polar

Chemical Reactions
1. Synthesis; 2. Anabolism; 3. Decomposition; 4. Catabolism; 5. Metabolism; 6. Exchange; 7. Dehydration; 8. Hydrolysis; 9. Oxidation 10. Reduction; 11. Reversible; 12. Equilibrium

Rate of Chemical Reactions
1. Increases; 2. Decreases; 3. Enzymes; 4. Catalyst; 5. Concentration

Energy
A. 1. Potential energy; 2. Kinetic energy
B. 1. Electrical; 2. Electromagnetic; 3. Chemical; 4. Heat
C. 1. Exergonic; 2. Exergonic; 3. Endergonic; 4. Endergonic

Inorganic Molecules
A. 1. Water; 2. Polar covalent bonding; 3. Hydrogen bonding
B. 1. Specific heat; 2. Lubricates 3. Dissolved; 4. Ions; 5. Hydrolysis and dehydration; 6. Mixes; 7. Transports

C. 1. Solution; 2. Suspension; 3. Colloid

Solution Concentration
1. Solvent; 2. Solute; 3. Percent; 4. Osmole; 5. Milliosmole

Acids and Bases
A. 1. Acids; 2. Bases; 3. Strong; 4. Salts; 5. Buffers
B. 1. Neutral solution; 2. Acidic solution; 3. Acidic solution

Oxygen and Carbon Dioxide
1. Oxygen; 2. Oxygen; 3. Carbon dioxide

Carbohydrates
A. 1. Monosaccharides; 2. Disaccharides; 3. Polysaccharides
B. 1. Glucose; 2. Glycogen; 3. Cellulose

Lipids
A. 1. Glycerol, fatty acid; 2. Fatty acid; 3. Carboxyl group; 4. Saturated; 5. Monounsaturated; 6. Polyunsaturated
B. 1. Fats; 2. Phospholipids; 3. Prostaglandins, leukotrienes, and thromboxanes; 4. Steroids; 5. Fat-soluble vitamins

Proteins
A. 1. Amino acid; 2. Peptide; 3. Polypeptide; 4. Primary; 5. Secondary; 6. Denaturation; 7. Tertiary; 8. Quaternary
B. 1. Chemical reactions; 2. Catalysts; 3. Activation energy; 4. Active site; 5. Lock-and-key model; 6 Induced fit model; 7. Cofactor; 8. Coenzymes; 9. Protease

Nucleic Acids and ATP
1. Nucleotide; 2. Deoxyribonucleic acid (DNA); 3. Ribonucleic acid (RNA); 4. Chromosome; 5. Adenosine Triphosphate (ATP)

Molecular Diagrams
1. Monosaccharide; 2. Polysaccharide; 3. Glycerol; 4. Fatty acid; 5. Triacylglycerol; 6. Amino acid; 7. Polypeptide; 8. ATP; 9. Nucleotide

1. Protons: positive charge; neutrons: no charge; and electrons: negative charge
2. Ionic, covalent, and hydrogen bonds
3. Synthesis, decomposition, exchange , and oxidation-reduction reactions
4. How easily substances react, concentration of react-ants, temperature of reactants, and presence of catalyst

5. Electrical energy: nerve impulses; electromagnetic energy: visible light; chemical energy: muscle contraction; heat energy: body temperature
6. High specific heat, effective lubricant, necessary for chemical reactions, and a good mixing medium

7. Carbohydrates: monosaccharides; fats: glycerol and fatty acids; proteins: amino acids; nucleic acids: nucleotides
8. Monosaccharides: glucose, fructose, galactose; Disaccharides: Sucrose, lactose; maltose; Polysaccharides: glycogen
9. Fats: energy storage, padding, insulation; Phospholipids: cell membrane; Prostaglandins, thromboxanes, or leukotrienes: response of tissue to injuries; Steroids: sex hormones; Fat soluble vitamins: vitamins A, D, E, and K.
10. DNA: adenine, thymine, guanine, cytosine; RNA: adenine, uracil, guanine, cytosine; ATP: adenine

MASTERY LEARNING ACTIVITY

1. E   An element consists of atoms of only one kind, and the atoms retain the properties of the element. A molecule is composed of two or more atoms joined together. Neutrons, protons, and electrons are parts of atoms.

2. C   The atomic number is the number of protons in an atom. Because the number of protons is equal to the number of electrons, it is also the number of electrons. The mass number is the number of protons and neutrons in an atom, and Avogadro's number is the number of atoms in a mole.

3. E   When a carbon atom shares four electrons, it increases the number of electrons in its outermost shell to eight, the number necessary for a stable outermost energy shell.

4. B   A cation is positively charged as a result of the loss of electrons.

5. A   A polar covalent bond is caused by unequal sharing of electrons, with one end of the molecule being more negative than the other end. Nonpolar covalent bonds share electrons equally between atoms, whereas ionic bonds involve complete transfer of electrons.

6. C   Decomposition reactions involve the breakdown of molecules. The sum of all synthesis reactions is called anabolism; amino acids are combined in synthesis reactions to produce proteins.

7. D   Reversible reactions can be synthesis, decomposition, or exchange reactions. Reversible reactions have reactants forming products and products forming reactants; the reactants form products and vice versa until an equilibrium is reached in which the ratio of products to reactants remains constant.

8. E   The rate of chemical reactions is influenced by how easily substances react with each other, by temperature, by concentration of the reactants, and by enzymes that catalyze the reaction in question.

9. C   Potential energy is energy that could do work but is not doing so. Chemical energy is a form of potential energy within the electrons forming a chemical bond. Kinetic energy is energy caused by an object's moving and is the form of energy that actually does work.

10. D   Heat energy results from the random movement of atoms, ions, or molecules; it can be produced from all other forms of energy and is released during exergonic reactions.

11. C   The polar nature of water molecules makes it possible for polar molecules and ions to dissolve in water. Water has one oxygen and two hydrogen atoms, has a high specific heat, and is used up in hydrolysis reactions (and is produced in dehydration reactions).

12. B   A solution is a solvent into which a solute is dissolved. A sugar solution consists of water (the solvent) into which sugar (the solute) is dissolved.

13. C   A suspension is a liquid that contains nondissolved materials that will settle out of the liquid unless it is continuously shaken. Sweat and sugar dissolved in water would be examples of solutions, whereas water and proteins inside cells form a colloid.

14. D   A 1-osmolal solution of one substance by definition has the same number of solute particles (but not necessarily the same number of molecules) and water molecules as a 1-osmolal solution of another substance. Because the sodium chloride dissociates into sodium and chloride ions but the glucose does not dissociate, the 1-osmolal solution of sodium chloride has one half the number of molecules as the 1-osmolal glucose solution.

15. C   Solutions with a pH of less than seven are acidic (above seven they are basic). Acid solutions contain more hydrogen ions than do neutral solutions.

16. D   Buffers resist a change in pH. Enzymes speed up chemical reactions they catalyze.

34

17. A Oxygen is an inorganic molecule (does not contain carbon). Carbon dioxide is released from glucose as glucose is broken down for energy.

18. A Glycogen is composed of long chains of glucose. Waxes and steroids are lipids, and DNA is a nucleic acid.

19. D. The basic building blocks of triacylglycerols are glycerol and fatty acids, the basic building blocks of carbohydrates are monosaccharides, the basic building blocks of proteins are amino acids, and the basic building blocks of nucleic acids are nucleotides.

20. A Peptide bonds are covalent bonds formed between amino acids during dehydration reactions.

21. E. Activation energy is required to start chemical reactions. Exergonic reactions give off energy, and endergonic reactions require the addition of energy.

22. E. According to the lock-and-key model of enzyme action, reactants must bind to a specific active site on the enzyme; therefore enzymes are highly specific for certain reactants and products. The three-dimensional shape of enzymes determines the structure of the active site. Because enzymes are sensitive to temperature change, heating could damage the enzyme.

23. C. Enzymes are often named by adding the suffix -ase to the name of the molecules on which they act. Lipase is an enzyme that catalyzes lipid breakdown.

24. A DNA, the genetic material, is a double strand of nucleotides. RNA is a single strand of nucleotides and contains the base uracil (DNA has thymine).

25. E. Energy from catabolism reactions is used to add a phosphate group to ADP (which has two phosphate groups) and form ATP (which has three phosphate groups). When ATP is broken down to ADP, energy is released that may be used in anabolism reactions.

 FINAL CHALLENGES

1. The atomic number is equal to the number of protons (7), and the number of protons equals the number of electrons (7). The number of neutrons (7) is equal to the mass number minus the atomic number (14 - 7 = 7).

2. Three hydrogen atoms can form covalent bonds with a nitrogen atom, so that the outermost shells of the nitrogen and hydrogen atoms are stable or complete. After the chemical reaction between the nitrogen and the three hydrogen atoms has taken place, the nitrogen atom has eight electrons in its outermost shell (its five original electrons plus three electrons shared with the three hydrogen atoms), and each hydrogen atom has two electrons in its outermost shell (its original electron plus one electron shared with nitrogen). The molecule formed is ammonia.

3. A Substances A and B require activation energy for the reaction to proceed, and this activation energy is provided by the heat.
    B. Substance D is a catalyst that lowers the activation energy requirement for the reaction, but it is not used up or altered by the reaction.

4. The osmolality of a solution is equal to the molality of the solution times the number of particles into which the solute dissociates. A 1-osmolal solution of KCl contains one half a mole of KCl (1-osmolal = 1/2 -molal X 2) and a 1-osmolal solution of fructose contains a mole of fructose (1-osmolal = 1-molal X 1). Therefore fructose has one half a mole more molecules than KCl. Because a mole contains 6.023×10^{23} molecules (Avogadro's number), the fructose solution has 3.012×10^{23} more molecules than the KCl solution. Note, however, that both solutions contain the same number of particles; i.e., the number of K^+ and Cl^- ions combined is equal to the number of fructose molecules.

5. Holding one's breath would cause an increase in blood carbon dioxide levels because carbon dioxide would not be eliminated by the respiratory system. As a result, more carbonic acid (H_2CO_3) would form and then dissociate into hydrogen ions (H^+) and bicarbonate ions (HCO_3^-). The increased number of hydrogen ions would make the blood more acidic, resulting in a decrease in blood pH.

6. Carbohydrates are usually polar molecules that dissolve in water, whereas lipids usually do not dissolve in water but will dissolve in organic solvents such as alcohol or acetone. One could try to dissolve each substance in water, alcohol (medicinal supply), or acetone (fingernail polish remover). Because many carbohydrates are sugars, one could also taste each substance (not really a good idea for an unknown substance, however!).

35

"I was gratified to answer promptly. I said I
don't know."

Mark Twain

Structure and Function of the Cell

FOCUS: The basic unit of the human body is the cell, which consists of organelles that are specialized to perform a variety of functions. The plasma membrane regulates the movement of materials into and out of the cell by diffusion, facilitated diffusion, active transport, and other mechanisms. Chromosomes in the nucleus contain DNA, which controls the activities of the cell through the production of mRNA (transcription), which leaves the nucleus and at ribosomes causes the production of enzymes (translation). The chemical reactions that occur within the cell, such as the production of ATP in mitochondria, are regulated by the enzymes. Also produced at ribosomes are proteins, such as microtubules and microfilaments, that provide structural support to the cell. Some of these proteins, the spindle fibers, are involved with the separation of chromosomes during cell division. Cells divide by mitosis during growth or repair of tissues and divide by meiosis to produce gametes.

CONTENT LEARNING ACTIVITY

Structure of the Plasma Membrane

" The plasma, or cell, membrane is the outermost component of a cell. "

A. Match these terms with the correct statement or definition:

Extracellular
Cholesterol
Fluid mosaic model
Intercellular

Intracellular
Hydrophilic
Hydrophobic
Phospholipids

_____ 1. This term refers to substances inside the cell membrane.

_____ 2. Double layer of molecules that form the plasma membrane.

_____ 3. The polar ends of phospholipids, which are exposed to water inside and outside of the cell.

_____ 4. Helps to determine the fluid nature of the plasma membrane.

_____ 5. Other molecules such as proteins float in the lipid bilayer; distributes molecules, allows repair of slight damage to the plasma membrane, and enables plasma membranes to fuse with one another.

B. Match these terms with
the correct plasma membrane
parts labeled in Figure 3-1:

Carbohydrate molecule
Cholesterol
Hydrophilic region of
  phospholipid
Hydrophobic region of
  phospholipid
Lipid bilayer
Protein

1. _____

2. _____

3. _____

4. _____

5. _____

6. _____

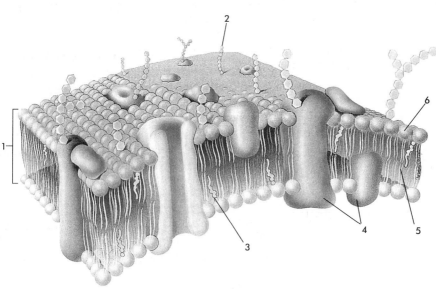

Figure 3-1

C. Match these terms with the
correct statement or definition:

Channel protein
Glycolipid
Glycoprotein
Integral protein

Marker molecule
Peripheral protein
Receptor molecule

_____ 1. General type of protein that penetrates the lipid bilayer from
one surface to the next.

_____ 2. Allows small molecules or ions to pass through the plasma
membrane.

_____ 3. Part of an intercellular communication system that enables
coordination of the activities of cells.

_____ 4. Allows cells to identify and attach to each other.

_____ 5. Protein with attached carbohydrates.

38

D. Match these terms with the correct statement or definition:

Binding site
G protein
Ligand

Ligand gated
Second messenger
Voltage gated

_____ 1. Part of receptor molecule to which other molecules attach.

_____ 2. Molecule that binds to receptor molecules.

_____ 3. Activated when a ligand attaches to a receptor molecule.

_____ 4. Stimulates cell functions following activation by G protein.

_____ 5. Channel protein that opens following a change in charge across the cell membrane.

_____ 6. Channel protein that opens after a molecule binds to its receptor.

Nucleus

66 *The information contained in the nucleus determines the chemical events that occur in the cell.* **99**

Using the terms provided, complete these statements:

Chromatin
Chromosomes
Histones
Messenger RNA

Nuclear envelope
Nuclear pores
Nucleolus
Nucleus

The _(1)_ is a large, membrane-bound structure usually located near the center of the cell. It is surrounded by a _(2)_ composed of two membranes separated by a space. At many points the inner and outer membranes fuse to form holes called the _(3)_. Except during cell division, DNA and associated protein are dispersed throughout the nucleus as thin strands called _(4)_. The proteins associated with DNA include _(5)_ that play a role in the regulation of DNA function. During cell division the chromatin condenses to form _(6)_. In the nucleus, DNA determines the structure of _(7)_, which moves out of the nucleus through nuclear pores into the cytoplasm. Within the cytoplasm mRNA is involved in the synthesis of proteins. A _(8)_ is a somewhat rounded, dense, well-defined nuclear body with no surrounding membrane. It is the site of ribosomal subunit manufacture.

1. _____

2. _____

3. _____

4. _____

5. _____

6. _____

7. _____

8. _____

Cytoplasm

"_Cytoplasm is the cellular material outside the nucleus but inside the plasma membrane._**"**

A. Match these terms with the correct statement or definition:

Cytoplasmic inclusion   Cytosol
Cytoskeleton   Organelles

_____ 1. Fluid portion of cytoplasm; contains enzymes.

_____ 2. Functions to support the cell and hold organelles in place; responsible for cell movements.

_____ 3. Aggregate of chemicals within cells.

_____ 4. General term for small structures within cells that are specialized for particular functions.

B. Match these terms with the correct statement or definition:

Actin filament   Microtubule
Intermediate filament

_____ 1. Composed primarily of tubulin; found in centrioles, spindle fibers, cilia, and flagella.

_____ 2. Provides support in microvilli; involved with cell movement such as muscle contraction.

_____ 3. Provides mechanical strength to cells.

Ribosomes and Endoplasmic Reticulum

"_The functional ribosome, composed of ribosomal RNA and protein, consists of a large subunit and a smaller subunit joined together; the endoplasmic reticulum forms broad, flattened sacs and tubules that interconnect._**"**

Match these terms with the correct statement or definition:

Cisternae   Rough endoplasmic reticulum
Ribosome   Smooth endoplasmic reticulum

_____ 1. Site where mRNA and tRNA come together and assemble amino acids to form proteins; free or attached to endoplasmic reticulum.

_____ 2. Tubules and flattened sacs with many ribosomes attached.

_____ 3. Large amounts of this structure are found in cells that secrete proteins.

_____ 4. Found in the cell when lipid synthesis, detoxification processes, or storage of calcium ions occurs.

☞ The outer membrane of the nuclear envelope is continuous with the endoplasmic reticulum.

Golgi Apparatus and Secretory Vesicles

"The Golgi apparatus is specialized smooth endoplasmic reticulum.**"**

Match these terms with the
correct statement or definition:

Golgi apparatus   Vesicle
Secretory vesicle

_____ 1. Modifies, packages, and distributes proteins and lipids
manufactured by the rough and smooth endoplasmic reticulum.

_____ 2. General term for a sac that pinches off from the Golgi
apparatus.

_____ 3. Releases its contents to the exterior of the cell by exocytosis.

☞ Vesicles contain proteins that are secreted, become part of the plasma membrane, or
function as enzymes.

Lysosomes and Peroxisomes

"There are several types of membrane-bound vesicles in the cell.**"**

Match these terms with the
correct statement or definition:

Autophagia   Lysosomes
Catalase   Peroxisomes

_____ 1. Vesicles that function as intracellular digestive systems.

_____ 2. Process of digesting cell organelles that are no longer
functional.

_____ 3. Membrane-bound bodies containing enzymes that produce
hydrogen peroxide while breaking down fatty acids and amino
acids.

_____ 4. Enzyme in peroxisomes that breaks down hydrogen peroxide.

Mitochondria

" *Mitochondria are small, spherical, rod-shaped, or thin filamentous* **"** *structures found throughout the cytoplasm.*

Using the terms provided, complete the following statements:

ATP   Electron-transport chain
Cristae   Matrix
DNA   Oxidative metabolism

Mitochondria are the major sites of __(1)__ production within cells. Mitochondria have inner and outer membranes separated by a space. The inner membranes have numerous infoldings called __(2)__ that project like shelves into the interior of the mitochondria. A complex series of mitochondrial enzymes forms two major enzyme systems, which are responsible for __(3)__ and most ATP synthesis. The enzymes of the citric acid (Krebs') cycle are found in the __(4)__, and the enzymes of the __(5)__ are embedded with the inner membrane. The information for making some mitochondrial proteins is contained with __(6)__ found within mitochondria.

1. _____

2. _____

3. _____

4. _____

5. _____

6. _____

Centrioles, Spindle Fibers, Cilia, Flagella, and Microvilli

" *Microtubules form essential components of centrioles, spindle fibers, cilia, and flagella.* **"**

Match these terms with the correct statement or definition:

Centrioles   Flagella
Centrosome   Microvilli
Cilia   Spindle fibers

_____ 1. Specialized zone of cytoplasm containing two centrioles.

_____ 2. Small cylindrical organelles oriented perpendicular to each other.

_____ 3. Attach to chromosomes; aid chromosome separation in cell division.

_____ 4. Short appendages that are capable of moving.

_____ 5. Increase cell surface area.

Cell Diagram

Match these terms with the
cell parts labeled on Figure 3-2:

Centrioles
Cilia
Golgi apparatus
Lysosome
Microtubule
Microvilli
Mitochondrion

Nucleolus
Plasma membrane
Ribosome
Rough endoplasmic reticulum
Secretory vesicle
Smooth endoplasmic reticulum

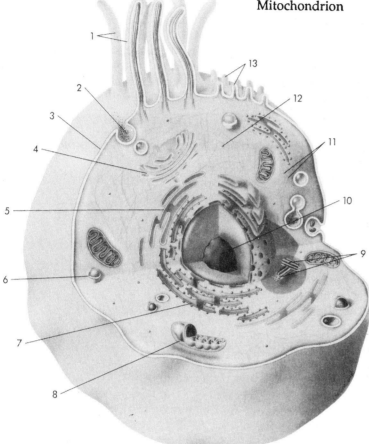

Figure 3-2

1. _____

2. _____

3. _____

4. _____

5. _____

6. _____

7. _____

8. _____

9. _____

10. _____

11. _____

12. _____

13. _____

Movement Through the Plasma Membrane

The plasma membrane separates the extracellular material from the intracellular material and is selectively permeable, allowing only certain substances to pass through it.

Match these terms with the correct statement or definition:

Carrier molecules   Lipid bilayer
Membrane channels   Vesicle

_____ 1. Molecules that are soluble in lipid dissolve in this layer; acts as a barrier to most polar substances.

_____ 2. Allow molecules of only a certain size range to pass through.

_____ 3. Positively charged, these structures cause positive ions to pass through less readily than neutral or negatively charged ions.

_____ 4. Large polar molecules cannot pass through the cell membrane in significant amounts unless transported by these proteins.

_____ 5. Transports large polar molecules, small pieces of matter, and even entire cells.

Diffusion

Diffusion of molecules is an important means by which substances move through the extracellular and intracellular fluids in the body.

A. Match these terms with the correct statement or definition:

Diffusion   Solvent
Solute

_____ 1. Predominant liquid or gas in a solution.

_____ 2. Product of the constant random motion of all atoms, molecules, or ions in a solution.

_____ 3. Tendency for solute molecules to move from an area of higher concentration to an area of lower concentration in a solution.

☞ The concentration gradient is the concentration difference between two points, divided by the distance between those two points. Viscosity is a measure of how easily a liquid flows.

B. Match these terms with the correct statement or definition:

Decrease
Increase

_____ 1. Change in the rate of diffusion when there is an increase in the concentration gradient or an increase in temperature.

_____ 2. Change in the rate of diffusion when there is a decrease in the size of diffusing molecules.

_____ 3. Change in the rate of diffusion when there is an increase in viscosity.

Osmosis

Osmosis is important to cells because large volume changes caused by water movement disrupt normal cell function.

Match these terms with the correct statement or definition:

Crenation   Isosmotic
Hyperosmotic   Isotonic
Hypertonic   Lysis
Hyposmotic   Osmosis
Hypotonic   Osmotic pressure

_____ 1. Diffusion of water across a selectively permeable membrane.

_____ 2. Force required to prevent the movement of water by osmosis across a selectively permeable membrane.

_____ 3. Solution with a lower osmotic pressure and fewer solute particles than another solution.

_____ 4. When a cell is placed in this type of solution, the cell, by definition, shrinks.

_____ 5. Rupture of the cell, which may occur from placing a cell into a hypotonic solution.

Filtration

Filtration results when a partition containing small holes is placed in a stream of moving liquid.

Using the terms provided, complete these statements:

Concentration difference   Lower
Higher   Pressure difference
Larger   Smaller

1. _____

2. _____

3. _____

Filtration depends upon a (1) on either side of a partition. Liquid moves through the partition from the side with the (2) pressure. Holes in the partition prevent particles that are (3) than the holes from moving through the partition.

Mediated Transport Mechanisms

Mediated transport mechanisms involve carrier molecules within the plasma membrane.

A. Match these terms with the correct statement or definition:

Competition   Specificity
Saturation

_____ 1. Carrier molecule transports only a single type of molecule.

_____ 2. Result of similar molecules binding to the carrier molecule.

_____ 3. Rate of transport is limited by the number of carrier molecules.

B. Match these terms with the correct statement or definition:

Active transport
Facilitated diffusion

_____ 1. Does not require metabolic energy (ATP).

_____ 2. Always moves substances from a higher to a lower concentration.

_____ 3. Can move substances against a concentration gradient.

C. Using the terms provided, complete these statements:

Against
Cotransport
Countertransport

Into
Out of
With

1. _____

2. _____

3. _____

4. _____

5. _____

In secondary active transport, a sodium-potassium exchange pump moves Na^+ ions _(1)_ a cell, establishing a concentration gradient for Na^+ ions. As Na^+ ions move _(2)_ the concentration gradient, Na^+ ions and glucose molecules can be transported _(3)_ cells by carrier molecules. The energy provided by Na^+ ion movement can result in the movement of glucose _(4)_ its concentration gradient. The movement of Na^+ ions and glucose in the same direction across the cell membrane is called _(5)_.

Endocytosis and Exocytosis

66 Endocytosis and exocytosis involve bulk movement of material into and out of the cell. 99

Match these terms with the correct statement or definition:

Endocytosis
Exocytosis

Phagocytosis
Pinocytosis

_____ 1. Includes both phagocytosis and pinocytosis.

_____ 2. Means "cell eating" and is the ingestion of solid particles.

_____ 3. Vesicle fuses with the plasma membrane and the content of the vesicle is expelled from the cell.

Cell Metabolism

66 Cell metabolism is the sum of all the catabolic and anabolic reactions in the cell. 99

Match these terms with the correct statement or definition:

Aerobic respiration
Anaerobic respiration

Glycolysis

_____ 1. The conversion of glucose to pyruvic acid; occurs in the cytosol.

_____ 2. Occurs when oxygen is available.

_____ 3. Converts glucose to carbon dioxide and water, producing 36 to 38 ATP molecules.

_____ 4. Results in the formation of lactic acid and two ATP molecules.

Protein Synthesis

66_Events that lead to protein synthesis begin in the nucleus and end in the cytoplasm._**99**

A. Match these terms with the correct statement or definition:

mRNA   Transcription
rRNA   Translation
tRNA

_____ 1. This process occurs when double strands of a DNA segment separate and RNA nucleotides pair with DNA nucleotides.

_____ 2. This type of RNA carries information in groups of three nucleotides, called codons, and each codon codes for a specific amino acid.

_____ 3. Type of RNA with an anticodon; binds to a specific amino acid.

_____ 4. This process involves the synthesis of polypeptide chains at the ribosome in response to information contained in mRNA molecules.

The proteins produced in a cell function as structural components, transport molecules, receptors, and enzymes.

B. Match these terms with the correct statement or definition:

Adenine   Guanine
Codon   Thymine
Cytosine   Triplet
Gene   Uracil

_____ 1. A sequence of three nucleotides in DNA.

_____ 2. The triplets in DNA required for the synthesis of a specific protein.

_____ 3. A term for the nucleotides responsible for the genetic code.

_____ 4. Two nucleotides that can pair with adenine.

_____

_____ 5. The nucleotide that pairs with cytosine.

C. Match these terms with the correct statement or definition:

Exon
Gene
Intron

Pre-RNA
Post-transcriptional processing
Transcriptional unit

_____ 1. Region of DNA between a start and stop codon.

_____ 2. Region of DNA that codes for a portion of a protein.

_____ 3. Region of DNA that does not code for a portion of a protein.

_____ 4. An mRNA containing introns.

_____ 5. Removal of introns and splicing together of extrons.

_____ 6. All of the nucleic acid sequences necessary to make a functional RNA or protein.

D. Match these terms with the correct statement or definition:

Polyribosome
Post-translational processing

Proenzyme
Proprotein

_____ 1. Protein with extra pieces that are removed by enzymes to form a functional protein.

_____ 2. Proproteins that can be modified to form functional enzymes.

_____ 3. Adding polysaccharides to a protein or combing of amino acid chains to form a protein.

_____ 4. A cluster of ribosomes attached to an mRNA; the ribosomes produce identical proteins.

Mitosis

66 *Nearly all cell divisions in the body occur by the same process, and the resultant* **99** *"daughter" cells have the same amount and type of DNA as the "parent" cells.*

A. Match these terms with the correct statement or definition:

Autosome
Cytokinesis
Homologous

Interphase
Mitosis
Sex chromosomes

_____ 1. Period between cell divisions when DNA and organelles are replicated.

_____ 2. In human cells, 22 pairs of chromosomes are included in this group.

_____ 3. Members of each pair of autosomal chromosomes, which look structurally alike, are described by this term.

_____ 4. Division of the nucleus into two nuclei.

_____ 5. Division of the cell's cytoplasm to produce two new cells.

👉 During replication of DNA, each strand of DNA serves as a template for the production of a new strand of DNA. This results in the production of two identical DNA molecules.

B. Match these terms with the correct statement or definition:

Astral fibers   Chromatids
Centromere   Spindle fibers

_____ 1. Identical pieces of DNA

_____ 2. Specialized region that joins two chromatids together.

_____ 3. Microtubules that radiate from the centrioles and end blindly.

_____ 4. Microtubules that project toward the equator and may attach to the centromeres of the chromosomes.

C. Match these terms with the phases of mitosis and the cell parts involved in mitosis labeled in Figure 3-3:

Anaphase   Prophase
Centriole   Spindle fiber
Centromere   Telophase
Metaphase

1. _____   4. _____   6. _____

2. _____   5. _____   7. _____

3. _____

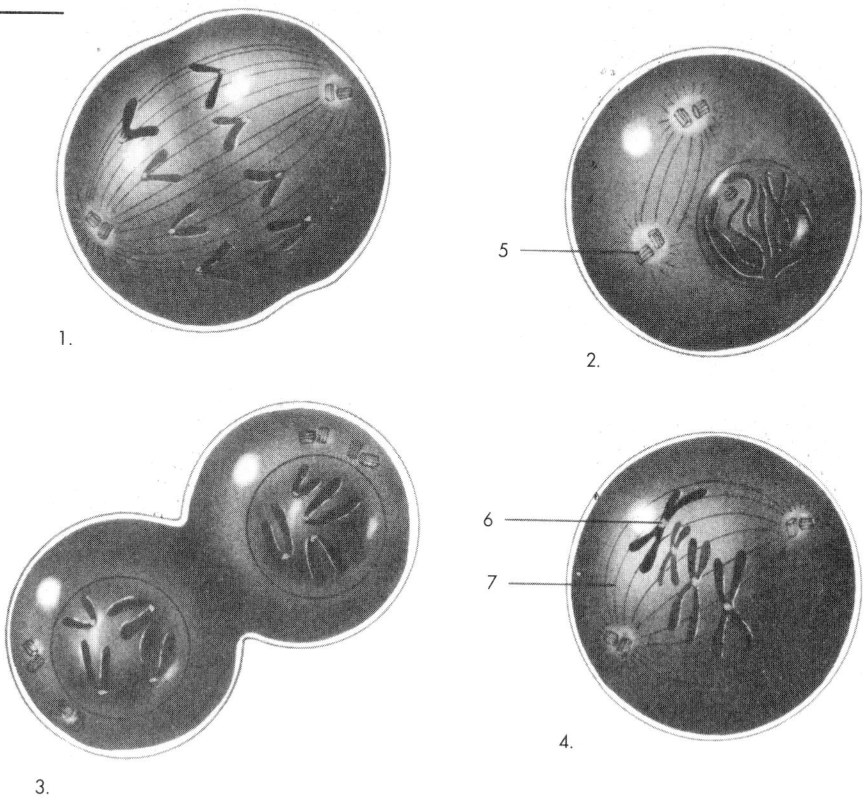

Figure 3-3

D. Match these terms with the correct statement or definition as they relate to the events in mitosis:

Anaphase   Prophase
Metaphase   Telophase

_____ 1. Chromosomes become visible.

_____ 2. Chromosomes (each with two chromatids) align along the equator with spindle fibers attached to their centromeres.

_____ 3. Separation of chromatids occurs.

_____ 4. Migration of each set of chromosomes toward the centrioles is completed.

Meiosis

"Meiosis is the special process in which the nucleus undergoes two divisions to produce gametes, each containing half the number of chromosomes in the parent cell."

Match these terms with the correct statement or definition:

Crossing over   Haploid
Diploid   Interkinesis
Gamete   Tetrad

_____ 1. Reproductive cells; a sperm cell in a male or an oocyte in a female.

_____ 2. Refers to the complement of chromosomes contained in a gamete.

_____ 3. In prophase I, the four chromatids of a homologous pair of chromosomes join together, or synapse, to form this structure.

_____ 4. Time between the formation of daughter cells from the first division of meiosis and the second meiotic division.

_____ 5. When tetrads are formed, some of the chromosomes break apart, and part of one chromatid is exchanged for part of another chromatid.

 Genetic diversity occurs for two reasons. First, crossing over produces chromatids with different DNA content; and second, there is a random distribution of the genetic material received from each parent.

Complete the following chart by writing in the organelle described by the structures and functions listed.

ORGANELLE FUNCTION

STRUCTURE

1. _____
Membrane-bound vesicle pinched off from the Golgi apparatus.
Contents released to the exterior of the cell by exocytosis.

2. _____
Broad, flattened sacs and tubules that interconnect, no ribosomes attached.
Lipid synthesis, detoxification, and calcium ion storage.

3. _____
Surrounded by double-layered envelope with pores.
Contains DNA in the form of chromatin (chromosomes) which produces RNA.

4. _____
Composed primarily of protein units called tubulin.
Support for the cytoplasm of the cell; involved in cell division; essential component of centrioles, spindle fibers, cilia, and flagella.

5. _____
Cylindrical extensions of the cell membrane supported by microfilaments.
Increase cell surface area for absorption.

6. _____
Closely packed stacks of curved cisternae composed of smooth endoplasmic reticulum.
Modifies, packages, and distributes proteins and lipids.

7. _____
Two subunits composed of ribosomal RNA and protein.
Site where mRNA and tRNA come together to assemble amino acids into proteins.

8. _____
Membrane-bound vesicle that contains hydrolytic enzymes.
Breakdown of phagocytized particles; autophagia.

9. _____ Small, spherical, rod-shaped or filamentous organelle; double membrane with infoldings of the inner membrane called cristae. | Most ATP synthesis in the cell.

10. _____ Nine evenly spaced, longitudinally oriented, parallel units; each unit consists of three parallel microtubules joined together. | Moves to each side of the nucleus during cell division; special microtubules called spindle fibers develop from the surrounding region.

11. _____ Broad, flattened sacs and tubules that interconnect; ribosomes attached. | Synthesis of proteins.

12. _____ One to four rounded, dense, well-defined nuclear bodies. | Production of ribosomal subunits.

13. _____ Granule in the cytoplasm. | Aggregates of chemicals include hemoglobin and glycogen.

14. _____ Appendages from the surface of the cell with two centrally located microtubules and nine peripheral pairs of microtubules joined together. | Movement of materials over the surface of the cells.

15. _____ Membrane-bound vesicle that contains a variety of enzymes such as catalase. | Detoxifies harmful molecules, breaks down hydrogen peroxide.

16. _____ Small protein fibrils that form bundles, sheets, or networks. | Provide structure to cytoplasm and mechanical support to microvilli; responsible for muscle's contractile capabilities.

17. _____ Protein fibrils, intermediate in size between actin filaments and microtubules. | Provide mechanical strength to cells.

18. List the three major types of molecules found in the cell membrane and give their functions.

19. List the factors that affect the rate and direction of diffusion in a solution.

20. List eight types of movement of materials across cell membranes.

21. List three characteristics of mediated transport mechanisms.

22. List three terms used to describe the tendency of cells to shrink or swell when placed in a solution.

23. List three types of RNA.

24. Name the four phases of mitosis.

MASTERY LEARNING ACTIVITY

Place the letter corresponding to the correct answer in the space provided.

_____1. Which of the following are functions of the proteins found in cell membranes?
a. membrane channels
b. carrier molecules
c. enzymes
d. receptor molecules
e. all of the above

_____2. Cytoplasm is found
a. in the nucleus.
b. between the plasma membrane and nuclear membrane.
c. inside the plasma membrane.
d. throughout the cell.

_____3. A large structure normally visible in the nucleus of a cell is the
a. endoplasmic reticulum.
b. mitochondria.
c. nucleolus.
d. lysosome.

_____4. Which cell organelle is involved in the modification and packaging of secretory materials?
a. nucleolus
b. ribosome
c. mitochondria
d. Golgi apparatus
e. chromosomes

_____5. Which of the following organelles produces large amounts of ATP?
a. nucleus
b. mitochondria
c. ribosomes
d. endoplasmic reticulum
e. lysosomes

_____6. Which of the following organelles would one expect to be present in large amounts in a cell responsible for producing lipids?
a. rough endoplasmic reticulum
b. lysosomes
c. smooth endoplasmic reticulum
d. ribosomes
e. a and c

_____7. Mature red blood cells cannot
a. synthesize ATP.
b. transport oxygen.
c. synthesize new protein.
d. use glucose as a nutrient.

_____8. If you observed the following characteristics in an electron micrograph of a cell:
1. microvilli
2. many mitochondria
3. the cell lines a cavity
4. no lysosome
5. little smooth endoplasmic reticulum
6. few vesicles

On the basis of these observations alone, which of the following is likely to be a major function of that cell?
a. active transport
b. secretion of protein
c. intracellular digestion
d. secretion of a lipid

_____9. In general, lipid-soluble molecules diffuse through the _____; small water-soluble molecules diffuse through the _____.
a. membrane channels; membrane channels
b. membrane channels; cell membranes
c. cell membranes; membrane channels
d. cell membranes; cell membranes

_____10. The end result of diffusion is
a. all net movement of molecules ceases.
b. some molecules become more concentrated.
c. determined by the amount of energy expended in the diffusion process.
d. determined by the rate at which diffusion took place.

_____ 11. Which of the following statements is true about osmosis?
   a. Always involves a membrane that allows water but not all solutes to diffuse through it.
   b. The greater the solute concentration, the greater the osmotic pressure of a solution.
   c. Osmosis always involves difference in solvent concentration.
   d. The greater the osmotic pressure of a solution, the greater is the tendency for water to move into the solution.
   e. all of the above

_____ 12. Cells placed in a hypertonic solution will
   a. swell.
   b. shrink.
   c. neither swell nor shrink.
   d. this term does not describe the tendency of cells to shrink or swell.

_____ 13. Container A contains a 10% salt solution and container B a 20% salt solution. If the two solutions were connected, the net movement of water by diffusion would be from to _____, and the net movement of the salt by diffusion would be from to _____.
   a. A,B,A,B
   b. A,B,B,A
   c. B,A,A,B
   d. B,A,B,A

_____ 14. Suppose that a woman was running a long distance race. During the race she lost a large amount of hyposmotic sweat. You would expect her cells to
   a. shrink.
   b. swell.
   c. remain the same.

_____ 15. Suppose that a man is doing heavy exercise in the hot summer sun. He sweats profusely. He then drinks a large amount of distilled water. You would expect his tissue cells to
   a. shrink.
   b. swell.
   c. remain the same.

_____ 16. The basic principle of the artificial kidney is to pass blood through very small blood channels made up of thin membranes. On the other side of the membrane is dialyzing fluid. The membrane is porous enough to allow all materials in the blood, except the plasma proteins and blood cells, to diffuse through. It is known that normal blood plasma has the following concentrations.

Sodium	Potassium	Urea	Glucose
142	5	26	100

The artificial kidney removes waste products such as urea while maintaining important constituents of the blood plasma such as sodium, potassium, and glucose as close as possible to their normal levels. Which of the proposed dialyzing fluids would accomplish this task?

	Sodium	Potassium	Urea	Glucose
a.	0	0	0	0
b.	142	5	0	125
c.	0	0	26	100
d.	148	10	0	90
e.	152	10	0	125

_____ 17. Which of the following statements is true about facilitated diffusion?
   a. Net movement is with the concentration gradient.
   b. It requires the expenditure of energy.
   c. It does not require a carrier molecule.
   d. It moves materials through membrane channels.
   e. a and b

_____ 18. The concentration of fluids in the human intestine tends to be isosmotic with the blood plasma. The intestinal membrane is essentially impermeable to magnesium salts. If a person ingested a large amount of magnesium salts for treatment of constipation, you would expect his tissue cells to
   a. shrink.
   b. swell.
   c. remain the same.

19. In diabetes mellitus, because of insufficient insulin production, glucose cannot enter cells. Instead it accumulates in the blood plasma. Which of the following statements would be true under these conditions?
   a. The concentration (osmolality) of the plasma would decrease.
   b. The osmotic pressure of the plasma would remain the same.
   c. Water would move out of the tissues into the plasma.
   d. Tissue cells would swell.

20. Given the following characteristics:
   1. specificity
   2. competition
   3. saturation

   Which are characteristics of active transport?
   a. 1
   b. 1,2
   c. 1,3
   d. 2,3
   e. 1,2,3

21. Which of the following statements concerning the cotransport of glucose into cells is true?
   a. The sodium-potassium exchange pump moves Na^+ ions into cells.
   b. The concentration of Na^+ ions outside cells is less than inside cells.
   c. A carrier molecule moves Na^+ ions into cells and glucose out of cells.
   d. The concentration of glucose inside cells can be greater than outside cells.

22. A process that requires energy and moves large molecules in solution (i.e., not particulate matter) into cells is called
   a. diffusion.
   b. facilitated diffusion.
   c. pinocytosis.
   d. phagocytosis.
   e. exocytosis.

23. An experimenter made up a series of solutions of different concentrations of a substance. Cells were then placed in each solution, and after 1 hour the amount of the substance that had moved into the cells was determined. The results are graphed below:

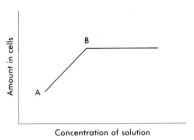

Concerning the part of the graph from A to B:
   a. the transport process appears to require membrane channels.
   b. the substance moving into the cells must be lipid soluble.
   c. movement is against the concentration gradient.
   d. concentration of the substance within the cell fails to increase.
   e. carriers are unsaturated.

24. In an experiment the rate of transfer of an amino acid into a cell was carefully monitored. Part way through the experiment, at time A, a metabolic inhibitor was introduced. The results of this experiment are:

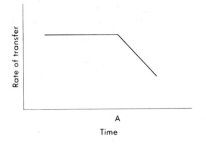

On the basis of these data it can be concluded that the mechanism responsible for the movement of the amino acid was
   a. diffusion.
   b. facilitated diffusion.
   c. active transport.
   d. there is insufficient information to reach a conclusion.

25. Cells that initially contained NO intracellular glucose were placed in a series of glucose solutions of different concentrations. After 5 minutes, the amount of intracellular glucose was determined. The results of the experiment are graphed below. (Hint: note the concentration indicated on each axis of the graph).

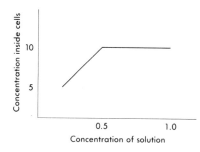

Which transport mechanism is responsible for moving glucose into the cell?
a. diffusion
b. facilitated diffusion
c. active transport
d. inadequate data to support a conclusion

26. Two different molecules, A and B, can both move easily across cell membranes when only one of them is present. However, when both types are present, the movement of A is greatly reduced, whereas the movement of B is only slightly reduced. This information suggests that the movement of molecule A is by _____, and the movement of molecule B is by _____.
a. diffusion, diffusion
b. diffusion, active transport
c. active transport, diffusion
d. active transport, active transport

27. A portion of an mRNA molecule that determines one amino acid in a polypeptide chain is called a
a. nucleotide.
b. gene.
c. codon.
d. exon.
e. intron.

28. In which of the following organelles is mRNA synthesized?
a. nucleus
b. ribosome
c. endoplasmic reticulum
d. nuclear envelope
e. peroxisome

29. Which organelle of the cell serves as the site of protein synthesis?
a. ribosome
b. lysosome
c. Golgi apparatus
d. nucleolus
e. vacuole

30. Transfer RNA
a. specifies the structure of enzymes.
b. duplicates itself at mitosis.
c. provides a template for mRNA.
d. a and c
e. none of the above

31. Choose the consequence that most specifically predicts the response of a cell to a substance that inhibits messenger RNA synthesis.
a. inhibits protein synthesis
b. inhibits DNA synthesis
c. inhibits fat synthesis
d. stimulates protein synthesis
e. stimulates fat synthesis

32. Given the following activities:
1. repair
2. growth
3. gamete production
4. differentiation

Which of the activities are the result of mitosis?
a. 2
b. 3
c. 1, 2
d. 3, 4
e. 1, 2, 4

33. Chromosomes disperse to become chromatin during
a. anaphase.
b. metaphase.
c. prophase.
d. telophase.

_____ 34. The major function of meiosis is to ensure that each of the resultant daughter cells
a. has the same number and kind of chromosomes as the parent cell.
b. has one half the number of chromosomes as the parent cell.
c. has one half the number and one of each pair of homologous chromosomes from the parent cell.
d. none of the above

_____ 35. The sex of individuals is determined by their sex chromosomes. An individual is male if he has chromosomes
a. MM.
b. MF.
c. XX.
d. XY.

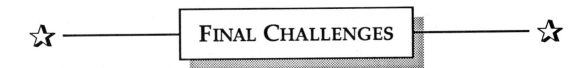

FINAL CHALLENGES

Use a separate sheet of paper to complete this section.

1. Given that you can make a solution of different concentrations (using a solute that does not cross the cell membrane), design an experiment that could be used to determine the concentration (osmolality) of the cytoplasm of red blood cells.

2. Sometimes isotonic glucose solutions are administered intravenously to patients who cannot otherwise take adequate amounts of nourishment. A nursing student hypothesized that such a treatment would result in the patient producing a large amount of dilute urine. Explain why you agree or disagree with this hypothesis.

3. Two different molecules, A and B, can both easily cross cell membranes when only one of them is present. Furthermore, even when both molecules are present, the rate of movement of each molecule is unaffected by the presence of the other molecule.

   A. Given that molecules A and B move by either diffusion or facilitated diffusion, what additional information or experiments would be necessary to determine the method of transport?

   B. Given that molecules A and B move by either facilitated diffusion or active transport, what additional information or experiments would be necessary to determine the method of transport?

4. A suspension of cells was prepared from two different types of tissue, A and B. The oxygen consumption (milliliters of oxygen used per minute) of the cells was determined with the following results:

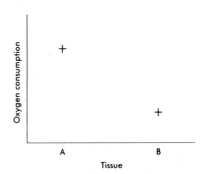

Can it be concluded that tissue A has a greater number of mitochondria? Explain.

5. Suppose that you have available for use a light microscope and an electron microscope. Design an experiment that would adequately test the hypothesis that cells synthesize enzymes for the purpose of secretion.

6. Suppose you have the following results from an experiment: (1) a stimulus caused a cell to secrete a lipid, and (2) treatment of the cell with an inhibitor of mRNA did not affect the response of the cell to the stimulus. Describe the events that lead to lipid secretion in this cell in response to a stimulus.

ANSWERS TO CHAPTER 3

Structure of the Plasma Membrane
A. 1. Intracellular; 2. Phospholipid;
3. Hydrophilic; 4. Cholesterol; 5. Fluid mosaic model
B. 1. Lipid bilayer; 2. Carbohydrate;
3. Cholesterol; 4. Protein; 5. Hydrophobic;
6. Hydrophilic
C. 1. Integral protein; 2. Channel protein;
3. Receptor molecule; 4. Marker molecule;
5. Glycoprotein
D. 1. Binding site; 2. Ligand; 3. G protein;
4. Second messenger; 5. Voltage gated;
5. Ligand gated

Nucleus
1. Nucleus; 2. Nuclear envelope; 3. Nuclear pores; 4. Chromatin; 5. Histones;
6. Chromosomes;
7. Messenger RNA; 8. Nucleolus

Cytoplasm
A. 1. Cytosol; 2. Cytoskeleton; 3. Cytoplasmic inclusions; 4. Organelles
B. 1. Microtubule; 2. Actin filament;
3. Intermediate filament

Ribosomes and Endoplasmic Reticulum
1. Ribosome; 2. Rough endoplasmic reticulum;
3. Rough endoplasmic reticulum; 4. Smooth endoplasmic reticulum

Golgi Apparatus and Secretory Vesicles
1. Golgi apparatus; 2. Vesicle; 3. Secretory vesicle

Lysosomes and Peroxisomes
1. Lysosomes; 2. Autophagia; 3. Peroxisomes;
4. Catalase

Mitochondria
1. ATP; 2. Cristae; 3. Oxidative metabolism;
4. Matrix; 5. Electron transport chain; 6. DNA

Centrioles, Spindle Fibers, Cilia, Flagella, and Microvilli
1. Centrosome; 2. Centrioles; 3. Spindle fibers;
4. Cilia; 5. Microvilli

Cell Diagram
1. Cilia; 2. Secretory vesicle; 3. Plasma membrane; 4. Golgi apparatus; 5. Smooth endoplasmic reticulum; 6. Lysosome; 7. Rough endoplasmic reticulum; 8. Mitochondrion;
9. Centrioles; 10. Nucleolus; 11. Microtubule;
12. Ribosome; 13. Microvilli

Movement Through the Plasma Membrane
1. Lipid bilayer; 2. Membrane channels;
3. Membrane channels; 4. Carrier molecules;
5. Vesicle

Diffusion
A. 1. Solvent; 2. Diffusion; 3. Diffusion
B. 1. Increase; 2. Increase; 3. Decrease

Osmosis
1. Osmosis; 2. Osmotic pressure; 3. Hyposmotic;
4. Hypertonic; 5. Lysis

Filtration
1. Pressure difference; 2. Higher; 3. Larger

Mediated Transport Mechanisms
A. 1. Specificity; 2. Competition; 3. Saturation
B. 1. Facilitated diffusion; 2. Facilitated diffusion; 3. Active transport
C. 1. Out of; 2. With; 3. Into; 4. Against;
5. Cotransport

Endocytosis and Exocytosis
1. Endocytosis; 2. Phagocytosis; 3. Exocytosis

Cell Metabolism
1. Glycolysis; 2. Aerobic respiration; 3. Aerobic respiration; 4. Anaerobic respiration

Protein Synthesis
A. 1. Transcription; 2. mRNA; 3. tRNA;
4. Translation
B. 1. Triplet; 2. Gene; 3. Codon; 4. Thymine, uracil; 5. Guanine
C. 1. Transcriptional unit; 2. Exon; 3. Intron;
4. Pre-RNA; 5. Post-transcriptional processing; 6. Gene
D. 1. Proprotein; 2. Proenzyme; 3. Post-translational processing; 4. Polyribosome

Mitosis
A. 1. Interphase; 2. Autosome; 3. Homologous;
4. Mitosis; 5. Cytokinesis
B. 1. Chromatids; 2. Centromere; 3. Astral fibers;
4. Spindle fibers
C. 1. Anaphase; 2. Prophase; 3. Telophase;
4. Metaphase; 5. Centriole; 6. Centromere;
7. Spindle fiber
D. 1. Prophase; 2. Metaphase; 3. Anaphase;
4. Telophase

Meiosis
1. Gamete; 2. Haploid; 3. Tetrad; 4. Interkinesis;
5. Crossing over

1. Secretory vesicle
2. Smooth endoplasmic reticulum
3. Nucleus
4. Microtubule
5. Microvillus
6. Golgi apparatus
7. Ribosome
8. Lysosome

9. Mitochondrion
10. Centriole
11. Rough endoplasmic reticulum
12. Nucleoli
13. Cytoplasmic inclusion
14. Cilia
15. Peroxisome
16. Actin filaments
17. Intermediate filaments
18. Lipids: phospholipids form the lipid bilayer that separates the inside of the cell from the outside; cholesterol: contributes to the fluid nature of the membrane; proteins: membrane channels, carrier molecules, receptor molecules, marker molecules, enzymes, or structural support
19. Magnitude of concentration gradient, temperature of solution, size of diffusing molecules, and viscosity of the solvent
20. Diffusion, osmosis, filtration, facilitated diffusion, active transport, secondary active transport, phagocytosis, pinocytosis, and exocytosis
21. Specificity, saturation, and competition
22. Isotonic, hypotonic, and hypertonic
23. Messenger RNA, transfer RNA, and ribosomal RNA
24. Prophase, metaphase, anaphase, and telophase

MASTERY LEARNING ACTIVITY

1. E. Proteins act as carrier molecules, function as enzymes, form membrane channels, and function as receptor and marker molecules.

2. B. The nucleus is defined as everything inside the nuclear membrane. Everything between the nuclear membrane and the plasma membrane is cytoplasm.

3. C. Normally the only large structure visible in the nucleus is the nucleolus. It is made up of RNA. The chromosomal material is dispersed, so the chromosomes are not visible except during cell division.

4. D. The Golgi apparatus is an extension of the endoplasmic reticulum (ER). Rough ER (with ribosomes) produces proteins, and smooth ER (no ribosomes) produces lipids that are transferred to the Golgi apparatus. The Golgi apparatus modifies, packages, and distributes the proteins and lipids.

5. B. The enzymes of the citric acid cycle (Krebs' cycle) and the electron transport system are located within the mitochondria. Therefore it is the site of oxidative metabolism and the major site of ATP synthesis within the cell. To answer this question and similar questions about the cellular organelles that could be asked, one must memorize their functions.

6. C. Smooth endoplasmic reticulum produces lipids that pass into the Golgi apparatus, where they are packaged into vesicles and then secreted. Rough endoplasmic reticulum (ER with ribosomes) produces proteins, which are then packaged by the Golgi apparatus. Lysosomes contain digestive enzymes.

7. C. The red blood cell has no nucleus and therefore cannot synthesize mRNA, which is required for new protein synthesis. Red blood cells can synthesize ATP, transport oxygen, and use glucose because of enzymes that were produced in the red blood cell before it lost its nucleus.

8. A. The cell lines a cavity, indicating that substances entering or leaving the cavity must be transported or must diffuse across that cellular barrier. The microvilli increase the surface area of the cell exposed to the lumen of the cavity, and numerous mitochondria suggest that energy, in the form of ATP, is used by the cell. Thus the data suggest that the cell is involved in active transport.

Lipid secretion is unlikely with the small amount of smooth endoplasmic reticulum. Without secretory vesicles, protein secretion is unlikely.

9. C. Water-soluble molecules usually do not diffuse through the membrane because of its lipid nature. If small enough, water-soluble molecules diffuse through membrane channels. Lipid-soluble molecules can diffuse through the membrane proper.

10. A. When net movement of molecules ceases, an equilibrium condition of a uniform distribution of molecules is achieved. This is the end result of diffusion.

b is incorrect, because with a uniform distribution molecules will not be more concentrated.

c is incorrect. Diffusion is a passive transport mechanism requiring no expenditure of energy.

d is incorrect, because the rate of diffusion has no effect on the equilibrium condition. A uniform distribution may be achieved quickly or slowly, but the end result is the same.

11. E   When a membrane does not permit the passage of at least one solute molecule, the result is an unequal concentration of solvent that diffuses through the membrane (osmosis). The greater the solute concentration of a solution, the greater is its osmotic pressure, and the greater is the tendency for water to move into the solution by osmosis.

12. B.   Cells placed in a hypertonic solution shrink, in a hypotonic solution they swell, and in an isotonic solution they neither swell nor shrink.

13. B.   Because A has 10% salt, it has 90% water; B has 20% salt, and thus 80% water. Substances diffuse from areas of higher concentration to low. Thus water will diffuse from A (90%) to B (80%), and salt will diffuse from B (20%) to A (10%).

14. A   Sweat is less concentrated than blood plasma. Thus, compared with blood plasma, more water is lost than salts. This increases the osmotic concentration of the blood plasma. Water then moves from the tissues into the plasma, resulting in shrinkage of the cells.

15. B.   Replacing the lost water and salts with distilled water results in a more dilute blood plasma. Relatively speaking, then, the blood plasma has more water than previously, and thus water moves from the blood plasma into the tissues, causing them to swell.

16. B.   There is no net movement of sodium or potassium, because the concentrations of these substances are the same in both the dialyzing fluid and the blood plasma. Urea definitely moves from the blood plasma (26) into the dialyzing fluid (0). There is a slight movement of glucose from the dialyzing fluid into the blood plasma, providing the patient with a small amount of glucose for energy.

Substances diffuse from areas of higher concentration to areas of lower concentration. A comparison of each proposed dialyzing fluid with normal blood plasma must be made to find one that maintains sodium, potassium, and glucose levels as close to the normal levels as possible, yet still remove urea.

a is incorrect. All of the substances involved are removed from the blood plasma.

c is incorrect. There is a loss of sodium and potassium, but no net exchange of urea.

d is incorrect. There is a gain of sodium and potassium and a loss of glucose.

e is incorrect. There is a gain of sodium, potassium, and glucose.

17. A   Facilitated diffusion moves materials with the concentration gradient, does not require the expenditure of energy, and does use a carrier molecule. Materials move through membrane channels by simple diffusion.

18. A   Because the magnesium salts tend to remain in the intestine, they cause the intestinal contents to be more concentrated (hyperosmotic) with respect to the blood plasma. Relatively speaking, then, the gut would contain less water than the blood plasma. Water moves from the blood plasma into the intestine. By a chain reaction, as water leaves the blood plasma, the concentration of the plasma relative to the tissues changes. As a result, water moves from the tissues (causing them to shrink) into the blood plasma.

19. C.   When glucose concentrations increase, the blood plasma contains proportionately less water. Thus water moves from the tissues into the blood plasma.

a is incorrect. As glucose increases, concentration (osmolality) of the plasma increases.

b is incorrect. As concentration (osmolality) increases, the osmotic pressure of the blood plasma increases.

d is incorrect. As water moves from the tissues (see answer C) into the blood plasma, the tissue cells shrink.

20. E   Active transport has specificity because molecules are transported by means of a carrier molecule. Because a carrier is involved, similar molecules may compete for the carrier. Also, once all the carrier molecules are in use, saturation occurs.

21. D.   The sodium-potassium exchange pump moves Na^+ ions out of cells, making the concentration of Na^+ ions outside of cells greater than inside of cells. As the Na^+ ions diffuse down their concentration gradient, carrier molecules move Na^+ ions and glucose molecules into cells. The energy from the Na^+ ion movement can be used to move glucose molecules against their concentration gradient, resulting in a greater concentration of glucose inside of cells than outside of cells.

22. C.   Diffusion and facilitated diffusion do not require energy. Phagocytosis moves particulate matter into cells, pinocytosis moves large molecules in solution into cells, and exocytosis moves substances out of cells.

23. E   Because the amount transferred after point B levels off despite increasing concentrations, it can be concluded that carrier-mediated transport is involved. In section A - B of the graph the carrier system is not saturated: thus increasing concentrations result in increased transport of the substance as more and more carrier molecules are used. From these data there is no way to tell if the carrier-mediated process is facilitated diffusion or active transport.

24. C   Because the metabolic inhibitor greatly reduced the rate of transfer of the amino acid, it can be concluded that the transport mechanism involved requires energy. Of the three transport mechanisms listed, only active transport uses energy.

25. C   Because the curve levels off (exhibits saturation), simple diffusion can be ruled out. Note that the cells started with NO intracellular glucose. After 5 minutes, however, the intracellular concentration of glucose was always greater than the extracellular concentration. Clearly this could occur only if active transport was involved.

26. D   The data suggest that both types of molecules are competing for the same carrier molecule. Apparently molecule B is a much better competitor, because its rate of transport is hardly affected by the presence of molecule A. Diffusion does not require carrier molecules, so both molecules must be moved by active transport.

27. C   A codon refers to a sequence of three organic bases (nucleotides) in a mRNA molecule. At the ribosome the codon determines which tRNA can combine with the mRNA. The particular tRNA that combines with the organic bases of mRNA is attached to a specific amino acid. Therefore a codon determines one amino acid in a polypeptide chain.

28. A   mRNA is synthesized within the nucleus. The sequence of organic bases that make up mRNA is determined by DNA within the nucleus. That is, DNA acts as a template for mRNA. The sequence of organic bases in mRNA (codons) subsequently determines the amino acid sequence of a protein.

29. A   At the ribosomes, mRNA and tRNA interact to form proteins.

30. E   Transfer RNA has as a component a sequence of three organic bases (an anticodon). These organic bases can pair with a unique sequence of three organic bases on mRNA (a codon), which was determined by the sequence of organic bases in the nuclear DNA. There are in excess of 20 different types of tRNA. Each type of tRNA is bound to a specific amino acid. Therefore the function of tRNA is to combine with a specific sequence of organic bases on mRNA and participate in determining the sequence of amino acids in a polypeptide.

31. A   Messenger RNA carries "instructions" for protein synthesis from the chromosomes (DNA) to the ribosomes. A substance that prevented the production of mRNA would prevent protein synthesis.

32. E   All of the characteristics listed except gamete production occur by mitosis. Gamete production is the result of meiosis.

33. D   During telophase the cell returns to the interphase condition. The chromosomes unravel to form chromatin, the nuclear membrane reforms, and the nucleolus reappears.

34. C   Meiosis results in the daughter cells having one half of the number of chromosomes as the parent cell. It also results in the daughter cells having one of each homologous pair of chromosomes (one of each type of chromosome found in the parent cell).

35. D   XY combination produces a male, and XX produces a female.

☆ FINAL CHALLENGES ☆

1. One approach is to make up different solutions of known concentrations (osmolalities). Red blood cells could be placed in each solution and observed to see if they change shape. The solution in which the red blood cells neither swell nor shrink is an isotonic solution. Because the solute molecules do not cross the cell membrane, this would also be an isosmotic solution. The concentration (osmolality) of the cytoplasm of the red blood cell would be equal to the concentration of this solution.

2. Agree. Initially, because the glucose solution is isotonic, there is only an increase in blood volume. However, as the glucose is used up (metabolized), the concentration (osmolality) of the blood decreases. This means that the blood is more dilute than usual; the kidneys compensate by producing large quantities of dilute urine. This removes the excess water from the blood, restoring it to its normal volume and concentration (osmolality).

3. A. Diffusion of molecules A and B could be distinguished from facilitated diffusion in two ways:

   1. If the process is facilitated diffusion, there could be competition for a carrier molecule. The fact that neither molecule affects the rate of transport of the other could mean that the process is not facilitated diffusion or that two different carrier molecules are involved. Introduction of other molecules similar to A or B could resolve this issue.

   2. If the process is facilitated diffusion, then increasing the concentration gradient for molecules A and B should result in saturation of the carrier molecule and a leveling off of the rate of transport. If the process is diffusion, increasing the concentration gradient results in increasing rates of transport.

B. Active transport of molecules A and B could be distinguished from facilitated diffusion (or simple diffusion) in two ways:

   1. Determine the concentration gradient and, if movement is against the concentration gradient, the transport mechanism is active transport.

   2. Introduce a metabolic inhibitor. Because active transport requires energy, this stops the movement of molecules A and B.

4. It is logical to conclude that tissue A has the greatest number of mitochondria (which use oxygen to produce ATP). However, it is not necessary to make that conclusion. Other possibilities are that tissue A has larger mitochondria or mitochondria that have a greater enzymatic activity than tissue B.

5. Enzymes are proteins. The presence of well-developed rough endoplasmic reticulum, Golgi apparatuses, and secretory vesicles is strong evidence for the secretion of a protein.

6. The fact that an inhibitor of mRNA synthesis does not inhibit the secretion of lipid suggests that the response of the cell involves the release of stored lipids or the activation of enzymes that already exist. Neither of these processes requires protein synthesis, which is blocked by an mRNA inhibitor.

A Pun Poem (untitled)

Where can a man buy a cap for his knee,
Or a key to the lock of his hair?
Should your eyes be called an academy
Because there are pupils there?

In the crown of your head, what jewels are found?
Who travels the bridge of your nose?
Could you use in shingling the roof of your mouth
The nails on the ends of your toes?

Could the crook in your elbow be sent to jail?
If so, what did he do?
How can you sharpen your shoulder blades,
I'll be darned if I know, do you?

Can you sit in the shade of the palm of your hand
And play on the drum of your ear?
Do the calves of our legs eat the corn on our toes,
Then why does it grow on the ear?

William Dunkle

64

Histology: The Study of Tissues

FOCUS: The cells of the body are specialized to form four basic types of tissues. Epithelial tissue consists of a single or a multiple layer of cells with little extracellular material between the cells. It covers the free surfaces of the body, providing protection, regulating the movement of materials, and producing secretions (glands). Connective tissue holds structures together (e.g., tendons and ligaments), is a site of fat storage and blood cell production, and supports other tissues (e.g., cartilage and bone). The extracellular matrix produced by connective tissue cells is responsible for many of the functional characteristics of connective tissue. Muscle tissue has the ability to contract and is responsible for body movement (skeletal muscle), blood movement (cardiac muscle), and movement of materials through hollow organs (smooth muscle). Nervous tissue is specialized to conduct electrical impulses and functions to control and coordinate the activities of other tissues. The four tissue types develop from three primary germ layers (ectoderm, mesoderm, and endoderm) in the embryo. In some locations epithelial and connective tissues form serous, mucous, or synovial membranes. Inflammation is a process that isolates and destroys injurious agents. Some tissues recover from injury by replacing lost cells with cells of the same type (regeneration), but other tissues can repair only by the addition of cells of a different type (replacement), leading to scar formation and loss of function.

CONTENT LEARNING ACTIVITY

Epithelial Tissue

❝*Epithelial tissues are components of every organ.*❞

Using the terms provided, complete these statements:

Basement membrane	Free
Diffusion	Mitosis
Extracellular	Surfaces

A major characteristic of epithelial cells is that they have very little _(1)_ material between them. Epithelium covers _(2)_, or forms structures that are derived developmentally from surfaces. Most epithelial tissues have one _(3)_ surface and a _(4)_, which is extracellular material secreted primarily on the side opposite their free surface. Blood vessels do not penetrate the basement membrane; all gases and nutrients must reach the epithelium by _(5)_. Because epithelial cells retain the ability to undergo _(6)_, damaged cells can be replaced with new epithelial cells.

1. _____

2. _____

3. _____

4. _____

5. _____

6. _____

Classification of Epithelium

"Epithelia are classified according to the number of cell layers and the shape of the cells."

A. Match these types of
epithelium with the
correct statement or definition:

Pseudostratified columnar   Stratified columnar
Simple columnar   Stratified cuboidal
Simple cuboidal   Stratified squamous
Simple squamous   Transitional

_____   1. Single layer of cube-shaped cells.

_____   2. Multiple layers of tall, thin cells.

_____   3. Layers of cells that appear cubelike when an organ is relaxed
                              and flattened when the organ is distended by fluid.

_____   4. Single layer of flat, often hexagonal cells.

_____   5. Single layer of cells; all cells are attached to the basement
                              membrane, but only some of them reach the free surface.

_____   6. Multiple layers of cells in which the basal layer is cuboidal
                              and surface layers become flattened ; moist or keratinized.

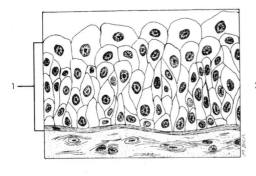
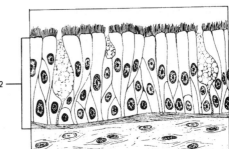
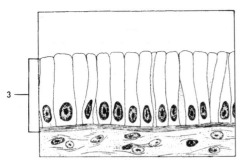
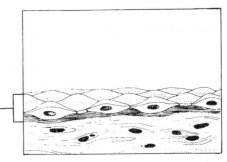

Figure 4-1

B. Match these terms with
the types of epithelial
tissue in Figure 4-1:

Pseudostratified columnar   1. _____
Simple columnar
Simple squamous   2. _____
Transitional
                             3. _____

                             4. _____

Functional Characteristics

"*Epithelial tissues perform many functions.***"**

A. Match these terms with the correct statement or definition:

Simple epithelium
Stratified epithelium

———————————————— 1. Found in organs where principal functions are diffusion, filtration, secretion, or absorption.

———————————————— 2. Found in areas where protection is a major function.

———————————————— 3. Found in areas such as the mouth, skin, throat, anus, and vagina.

B. Match these terms with the correct statement or definition:

Cuboidal or columnar
Squamous

———————————————— 1. Epithelial cells involved with secretion or absorption.

———————————————— 2. Epithelial cells involved with diffusion or filtration.

C. Match these terms with the correct statement or definition:

Ciliated   Smooth
Folded   Microvillar

———————————————— 1. Cell surface that reduces friction.

———————————————— 2. Cell surface that greatly increases surface area.

———————————————— 3. Propels materials along the cell surface.

———————————————— 4. Cell surface with rigid sections alternating with flexible sections.

D. Using the terms provided, complete these statements:

Desmosomes   Intercalated disks
Gap junctions   Zonula adherens
Glycoproteins   Zonula occludens
Hemidesmosomes

1. ————————————————

2. ————————————————

3. ————————————————

4. ————————————————

5. ————————————————

6. ————————————————

7. ————————————————

Epithelial cells secrete __(1)__ that attach the cells to the basement membrane and to each other. This relatively weak binding is reinforced by __(2)__, disk-shaped structures with especially adhesive glycoproteins that bind cells to each other. __(3)__, similar to one half of a desmosome, attach epithelial cells to the basement membrane. Tight junctions consist of the __(4)__, which acts like a weak glue to hold cells together, and the __(5)__, which acts as a permeability barrier. __(6)__ are small channels that allow the passage of ions and small molecules between cells as a means of intercellular communication. Specialized gap junctions between cardiac cells are called __(7)__.

Glands

"*Glands are secretory organs composed primarily of epithelium and connective tissue.*"

A. Match these terms with the correct statement or definition:

Endocrine
Exocrine

_____

1. Glands with a duct.

_____

2. Ductless glands that secrete hormones.

B. Match these terms with the correct statement or definition:

Acinar or alveolar   Straight
Coiled   Tubular
Compound   Unicellular
Simple

_____

1. Exocrine glands composed of one cell, e.g., goblet cells.

_____

2. Exocrine glands with ducts that branch repeatedly.

_____

3. Exocrine glands that end in saclike structures.

_____

4. Tubular exocrine glands with no coiling.

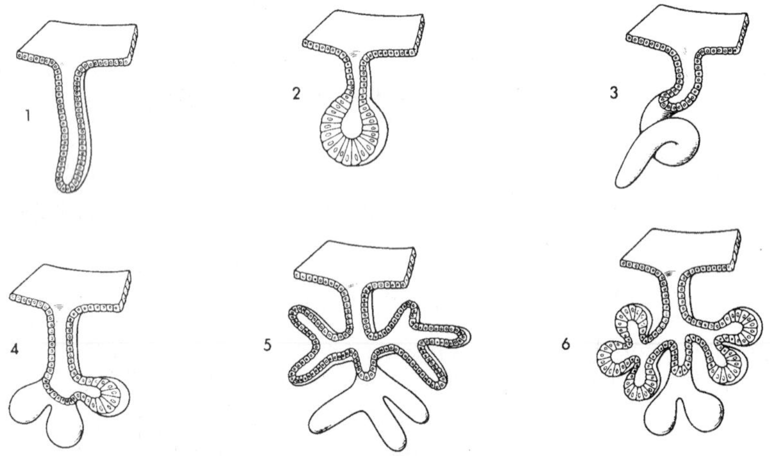

Figure 4-2

C. Match these terms with the types of exocrine glands in Figure 4-2:

Compound acinar
Compound tubular
Simple acinar
Simple branched acinar
Simple coiled tubular
Simple straight tubular

1. _____

2. _____

3. _____

4. _____

5. _____

6. _____

D. Match these terms with the correct statement or definition:

Apocrine   Merocrine
Holocrine

_____ 1. Exocrine glands that secrete products with no loss of actual cellular material, e.g., water-producing sweat glands.

_____ 2. Exocrine glands that discharge fragments of the gland's cells into the secretion, e.g., mammary glands.

_____ 3. Exocrine glands that shed entire cells, e.g., sebaceous glands.

☞ Endocrine glands are so varied in their structure that they are not easily classified.

Connective Tissue

66 *The essential characteristic that distinguishes connective tissue from the other three tissue* 99 *types is that it consists of cells separated from each other by abundant extracellular matrix.*

Match these terms with the correct statement or definition:

Blasts   Cytes
Clasts

_____ 1. Suffix for connective tissue cells that create the extracellular matrix.

_____ 2. Suffix for connective tissue cells that maintain the extracellular matrix.

_____ 3. Suffix for connective tissue cells that break down the extracellular matrix for remodeling.

☞ The extracellular matrix has three major components: (1) protein fibers, (2) ground substance consisting of nonfibrous protein and other molecules, and (3) fluid.

Protein Fibers of the Matrix

66 *Three types of protein fiber are in connective tissue.* 99

Match these terms with the correct statement or definition:

Collagen fibers   Reticular fibers
Elastin fibers

_____ 1. These protein fibers are the most common protein in the body and are strong and flexible, but inelastic.

_____ 2. These protein molecules are very short, thin collagen fibers that branch to form a network.

_____ 3. This protein gives the tissue in which it is found an elastic quality. The structure of this molecule is similar to that of a coiled metal spring.

Nonprotein Matrix Molecules

❝_There are two types of large, nonprotein molecules of the extracellular matrix._**❞**

Using the terms provided, complete these statements:

Aggregates   Monomers
Ground substance   Proteoglycans
Hyaluronic acid   Water

The __(1)__ is the shapeless background against which
collagen fibers are seen through the microscope. One type
of ground substance is __(2)__, a long, unbranched polysaccharide that is very slippery, whereas __(3)__ are formed
from proteins and polysaccharides. Proteoglycan
__(4)__ resemble minute pine tree branches. If the protein
cores of proteoglycan monomers attach to hyaluronic acid,
proteoglycan __(5)__ are formed. Proteoglycans trap large
quantities of __(6)__, which gives them the capacity to return
to their original shape when compressed.

1. _____

2. _____

3. _____

4. _____

5. _____

6. _____

Matrix with Fibers as the Primary Feature

❝_Connective tissue with a matrix of primarily fibers has two subtypes: fibrous and special._**❞**

A. Match these terms with the
   correct statement or definition:

   Dense   Loose (areolar)
   Fibroblasts   Macrophages

   _____

   _____

   _____

   _____

1. Fibrous connective tissue in which protein fibers fill nearly all
   the extracellular space.

2. Fibrous connective tissue in which protein fibers form a lacy
   network with numerous fluid-filled spaces.

3. Cells that produce the fibers of connective tissue.

4. Cells that engulf bacteria and cell debris within connective
   tissue.

B. Match these terms with the
   correct statement or definition:

   Dense irregular
   Dense regular

   _____

   _____

1. Dense connective tissue that contains protein fibers
   predominantly oriented in the same direction; strong in one
   direction.

2. Dense connective tissue that contains protein fibers that can be
   arranged as a network of randomly oriented fibers; strong in
   many directions.

C. Match these terms with the correct statement or definition:

Dense irregular collagenous   Dense regular collagenous
Dense irregular elastic   Dense regular elastic

_____ 1. Connective tissue found in tendons and ligaments; collagen fibers oriented in the same direction.

_____ 2. Connective tissue found in special ligaments such as the nuchal ligament; elastin fibers oriented in the same direction.

_____ 3. Dense connective tissue with randomly oriented collagen fibers; found in the skin (dermis).

_____ 4. Dense connective tissue with randomly oriented fibers of elastin; found in walls of large arteries.

☞ The nuchal ligament lies along the posterior neck and helps hold the head upright.

D. Match these terms with the correct statement or definition:

Adipose   Reticular tissue
Dendritic cells   Yellow bone marrow
Red bone marrow

_____ 1. Consists of adipocytes (fat cells), which contain large amounts of lipid and can be either brown or yellow (white) in color.

_____ 2. Forms the framework of lymphatic tissue, bone marrow, and the liver.

_____ 3. Cells that occur between reticular fibers and look very much like reticular cells, but are part of the immune system.

_____ 4. Found in bone; contains adipose tissue.

_____ 5. Hemopoietic tissue surrounded by reticular fibers.

Matrix with Protein Fibers and Ground Substance

66 _Two types of connective tissue have a matrix with both protein fibers and ground substance._ 99

A. Match these terms with the correct statement or definition:

Chondrocytes   Hyaline cartilage
Elastic cartilage   Lacunae
Fibrocartilage

_____ 1. Cartilage cells.

_____ 2. Spaces in which cartilage cells are located.

_____ 3. Very smooth tissue with a glassy, translucent matrix. It is found in areas where strong support and some flexibility are needed (e.g., rib cage and rings of bronchi and trachea).

_____ 4. Cartilage with thick bundles of collagen dispersed through the matrix. Found in areas that must withstand a great deal of pressure, such as the knee.

B. Match these terms with the correct statement or definition:

Cancellous bone
Compact bone
Hydroxyapatite

Lacunae
Osteocytes
Trabeculae

_____ 1. Complex salt crystal; the mineral (inorganic) portion of bone.

_____ 2. Bone cells.

_____ 3. Spaces occupied by bone cells.

_____ 4. Plates of bone with spaces between them.

_____ 5. Essentially solid bone, with almost no space between lamina.

Identification of Connective Tissue

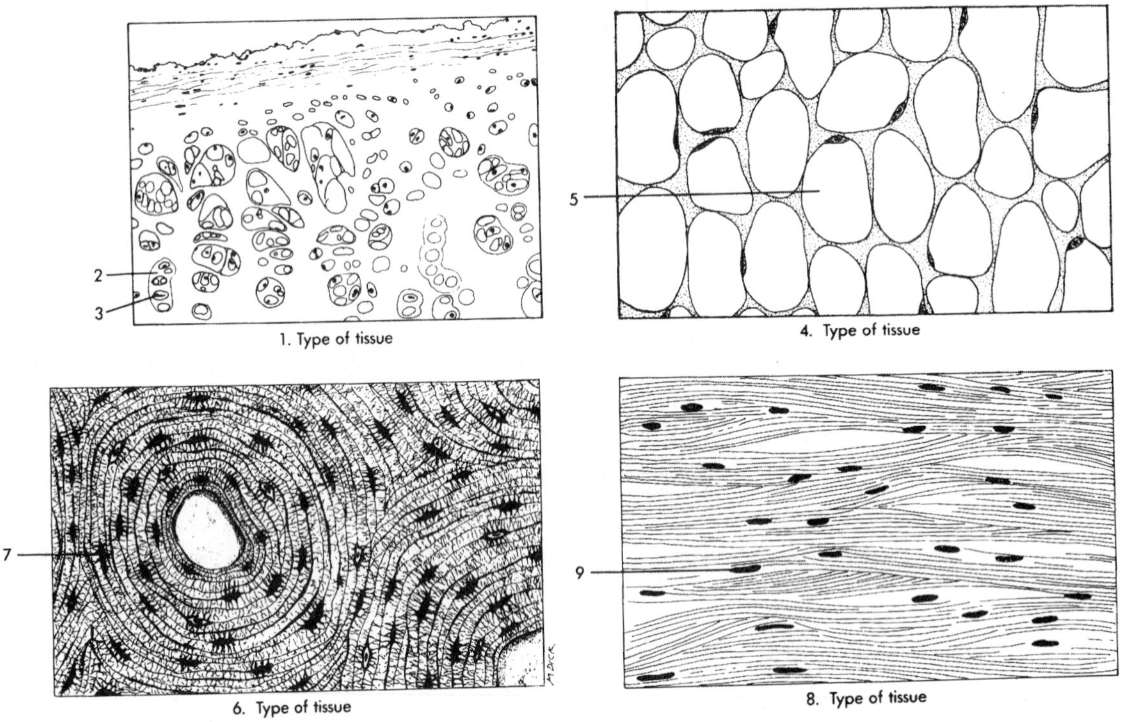

1. Type of tissue

4. Type of tissue

6. Type of tissue

8. Type of tissue

Figure 4-3

Match these terms with the correct tissue type or structure labeled in Figure 4-3:

Adipose
Bone
Cartilage
Chondrocyte
Dense regular
  connective tissue

Fat droplet
Fibroblast
Lacuna
Osteocyte

1. _____   4. _____   7. _____

2. _____   5. _____   8. _____

3. _____   6. _____   9.

 Blood is unique among the connective tissues because the matrix between the cells is liquid.

Muscular Tissue

"The main characteristics of muscle tissue are that it is contractile and responsible for movement."

A. Match these terms with the correct statement or definition:

Cardiac   Smooth
Skeletal

_____ 1. Striated, voluntary muscle cells.

_____ 2. Striated, involuntary muscle cells.

_____ 3. Nonstriated, involuntary muscle cells.

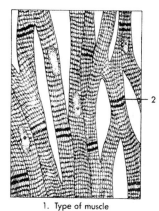

1. Type of muscle

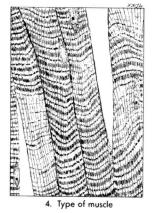

3. Type of muscle

4. Type of muscle

Figure 4-4

B. Match these terms with the correct tissue type or structure labeled in Figure 4-4:

Cardiac muscle   Skeletal muscle
Intercalated disk   Smooth muscle

1. _____   3. _____

2. _____   4. _____

Nervous Tissue

Nervous tissue has the ability to conduct electrical signals called action potentials.

Match these terms with the correct statement or definition:

Axon
Bipolar
Cell body
Dendrite
Multipolar
Neuroglia
Unipolar

_____ 1. Part of a neuron containing the nucleus.

_____ 2. Nerve cell process that conducts action potentials away from the cell body.

_____ 3. Neuron with several dendrites and one axon.

_____ 4. Neuron with one dendrite and one axon.

_____ 5. Support cells of the brain, spinal cord, and peripheral nerves that nourish, protect, and insulate neurons.

Embryonic Tissue

Four tissue types are derived from three germ layers during early development of the embryo.

Match these terms with the correct statement or definition:

Ectoderm
Endoderm
Mesoderm
Neural crest cells
Neuroectoderm

_____ 1. Inner germ layer; forms the lining of the digestive tract and its derivations.

_____ 2. Middle germ layer; forms tissues such as muscle, blood vessels, and bone.

_____ 3. Outer germ layer; forms skin.

_____ 4. Becomes the nervous system.

_____ 5. Groups of cells that break away from neuroectoderm; become peripheral nerves, skin pigment, and many tissues of the face.

Membranes

A membrane is a thin sheet or layer of tissue that covers a structure or lines a cavity.

A. Match these terms with the correct statement or definition:

Mucous membranes   Synovial membranes
Serous membranes

_____ 1. Consist of simple squamous epithelium (mesothelium), its basement membrane, and a delicate layer of loose connective tissue; line cavities that do not open to the exterior.

_____ 2. Consist of epithelial cells and their basement membrane, which rests on a thick layer of loose connective tissue called the lamina propria; line cavities that open to the outside of the body.

_____ 3. Modified connective tissue cells associated with the dense connective tissue of the joint capsule.

B. Match these terms with the correct statement or definition:

Mucus   Synovial fluid
Serous fluid

_____ 1. Fluid that lubricates serous membranes.

_____ 2. Viscous substance produced by goblet cells or multicellular glands.

_____ 3. Fluid rich in hyaluronic acid; lubricates freely movable joints.

Inflammation

The inflammatory response mobilizes the body's defenses, isolates and destroys microorganisms and other infectious agents, and removes foreign materials and damaged cells so that tissue repair can proceed.

Match these terms with the correct statement or definition:

Coagulation   Mediators of inflammation
Disturbance of function   Pain
Edema   Vasodilation

_____ 1. Chemical substances that are released or activated in tissues and adjacent blood vessels after a person is injured.

_____ 2. Expansion of blood vessels, which produces symptoms of redness and heat.

_____ 3. Swelling of a tissue because of fluid accumulation.

_____ 4. Clotting of blood and other proteins, which "walls off" the site of the injury from the rest of the body.

_____ 5. Result of edema and some mediators stimulating nerves.

_____ 6. Result of pain, limitation of movement resulting from edema, and tissue destruction.

Tissue Repair

66 *Tissue repair is the substitution of viable cells for dead cells, and it can occur by regeneration or* **99** *replacement.*

A. Using the terms provided, complete these statements:

Labile	Replacement
Permanent	Scar
Regeneration	Stable

In _(1)_ , the new cells are of the same type as those that were destroyed, whereas in _(2)_ , a new type of tissue develops that eventually causes _(3)_ production and the loss of some tissue function. _(4)_ cells continue to divide throughout life. Damage to these cells can be completely repaired by regeneration. _(5)_ cells do not actively replicate after growth ceases, but they do retain the ability to divide if necessary and are capable of regeneration. _(6)_ cells cannot replicate, and, if killed, they are replaced by a different type of cell.

1. _____
2. _____
3. _____
4. _____
5. _____
6. _____

B. Using the terms provided, complete these statements:

Fibrin	Scab
Granulation tissue	Scar
Phagocytic	Secondary union
Primary union	Wound contraction
Pus	

If the edges of a wound are close together such as in a surgical incision, the wound heals by a process called _(1)_ . The wound fills with blood and a clot forms. The clot is filled with a threadlike protein, _(2)_ , and the surface of the clot dries to form a _(3)_ . Some of the white blood cells that move into the damaged tissue are _(4)_ cells called neutrophils. As these cells ingest bacteria and tissue debris, they may be killed and accumulate as a mixture of dead cells and fluid called _(5)_ . Eventually the clot is replaced by _(6)_ , which consists of fibroblasts, collagen, and capillaries. A large amount of granulation tissue sometimes persists as a _(7)_ . If the edges of the wound are not close together or if there has been extensive loss of tissue, the process is called _(8)_ . In this process, wound regeneration takes longer, _(9)_ occurs when fibroblasts in the granulation tissue contract, and disfiguring and debilitating scars may result.

1. _____
2. _____
3. _____
4. _____
5. _____
6. _____
7. _____
8. _____
9. _____

1. List eight kinds of epithelium based on numbers of cell layers and shape of cells.

2. List four functions performed by epithelial cells.

3. List the four types of free cell surfaces of epithelial cells.

4. Name six ways that cells are mechanically bound together.

5. List three types of exocrine glands based on how products leave the cell.

6. Name three types of protein fibers found in connective tissue.

7. List four types of fibrous tissue found in the human body.

8. List three types of special connective tissue.

9. Name three types of cartilage found in the human body.

10. Name two types of bone found in the human body.

11. List three types of muscle cells found in the human body.

12. List the three primary germ layers and one derivative of each.

13. List the three major categories of membranes found in the human body.

14. List the five major symptoms of inflammation.

15. List the three categories into which cells can be classified according to their regenerative ability.

MASTERY LEARNING ACTIVITY

Place the letter corresponding to the correct answer in the space provided.

_____ 1. Given these characteristics:
   1. capable of contraction
   2. covers all free body surfaces
   3. lacks blood vessels
   4. composes various glands
   5. anchored to connective tissue by a basement membrane

   Which of these are characteristics of epithelial tissue?
   a. 1,2,3
   b. 2,3,5
   c. 3,4,5
   d. 1,2,3,4
   e. 2,3,4,5

_____ 2. A tissue that covers a surface, is one cell layer thick, and is composed of flat cells is
   a. simple squamous epithelium.
   b. simple cuboidal epithelium.
   c. simple columnar epithelium.
   d. stratified squamous epithelium.
   e. transitional epithelium.

_____ 3. Epithelium composed of two or more layers of cells with only the deepest layer in contact with the basement membrane is known as
   a. stratified epithelium.
   b. simple epithelium.
   c. pseudostratified epithelium.
   d. columnar epithelium.

_____ 4. Stratified epithelium is usually found in areas of the body where the principal activity is
   a. filtration.
   b. protection.
   c. absorption.
   d. diffusion.

_____ 5. An epithelial cell with microvilli would most likely be found
   a. lining blood vessels.
   b. lining the lungs.
   c. in serous membranes.
   d. lining the digestive tract.

_____ 6. A tissue that contains cells with these characteristics:
   1. covers a surface
   2. one layer of cells
   3. cells are flat

   Performs which of the functions listed below?
   a. phagocytosis
   b. active transport
   c. secretion of many complex lipids and proteins
   d. is adapted to allow certain substances to diffuse across it
   e. a and b

_____ 7. A type of junction between epithelial cells whose ONLY function is to prevent the cells from coming apart (provides mechanical strength) is the
   a. desmosome.
   b. gap junction.
   c. dermatome.
   d. tight junction.

_____ 8. Pseudostratified ciliated epithelium can be found lining the
   a. digestive tract.
   b. trachea.
   c. thyroid gland.
   d. kidney tubules.
   e. a and b

_____ 9. Cuboidal epithelium can be found in the
   a. mesothelium.
   b. endothelium.
   c. vagina.
   d. oviducts.
   e. salivary glands.

_____ 10. In parts of the body such as the urinary bladder, where considerable distention occurs, one can expect to find which type of epithelium?
   a. cuboidal
   b. pseudostratified
   c. transitional
   d. squamous

11. A histologist observed the characteristics listed below while viewing a tissue with a microscope:
    1. consists of more than one layer of cells
    2. surface cells are alive
    3. cells near the free surface are flat
    4. cells in the basal layers appear to be undergoing mitosis
    5. there appear to be many cell-to-cell attachments in all layers

    Choose the MINIMUM number of characteristics that would allow one to classify the tissue as moist, stratified squamous epithelium.
    a. 1
    b. 1,2
    c. 1,2,3
    d. 1,2,3,4
    e. 1,2,3,4,5

12. Those glands that lose their connection with the epithelium during embryonic development and thus empty their secretions directly into the blood are called
    a. exocrine glands.
    b. apocrine glands.
    c. endocrine glands.
    d. merocrine glands.
    e. holocrine glands.

13. Glands that accumulate secretions and release them only when the individual secretory cells rupture and die are
    a. merocrine glands.
    b. holocrine glands.
    c. apocrine glands.

14. The fibers in connective tissue are formed by
    a. fibroblasts.
    b. adipocytes.
    c. osteoblasts.
    d. macrophages.

15. Extremely delicate fibers that make up the framework for organs such as the liver, spleen, and lymph nodes are
    a. collagen fibers.
    b. elastic fibers.
    c. reticular fibers.
    d. microvilli.
    e. cilia.

16. A tissue that contains a large amount of extracellular collagen organized as parallel fibers would probably be found in
    a. a muscle.
    b. a tendon.
    c. adipose tissue.
    d. bone.
    e. cartilage.

17. Which of these is true of adipose tissue?
    a. site of energy storage
    b. a type of connective tissue
    c. acts as a protective cushion
    d. functions as a heat insulator
    e. all of the above

18. Fibrocartilage is found in the
    a. cartilage rings of the trachea.
    b. costal cartilages.
    c. intervertebral disks.
    d. a and b
    e. all of the above

19. A tissue in which cells are located in lacunae surrounded by a hard matrix of hydroxyapatite is
    a. hyaline cartilage.
    b. bone.
    c. nervous tissue.
    d. dense fibrous connective tissue.
    e. fibrocartilage.

20. Blood is an example of
    a. epithelial tissue.
    b. connective tissue.
    c. muscle tissue.
    d. nervous tissue.
    e. none of the above

21. Which of these is (are) characteristic of smooth muscle?
    a. under voluntary control
    b. striated
    c. under involuntary control
    d. a and b
    e. b and c

22. Which of these statements about nervous tissue is NOT true?
    a. Neurons have cytoplasmic extensions called axons.
    b. Electrical signals (action potentials) are conducted along axons.
    c. Bipolar neurons have two axons.
    d. Neurons are protected and nourished by neuroglia.

23. A bullet that passes through one's arm without hitting bone could contact which of these tissue types?
    a. nervous
    b. muscle
    c. connective
    d. epithelial
    e. all of the above

24. Concerning germ layers:
    a. There are four germ layers.
    b. The germ layers give rise to four tissue types.
    c. Muscle and bone are derived from endoderm.
    d. all of the above

25. Linings of the digestive, respiratory, urinary, and reproductive tracts are composed of
    a. serous membranes.
    b. mesothelium.
    c. mucous membranes.
    d. endothelium.
    e. a and b

26. The lamina propria
    a. is a layer of smooth muscle.
    b. is located under (deep to) mucous membranes.
    c. binds the skin (integument) to underlying tissues.
    d. all of the above

27. Chemical mediators of inflammation
    a. stimulate nerve endings to produce the symptom of pain.
    b. increase the permeability of blood vessels.
    c. cause vasodilation (expansion) of blood vessels.
    d. are released into or activated in tissues following injury.
    e. all of the above

28. Which of these tissues are capable of division throughout life and thus have the ability to undergo repair by producing new cells to replace damaged cells?
    a. connective tissue
    b. epithelial tissue
    c. nervous tissue (neurons)
    d. a and b
    e. all of the above

29. Permanent cells
    a. divide and replace damaged cells in replacement tissue repair.
    b. form granulation tissue.
    c. are responsible for removing scar tissue.
    d. are replaced by a different cell type if they are destroyed.
    e. are replaced during regeneration tissue repair.

30. Which of these is true about granulation tissue?
    a. produced by fibroblasts
    b. a type of connective tissue
    c. may turn into scar tissue
    d. a and b
    e. all of the above

☆ ——————— FINAL CHALLENGES ——————— ☆

Use a separate sheet of paper to complete this section.

1. On a histology exam, Slide Mann was asked to identify the types of epithelial tissue lining the surface of an organ. He identified the first tissue as stratified squamous epithelium and the second tissue as stratified cuboidal epithelium. In both cases he was wrong. If the tissues both came from the same organ, what was the epithelial type?

2. Ciliated cells move materials over their surfaces. Explain why ciliated epithelial cells are found in the tube system of the respiratory system, but are not found in the duct system of the urinary system.

3. Slide Mann was examining a ligament under the microscope. Slide knew that ligaments attached bones to bones, so he was surprised to observe a large number of elastic fibers in the ligament. Why did it seem inappropriate for a ligament to have elastic fibers?

4. When Slide Mann asked his instructor about the ligament with elastic fibers, the instructor asked, "Do you think the ligament joined the vertebrae to each other, or did it join the thigh bone (femur) to the hip bone (coxa)?" Explain why this question should make everything clear to Slide.

5. Cardiac muscle is striated, has a single nucleus, is under involuntary control, and is capable of spontaneous contractions. List all the pairs of traits (e.g., striated, involuntary control) from this list that would identify the muscle as cardiac muscle.

ANSWERS TO CHAPTER 4

Epithelial Tissue
1. Extracellular; 2. Surfaces; 3. Free;
4. Basement membrane; 5. Diffusion; 6. Mitosis

Classification of Epithelium
A. 1. Simple cuboidal; 2. Stratified columnar; 3. Transitional; 4. Simple squamous; 5. Pseudostratified columnar; 6. Stratified squamous
B. 1. Transitional; 2. Pseudostratified columnar; 3. Simple columnar; 4. Simple squamous

Functional Characteristics
A. 1. Simple epithelium; 2. Stratified epithelium; 3. Stratified epithelium
B. 1. Cuboidal or columnar; 2. Squamous
C. 1. Smooth; 2. Microvillar; 3. Ciliated; 4. Folded
D. 1. Glycoprotein; 2. Desmosomes; 3. Hemidesmosomes; 4. Zonula adherens; 5. Zonula occludens; 6. Gap junctions; 7. Intercalated disks

Glands
A. 1. Exocrine; 2. Endocrine
B. 1. Unicellular; 2. Compound; 3. Acinar or alveolar; 4. Straight
C. 1. Simple straight tubular; 2. Simple acinar; 3. Simple coiled tubular; 4. Simple branched acinar; 5. Compound tubular; 6. Compound acinar
D. 1. Merocrine; 2. Apocrine; 3. Holocrine

Connective Tissue
1. Blasts; 2. Cytes; 3. Clasts

Protein Fibers of the Matrix
1. Collagen fibers; 2. Reticular fibers; 3. Elastin fibers

Nonprotein Matrix Molecules
1. Ground substance; 2. Hyaluronic acid; 3. Proteoglycans; 4. Monomers; 5. Aggregates; 6. Water

Matrix with Fibers as the Primary Feature
A. 1. Dense; 2. Loose (areolar); 3. Fibroblasts; 4. Macrophages
B. 1. Dense regular; 2. Dense irregular

C. 1. Dense regular collagenous; 2. Dense regular elastic; 3. Dense irregular collagenous; 4. Dense irregular elastic
D. 1. Adipose; 2. Reticular tissue; 3. Dendritic cells; 4. Yellow bone marrow; 5. Red bone marrow

Matrix with Protein Fibers and Ground Substance
A. 1. Chondrocytes; 2. Lacunae; 3. Hyaline cartilage; 4. Fibrocartilage
B. 1. Hydroxyapatite; 2. Osteocytes; 3. Lacunae; 4. Trabeculae; 5. Compact bone

Identification of Connective Tissue
1. Cartilage; 2. Lacuna; 3. Chondrocyte; 4. Adipose; 5. Fat droplet; 6. Bone; 7. Osteocyte; 8. Dense regular connective tissue; 9. Fibroblast

Muscular Tissue
A. 1. Skeletal; 2. Cardiac; 3. Smooth
B. 1. Cardiac muscle; 2. Intercalated disk; 3. Smooth muscle; 4. Skeletal muscle

Nervous Tissue
1. Cell body; 2. Axon; 3. Multipolar; 4. Bipolar; 5. Neuroglia

Embryonic Tissue
1. Endoderm; 2. Mesoderm; 3. Ectoderm; 4. Neuroectoderm; 5. Neural crest cells

Membranes
A. 1. Serous membranes; 2. Mucous membranes; 3. Synovial membranes
B. 1. Serous fluid; 2. Mucus; 3. Synovial fluid

Inflammation
1. Mediators of inflammation; 2. Vasodilation; 3. Edema; 4. Coagulation; 5. Pain; 6. Disturbance of function

Tissue Repair
A. 1. Regeneration; 2. Replacement; 3. Scar; 4. Labile; 5 Stable; 6. Permanent
B. 1. Primary union; 2. Fibrin; 3. Scab; 4. Phagocytic; 5. Pus; 6. Granulation tissue; 7. Scar; 8. Secondary union; 9. Wound contraction

1. Simple squamous, simple cuboidal, simple columnar, stratified squamous, stratified cuboidal, stratified columnar, pseudostratified, and transitional epithelium
2. Protection, secretion, transportation, absorption, filtration, and diffusion
3. Smooth, microvillar, ciliated, and folded
4. Glycoproteins, desmosomes, hemidesmosomes, zonula adherens, tight junctions, (zonula adherens and zonula occludens), and gap junctions
5. Merocrine, apocrine, and holocrine
6. Collagen, reticular fibers, and elastin

7. Dense regular collagenous, dense irregular collagenous, dense regular elastic, and dense irregular elastic
8. Reticular, adipose, and bone marrow
9. Hyaline, fibrocartilage, and elastic cartilage
10. Compact and cancellous
11. Skeletal, cardiac, and smooth muscle
12. Ectoderm: brain, spinal cord, peripheral nerves; endoderm: lining of digestive system, trachea, bronchi, lungs, liver, thyroid; mesoderm: bone, cartilage, tendons, muscle, and blood
13. Serous, mucous, and synovial membranes
14. Redness, heat, swelling, pain, and disturbance of function
15. Labile, stable, and permanent

MASTERY LEARNING ACTIVITY

1. E. Muscle tissue, not epithelial tissue, is capable of contraction. The other statements are true for epithelial tissue.

2. A. The tissue is simple (one cell layer thick), squamous (flat cells) epithelium (covers a surface).

3. A. Because the description specified two or more layers, only stratified and pseudostratified epithelium are possible candidates for a correct answer. Of these two choices, stratified epithelium has only the deepest layer in contact with the basement membrane. Pseudostratified epithelium has all cells contacting the basement membrane.

4. B. Stratified epithelium (many layers of cells) is protective in function and is found in areas subjected to friction (mouth, pharynx, esophagus, anus, vagina, and skin). Simple squamous epithelium would be associated with filtration and diffusion. Secretion is typically a function of cuboidal or columnar shaped cells.

5. D. Microvilli greatly increase the surface area of epithelial cells. This allows greater absorption at the surface of the cell, an ideal trait for cells that must absorb materials from the intestine.

6. D. One layer of flat cells covering a surface describes simple squamous epithelium. This tissue forms a thin sheet through which diffusion can occur. Examples are found in blood vessels (endothelium), air sacs (alveoli) in the lungs, and in the kidneys (Bowman's capsule). Secretion, phagocytosis, active transport, and absorption would be expected of larger cells (cuboidal or columnar).

7. A. Desmosomes provide mechanical strength. Gap junctions provide a way for materials to be exchanged between cells. Tight junctions prevent extracellular materials from passing between cells.

8. B. Pseudostratified ciliated epithelium is found in the nasal cavity (upper respiratory tract) and in the trachea (lower respiratory tract). The cilia move mucus with entrapped particulate matter and microorganisms to the back of the throat, where it is swallowed. This helps to keep the respiratory tract clear. The digestive tract has simple columnar epithelium; it is not ciliated. The kidney tubules and the thyroid gland have cuboidal epithelium.

9. E. Cuboidal epithelial cells are often involved in secretion. They are found in the salivary glands, sweat glands, thyroid gland, and meninges. They also are found in the ducts of glands, the kidney tubules, and the covering of the ovaries.

10. C. Transitional epithelium is composed of cells that can flatten and slide over each other. This allows stretching of the tissue and makes transitional epithelium an ideal lining for the urinary bladder.

11. C. To identify a tissue as being stratified squamous epithelium one must know that the tissue consists of more than a single layer (1), that the tissue lines a surface (2 and 3), and that the cells near the surface are flat (3).

Next, one must distinguish between moist, stratified squamous epithelium (lining of the mouth, esophagus, and vagina) and keratinized, stratified squamous epithelium (outer layer of the skin). In moist, stratified squamous epithelium the surface cells are alive (2). Keratinized squamous epithelium has surface cells that do not have nuclei and are dead.

The first three characteristics are the MINIMUM necessary to identify the tissue as moist, stratified squamous epithelium.

12. C. Endocrine glands empty their secretions directly into the blood. Exocrine glands (apocrine, merocrine, and holocrine glands) empty their secretions by means of ducts.

13. B.   Holocrine glands accumulate secretions, then release the secretions by rupturing. This kills the cell. Apocrine glands also accumulate secretions, but they release the secretions by pinching off a small portion of the cell. The remainder of the cell lives and continues to accumulate secretions. Merocrine glands form and discharge their secretions in a cyclic fashion. There is no cell destruction, as occurs in holocrine and merocrine glands, during this process.

14. A.   Fibroblasts secrete the ground substance and fibers of connective tissue. Adipocytes are cells within loose connective tissue that store fat droplets within their cytoplasm. Osteoblasts secrete the ground substance, fibers, and mineral salts that compose bone. Macrophages are cells that are capable of phagocytosis.

15. C.   The internal framework of some organs and many glands consists of reticular fibers. Collagen fibers are typically found in tendons and ligaments. Elastic fibers are found in the external ear (auricle), elastic cartilage, elastic ligaments, and the walls of arteries.

16. B.   A large amount of parallel collagen fibers would be found in a tissue resistant to pull in one direction, e.g., a tendon or ligament.

17. E.   Adipose tissue is a type of connective tissue. It stores energy (lipids), acts as a protective cushion (around the kidneys, for example), and functions as a heat insulator (under the skin).

18. C.   There are three types of cartilage: hyaline cartilage (embryonic skeleton, costal cartilage, cartilage rings of the trachea, and articular cartilage on the ends of bones), elastic cartilage (external ear, epiglottis, and auditory tubes), and fibrocartilage (intervertebral disks and articular disks).

19. B.   Bone is formed of a mineralized matrix called hydroxyapatite. Cells within the matrix are surrounded by small spaces called lacunae.

20. B.   You simply had to know that blood is an example of connective tissue.

21. C.   Smooth muscle is under involuntary control, and is not striated.

22. C.   Bipolar neurons have one axon and one dendrite.

23. E.   The question requires that one know what tissue types are components of the upper arm. Epithelium (part of the skin), connective tissue (part of the skin and material around muscle), muscle, and nervous tissue (especially if pain sensations are a result) would be penetrated.

24. B.   The germ layers are so named because they give rise to four tissue types. There are three germ layers: the ectoderm (outer layer), mesoderm (middle layer), and endoderm (inner layer). Skin is derived from ectoderm, muscle and bone are derived from mesoderm, and the lining of the digestive tract is derived from endoderm.

25. C.   Cavities or organs that open to the exterior are lined by mucous membranes. Serous membranes (mesothelium) line the ventral body cavity. Endothelium lines blood vessels.

26. B.   The lamina propria is a loose layer of connective tissue that lies under the epithelial cells of mucous membranes.

27. E.   Chemical mediators, such as histamine, kinins, prostaglandins, and leukotrienes, are released or activated in tissue following injury to the tissue. The mediators cause vasodilation, increase vascular permeability, and stimulate neurons.

28. D.   Essentially an individual is born with all the nerve cells he or she will ever have. Epithelial and connective tissue cells are capable of division throughout life.

29. D.   Permanent cells cannot replicate and are replaced by other types of cells if destroyed (replacement tissue repair). Labile and stable cells can replicate (regeneration tissue repair).

30. E.   Following tissue damage, a blood clot forms that temporarily binds the damaged tissue. This eventually is invaded by fibroblasts and blood vessels. The fibroblasts produce fibers that replace the clot, and the blood vessels provide revascularization of the area. The fibroblasts and blood vessels form granulation tissue. The granulation tissue will form a scar if it is not reabsorbed and replaced by the original tissue at the damage site.

1. The tissue is transitional epithelium, a stratified epithelium that lines organs such as the urinary bladder and ureters. When the organ is stretched, the cells of transitional epithelium become squamouslike; and when the organ is not stretched, the cells are roughly cuboidal in shape.

2. One possible factor is the amount of material moved. Cilia can move a small amount of mucus each day out of the lungs, but are not effective in moving the large amount of urine excreted each day. In the urinary system movement of urine results from contraction of smooth muscle and gravity. Movement of mucus by contraction of the tube system of the lungs could interfere with breathing, and gravity does not move mucus in a superior direction. In addition, the tube system of the lungs is always open, creating a space that allows the cilia to beat and move materials.

3. Ligaments join bones to bones and usually do not stretch. This holds the bones in proper relationship to each other. One would not expect elastic fibers because they would allow the ligaments to stretch, possibly causing misalignment of the bones.

4. The elastic fibers would be expected in the ligaments connecting the bones of the back (vertebrae). These ligaments, called the ligamentum flavae, are similar to the ligamentum nuchae that help to hold the head in place. They allow the spine to stretch and then return to its original position. Thus they allow flexibility and help to maintain the position of the vertebrae by pulling them back into the correct position. The vertebrae have a number of bony processes that help to prevent misalignment (see Chapter 7).

5. There are three pairs of traits that would identify the muscle as cardiac muscle:
   1. striated, involuntary control
   2. striated, spontaneous contraction
   3. striated, single nucleus

   The other three possible pairs of traits describe cardiac muscle or smooth muscle:
   4. single nucleus, involuntary control
   5. single nucleus, spontaneous contraction
   6. involuntary control, spontaneous contraction

Integumentary System

FOCUS: The integumentary system consists of the skin, hair, nails, and a variety of glands. The epidermis of skin provides protection against abrasion, ultraviolet light, and water loss and produces vitamin D. The dermis provides structural strength and contains blood vessels involved in temperature regulation. The skin is attached to underlying tissue by the hypodermis, which is a major site of fat storage. Hair and nails consist of dead, keratinized epithelial cells.

CONTENT LEARNING ACTIVITY

Hypodermis

❝The skin is attached to underlying bone or muscle by the hypodermis.❞

Match these terms with the correct statement or definition:

Fibroblasts, adipose cells, and macrophages
Loose connective tissue

_____ 1. The type of tissue found in the hypodermis.

_____ 2. The main types of cells within the hypodermis.

☞ The hypodermis is sometimes called subcutaneous tissue or superficial fascia.

Skin

❝The integumentary system includes skin, hair, nails, and glands.❞

Match these terms with the correct statement or definition:

Dermis
Epidermis

_____ 1. Dense, irregular connective tissue

_____ 2. Layer of epithelial tissue.

_____ 3. Most superficial layer of the skin.

Dermis

" *The dermis is responsible for most of the structural strength of the skin.* **"**

Match these terms with the
correct statement or definition:

Cleavage lines   Reticular layer
Papillae   Striae
Papillary layer

_____ 1. Deep layer of dermis; fibrous layer that blends into the
hypodermis.

_____ 2. Lines visible through the epidermis produced by rupture of the
dermis.

_____ 3. Projections from the dermis into the epidermis.

Epidermis

" *The epidermis is a stratified squamous epithelium separated* **"**
from the dermis by a basement membrane.

A. Match these terms with the
correct statement or definition:

Desquamate   Langerhans cells
Keratinization   Melanocytes
Keratinocytes   Strata

_____ 1. Cells that produce a protein called keratin; the most abundant
epidermal cell.

_____ 2. Cells in the epidermis that are part of the immune system.

_____ 3. To slough or be lost from the surface of the epidermis.

_____ 4. A process that occurs in epidermal cells during their movement
from deeper epidermal layers to the surface.

_____ 5. Layers of cells within the epidermis.

B. Match these terms with the
correct statement or definition:

Keratohyalin   Stratum germinativum
Lamellar bodies   Stratum granulosum
Stratum basale   Stratum lucidum
Stratum corneum   Stratum spinosum

_____ 1. Deepest portion of the epidermis; a single layer of cells; the
site of production of most epidermal cells.

_____ 2. Epidermal layer superficial to the stratum basale, consisting
of eight to ten layers of many-sided cells.

_____ 3. Name for the stratum of the epidermis that includes both the
stratum basale and the stratum spinosum.

_____ 4. Derives its name from protein granules contained in the cells and is superficial to the stratum spinosum.

_____ 5. Nonmembrane-bound protein granules found in the cells of the stratum granulosum.

_____ 6. Structures that move to the cell membrane and release their lipid contents into the intercellular space; the lipids are responsible for the permeability characteristics of the epidermis.

_____ 7. Clear, thin zone above the stratum granulosum; absent in most skin.

_____ 8. Most superficial stratum of the epidermis; dead cells with hard protein envelope and filled with keratin, which provides structural strength.

C. Match these terms with the parts labeled in Figure 5-1:

Dermis	Stratum germinativum
Epidermis	Stratum granulosum
Stratum basale	Stratum lucidum
Stratum corneum	Stratum spinosum

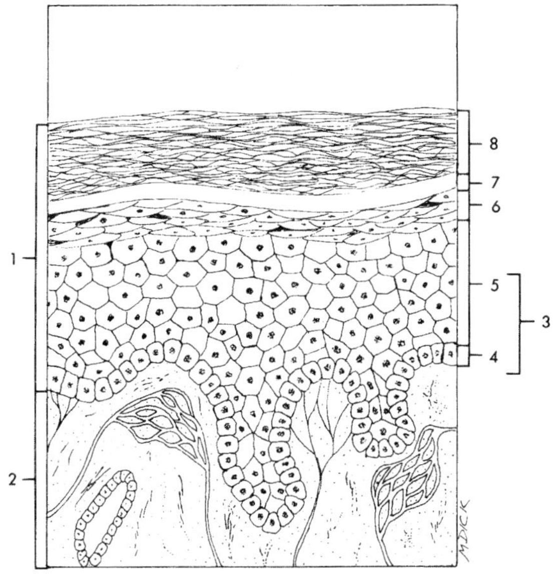

Figure 5-1

1. _____

2. _____

3. _____

4. _____

5. _____

6. _____

7. _____

8. _____

89

Thick Skin and Thin Skin

"Skin is classified as thick or thin based on the structure of the epidermis.**"**

Match these terms with the correct statement or definition:

Callus
Corn

Thick skin
Thin skin

_____

1. The papillae of the dermis of this type of skin compose curving ridges that produce fingerprints and footprints.

_____

2. In this type of skin the stratum lucidum is usually absent; hair is found in this type of skin.

_____

3. Thickened area of thin or thick skin resulting from greatly increased number of layers of stratum corneum.

_____

4. Cone-shaped structure that develops in thin or thick skin over a bony prominence.

Skin Color

"Skin color is determined by pigments in the skin, by blood circulating through the skin, and by the thickness of the stratum corneum.**"**

Using the terms provided, complete the following statements:

Albinism
Carotene
Cyanosis
Melanocytes

Melanin
Melanosomes
Ultraviolet light

1. _____

2. _____

3. _____

4. _____

5. _____

6. _____

7. _____

 (1) , a brown to black pigment, is responsible for most skin color. It is produced by (2) , irregularly shaped cells with many long processes that extend between the keratinocytes of the stratum basale and stratum spinosum. Melanin is packaged into vesicles called (3) , which are released from the cell processes by exocytosis. A single mutation can prevent the manufacture of melanin, resulting in (4) . Exposure to (5) increases melanin production. (6) is a yellow pigment found in plants such as carrots. When large amounts of this pigment are consumed, the excess accumulates in the stratum corneum and fat cells of the dermis and hypodermis, causing the skin to develop a yellowish tint. A decrease in blood oxygen content produces (7) , a bluish skin color, whereas an abundant supply of oxygenated blood produces a reddish hue.

Hair

"The presence of hair is one of the characteristics common to all mammals."

A. Match these terms with the correct statement or definition:

Lanugo   Vellus hairs
Terminal hairs

_____ 1. Delicate unpigmented hair that covers the fetus.

_____ 2. Long, coarse, pigmented hairs that replace lanugo on the scalp, eyebrows, and eyelids.

_____ 3. Short, fine, unpigmented hairs that replace lanugo over most of the body.

_____ 4. The type of hair that replaces vellus hairs at puberty.

B. Match these terms with the correct statement or definition:

Cortex   Medulla
Cuticle   Root
Hair bulb   Shaft

_____ 1. Portion of hair protruding above the skin.

_____ 2. An expanded knob at the base of the hair root.

_____ 3. Central axis of the hair that consists of two or three layers of cells containing soft keratin.

_____ 4. Forms the bulk of the hair and consists of cells containing hard keratin.

_____ 5. Outermost layer of the hair shaft and root, composed of a single overlapping layer of cells containing hard keratin.

C. Match these terms with the correct statement or definition:

Arrector pili   Epithelial root sheath
Dermal root sheath   Matrix

_____ 1. Surrounds the epithelial root sheath.

_____ 2. Layers of epithelial cells immediately surrounding the root of the hair.

_____ 3. Mass of undifferentiated epithelial cells inside the hair bulb that produces the hair and internal epithelial root sheath.

_____ 4. Smooth muscle cells that attach to the hair follicle dermal root sheath and the papillary layer of the dermis; cause hair to "stand on end."

☞ Hair is produced in cycles that involve a growth stage alternating with a resting stage.

D. Match these terms with
the correct parts labeled
in Figure 5-2:

Arrector pili
Dermal root sheath
Hair bulb
Hair root
Hair shaft
Matrix
Papilla

1. _____

2. _____

3. _____

4. _____

5. _____

6. _____

7. _____

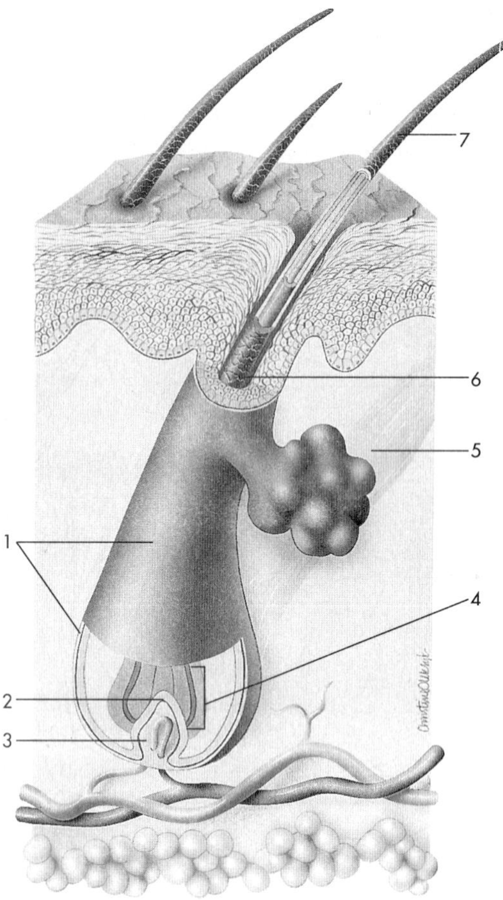

Figure 5-2

Glands

66 *The major glands of the skin consist of epidermal tissue that extends into the dermis* 99
or through the dermis into the hypodermis.

A. Match these terms with the
correct statement or definition:

Apocrine sweat gland   Sebaceous gland
Ceruminous gland   Sebum
Merocrine sweat gland

_____ 1. White substance rich in lipids, which oils the hair and skin
surface, prevents drying, and protects against bacteria.

_____ 2. Gland that opens into a hair follicle; produces sebum.

_____ 3. Gland that opens to the surface of the skin and secretes an
isotonic fluid that is mostly water; involved with temperature
regulation.

_____ 4. Gland that usually opens into a hair follicle; secretes an
organic substance that is metabolized by bacteria to produce
body odor.

92

B. Match these terms with the types of glands labeled in Figure 5-3:

Apocrine sweat gland
Merocrine sweat gland
Sebaceous gland

1. _____

2. _____

3. _____

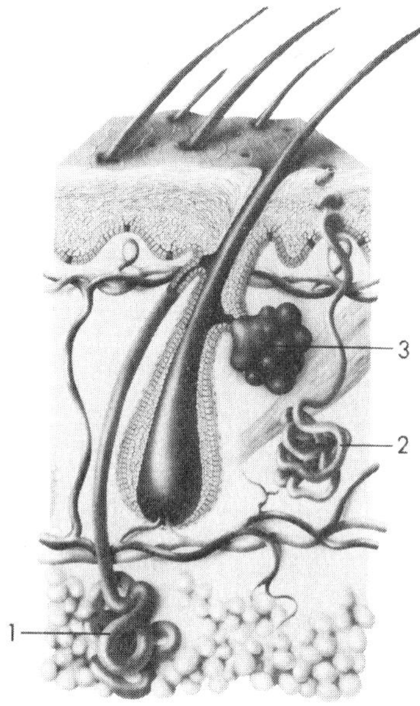

Figure 5-3

Nails

66 *Nails protect the ends of the digits, aid in manipulation and grasping of small objects,* **99** *and are used for scratching.*

A. Match these terms with the correct statement or definition:

Eponychium   Nail fold
Hyponychium   Nail groove
Lunula   Nail matrix
Nail bed   Nail root
Nail body

_____ 1. Proximal portion of the nail that is covered by skin.

_____ 2. Portion of the skin that covers the lateral and proximal edges of the nail.

_____ 3. Holds the edges of the nail in place.

_____ 4. Cuticle; the stratum corneum of the nail fold that grows onto the nail body.

_____ 5. Nail root and nail body attach to this structure.

_____ 6. Proximal portion of the nail bed; produces most of the nail.

_____ 7. Whitish, crescent-shaped area at the base of the nail; part of the nail matrix.

B. Match these terms with
   the parts of the nail labeled
   in Figure 5-4:

Nail body
Eponychium
Lunula
Nail root

1. _____

2. _____

3. _____

4. _____

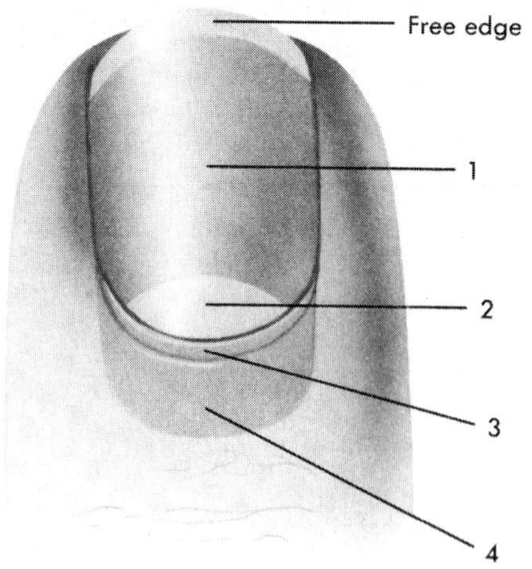

Figure 5-4

Functions of the Integumentary System

"The integumentary system has many functions in the body."

Match these terms with the
correct statement or definition:

Excretion
Protection
Sensation

Temperature regulation
Vitamin D production

_____ 1. Accomplished by the skin as a physical barrier, as a
permeability barrier, as a barrier against ultraviolet light,
and as a barrier against abrasion.

_____ 2. Carried out by producing sweat and increasing or decreasing
blood vessel diameter

_____ 3. Begins when a precursor molecule in the skin is exposed to
ultraviolet light and is converted to cholecalciferol.

_____ 4. Detection of touch, temperature, and pain.

_____ 5. Occurs to a very slight degree with sweat production when
some urea, uric acid, and ammonia are lost.

The Effects of Aging on the Integumentary System

66 *As the body ages, many changes occur in the integumentary system.* 99

Using the terms provided, complete the following statements:

Decrease(s)   Increase(s)

As the body ages, blood flow to the skin __(1)__, and the
thickness of skin __(2)__. Elastic fibers in the dermis __(3)__,
and the skin tends to sag. A __(4)__ in the activity of
sebaceous and sweat glands results in dry skin and a __(5)__ in
thermoregulatory ability. The __(6)__ in ability to sweat can
contribute to death from heat prostration in elderly
individuals. The number of functioning melanocytes __(7)__,
but in some localized areas, especially the hands and face,
melanocytes __(8)__ to produce age spots. White or gray
hairs occur because of a __(9)__ in melanin production.

1. _____

2. _____

3. _____

4. _____

5. _____

6. _____

7. _____

8. _____

9. _____

Integumentary Disorders

66 *The integumentary system is directly affected by many disorders.* 99

A. Match these terms with the
   correct statement or definition:

Acne   Ringworm and athlete's foot
Decubitus ulcers   Warts

_____

_____

_____

_____

1. Disorder of the hair follicles and sebaceous glands that
   involves testosterone and bacteria.

2. Viral infection of the epidermis.

3. Fungal infections that affect the keratinized portion of the
   skin.

4. Disorder caused by ischemia and necrosis of the hypodermis.

B. Match these terms with the
   correct statement or definition:

Basal cell carcinoma   Squamous cell carcinoma
Malignant melanoma

_____

_____

_____

1. Cancer that begins in the stratum basale and extends into the
   dermis to produce an open ulcer; the most frequent type of skin
   cancer.

2. Cancer that typically produces a nodular, keratinized tumor
   confined to the epidermis.

3. Less common form of skin cancer that usually arises from a
   preexisting mole; the skin cancer that is most often fatal if not
   diagnosed and treated early.

1. Name the two layers of the dermis.

2. List the five strata of the epidermis from the deepest to the most superficial.

3. List three types of hair found at different stages of development in humans.

4. Name the two stages in the hair production cycle.

5. List the three major types of glands associated with the skin.

6. List four protective functions of the integumentary system.

7. List two ways the integumentary system functions to regulate the temperature of the body.

8. List four effects that aging has on the integumentary system.

Place the letter corresponding to the correct answer in the space provided.

_____ 1. The hypodermis
   a. is the layer of skin where hair is produced.
   b. is the layer of skin where nails are produced.
   c. connects the dermis and the epidermis.
   d. is dense irregular connective tissue.
   e. none of the above

_____ 2. Fingerprint and footprint patterns are a result of the development of the
   a. stratum corneum.
   b. dermis.
   c. hypodermis.
   d. stratum germinativum.

_____ 3. The layer of the skin (where mitosis occurs) that replaces cells lost from the outer layer of the epidermis is the
   a. stratum corneum.
   b. stratum basale.
   c. reticular layer of the dermis.
   d. hypodermis.

_____ 4. If a splinter penetrated the skin of the sole of the foot to the second epidermal layer from the surface, the last layer to be damaged would be the
   a. stratum granulosum.
   b. stratum basale.
   c. stratum corneum.
   d. stratum lucidum.

_____ 5. In what area of the body would you expect to find an especially thick stratum corneum?
   a. back of the hand
   b. abdomen
   c. over the shin
   d. bridge of the nose
   e. heel of the foot

_____ 6. The function of melanin in the skin is
   a. lubrication of the skin.
   b. prevention of skin infections.
   c. protection from ultraviolet light.
   d. to reduce water loss.
   e. to help regulate body temperature.

_____ 7. Concerning skin color, which of the following statements is NOT correctly matched?
   a. skin appears yellow - carotene present
   b. no skin pigmentation (albinism) - genetic disorder
   c. skin tans - increased melanin production
   d. skin appears blue (cyanosis) - oxygenated blood
   e. Afro-Americans darker than Caucasians - more melanin in Afro-Americans

_____ 8. Hair
   a. is produced by the matrix of the hair bulb.
   b. consists of dead epithelial cells.
   c. is colored by melanin.
   d. is the same thing as fur.
   e. all of the above

_____ 9. A hair follicle is
   a. an extension of the epidermis deep into the dermis.
   b. an extension of the dermis deep into the epidermis.
   c. often associated with merocrine sweat glands.
   d. a and c
   e. b and c

_____ 10. Smooth muscles that produce "goose flesh" when they contract and are attached to hair follicles are called
   a. external root sheaths.
   b. arrector pili.
   c. dermal papillae.
   d. internal root sheaths.

11. Sebum
    a. oils the hair.
    b. is produced by sweat glands.
    c. consists of dead cells from hair follicles.
    d. all of the above.

12. A congenital lack of merocrine sweat glands would primarily affect one's ability to
    a. secrete waste products.
    b. flush out secretions that accumulate in the hair follicle.
    c. oil the skin.
    d. prevent bacteria from growing on the skin.
    e. control one's body temperature in warm environments.

13. An experimenter wishes to determine the relationship between body temperature and water loss in an animal. At different air temperature, the amount of water lost and the body temperature of the animal was measured. The results are graphed below:

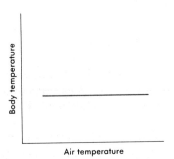

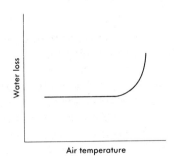

Given the following statements:
1. The animal has sweat glands.
2. The animal conserves water at low air temperatures.
3. The skin of the animal is resistant to water loss.

4. The animal uses water loss to maintain body temperature at a constant level at all air temperatures.

Using the experimental results, which of the statements can you conclude are true?
a. 2, 3
b. 1, 2, 3
c. 1, 3, 4
d. 2, 3, 4
e. 1, 2, 3, 4

14. Nails
    a. appear pink because of a special pigment within nail cells.
    b. are an outgrowth of the dermal papillae.
    c. contain lots of hard keratin.
    d. all of the above

15. While building the patio deck to his house, an anatomy and physiology instructor hit his finger with a hammer. He responded to this stimulus by saying, "Gee, I hope I didn't irreversibly damage the _____, because if I did, my fingernail will never grow back."
    a. nail bed beneath the nail body
    b. stratum corneum
    c. nail matrix
    d. eponychium
    e. a and b

16. To increase heat loss from the body, one would expect
    a. dilation of dermal capillaries.
    b. constriction of dermal capillaries.
    c. increased sweating.
    d. a and c
    e. b and c

17. Body odor
    a. results from special scent glands located under the arm pits.
    b. occurs when bacteria break down the organic secretions of apocrine glands.
    c. results from a chemical reaction of sweat with the air.
    d. results from sebum.
    e. results from keratin.

_____18. Skin aids in maintaining the calcium and phosphate levels of the body at the optimum levels by participating in the production of
  a. vitamin A.
  b. vitamin B.
  c. vitamin D.
  d. melanin.
  e. keratin.

_____19. On a sunny spring day a student decided to initiate her annual tanning ritual. However, in doing so she fell asleep while sunbathing. After awaking she noticed that the skin on her back was burned. She experienced redness, blisters, edema, and significant pain. The burn was nearly healed about 10 days later.

The burn was best classified as a
  a. first degree burn.
  b. second degree burn.
  c. third degree burn.

_____20. A burn patient is admitted into the hospital while you are present in the emergency room. Someone asks for an estimate of the percentage of surface area burned. Remembering the formula for assessing the extent of burns, you take a shot at it. If the patient was burned on the entire left upper and lower limbs and over three quarters of the anterior and posterior trunk, what would be the estimate?
  a. 54%
  b. 63%
  c. 45%
  d. 72%
  e. none of the above

FINAL CHALLENGES

Use a separate sheet of paper to complete this section.

1. The rate of water loss from the skin of the hand was measured. Following the measurement the hand was soaked in alcohol for 15 minutes. After all the alcohol was removed from the hand, the rate of water loss was again determined. Compared to the rate of water loss before soaking the hand in alcohol, what difference, if any, would you expect in the rate of water loss after soaking the hand in alcohol?

2. It has been several weeks since Goodboy Player has competed in a tennis match. After the match he discovers that a blister has formed beneath an old callus on his foot and the callus has fallen off. When he examines the callus he discovers that it appears yellow. Can you explain why?

3. Why is it difficult to surgically remove a large tattoo without causing scar tissue to form? (Hint: why do tattoos appear bluish in color?)

4. Are dark-skinned children more susceptible to rickets (insufficient calcium in the bones) than fair-skinned children? Explain.

5. Given what you know about the cause of body odor, propose some ways to prevent the condition.

6. Pulling hair can be quite painful, yet cutting hair is not painful. Explain.

7. A man is using gasoline to clean automobile parts. There is an explosion, and he is burned over the entire anterior surface of the trunk and the entire anterior surface of both lower limbs. Approximately how much of the man's surface area was damaged?

8. Dandy Chef has been burned on the arm. The doctor, using a forceps, pulls on a hair within the area that was burned. The hair easily pulls out. What degree of burn did the patient have and how do you know?

ANSWERS TO CHAPTER 5

Hypodermis
1. Loose connective tissue; 2. Fibroblasts, adipose cells, and macrophages

Skin
1. Dermis; 2. Epidermis; 3. Epidermis

Dermis
1. Reticular layer; 2. Striae; 3. Papillae

Epidermis
A. 1. Keratinocytes; 2. Langerhans cells; 3. Desquamate; 4. Keratinization; 5. Strata
B. 1. Stratum basale; 2. Stratum spinosum; 3. Stratum germinativum; 4. Stratum granulosum; 5. Keratohyalin; 6. Lamellar bodies; 7. Stratum lucidum; 8. Stratum corneum
C. 1. Epidermis; 2. Dermis; 3. Stratum germinativum; 4. Stratum basale; 5. Stratum spinosum; 6. Stratum granulosum; 7. Stratum lucidum; 8. Stratum corneum

Thick Skin and Thin Skin
1. Thick skin; 2. Thin skin; 3. Callus; 4. Corn

Skin Color
1. Melanin; 2. Melanocytes; 3. Melanosomes; 4. Albinism; 5. Ultraviolet light; 6. Carotene; 7. Cyanosis

Hair
A. 1. Lanugo; 2. Terminal hairs; 3. Vellus hairs; 4. Terminal hairs
B. 1. Shaft; 2. Hair bulb 3. Medulla; 4. Cortex; 5. Cuticle;

C. 1. Dermal root sheath; 2. Epithelial root sheath; 3. Matrix; 4. Arrector pili
D. 1. Dermal root sheath; 2. Matrix; 3. Papilla; 4. Hair bulb; 5. Arrector pili; 6. Hair root; 7. Hair shaft

Glands
A. 1. Sebum; 2. Sebaceous gland; 3. Merocrine sweat gland; 4. Apocrine sweat gland
B. 1. Apocrine sweat gland; 2. Merocrine sweat gland; 3. Sebaceous gland

Nails
A. 1. Nail root; 2. Nail fold; 3. Nail groove; 4. Eponychium; 5. Nail bed; 6. Nail matrix; 7. Lunula
B. 1. Nail body; 2. Lunula; 3. Eponychium; 4. Nail root

Functions of the Integumentary System
1. Protection; 2. Temperature regulation; 3. Vitamin D production; 4. Sensation; 5. Excretion

The Effects of Aging on the Integumentary System
1. Decreases; 2. Decreases; 3. Decrease; 4. Decrease; 5. Decrease; 6. Decrease; 7. Decreases; 8. Increase; 9. Decrease

Integumentary Disorders
A. 1. Acne; 2. Warts; 3. Ringworm and athlete's foot; 4. Decubitus ulcers
B. 1. Basal cell carcinoma; 2. Squamous cell carcinoma; 3. Malignant melanoma

1. Reticular and papillary layer
2. Stratum basale, stratum spinosum, stratum granulosum, stratum lucidum, and stratum corneum
3. Lanugo, vellus hairs, and terminal hairs
4. Growth stage and resting stage
5. Sebaceous glands, apocrine sweat glands, and merocrine sweat glands
6. Prevents water loss, physical barrier to microorganisms, protection against mechanical damage, protection against ultraviolet light, protection against damage to eyes and digits
7. Vasodilation/vasoconstriction and sweat production
8. Decreased blood flow to skin, skin thinner, decreased elastic fibers, decreased activity of sweat and sebaceous glands, decreased melanocyte activity, and age spots in some areas

1. E. The hypodermis, which is a layer of loose connective tissue (areolar tissue), connects the dermis to underlying tissue such as bone or muscle. It is not really part of the skin. The hypodermis is sometimes called subcutaneous tissue and is the site for some injections. The epidermis produces hair and nails.

2. **B.** The dermis consists of two layers, the papillary layer and the reticular layer. The papillary layer fits closely to the basement membrane of the epidermis and has projections (papillae) that extend into the epidermis. These contain capillaries and sensory receptors. On the hands and feet they cause the overlying epidermis to form fingerprint and footprint patterns.

3. **B.** The epidermis is divided into five layers. The innermost of these layers, the stratum basale (and to a lesser extent, the stratum spinosum) is capable of mitosis. The other layers consist of dying or dead cells. This occurs because the stratum basale is close enough to blood vessels (found in the dermis) to receive adequate oxygen and nutrients. The dermis and the hypodermis do not contribute to the cells that eventually reach the surface of the skin.

4. **D.** In thick skin (sole of foot) from the outside in, the layers of the epidermis are the stratum corneum, stratum lucidum, stratum granulosum, stratum spinosum, and stratum basale.

5. **E.** The stratum corneum consists of many rows of dead, cornified, squamous epithelium. Areas of the body subjected to abrasion such as the heel of the foot or calluses have especially thick stratum corneums. This provides protection for the underlying tissues. As cells of the stratum corneum are worn off, they are replaced by cells from deeper layers of the epidermis.

6. **C.** Melanin, produced by the melanocytes in the epidermis, is a dark pigment that prevents the penetration of ultraviolet light. Lipids, found in the stratum corneum of the epidermis, help to reduce water loss. Lubrication of the skin and prevention of skin infections result from secretions of skin glands. Body temperature regulation occurs by vasodilation and vasoconstriction of the vessels within the dermis of the skin.

7. **D.** The skin appears blue because of deoxygenated blood. This could indicate respiratory and/or circulatory dysfunction resulting in inadequate oxygen delivery to tissues. Another cause of cyanosis is cold. In this case blood flow to the surface of the body is reduced (vasoconstriction) to conserve heat, and the skin appears pale or bluish.

8. **E.** The matrix of the hair bulb produces the hair. New cells formed at the base of the hair quickly die, so the hair shaft is composed of dead epithelial cells. Just as skin is colored because of melanin produced by melanocytes, varying amounts of melanin account for hair color. Fur is just lots of hair.

9. **A.** A hair follicle is an extension of the epidermis deep into the dermis. It consists of the epidermis, which gives rise to the hair, and an outer layer of connective tissue from the dermis. Merocrine sweat glands open directly onto the surface of the skin. Sebaceous glands and apocrine sweat glands are associated with hair follicles.

10. **B.** The arrector pili muscle is attached to the hair follicle and the papillary layer of the dermis. When it contracts, the hair stands up and a small section of skin bulges, producing "goose flesh." This same response is seen in many mammals in response to cold. Causing the hair to stand up produces a thicker layer of fur, which provides greater insulation.

11. **A.** Sebum is an oily substance rich in lipids that is produced by sebaceous glands. With few exceptions, sebaceous glands are associated with hair follicles. The sebum oils the hair and surrounding skin, preventing drying.

12. **E.** Merocrine sweat glands help to regulate body temperature in warm temperatures. Although it is true that a small amount of waste products is excreted in sweat, lack of sweat glands is not harmful in this regard. The kidneys are responsible for waste product excretion. Merocrine sweat glands are not associated with hair follicles but open directly onto the skin surface. Sebaceous glands are associated with hair follicles and produce sebum. Sebum oils hair and the skin and prevents the growth of some bacteria.

13. **A.** The animal maintained a constant body temperature (homeostasis) for all air temperatures used in the experiment. This was not caused by water loss alone, because the rate of water loss remained constant at several of the lower air temperatures. Undoubtedly other mechanisms such as vasodilation or vasoconstriction were involved in maintaining body temperature. Once air temperature increased enough, however, these other mechanisms could not prevent a rise in body temperature. At this point water loss increased dramatically and was responsible for keeping body temperature from increasing.

The fact that water loss was constant at the lower air temperatures suggests that the animal conserves water at these temperatures. This is accomplished because the skin is resistant to water loss. When the need arises, however, large amounts of water can be lost. There are two reasonable ways this can happen: from sweat glands (as in humans) or by panting (as in dogs). The data presented do not allow one to conclude which mechanism was used.

14. C. The nail is stratum corneum. The hardness of the nail results from the large amounts of hard keratin. The nail appears pink because the nail is semitransparent and the underlying vascular tissue can be seen. Nails grow from the nail matrix at the proximal end of the nail.

15. C. The nail bed upon which the nail body rests is formed by the stratum germinativum of the epidermis. However, the nail is produced by the stratum germinativum in the nail matrix at the proximal end of the nail (the nail root). Part of the nail matrix can be seen as a whitish area at the base of the nail, the lunula. The eponychium (cuticle) does not affect nail growth.

16. D. Increased sweating cools the body as the sweat evaporates from the skin surface. Dilation of dermal capillaries produces increased blood flow through the skin, moving heat through warm blood from the body core to the surface and then to the environment.

17. B. Located in the axilla, around the anus, and on the labia majora of females and the scrotum of males are modified sweat glands (apocrine glands) that produce organic secretions. When these secretions are decomposed by bacteria, which use them as a source of energy, body odor develops. Because normal sweat does not contain the organic matter, sweat in other areas of the body does not smell. The apocrine sweat glands do not become active until puberty, so body odor is usually not a problem in children.

18. C. When exposed to sunlight (ultraviolet light), the skin produces cholecalciferol, which is absorbed into the blood. Cholecalciferol is metabolically transformed, first in the liver, then in the kidneys, into active vitamin D, which assists in the absorption of calcium and phosphate from the small intestine.

19. B. Sunburns are often first degree burns. Because of the blistering, it is a second degree burn.

20. A. The estimate:

Left upper limb:	9%
Left lower limb:	18%
3/4 of trunk	27%
Total	54%

 FINAL CHALLENGES

1. Alcohol is an organic solvent and removes the lipids from skin, especially in the stratum corneum. Because the lipids normally prevent water loss, after soaking the hand in alcohol, the rate of water loss can be expected in increase.

2. Carotene, a yellow pigment from ingested plants, accumulates in lipids. The stratum corneum of a callus has more layers of cells than other noncallused parts of the skin. The carotene in the lipids surrounding these cells makes the callus appear yellow.

3. The tattoo is located in the dermis; to remove the tattoo, the overlying epidermis and most of the dermis must be removed. The wound produced is like a second degree burn. After removal of a small tattoo the edges of the wound can be sutured together, producing a small scar (primary union). After removal of a large tattoo, more scar tissue forms because most of the epithelial tissue regeneration occurs from the edge of the wound (secondary union).

4. Rickets is a children's disease caused by a lack of vitamin D. Without vitamin D there is insufficient absorption of calcium from the small intestine, resulting in soft bones. Ingested vitamin D can prevent rickets. If dietary vitamin D is lacking, when the skin is exposed to ultraviolet light it can produce a precursor molecule that can be transformed into vitamin D. Dark-skinned children are more susceptible to rickets because the additional melanin in their skins screens out the ultraviolet light and they produce less vitamin D.

5. Body odor results from the breakdown of the organic secretions of apocrine sweat glands by bacteria; remove the secretions by washing or kill the bacteria. The aluminum salts in some antiperspirants kill bacteria. Antiperspirants reduce the watery secretions of merocrine sweat glands, but these secretions are not the cause of body odor. Deodorants mask body odor with another scent.

6. The hair follicle (but not the hair) is well supplied with nerve endings that can detect movement or pulling of the hair. The hair itself is dead, keratinized epithelium so cutting the hair is not painful.

7. About 36% was damaged: 18% from the trunk and 18% from the lower limbs.

8. Because the hair follicle pulled out easily, the hair follicle has been destroyed. Therefore repair can occur only from the edge of the wound and this is a third degree burn.

Skeletal System: Histology and Development

FOCUS: The skeletal system consists of ligaments, tendons, cartilage, and bones. Ligaments attach bones to bones, and tendons attach muscles to bones. Cartilage forms the articular surfaces of bone, is the site of bone growth in length, and provides flexible support within the ear, nose, and ribs. Bone protects internal organs and contraction of skeletal muscle moves the bones, producing body movements. Bone can form by intramembranous (within a membrane) or endochondral (within cartilage) ossification, grows in length at the epiphyseal plate, and increases in diameter beneath the periosteum. Bone is a dynamic tissue constantly undergoing remodeling to adjust to stress and is capable of repair.

CONTENT LEARNING ACTIVITY

Functions of the Skeletal System

66 *The skeletal system consists of bones and their associated connective tissues, including cartilage, tendons, and ligaments.* 99

Match these terms with the correct statement or definition:

Bone　　　　　　　Ligament
Cartilage　　　　　Tendon

_____ 1. Very rigid tissue that functions to maintain body shape, protects internal organs, and provides attachment for muscles to produce body movement.

_____ 2. Somewhat rigid tissue that forms a smooth surface at some joints, provides flexible support, and provides a model for most adult bones.

_____ 3. Strong band of fibrous connective tissue that attaches muscle to bone.

_____ 4. Strong band of fibrous connective tissue that attaches bone to bone.

_____ 5. Tissue that stores minerals and fat.

_____ 6. Contains marrow that gives rise to blood cells and platelets.

Tendons and Ligaments

66 *Tendons and ligaments are dense, regular connective tissue, consisting mostly of collagen fibers.* 99

Using the terms provided, complete these statements:

Appositional   Less
Fibroblasts   More
Fibrocytes   Sheets
Interstitial

1. _____

2. _____

3. _____

4. _____

5. _____

6. _____

7. _____

8. _____

Although the general structure of tendons and ligaments is similar, ligaments often have __(1)__ compact collagen fibrils, have __(2)__ parallel fibrils, usually are __(3)__ flattened than tendons, and tend to form __(4)__ or bands of tissue. Cells of developing tendons and ligaments are spindle-shaped __(5)__; once surrounded by matrix, they are called __(6)__. Tendons and ligaments grow by __(7)__ growth when surface fibroblasts produce additional fibroblasts and secrete matrix outside existing fiber. __(8)__ growth occurs when fibroblasts proliferate and secrete matrix inside the tissue.

Hyaline Cartilage

66 *Hyaline cartilage is the cartilage most closely associated with bone function and development.* 99

Match these terms with the correct statement or definition:

Chondroblast   Lacuna
Chondrocyte   Perichondrium

_____ 1. Cell that produces new matrix on the outside of cartilage.

_____ 2. Mature cartilage cell that is surrounded by matrix.

_____ 3. Double-layered connective tissue sheath around cartilage.

☞ Like tendons and ligaments, cartilage grows by both appositional and interstitial growth.

Bone Shape

66 *Individual bones can be classified according to their shape.* 99

Match these bone shapes with the correct bones:

Flat bones   Long bones
Irregular bones   Short bones

_____ 1. Limb bones.

_____ 2. Carpals and tarsals.

_____ 3. Some skull bones, ribs, sternum, and scapula.

_____ 4. Vertebrae and facial bones.

Bone Anatomy

66 *About half the bones of the body are 'long bones' and the remaining half have some other shape.* 99

A. Match these terms with the correct statement or definition:

Diaphysis
Endosteum
Epiphyseal line
Epiphyseal plate
Epiphysis

Marrow
Medullary cavity
Perforating (Sharpey's) fibers
Sinus

_____ 1. Shaft of a long bone; composed of compact bone.

_____ 2. End of a long bone; composed of cancellous (spongy) bone.

_____ 3. Area of growth between the diaphysis and epiphysis.

_____ 4. Ossified epiphyseal plate.

_____ 5. Large cavity within the diaphysis.

_____ 6. Air space located in some skull bones.

_____ 7. Substance in medullary cavity and cavities of cancellous bone.

_____ 8. Fibers that penetrate the periosteum and outer portion of bone.

_____ 9. Membrane that lines the medullary cavity.

 Flat bones usually have no diaphysis or epiphysis, and they contain an interior framework of cancellous bone sandwiched between two layers of compact bone. Short and irregular bones have a composition similar to the epiphyses of long bones.

B. Match these terms with the correct parts labeled in Figure 6-1:

Articular cartilage
Cancellous bone
Compact bone
Diaphysis

Epiphyseal line
Epiphysis
Medullary cavity
Periosteum

1. _____

2. _____

3. _____

4. _____

5. _____

6. _____

7. _____

8. _____

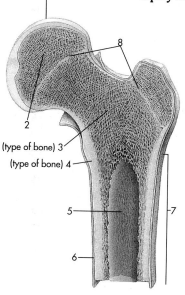

Figure 6-1

105

Bone Matrix

" Bone matrix is normally about 35% organic and 65% inorganic ."

Match these terms with the correct statement or definition:

Canaliculi
Collagen
Hydroxyapatite
Lacunae

Lamellae
Osteoblast
Osteoclast
Osteocyte

_____ 1. Major organic component of bone; lends flexible strength to bones.

_____ 2. The primary mineral in bone; gives bone matrix compression (weight-bearing) strength.

_____ 3. Bone cell that produces, but is not surrounded by, bone matrix.

_____ 4. Large cell with several nuclei that breaks down bone matrix.

_____ 5. Thin sheets or layers of bone matrix.

_____ 6. Spaces in the matrix occupied by osteocytes.

_____ 7. Spaces occupied by long, thin cell processes between lacunae.

Cancellous Bone and Compact Bone

" Compact bone is mostly solid matrix and cells with few spaces; cancellous bone consists of a lacy network of bony plates and beams.

A. Using the terms provided, complete these statements:

Circumferential lamellae
Concentric lamellae
Haversian (central) canals
Haversian system (osteon)

Trabeculae
Volkmann's
(perforating)
canals

Cancellous bone consists of interconnecting rods or plates of bone called _(1)_. Compact bone is more dense, with blood vessels that run parallel to the long axis of the bone through _(2)_, surrounded by _(3)_. A _(4)_ consists of a single haversian canal, its contents, and associated concentric lamellae and osteocytes. The blood vessels of the haversian canals are interconnected by a network of vessels contained within _(5)_, which run perpendicular to the long axis of the bone. The outer surfaces of compact bone are covered by flat plates of bone called _(6)_.

1. _____
2. _____
3. _____
4. _____
5. _____
6. _____

B. Match these terms with the correct parts labeled in Figure 6-2:

Circumferential lamellae
Concentric lamellae
Haversian (central) canal
Haversian system (osteon)
Interstitial lamellae
Volkmann's (perforating) canal

1. _____

2. _____

3. _____

4. _____

5. _____

6. _____

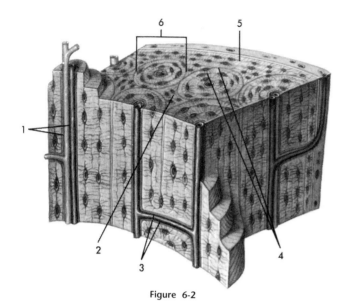

Figure 6-2

Bone Ossification

❝Ossification is the formation of bone by osteoblasts.❞

Match these terms with the correct statement or definition:

Endochondral   Remodeling
Intramembranous   Woven bone

_____ 1. First bone matrix formed during fetal development or during repair of a fracture.

_____ 2. Breakdown of bone matrix by osteoclasts, followed by the formation of new matrix by osteoblasts.

_____ 3. Bone formation that occurs within connective tissue membranes.

_____ 4. Ossification process that produces many skull bones and the clavicle.

_____ 5. Ossification process that produces most of the skeletal system.

107

Cancellous bone and compact bone are formed by both endochondral and intramembranous ossification.

Intramembranous Ossification

❝*Bone formation that occurs within connective tissue membranes is intramembranous.*❞

Using the terms provided, complete these statements:

Centers of ossification	Remodeling
Fontanels	Skull
Membrane	Woven bone
Osteoprogenitor cells	

At the site of intramembranous ossification, unspecialized cells aggregate to form a connective tissue __(1)__. __(2)__ cells differentiate to become osteoblasts that produce delicate trabeculae of __(3)__. These locations of bone formation are called __(4)__. __(5)__ converts woven bone to mature bone and contributes to the final shape of the bone. Intramembranous ossification results in the formation of the flat bones of the __(6)__. At birth some membrane is not ossified; these regions are called __(7)__.

1. _____
2. _____
3. _____
4. _____
5. _____
6. _____
7. _____

Endochondral Ossification

❝*Bone formation in association with cartilage is endochondral.*❞

Using the terms provided, complete these statements:

Bone collar	Osteoblasts
Calcified cartilage	Osteoclasts
Chondrocytes	Osteoprogenitor
Epiphyses	Primary
Hyaline cartilage	ossification center
Hypertrophy	

The first phase of endochondral ossification is the formation of a __(1)__ model by chondroblasts, which become __(2)__ when they are surrounded by cartilage matrix. When blood vessels invade the perichondrium surrounding the cartilage, __(3)__ cells within the perichondrium become osteoblasts. The osteoblasts produce compact bone on the surface of the cartilage model, forming a __(4)__. Chondrocytes in the center of the cartilage model __(5)__, or enlarge, and the matrix between the enlarged cells becomes mineralized to form __(6)__. When chondrocytes in this calcified area die, blood vessels grow into the enlarged lacunae and bring in __(7)__, which produce bone trabeculae. This area of bone formation is called a __(8)__. At the same time, __(9)__ remove bone from the diaphysis to produce the medullary cavity. Secondary ossification centers appear in the __(10)__ about one month before birth.

1. _____
2. _____
3. _____
4. _____
5. _____
6. _____
7. _____
8. _____
9. _____
10. _____

Bone Growth

"Unlike tendons, ligaments, and cartilage, bones cannot grow by interstitial growth."

A. Using the terms provided, complete these statements:

Appositional growth   Epiphyses
Diameter   Length
Endochondral growth

1. _____

2. _____

3. _____

4. _____

5. _____

Bone growth can occur either by __(1)__, the formation of new bone on the surface of bone, or by __(2)__, the growth of cartilage in the epiphyseal plate or articular cartilage and its eventual replacement by bone. Appositional growth is responsible for increase in __(3)__ of long bones and most growth of other bones, whereas endochondral growth is responsible for the increase in __(4)__ of bones, or elongation of projections on some bones. Endochondral growth also occurs in articular cartilage, and is responsible for the growth of __(5)__.

B. Using the terms provided, complete these statements:

Zone of calcification   Zone of resting
Zone of hypertrophy   cartilage
Zone of proliferation

1. _____

2. _____

3. _____

4. _____

In the epiphyseal plate, randomly arranged chondrocytes nearest the epiphysis are in the __(1)__ and divide slowly. Chondrocytes in the __(2)__ divide rapidly and produce new cartilage by interstitial growth. Chondrocytes produced in the zone of proliferation mature and enlarge in the __(3)__. Hypertrophied chondrocytes die, and osteoblasts deposit bone in the __(4)__, a thin layer close to the diaphysis.

☞ Eventually, the epiphyseal plate is completely ossified and becomes the epiphyseal line.

Factors Affecting Bone Growth

"The potential shape and size of a bone as well as final adult height are genetically determined, but factors such as nutrition and hormones may greatly modify the expression of genetic factors."

A. Match these terms with the correct statement or definition:

Osteomalacia   Vitamin C
Rickets   Vitamin D
Scurvy

_____ 1. Necessary for normal absorption of calcium from the intestines.

_____ 2. Can occur when children have insufficient vitamin D.

_____ 3. Softening of the bones because of calcium depletion in adults.

_____ 4. Necessary for normal collagen synthesis by osteoblasts.

B. **Match these terms with the correct statement or definition:**

Calcitonin
Growth hormone
Parathyroid hormone

Sex hormone
Thyroid hormone

_____ 1. Increase general tissue growth (two).

_____

_____ 2. Stimulates bone growth, but also ossifies epiphyseal plate.

☞ Bone is the major storage site for calcium in the body.

Calcium Homeostasis

66 *Bones play an important role in regulating blood calcium levels.* 99

Match these terms with the correct statement or definition:

Calcitonin
Osteoblasts

Osteoclasts
Parathyroid hormone

_____ 1. Activity of these cells increases if blood calcium is low.

_____ 2. Cells that take calcium from blood to produce new bone.

_____ 3. Major regulator of blood calcium levels; stimulates osteoclast activity and increases blood calcium levels.

_____ 4. Increases calcium uptake in the small intestine and decreases calcium loss from the kidneys.

_____ 5. Hormone that inhibits osteoclast activity.

Bone Remodeling

66 *Bone remodeling is involved in bone growth, changes in bone shape, adjustment of bone to stress,* 99 *and calcium ion regulation.*

Using the terms provided, complete these statements:

Interstitial lamellae
Medullary cavity
Osteons

Osteoblasts
Osteoclasts

Remodeling occurs constantly in bones. For example, as long bones increase in length and diameter, the size of the _(1)_ also increases. Remodeling is also responsible for the formation of new _(2)_ in compact bone. _(3)_ are constantly removing osteons and circumferential lamellae, and new osteons are formed by osteoblasts. However, this process leaves portions of older osteons and circumferential lamellae, called _(4)_, between the newly developed osteons. Mechanical stress increases the activity of _(5)_; in addition, pressure in bone causes an electrical change that decreases bone resorption by _(6)_ and speeds healing.

1. _____

2. _____

3. _____

4. _____

5. _____

6. _____

Bone Repair

"Bone is a living tissue capable of repair."

Using the terms provided, complete these statements:

Callus   Osteoblasts
Chondroblasts   Osteoclasts
Compact   Periosteum
Fibroblasts

When bone is damaged, blood vessels in the (1) and in the bone bleed, and a clot is formed. Uncommitted cells from surrounding tissue invade the clot and become (2) , which produce a fibrous network, and others become (3) , which produce islets of fibrocartilage. The area of fiber and cartilage formation is called a fibrocartilage (4) . From adjacent bone, (5) invade and form a bony callus through intramembranous and endochondral ossification. Finally, the repaired bone is remodeled to form (6) bone and cancellous bone.

1. _____

2. _____

3. _____

4. _____

5. _____

6. _____

QUICK RECALL

1. List three important components of the extracellular matrix of cartilage.

2. List the major organic and inorganic compounds found in the extracellular matrix of bone.

3. List three types of bone cells, depending upon their function.

4. List three types of lamellae visible in bone matrix.

5. Name, and distinguish between, two types of bone marrow.

6. List two types of bone depending upon its internal structure.

7. List three important components of a haversian system (osteon).

8. Name two types of bone ossification.

9. Name two types of bone growth.

10. List three factors that affect bone growth.

11. Name the abnormal bone conditions that can occur with a dietary deficiency of vitamin D.

12. List five hormones that influence bone growth.

MASTERY LEARNING ACTIVITY

Place the letter corresponding to the correct answer in the space provided.

_____1. Which of these is a function of bone?
a. internal support and protection
b. provides attachment for the muscles
c. calcium and phosphate storage
d. blood cell formation
e. all of the above

_____2. Tendons
a. increase in size by endochondral growth.
b. are composed of collagen fibers oriented in the same direction.
c. are produced by lacuna cells.
d. all of the above

_____3. Chondrocytes are mature cartilage cells found within the _____, and they are derived from _____.
a. perichondrium, fibroblasts
b. perichondrium, chondroblasts
c. lacunae, fibroblasts
d. lacunae, chondroblasts

_____4. Cartilage
a. often occurs in thin plates or sheets.
b. receives nutrients and oxygen by diffusion.
c. grows as a result of chondroblast formation by the perichondrium.
d. all of the above

_____5. A fracture in the shaft of a bone would be a break in the
a. epiphysis.
b. perichondrium.
c. diaphysis.
d. articular cartilage.

_____6. Yellow marrow is
a. mostly adipose tissue.
b. found within the medullary cavity in adults.
c. associated mostly with the long bones of the limbs in adults.
d. all of the above

_____7. Which of these connective tissue membranes covers the surface of mature bones?
a. periosteum
b. perichondrium
c. hyaline cartilage
d. b and c

_____8. Which of these substances make up the major portion of bone?
a. collagen
b. hydroxyapatite
c. proteoglycan aggregates
d. osteocytes
e. osteoblasts

_____9. The flexibility of bone is due to
a. osteoclasts.
b. ligaments.
c. calcium salts.
d. collagen fibers.

_____10. The prime function of osteoclasts is to
a. prevent osteoblasts from forming.
b. become osteocytes.
c. break down bone.
d. secret calcium salts and collagen fibers.
e. form bone.

_____11. Haversian canals
a. connect Volkmann's canals to canaliculi.
b. connect spongy bone to compact bone.
c. are where red blood cells are produced.
d. are found only in spongy bone.

_____12. The type of lamellae found in osteons is
a. circumferential lamellae.
b. concentric lamellae.
c. interstitial lamellae.
d. none of the above

13. Given these events:
    1. osteoprogenitor cells become osteoblasts
    2. connective tissue membrane formed
    3. osteoblasts produce woven bone

    Which sequence best describes intramembranous bone formation?
    a. 1,2,3
    b. 1,3,2
    c. 2,1,3
    d. 2,3,1
    e. 3,2,1

14. Given these processes:
    1. chondrocytes die
    2. calcification of cartilage matrix
    3. chondrocytes hypertrophy
    4. osteoblasts deposit bone
    5. blood vessels grow into lacunae

    Which of these sequences best represents the order in which they occur during endochondral bone formation?
    a. 3,2,1,4,5
    b. 3,2,1,5,4
    c. 5,2,3,4,1
    d. 3,2,5,1,4
    e. 3,5,2,4,1

15. Intramembranous bone formation
    a. occurs at the epiphyseal plate.
    b. gives rise to the flat bones of the skull.
    c. is responsible for growth in diameter of a bone.
    d. occurs within a hyaline cartilage model.
    e. all of the above

16. The ossification regions formed during early fetal development are
    a. secondary ossification centers.
    b. cartilage at the ends of bones.
    c. primary ossification centers.
    d. medullary cavities.

17. Articular cartilage is
    a. found on the ends of bone.
    b. hyaline cartilage.
    c. embryonic cartilage that does not ossify.
    d. all of the above

18. Growth in length of bone occurs
    a. at the primary ossification center.
    b. beneath the periosteum.
    c. at the diaphysis.
    d. at the epiphyseal plate.

19. During bone growth in length cartilage is formed and then ossified. The location of the ossification is the
    a. zone of proliferation.
    b. zone of resting cartilage.
    c. zone of hypertrophy.
    d. zone of calcification.

20. Chronic vitamin D deficiency would result in which of these?
    a. Bones would become brittle.
    b. Bones would become soft and pliable.
    c. The percentage of bone composed of collagen would increase.
    d. The percentage of bone composed of hydroxyapatite would increase.
    e. b and c

21. Osteoclast activity is stimulated by
    a. calcitonin.
    b. growth hormone.
    c. parathyroid hormone.
    d. sex hormones.

22. A 16-year-old female with delayed sexual maturation was found to have an adrenal tumor that produced large amounts of sex steroids (estrogen and testosterone). If that condition was allowed to go untreated, which of these symptoms could be expected?
    a. She would experience a short period of rapid growth.
    b. As an adult she would be taller than a normal adult.
    c. As an adult she would be shorter than a normal adult.
    d. She would grow normally.
    e. a and b

_____23. Destruction and remodeling of bone may occur
   a. as bones change shape during growth.
   b. as bones are subjected to varying patterns of stress.
   c. during and following the healing of fractures.
   d. all of the above

_____24. In a bone that was no longer growing, where would one be most likely to find an osteoclast?
   a. under the periosteum
   b. epiphyseal plate
   c. articular cartilage
   d. Volkmann's canal

_____25. In the healing of bone fractures
   a. a blood clot forms around the break.
   b. both intramembranous and endochondral ossification are involved.
   c. the callus may eventually disappear.
   d. all of the above

FINAL CHALLENGES

Use a separate sheet of paper to complete this section.

1. A doctor tells an elderly friend of yours that the cartilage in his joints is degenerating and has lost its resiliency. Your friend asks you to explain what that means and what the consequences of it might be. What do you say?

2. When an x-ray film was taken to check for a broken leg in a patient, the physician observed that there was a transverse region of very dense bone below the epiphyseal line. How might this dense region have developed, what effect might it have on the height of the patient, and would you expect to see this region in the other leg?

3. Is the formation of bone that occurs when a bone increases in length more like intramembranous or endochondral ossification? Explain.

4. Is the repair of bone tissue following a fracture more like intramembranous or endochondral ossification? Explain.

5. A dental assistant overheard part of a conversation between a dentist and her patient. She was explaining to the patient why a tooth implant had come loose and stated that the metallic components of the tooth implant were so strong that the underlying bone was "shielded." Can you complete the dentist's explanation why this would cause the implant to come loose?

6. An adult patient has a digestive system disorder that inhibits the absorption of fats. What is more likely to develop, soft or brittle bones? Explain.

ANSWERS TO CHAPTER 6

Functions of the Skeletal System
1. Bone; 2. Cartilage; 3. Tendon; 4. Ligament;
5. Bone; 6. Bone

Tendons and Ligaments
1. Less; 2. Less; 3. More; 4. Sheets; 5. Fibroblasts;
6. Fibrocytes; 7. Appositional; 8. Interstitial

Hyaline Cartilage
1. Chondroblast; 2. Chondrocyte;
3. Perichondrium

Bone Shape
1. Long bones; 2. Short bones; 3. Flat bones;
4. Irregular bones

Bone Anatomy
A. 1. Diaphysis; 2. Epiphysis; 3. Epiphyseal plate;
4. Epiphyseal line; 5. Medullary cavity;
6. Sinus; 7. Marrow; 8. Perforating (Sharpey's)
fibers; 9. Endosteum
B. 1. Articular cartilage; 2. Epiphysis;
3. Cancellous bone; 4. Compact bone;
5. Medullary cavity; 6. Periosteum;
7. Diaphysis; 8. Epiphyseal line

Bone Matrix
1. Collagen; 2. Hydroxyapatite; 3. Osteoblast;
4. Osteoclast; 5. Lamellae; 6. Lacunae;
7. Canaliculi

Cancellous Bone and Compact Bone
A. 1. Trabeculae; 2. Haversian (central) canals;
3. Concentric lamellae; 4. Haversian system
(osteon); 5. Volkmann's (perforating) canals;
6. Circumferential lamellae
B. 1. Haversian (central) canal; 2. Interstitial
lamellae; 3. Volkmann's (perforating) canal;
4. Concentric lamellae; 5. Circumferential
lamellae; 6. Haversian system (osteon)

Bone Ossification
1. Woven bone; 2. Remodeling;
3. Intramembranous; 4. Intramembranous;
5. Endochondral

Intramembranous Ossification
1. Membrane; 2. Osteoprogenitor cells; 3. Woven
bone; 4. Centers of ossification; 5. Remodeling;
6. Skull; 7. Fontanels

Endochondral Ossification
1. Hyaline cartilage; 2. Chondrocytes;
3. Osteoprogenitor; 4. Bone collar;
5. Hypertrophy; 6. Calcified cartilage;
7. Osteoblasts; 8. Primary ossification center;
9. Osteoclasts; 10. Epiphyses

Bone Growth
A. 1. Appositional growth; 2. Endochondral
growth; 3. Diameter; 4. Length; 5. Epiphyses
B. 1. Zone of resting cartilage; 2. Zone of
proliferation; 3. Zone of hypertrophy; 4. Zone
of calcification

Factors Affecting Bone Growth
A. 1. Vitamin D; 2. Rickets; 3. Osteomalacia;
4. Vitamin C
B. 1. Growth hormone, thyroid hormone;
2. Sex hormone

Calcium Homeostasis
1. Osteoclasts; 2. Osteoblasts; 3. Parathyroid
hormone; 4. Parathyroid hormone; 5. Calcitonin

Bone Remodeling
1. Medullary cavity; 2. Osteons; 3. Osteoclasts;
4. Interstitial lamellae; 5. Osteoblasts;
6. Osteoclasts

Bone Repair
1. Periosteum; 2. Fibroblasts; 3. Chondroblasts;
4. Callus; 5. Osteoblasts; 6. Compact

1. Collagen, proteoglycans, and water
2. Organic: collagen; inorganic: hydroxyapatite
3. Osteoblasts, osteocytes, and osteoclasts
4. Concentric, circumferential, and interstitial
5. Yellow marrow: mostly fat; red marrow: site of blood cell formation
6. Compact bone and cancellous (spongy) bone
7. Haversian (central) canal, concentric lamellae, and osteocytes in lacunae
8. Intramembranous and endochondral
9. Appositional and endochondral
10. Genetics, nutrition, and hormones
11. Rickets in children, osteomalacia in adults
12. Growth hormone, thyroid hormone, sex hormones, parathyroid hormone, and calcitonin

1. E. Bone performs all of the functions listed plus acts as a lever system and a site of fat storage (yellow marrow).

2. B. Tendons are dense, regular connective tissue. Fibroblasts produce tendons by appositional growth, and fibrocytes produce tendons by interstitial growth.

3. D. Chondroblasts become chondrocytes when they are surrounded by cartilage matrix and are found within spaces called lacunae.

4. D. Because most cartilage is avascular, nutrients reach chondrocytes only by diffusing into the cartilage from blood vessels in the perichondrium. Diffusion is effective only over short distances; thus the chondrocytes must be close to the perichondrium (surface). This is accomplished by arranging the cartilage into thin plates or sheets. Chondroblasts, derived from the perichondrium, lay down new matrix along the surface of the cartilage, resulting in appositional cartilage growth. Chondrocytes within the cartilage are responsible for interstitial cartilage growth.

5. C. The shaft of a long bone is the diaphysis. It is covered by the periosteum. The perichondrium covers cartilage. The ends of long bones are the epiphyses; they are covered by the articular cartilage.

6. D. Yellow marrow is mostly adipose tissue found within the medullary cavities of long bones within the limbs of adults. Red marrow is the site of blood cell formation and is found in cancellous bone of the axial skeleton and girdles of adults. Children have red marrow in the limbs but it is replaced by yellow marrow with age.

7. A. Whether mature bones are formed by intramembranous ossification or endochondral ossification, they are covered with periosteum. With intramembranous ossification, the periosteum is formed directly. In endochondral ossification the periosteum is derived from the perichondrium. Hyaline cartilage covers the articular surface of bones.

8. B. For the average compact bone, salts (hydroxyapatite) make up about 65% of the bone, whereas the remaining 35% is a tough organic matrix. The organic matrix is composed mainly of collagen (95%). Also present are ground substance, osteoblasts, and osteoclasts. Proteoglycan aggregates are found in cartilage.

9. D. Collagen fibers are responsible for bone flexibility. Calcium salts allow bones to withstand compression.

10. C. Osteoclasts break down bone, and osteoblasts build bone. The interaction between these two types of cells is responsible for bone growth and remodeling. Osteoblasts are derived from uncommitted cells in the periosteum and endosteum. Osteoblasts become osteocytes. Osteoclasts are derived from uncommitted cells in red bone marrow.

11. A. Blood vessels, lymph nodes, and nerves from the periosteum or the medullary cavity enter bone by Volkmann's canals. Volkmann's canals join with haversian canals, which are connected by canaliculi to osteocytes. This system functions to provide osteocytes with nutrients and to remove waste products.

12. B. Concentric lamellae around haversian canals form the "rings" within osteons. Circumferential lamellae form the outer surface of bone, and interstitial lamellae are the remnants of partially destroyed concentric or circumferential lamellae.

13 C. First, a connective tissue membrane is formed. Osteoprogenitor cells become osteoblasts that produce delicate trabeculae of woven bone, forming centers of ossification.

14. B. Chondrocytes within the cartilage model hypertrophy, the cartilage matrix is calcified, and the chondrocytes die. Blood vessels grow into the enlarged lacunae of the calcified cartilage, bringing osteoblasts that deposit bone on the surface of the calcified cartilage.

15. B. Intramembranous bone formation gives rise to the flat bones of the cranial roof, some facial bones, and the clavicle. The other terms describe endochondral bone formation.

16. C. Primary ossification centers are formed during early fetal development. In many bones this occurs by the third month of fetal life. Most secondary ossification centers begin to differentiate sometime between 1 month prior to birth and the age of 3 years. The last secondary ossification center appears in the clavicle at 18 to 20 years of age.

17. D. In endochondral bone formation the embryonic hyaline cartilage model is ossified, except for the articular cartilage (on the ends of bones within joints) and the epiphyseal plate.

18. D. The primary ossification center results in conversion of the diaphysis from hyaline cartilage into bone. Increase in length occurs at the epiphyseal plate. Increase in diameter occurs beneath the periosteum.

19. D. Cartilage cells, produced in the zone of proliferation, are arranged in columns resembling stacks of coins. Chondrocytes mature and enlarge in the zone of hypertrophy. In the zone of calcification, hypertrophied cells die, and the matrix is mineralized. The zone of resting cartilage contains cells that are not dividing rapidly.

20. E. Lack of vitamin D leads to demineralization and subsequent softening of bones. Vitamin D promotes the uptake of calcium and phosphorus from the intestine, making these elements available for bone formation. As the bone becomes demineralized, the percentage of bone composed of collagen increases.

21. C. Parathyroid hormone is the major regulator of blood calcium levels. If blood calcium decreases, the secretion of parathyroid hormone increases, stimulating osteoclast activity, resulting in increased bone breakdown and increased blood calcium. Calcitonin decreases osteoclast activity, whereas both growth hormone and sex hormones would stimulate osteoblast activity and stimulate bone growth.

22. E. Estrogen and testosterone both increase bone growth. However, early uniting of the epiphyses of long bones is also caused by both, especially estrogen (which explains why growth in women stops several years earlier than in men). With the adrenal tumor, it is expected that the girl would have a spurt of growth, followed by uniting of the epiphyses. Because sexual maturation was delayed (she was 16 years old), the girl is taller than normal because, during the period of delayed sexual maturation, she would have been slowly growing.

23. D. Bone is continually broken down and rebuilt in response to growth, changes in stress, repair of fractures, and other factors.

24. A. Once the epiphyseal plate has ossified, the formation of new bone does not cease. There is a continual resorption (by osteoclasts) and deposition (by osteoblasts) of bone. This occurs primarily on the outer surface of the bone beneath the periosteum and within haversian canals.

25. D. Following a fracture, bleeding from blood vessels in the haversian system and the periosteum produces a blood clot. The clot becomes a fibrocartilage callus when fibroblasts and chondroblasts invade the clot, the fibroblasts lay down fibers, and the chondroblasts form cartilage. The fibrocartilage callus is ossified (becomes a bony callus) by intramembranous (fibers) and endochondral (cartilage) ossification. Remodeling of the bony callus restores the medullary cavity and can remove all traces of the bony callus.

☆ FINAL CHALLENGES ☆

1. The major components of the hyaline (articular) cartilage found in joints are collagen and proteoglycan aggregates. The collagen provides structural strength, whereas the proteoglycan aggregates trap water and are responsible for the resilient property of cartilage. Resilience is the ability of the cartilage to resume its shape after being compressed. Aging results in a decrease in proteoglycan aggregates and a loss of resiliency. As a consequence, the cartilage is compressed, the collagen and other cartilage components are subjected to more wear and tear, and the cartilage degenerates. The destruction of cartilage within joints can result in joint pain and decreased mobility.

2. The dense area of bone below the epiphyseal line indicates that there was a period of time during which the bone did not increase in length (or increased very slowly) and the bone tissue became more ossified than normal. This often happens when a person suffers from an extended period of malnutrition. One expects the other leg to show a similar area and the height of the individual to be shorter than it might have been otherwise. Alternatively, a break in the bone resulting in the formation of a callus could produce an area of dense bone. In this case the other leg would not show a similar area, and height is probably not affected.

3. It is like endochondral ossification. Cartilage is formed in the zone of proliferation of the epiphyseal plate, the site of bone increase in length. The cartilage corresponds to the cartilage template formed during embryonic development. The chondrocytes (cartilage cells) hypertrophy, the matrix is calcified, the chondrocytes die, the area is invaded by blood vessels and osteoblasts, and bone is laid down and remodeled. These are the same events that occur in endochondral ossification.

4. Bone repair involves events that are similar to intramembranous ossification and endochondral ossification. Fibroblasts produce a fibrous collagen network in the fibrocartilage callus that is to be ossified, similar to the ossification of the collagen membrane in intramembranous ossification. In the fibrocartilage callus, chondrocytes produce cartilage islets that are later ossified. This is similar to the cartilage template of endochondral ossification.

5. Shielding the underlying bone reduces the stress on the bone. Without stress osteoblast activity decreases, but osteoclast activity continues. The result is a reduction in bone, weakening of the bone, and loss of the tooth implant.

6. Without absorption of fats there is inadequate absorption of vitamin D in the small intestine. Without vitamin D there is inadequate absorption of calcium in the small intestine. Therefore the bone is softer than normal.

Skeletal System: Gross Anatomy

FOCUS: The skeletal system consists of the axial skeleton (skull, hyoid bone, vertebral column, and rib cage) and the appendicular skeleton (limbs and their girdles). The skull surrounds and protects the brain and the organs responsible for the primary senses (seeing, hearing, smelling, and tasting); the vertebral column supports the head and trunk and protects the spinal cord; and the rib cage protects the heart and lungs. The pectoral girdle attaches the upper limbs to the trunk and allows a wide range of movement of the upper limbs. The pelvic girdle attaches the lower limbs to the trunk and is specialized to support the weight of the body. The pelvis in women is broader and more open than the pelvis in men to facilitate delivery of a baby. The upper limbs are capable of detailed movements such as grasping and manipulating objects, whereas the lower limbs are best suited for locomotion.

CONTENT LEARNING ACTIVITY

Introduction

66*The skeletal system includes bones, cartilages, ligaments, and tendons.*99

Match these skeletal subdivisions
with the correct bones:

Appendicular skeleton
Axial skeleton

_____ 1. Vertebral column

_____ 2. Skull

_____ 3. Hyoid bone

_____ 4. Limbs

_____ 5. Thoracic cage

_____ 6. Limb girdles

General Considerations

66 *Most bone features are based on the relationship between bone and associated soft tissues.* **99**

Match these bone features
with the correct definition:

Canal (meatus)   Fossa
Condyle   Notch
Crest (crista)   Trochanter
Facet   Tubercle
Foramen

_____ 1. Smooth, rounded articular surface

_____ 2. Small, flattened articular surface

_____ 3. Prominent ridge

_____ 4. Small, rounded process

_____ 5. A hole in a bone for passage of blood vessels or nerves

_____ 6. A tunnel running within a bone

_____ 7. General term for a depression

The Skull

66 *The skull is composed of 28 separate bones.* **99**

A. Match these terms with the
correct parts of the skull
labeled in Figure 7-1:

Coronal suture
External auditory meatus
Frontal bone
Lacrimal bone
Lambdoid suture
Mandible
Mastoid process
Maxilla
Occipital bone
Parietal bone
Sphenoid bone
Squamous suture
Styloid process
Temporal bone
Zygomatic arch
Zygomatic bone

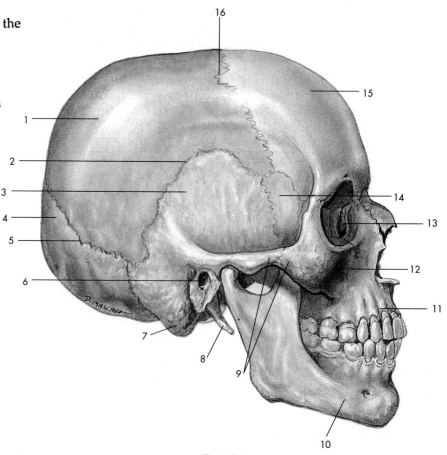

Figure 7-1

122

1. _____   7. _____   12. _____

2. _____   8. _____   13. _____

3. _____   9. _____   14. _____

4. _____   10. _____   15. _____

5. _____   11. _____   16. _____

6. _____

B. Match these structures with the correct description or definition:

Auditory ossicles	Hyoid bone
Cranial vault (brain case)	Nasal conchae
Cranium	Nasal septum
Crista galli	Orbit
Facial bones	Paranasal sinuses
Hard palate	Sella turcica

_____ 1. Six bones that function in hearing, found within cavities of the temporal bones.

_____ 2. Subdivision of the skull that protects the brain.

_____ 3. Neck and tongue muscles attach to it; "floats" in the neck.

_____ 4. Structure in the skull that encloses and protects the eye.

_____ 5. Bony shelves of the nasal cavity; warm and moisten the air.

_____ 6. Air-filled cavities attached to the nasal cavity.

_____ 7. Connective tissue membrane (one of the meninges) that holds the brain in place attaches to this prominent ridge.

_____ 8. Saddle-shaped structure occupied by the pituitary gland.

C. Match these bone parts with structures to which they contribute:

Horizontal plate of palatine bone	Perpendicular plate of ethmoid bone
Palatine process of maxilla	Vomer
Temporal process of zygomatic bone	Zygomatic process of temporal bone

_____ 1. Two parts that form the hard palate.

_____

_____ 2. Two parts that form the nasal septum.

_____

_____ 3. Two parts that form the zygomatic arch.

_____

123

D. Match these terms with the correct parts of the skull labeled in Figure 7-2A and 7-2B:

Hard palate
Horizontal plate of palatine bone
Palatine process of maxillary bone
Perpendicular plate of ethmoid bone
Septal cartilage

Temporal process of zygomatic bone
Vomer
Zygomatic arch
Zygomatic process of temporal bone

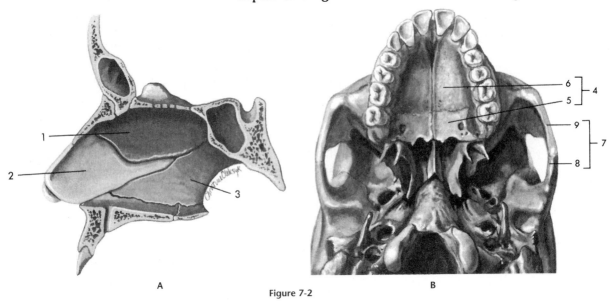

Figure 7-2

1. _____
2. _____
3. _____

4. _____
5. _____
6. _____

7. _____
8. _____
9. _____

E. Match these bones with the correct structures that make up part of that bone:

Ethmoid
Hyoid
Mandible

Occipital
Sphenoid
Temporal

_____ 1. Coronoid process

_____ 2. Cribriform plate

_____ 3. Foramen magnum

_____ 4. Greater cornu

_____ 5. Lateral pterygoid process

_____ 6. Mandibular fossa

_____ 7. Mastoid process

_____ 8. Medial nasal concha

_____ 9. Sella turcica

Muscles of the Skull

66*Many muscles attach to the skull.*99

Match these muscles with their correct attachments:

Eye movement muscles   Muscles of mastication

Muscles of facial expression   Neck muscles

_____ 1. Attach to the temporal bones, zygomatic arches, lateral pterygoid processes of the sphenoid, and mandible.

_____ 2. Attach to the nuchal lines, external occipital protuberance, and mastoid process.

_____ 3. Attach to the bones of the orbits.

Openings of the Skull

66*There are many openings through the bones of the skull.*99

Match the skull opening with its function or with the structures it contains:

Carotid canals   Jugular foramen

External auditory meatus   Nasolacrimal canal

Foramen magnum   Olfactory foramina

Internal auditory meatus   Optic foramen

_____ 1. Cribriform plate holes; contain nerves for the sense of smell.

_____ 2. Contains a duct carrying tears from the eye to the nasal cavity.

_____ 3. Contains the nerve for the sense of hearing.

_____ 4. Most blood leaves the skull through this opening.

_____ 5. The openings through which most blood reaches the brain.

_____

Joints of the Skull

66*The skull has several important joints.*99

Match these skull joints with their correct description:

Coronal suture   Occipital condyle

Lambdoid suture   Sagittal suture

Mandibular fossa   Squamous suture

_____ 1. Found between the parietal bones.

_____ 2. Found between the parietal bones and the occipital bone.

_____ 3. Found between the parietal bone and the temporal bone.

_____ 4. Articulation point between the skull and mandible.

_____ 5. Articulation point between the skull and the vertebral column.

Vertebral Column

The vertebral column provides flexible support and protects the spinal cord.

A. Match these terms with the correct statement or definition:

Cervical
Kyphosis
Lordosis
Lumbar

Sacral and coccygeal
Scoliosis
Thoracic

_____ 1. Two sections of the vertebral column that become convex anteriorly after birth.

_____

_____ 2. An exaggerated concave curve, especially in the thorax.

_____ 3. An abnormal bending of the spine to one side.

B. Match these terms with the correct statement or definition:

Anulus fibrosus
Intervertebral disk

Nucleus pulposus

_____ 1. Structure located between the bodies of adjacent vertebrae.

_____ 2. External portion of an intervertebral disk.

_____ 3. Inner portion of an intervertebral disk.

_____ 4. Ruptured (herniated) disk is due to damage to this structure.

Vertebra Structure

A typical vertebra consists of a body, an arch, and various processes.

A. Match these parts of a vertebra with their correct function or description:

Articular process
Body
Intervertebral foramina
Lamina

Pedicle
Spinous process
Transverse process
Vertebral foramen

_____ 1. Main weight-bearing portion of the vertebra.

_____ 2. Two parts that form the vertebral arch, which functions to protect the spinal cord.

_____

_____ 3. Contains the spinal cord; all of them together form the vertebral canal.

_____ 4. Where the spinal nerves exit the vertebrae.

_____ 5. Connects one vertebra to another, increasing the rigidity of the vertebral column while allowing movement.

_____ 6. Midline point of muscle attachment.

B. Match these terms with the correct parts of the vertebra labeled in Figure 7-3:

Articular facet for tubercle of rib
Body
Lamina
Pedicle

Spinous process
Superior articular facet
Transverse process
Vertebral arch
Vertebral foramen

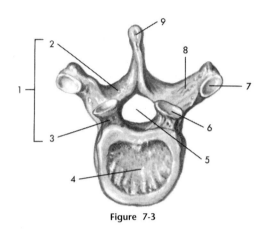

Figure 7-3

1. _____

2. _____

3. _____

4. _____

5. _____

6. _____

7. _____

8. _____

9. _____

Kinds of Vertebrae

❝*Several different kinds of vertebrae can be recognized.*❞

Match the type of vertebra with the characteristic unique to it:

Cervical vertebra
Coccygeal vertebra
Lumbar vertebra

Sacrum
Thoracic vertebra

_____

_____

_____

_____

_____

1. Transverse foramina through which the vertebral arteries going to the brain pass; spinous process is usually bifid (split).

2. Articular facet on the transverse process and on the body for the attachment of the rib.

3. Thick bodies and heavy, rectangular transverse and spinous processes; superior articular processes face medially, limiting movement of the vertebrae.

4. Transverse processes are fused to form the alae, which attach to the pelvic bones.

5. No vertebral foramen.

 The first cervical vertebra, the atlas, is specialized to articulate with the occipital condyle of the skull. The second cervical vertebra, the axis, has a modified process (the dens) around which the atlas rotates.

Thoracic Cage

66 *The thoracic cage, or rib cage, protects the vital organs within the thorax and prevents* **99** *the collapse of the lungs during respiration.*

A. Match these terms with the correct statement or definition:

Body   Manubrium
Head   True (vertebrosternal) ribs
False (vertebrochondral) ribs Tubercle
Floating (vertebral) ribs   Xiphoid process

_____ 1. First seven pairs of ribs, which articulate with both the vertebrae and the sternum.

_____ 2. Ribs that are attached to a common cartilage that, in turn, is attached to the sternum.

_____ 3. Eleventh and twelfth ribs, which have no attachment to the other ribs.

_____ 4. Rounded process on a rib that articulates with the transverse process of one vertebra.

_____ 5. Middle part of the sternum.

_____ 6. Most inferior portion of the sternum.

B. Match these terms with the parts of the thoracic cage labeled in Figure 7-4:

Body   Manubrium
Costal cartilage   Sternum
False (vertebrochondral) ribs True (vertebrosternal) ribs
Floating (vertebral) ribs   Xiphoid process

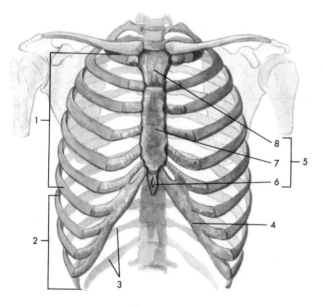

1. _____

2. _____

3. _____

4. _____

5. _____

6. _____

7. _____

8. _____

Figure 7-4

Pectoral Girdle

"The pectoral, or shoulder, girdle consists of the scapulae and clavicles which attach the upper limbs to the body."

A. Match these terms with the correct description:

Acromion process   Glenoid fossa
Clavicle   Scapular spine
Coracoid process

_____   1. Projection of scapula over the shoulder joint; articulates with the clavicle.

_____   2. Large ridge extending from the acromion process across the scapula.

_____   3. Projection on the anterior side of the scapula that provides a point of attachment for shoulder and arm muscles.

_____   4. Shallow depression in the scapula, where it articulates with the humerus.

_____   5. Functions to hold the upper limb away from the body.

B. Match these terms with the correct parts of the scapula labeled in Figure 7-5:

Acromion process   Lateral border
Coracoid process   Medial border
Glenoid fossa   Scapular spine
Infraspinous fossa   Supraspinous fossa

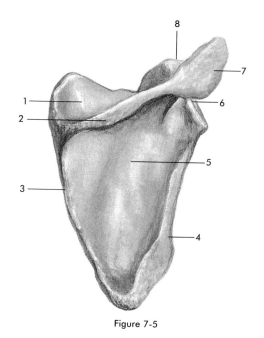

Figure 7-5

1. _____

2. _____

3. _____

4. _____

5. _____

6. _____

7. _____

8. _____

Upper Limb

66*The arm has one bone, the humerus, whereas the forearm has two bones, the radius and the ulna.*99

A. Match the bone parts that articulate with each other.

Head of humerus   Head of ulna
Head of radius   Trochlea

_____   1. Glenoid fossa

_____   2. Capitulum

_____   3. Trochlear (semilunar) notch

_____   4. Radial notch

_____   5. Carpals

B. Match the bony part with its function:

Bicipital groove   Lesser tubercle
Deltoid tuberosity   Olecranon process
Epicondyles   Radial tuberosity
Greater tubercle   Styloid processes

_____   1. Three places that shoulder muscles attach to the humerus.

_____

_____

_____   2. Groove between the greater and lesser tubercle where one tendon of the biceps brachii muscle passes.

_____   3. Location where forearm muscles attach to the humerus.

_____   4. Two places that arm muscles attach to the forearm.

_____

_____   5. Ligaments of the wrist attach to the forearm bones here.

C. Match these groups of bones with the correct descriptions:

Carpals   Phalanges
Metacarpals

_____   1. There are five of these bones; they form the hand.

_____   2. There are eight of these bones in two rows, forming the wrist.

_____   3. Each finger has three of these bones, and the thumb has two.

Pelvic Girdle

66*The pelvic girdle is formed by the coxae and the sacrum.*99

A. Using the terms provided, complete these statements:

Auricular surface   Iliac spines
Coxa   Ischial tuberosity
False pelvis   Obturator foramen
Iliac crest   Symphysis pubis
Iliac fossa   True pelvis

Each (1) is formed by the fusion of the ilium, ischium, and pubis. A large hole in the inferior half of the coxa is the (2) . The superior portion of the ilium is called the (3) , whereas the large depression on the medial side of the ilium is the (4) . The anterior and posterior ends of the iliac crest are referred to as (5) , which serve as points of muscle attachment. A heavy (6) on the ischium provides a place for muscle attachment and a place to sit. The two coxae are joined at the (7) , and the (8) of each coxa joins the sacrum (ala) to form the sacroiliac joint. The (9) is superior to the pelvic brim and is partially surrounded by bone, whereas the (10) is inferior to the pelvic brim and is completely surrounded by bone.

1. _____

2. _____

3. _____

4. _____

5. _____

6. _____

7. _____

8. _____

9. _____

10. _____

B. Match these terms with the correct parts of the coxa labeled in Figure 7-6:

Acetabulum
Anterior inferior iliac spine
Anterior superior iliac spine
Greater ischiadic (sciatic) notch
Iliac crest
Ilium
Ischial spine
Ischial tuberosity
Ischium
Obturator foramen
Posterior inferior iliac spine
Posterior superior iliac spine
Pubis

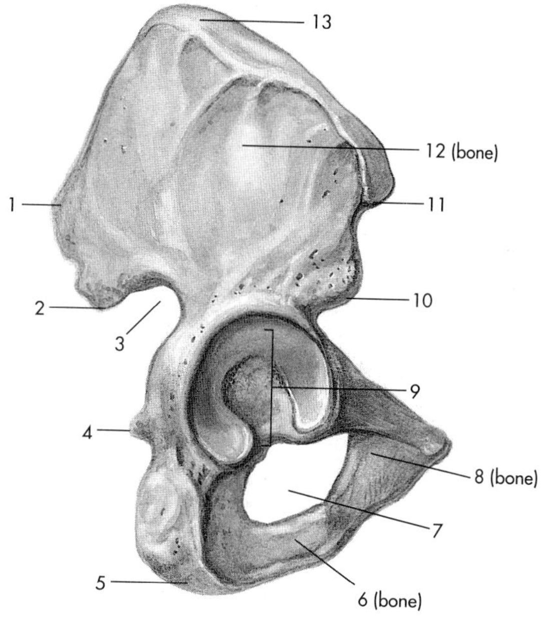

Figure 7-6

1. _____   6. _____   10. _____

2. _____   7. _____   11. _____

3. _____   8. _____   12. _____

4. _____   9. _____   13. _____

5. _____

C. Match these terms with the correct descriptions:

Female pelvis
Male pelvis

_____ 1. Pelvic inlet tends to be heart-shaped.

_____ 2. Subpubic angle is more than 90 degrees.

_____ 3. Ischial spines closer together.

_____ 4. Pelvic outlet is broader.

Lower Limb

66 *The thigh has one bone, the femur, whereas the leg has two bones, the tibia and the fibula.* **99**

A. Match the bone parts that articulate with each other:

Acetabulum   Patellar groove
Condyles   Talus
Head of fibula

_____ 1. Head of the femur articulates with the coxa.

_____ 2. Femur and tibia.

_____ 3. Tibia and proximal fibula.

_____ 4. Tibia and fibula with the ankle bone.

B. Match the bony part with its function:

Calcaneus   Patella
Condyles   Tibial tuberosity
Epicondyles   Trochanters (lesser and greater)

_____ 1. Site of hip muscle attachment lateral to the neck of the femur.

_____ 2. Muscle attachment sites next to the condyles of the femur.

_____ 3. Sesamoid bone within an anterior tendon of the thigh muscles.

_____ 4. Point of attachment on the tibia for anterior thigh muscles.

_____ 5. Attachment site for the calf muscles; the heel bone.

C. Match these terms with the correct description:

Lateral malleolus   Phalanges
Medial malleolus   Tarsals
Metatarsals

_____ 1. Found on the distal end of the tibia; forms the ankle.

_____ 2. Seven of these bones constitute the ankle.

_____ 3. Five of these bones are located in the foot.

_____ 4. Three of these bones are in each toe, except the big toe has two.

1. List the two major subdivisions of the skull.

2. List the four major sutures found in the cranial vault (braincase).

3. List the four major curvatures of the vertebral column of the adult, and give the direction in which they curve.

4. Name the five types of vertebrae, and give the number of each found in the vertebral column.

5. Name the three types of ribs according to their attachment, and give the number of each type.

6. List the three parts of the sternum.

7. Name the bones of the pectoral and pelvic girdles.

8. Give the number of carpals, metacarpals, and phalanges in the upper limb, and give the number of tarsals, metatarsals, and phalanges in the lower limb.

Name the bone or part of the bone responsible for the described bony landmark:

_____ 9. Bump posterior and inferior to the ear

_____ 10. Bump that can be felt anterior to the ear when the jaw is moved side to side

_____ 11. Bridge of the nose (two bones)

_____ 12. Tip of the shoulder

_____ 13. Elbow projection located on the medial side of the upper limb

_____ 14. Elbow projection located along the midline of the posterior surface of the upper limb

_____ 15. Bump at the distal end of the forearm on the medial, posterior surface

_____ 16. Knuckles

_____ 17. Ridge of bone felt when the hands are placed on the hips

_____ 18. The bony prominence that one sits upon

_____ 19. Kneecap

_____ 20. Shinbone

_____ 21. Bump on the anterior leg just distal to the kneecap

_____ 22. Large protuberance on the lateral surface of the ankle

MASTERY LEARNING ACTIVITY

Place the letter corresponding to the correct answer in the space provided.

_____ 1. Which of the following is part of the appendicular skeleton?
a. cranium
b. ribs
c. clavicle
d. sternum
e. vertebra

_____ 2. A knob like process on a bone?
a. spine
b. facet
c. tuberosity
d. sulcus
e. ramus

_____ 3. The superior and medial nasal conchae are formed by projections of this bone.
   a. sphenoid bone
   b. vomer bone
   c. palatine process of maxillae
   d. palatine bone
   e. ethmoid bone

_____ 4. The crista galli
   a. separates the nasal cavity into two parts.
   b. attaches the hyoid bone to the skull.
   c. holds the pituitary gland.
   d. is an attachment site for the membranes (meninges) that surround the brain.

_____ 5. The perpendicular plate of the ethmoid and the _____ form the nasal septum.
   a. palatine process of the maxilla
   b. horizontal plate of the palatine
   c. vomer
   d. a and b

_____ 6. Which of the following bones does NOT contain a paranasal sinus?
   a. ethmoid
   b. sphenoid
   c. temporal
   d. frontal
   e. maxillae

_____ 7. The mandible articulates with the skull at the
   a. styloid process.
   b. occipital condyle.
   c. mandibular fossa.
   d. zygomatic arch.
   e. medial pterygoid.

_____ 8. The nerves for the sense of smell pass through the
   a. nasolacrimal canal.
   b. internal auditory meatus.
   c. cribriform plate.
   d. optic foramen.
   e. orbital fissure.

_____ 9. The major blood supply to the brain enters through the
   a. foramen magnum.
   b. carotid canals.
   c. jugular foramina.
   d. a and b
   e. all of the above

_____ 10. Site of the sella turcica?
   a. sphenoid bone
   b. maxillae
   c. frontal bone
   d. ethmoid bone
   e. none of the above

_____ 11. Which of the following bones is NOT in contact with the sphenoid bone?
   a. maxilla
   b. inferior nasal concha
   c. ethmoid
   d. parietal
   e. vomer

_____ 12. The weight-bearing portion of a vertebra is the
   a. vertebral arch.
   b. articular process.
   c. body.
   d. transverse process.
   e. spinous process.

_____ 13. Transverse foramina are found only in
   a. cervical vertebrae.
   b. thoracic vertebrae.
   c. lumbar vertebrae.
   d. the sacrum.
   e. the coccyx.

_____ 14. Articular facets on the bodies and transverse processes are found only in
   a. cervical vertebrae.
   b. thoracic vertebrae.
   c. lumbar vertebrae.
   d. the sacrum.
   e. the coccyx.

_____ 15. Medially facing, superior articulating processes and laterally facing, inferior articulating processes are found in
   a. cervical vertebrae.
   b. thoracic vertebrae.
   c. lumbar vertebrae.
   d. the sacrum.
   e. the coccyx.

16. A ruptured (herniated) disk
    a. occurs when the anulus fibrosus ruptures.
    b. produces pain when the nucleus pulposus presses against spinal nerves.
    c. occurs when the disk slips out of place.
    d. a and b

17. Which of the following statements about vertebral column curvature is NOT true?
    a. The cervical curvature develops before birth.
    b. The thoracic curvature becomes exaggerated in kyphosis.
    c. The lumbar curvature becomes exaggerated in lordosis.
    d. The sacral curvature develops before birth.

18. Concerning ribs:
    a. the true ribs attach directly to the sternum.
    b. there are two pairs of floating ribs.
    c. the head of the rib attaches to the vertebra body.
    d. a and b
    e. all of the above

19. The point where the scapula and clavicle connect is the
    a. coracoid process.
    b. styloid process.
    c. glenoid fossa.
    d. acromion process.
    e. capitulum.

20. Distal medial process of the humerus to which the ulna joins?
    a. epicondyle
    b. deltoid tuberosity
    c. malleolus
    d. capitulum
    e. trochlea

21. Depression on the anterior surface of the humerus that receives part of the ulna when the forearm is flexed (bent)?
    a. epicondyle
    b. capitulum
    c. coronoid fossa
    d. olecranon fossa
    e. radial fossa

22. Which of the following is NOT a point of muscle attachment on the pectoral girdle or upper limb?
    a. epicondyles
    b. styloid processes
    c. radial tuberosity
    d. spine of scapula
    e. greater tubercle

23. Which of the following parts of the upper limb is NOT correctly matched with the number of bones in that part?
    a. arm: 1
    b. forearm: 2
    c. wrist: 10
    d. palm of hand: 5
    e. fingers: 14

24. The ankle bone that the tibia rests upon?
    a. talus
    b. calcaneus
    c. metatarsals
    d. navicular
    e. phalanges

25. Place(s) where nerves or blood vessels pass from the trunk to the lower limb?
    a. obturator foramen
    b. greater ischiadic (sciatic) notch
    c. ischial tuberosity
    d. a and b
    e. all of the above

26. When comparing the shoulder girdle to the pelvic girdle, which of the following statements is true?

	Pectoral Girdle	Pelvic Girdle
a. relative mass	more	less
b. security of axial attachment	less	more
c. security of limb attachment	more	less
d. relative flexibility	less	more

_____27. When comparing a male pelvis to a female pelvis, which of the following statements is true?

Part	Female	Male
a. pelvic inlet	small, heart-circular	large, shaped
b. pubic arch	sharp, rounded, acute	obtuse
c. ischial spine	farther apart	closer together
d. sacrum	narrow, more curved	broad, less curved

_____28. Site of muscle attachment on the proximal end of the femur?
a. greater trochanter
b. lesser trochanter
c. epicondyle
d. a and b
e. all of the above

_____29. Process forming the outer ankle?
a. lateral malleolus
b. lateral condyle
c. lateral epicondyle
d. lateral tuberosity
e. none of the above

_____30. Projection on the pelvic girdles used as a landmark for finding an injection site?
a. anterior superior iliac spine
b. posterior superior iliac spine
c. anterior inferior iliac spine
d. posterior inferior iliac spine

FINAL CHALLENGES

Use a separate sheet of paper to complete this section.

1. The length of the lower limbs of a 5-year-old was measured, and it was determined that one limb was 2 cm shorter than the other limb. Would you expect the little girl to exhibit kyphosis, scoliosis, or lordosis? Explain.

2. Suppose the vertebral column were forcefully bent backward. List the vertebrae (e.g., cervical, thoracic, and lumbar) in the order that they are most likely to be damaged. Explain.

3. In what region of the vertebral column is a ruptured disk most likely to develop? Explain.

4. Can you suggest a possible advantage in having the coccyx attached to the sacrum by a flexible cartilage joint?

5. Suppose you needed to compare the length of one upper limb to another in the same individual. Using bony landmarks, suggest an easy way to accomplish the measurements.

6. During a fist fight a man wildly swings at his opponent, striking him on the jaw. The force of the blow resulted in a broken hand. What bone of the hand was most likely damaged? Explain.

7. Seat belts should be fastened as low as possible across the hips. Defend or refute this statement.

ANSWERS TO CHAPTER 7

Introduction
1. Axial skeleton; 2. Axial skeleton; 3. Axial skeleton; 4. Appendicular skeleton; 5. Axial skeleton; 6. Appendicular skeleton

General Considerations
1. Condyle; 2. Facet; 3. Crest (crista); 4. Tubercle; 5. Foramen; 6. Canal (meatus); 7. Fossa

The Skull
A. 1. Parietal bone; 2. Squamous suture; 3. Temporal bone; 4. Occipital bone; 5. Lambdoid suture; 6. External auditory meatus; 7. Mastoid process; 8. Styloid process; 9. Zygomatic arch; 10. Mandible; 11. Maxilla; 12. Zygomatic bone; 13. Lacrimal bone; 14. Sphenoid bone; 15. Frontal bone; 16. Coronal suture

B. 1. Auditory ossicles; 2. Cranial vault (braincase); 3. Hyoid bone; 4. Orbit; 5. Nasal conchae; 6. Paranasal sinuses; 7. Crista galli; 8. Sella turcica

C. 1. Horizontal plate of palatine bone and palatine process of maxilla; 2. Perpendicular plate of ethmoid bone and vomer; 3. Temporal process of zygomatic bone and zygomatic process of temporal bone

D. 1. Perpendicular plate of ethmoid bone; 2. Septal cartilage; 3. Vomer; 4. Hard palate; 5. Horizontal plate of palatine bone; 6. Palatine process of maxillary bone; 7. Zygomatic arch; 8. Zygomatic process of temporal bone; 9. Temporal process of zygomatic bone

E. 1. Mandible; 2. Ethmoid; 3. Occipital; 4. Hyoid; 5. Sphenoid; 6. Temporal; 7. Temporal; 8. Ethmoid; 9. Sphenoid

Muscles of the Skull
1. Muscles of mastication; 2. Neck muscles; 3. Eye movement muscles

Openings of the Skull
1. Olfactory foramina; 2. Nasolacrimal canal; 3. Internal auditory meatus; 4. Jugular foramen; 5. Foramen magnum; carotid canals

Joints of the Skull
1. Sagittal suture; 2. Lambdoid suture; 3. Squamous suture; 4. Mandibular fossa; 5. Occipital condyle.

Vertebral Column
A. 1. Cervical; lumbar; 2. Kyphosis; 3. Scoliosis
B. 1. Intervertebral disk; 2. Anulus fibrosus; 3. Nucleus pulposus; 4. Anulus fibrosus

Vertebra Structure
A. 1. Body; 2. Lamina; pedicle; 3. Vertebral foramen; 4. Intervertebral foramina; 5. Articular process; 6. Spinous process
B. 1. Vertebral arch; 2. Lamina; 3. Pedicle;

4. Body; 5. Vertebral foramen; 6. Superior articular facet; 7. Articular facet for tubercle of rib; 8. Transverse process; 9. Spinous process.

Kinds of Vertebrae
1. Cervical vertebra; 2. Thoracic vertebra; 3. Lumbar vertebra; 4. Sacrum; 5. Coccygeal

Thoracic Cage
A. 1. True (vertebrosternal) ribs; 2. False (vertebrochondral) ribs; 3. Floating (vertebral) ribs; 4. Tubercle; 5. Body; 6. Xiphoid process
B. 1. True (vertebrosternal) ribs; 2. False (vertebrochondral) ribs; 3. Floating (vertebral) ribs; 4. Costal cartilage; 5. Sternum; 6. Xiphoid process; 7. Body; 8. Manubrium

Pectoral Girdle
A. 1. Acromion process; 2. Scapular spine; 3. Coracoid process; 4. Glenoid fossa; 5. Clavicle
B. 1. Supraspinous fossa; 2. Scapular spine; 3. Medial border; 4. Lateral border; 5. Infraspinous fossa; 6. Glenoid fossa; 7. Acromion process; 8. Coracoid process

Upper Limb
A. 1. Head of humerus; 2. Head of radius; 3. Trochlea; 4. Head of radius; 5. Head of ulna
B. 1. Greater tubercle; lesser tubercle; deltoid tuberosity; 2. Bicipital groove; 3. Epicondyles; 4. Olecranon process; radial tuberosity; 5. Styloid processes
C. 1. Metacarpals; 2. Carpals; 3. Phalanges

Pelvic Girdle
A. 1. Coxa; 2. Obturator foramen; 3. Iliac crest; 4. Iliac fossa; 5. Iliac spines; 6. Ischial tuberosity; 7. Symphysis pubis; 8. Auricular surface; 9. False pelvis; 10. True pelvis
B. 1. Posterior superior iliac spine; 2. Posterior inferior iliac spine; 3. Greater ischiadic (sciatic) notch; 4. Ischial spine; 5. Ischial tuberosity; 6. Ischium; 7. Obturator foramen; 8. Pubis; 9. Acetabulum; 10. Anterior inferior iliac spine; 11. Anterior superior iliac spine; 12. Ilium; 13. Iliac crest
C. 1. Male pelvis; 2. Female pelvis; 3. Male pelvis; 4. Female pelvis

Lower Limb
A. 1. Acetabulum; 2. Condyles; 3. Head of fibula; 4. Talus
B. 1. Trochanters (lesser and greater); 2. Epicondyles; 3. Patella; 4. Tibial tuberosity; 5. Calcaneus
C. 1. Medial malleolus; 2. Tarsals; 3. Metatarsals; 4. Phalanges

1. Cranial vault (braincase) and facial bones
2. Coronal, squamous, lambdoid, and sagittal sutures
3. Cervical: convex anteriorly; thoracic: concave anteriorly; lumbar: convex anteriorly; sacrum and coccyx: concave anteriorly
4. Cervical: 7; thoracic: 12; lumbar: 5; sacrum: 1; coccyx: 1
5. True (vertebrosternal) ribs: 7 pairs; False ribs: 5 pairs; 3 pairs of the false ribs are vertebrochondral ribs and 2 pairs are floating (vertebral) ribs
6. Manubrium, body, and xiphoid process
7. Pectoral girdle: scapula and clavicle; pelvic girdle: coxae and sacrum
8. Upper limb: carpals 8; metacarpals 5; phalanges

14; lower limb: tarsals 7; metatarsals 5; phalanges 14
9. Mastoid process
10. Mandibular condyle
11. Nasal bones and maxillae
12. Acromion process of scapula
13. Medial epicondyle of humerus
14. Olecranon process of ulna
15. Head of ulna
16. Distal end of metacarpals
17. Iliac crest
18. Ischial tuberosity
19. Patella
20. Anterior crest of tibia
21. Tibial tuberosity
22. Lateral malleolus of fibula

MASTERY LEARNING ACTIVITY

1. **C.** Appendicular skeleton: pectoral (clavicle, scapula) and pelvic (coxae) girdles plus the upper and lower limbs; Axial skeleton: skull (cranium and auditory ossicles), hyoid, vertebrae, and rib cage (ribs and sternum).

2. **C.** A tuberosity is a knoblike process. Other processes include tubercles (usually smaller than a tuberosity) and trochanters. A spine is a very high ridge, a facet is a small articular surface, a sulcus is a narrow depression, and a ramus is a branch off the main part (body) of a bone.

3. **E.** The superior and medial nasal conchae are projections of the ethmoid bone. The inferior nasal conchae are separate bones.

4. **D.** The crista galli serves as a point of attachment for the meninges. The nasal septum divides the nasal cavity, the hyoid bone does not have a bony attachment to the skull, and the sella turcica holds the pituitary gland.

5. **C.** The vomer and perpendicular plate of the ethmoid form the bony part of the nasal septum. The anterior portion of the nasal septum is hyaline cartilage. The palatine process of the maxilla and the horizontal plate of the palatine form the hard palate.

6. **C.** The temporal bone does not have a paranasal sinus (cavity connected to the nasal cavity). The mastoid process is filled with cavities called the mastoid air cells.

7. **C.** The mandibular condyle articulates with the mandibular fossa. The occipital condyle articulates with the vertebral column. The styloid process, zygomatic arch, and medial pterygoids are points of muscle attachment.

8. **C.** The nerves for the sense of smell pass through the olfactory foramina of the cribriform plate. The nasolacrimal duct passes from the eye to the nasal cavity through the nasolacrimal canal, the nerve for the sense of hearing and balance passes through the internal auditory meatus, the nerve for the sense of vision passes through the optic foramen, and nerves of the orbit and face pass through the orbital fissure.

9. **D.** The vertebral arteries enter through the foramen magnum, and the carotid arteries enter by means of the carotid canals. Blood leaves the brain through the jugular foramina.

10. **A.** The sella turcica is part of the sphenoid bone.

11. **B.** The sphenoid is in contact with all the bones listed except the inferior nasal concha.

12. **C.** The body is the weight-bearing portion. The vertebral arch protects the spinal cord; the articular processes connect vertebrae, allowing movement but providing stability; and the transverse and spinous processes are points of muscle attachment.

13. **A.** Only cervical vertebrae have transverse foramina.

14. **B.** The thoracic vertebrae have articular processes for the attachment of the ribs.

15. C.   The arrangement of the articular processes in lumbar vertebrae provides greater stability to the lumbar region, which is subjected to greater weight and pressure than the thoracic and cervical vertebrae.

16. D.   A herniated disk does not really involve a movement of the disk out of place. Instead, there is a rupture of the outer part of the disk, the anulus fibrosus. When the disk is subjected to pressure, its softer, inner portion, the nucleus pulposus, squeezes out of it. If the nucleus pulposus presses against the spinal cord or a spinal nerve, acute pain can result. Permanent damage to the nervous system is possible.

17. A.   The cervical curvature develops after birth as the infant raises its head.

18. E.   All of the statements are true.

19. D.   The scapula articulates with the clavicle at the acromion process and with the humerus at the glenoid fossa. The coracoid process is a point of muscle attachment.

20. E.   The humerus articulates with the ulna at the trochlea and with the radius at the capitulum.

21. C.   When the forearm is flexed, the coronoid process of the ulna fits into the coronoid fossa of the humerus. The head of the radius also fits into the radial fossa. When the forearm is extended (straight), the olecranon process of the ulna fits into the olecranon fossa of the humerus.

22. B.   The styloid processes of the ulna and radius are points of attachment for wrist ligaments.

23. C.   There are eight carpal bones in the wrist.

24. A.   The tibia and the fibula both articulate with the talus.

25. B.   The ischiadic (sciatic) nerve passes to the lower limb through the greater ischiadic (sciatic) notch. The obturator foramen is closed up by soft tissue, so it is not a major passageway for nerves and blood vessels. The ischial tuberosity is a site of posterior thigh muscle attachment.

26. B.   The pectoral girdle is attached to the axial skeleton only at the medial end of the clavicle. The pelvic girdle (coxa) is firmly attached to the sacrum at the sacroiliac joint.

27. C.   The pelvic outlet in females is larger than in males, in part because the ischial spines are further apart. This makes delivery possible.

28. D.   Hip muscles attach to the greater and lesser trochanter. The epicondyles are distal sites for the attachment of lower limb muscles.

29. A.   The lateral malleolus (fibula) forms the outer ankle and the medial malleolus (tibia) forms the inner ankle.

30. A.   The anterior superior iliac spine is used as a landmark for finding an injection site.

 # FINAL CHALLENGES

1. Scoliosis, because with one lower limb shorter than the other limb, the hips are abnormally tilted sideways. The vertebral column curves laterally in an attempt to compensate.

2. The most likely order of damage would be the cervical, lumbar, and thoracic vertebrae. The cervical vertebrae are the most movable vertebrae; but, once they have exceeded their normal range of movement, they are more likely to be damaged than other vertebrae. The lumbar vertebrae are more massive, and the thoracic vertebrae are supported by the ribs. In addition, because the ribs limit movement of the thoracic vertebrae, the cervical and lumbar vertebrae are more likely to be the region of the vertebral column that is bent and damaged.

3. A ruptured disk is most likely to develop in the lumbar region because it supports more weight than the cervical or thoracic regions. The sacral vertebrae are fused.

4. The coccyx can bend posteriorly to increase the size of the pelvic outlet during delivery.

5. Measure from the acromion process to the head of the ulna.

6. Inexperienced fighters do not punch in a straight line, often striking their opponent with the medial knuckles. The fifth metacarpal is the smallest and least supported (by muscles and other bones) metacarpal, and the distal end is fractured. A skilled boxer punches so that the second and third metacarpals absorb the force of the blow. These larger and better supported metacarpals are less likely to fracture.

7. The statement is true because the hip bones are strong. If the seat belt is fastened around the abdomen, internal organs are more likely to be damaged.

Articulations and Biomechanics of Body Movement

FOCUS: An articulation, or joint, is a place where two bones come together. Depending upon the type of connective tissue that binds the bones together, joints can be classified as fibrous, cartilaginous, and synovial. Fibrous joints exhibit little movement, cartilaginous joints are slightly movable, and synovial joints are freely movable. The range or type of movement in a synovial joint is determined by the shape of the articular surfaces. Plane and hinge joints allow movements in one plane, saddle and ellipsoid joints permit movement in two planes, ball-and-socket joints allow movement in all planes, and pivot joints restrict movement to rotation around a single axis.

CONTENT LEARNING ACTIVITY

Fibrous and Cartilaginous Joints

"Fibrous and cartilaginous joints are classified according to the tissue type that binds the bones together."

A. **Match the type of joint with its definition:**

Fontanel
Gomphosis
Periodontal ligament
Suture
Sutural ligament

Symphysis
Synchondrosis
Syndesmosis
Synostosis

_____ 1. Interdigitating bones joined by short, dense regular fibers.

_____ 2. Two layers of periosteum plus dense fibrous connective tissue between.

_____ 3. Wide suture found in the newborn.

_____ 4. Result when bones of a joint grow together.

_____ 5. Joint with bones joined by ligaments; farther apart than sutures.

_____ 6. Pegs that fit into sockets; bones joined by periodontal ligaments.

_____ 7. Connective tissue bundle between teeth and socket.

_____ 8. Joint with two bones joined by hyaline cartilage.

_____ 9. Joint with bones joined by fibrocartilage.

141

B. Match the type of joint
with the correct example:

Gomphosis Synchondrosis
Suture Syndesmosis
Symphysis Synostosis

_____ 1. Between the temporal and parietal bones.

_____ 2. Epiphyseal line.

_____ 3. Between the ulna and the radius.

_____ 4. Between the mandible and the teeth.

_____ 5. Between the ribs and the sternum.

_____ 6. Epiphyseal plate.

_____ 7. Intervertebral disks.

Synovial Joints

❝Synovial joints, those containing synovial fluid, allow considerable movement between bones.❞

A. Match these terms with the
correct statement or definition:

Articular cartilage Synovial fluid
Articular disk Synovial membrane
Bursa Tendon sheath
Fibrous capsule

_____ 1. Thin layer of hyaline cartilage that covers the articular surface of bones in synovial joints.

_____ 2. Fibrocartilage structures that provide extra strength and support for joints such as the knee and temporomandibular joint.

_____ 3. Portion of the joint capsule that is continuous with the fibrous layer of the periosteum.

_____ 4. Portion of the joint capsule; an inner membrane that secretes fluid.

_____ 5. Fluid that consists of serum filtrate and secretions from the synovial cells.

_____ 6. Pocket or sac of the synovial membrane extending away from the rest of the joint cavity.

_____ 7. Pocket or sac containing synovial fluid extending along a tendon for some distance.

 Bursitis is an inflammation of a bursa and may cause considerable pain around the joint and inhibition of movement.

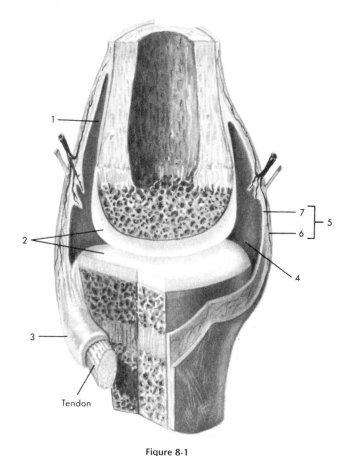

Figure 8-1

B. **M**atch these terms with the correct parts labeled in Figure 8-1:

Articular cartilage
Bursa
Fibrous capsule
Joint capsule

Joint cavity
Synovial membrane
Tendon sheath

1. _____

2. _____

3. _____

4. _____

5. _____

6. _____

7. _____

C. **M**atch the type of synovial joint with the correct definition and type of movement:

Ball-and-socket joint
Ellipsoid joint
Hinge joint

Pivot joint
Plane joint
Saddle joint

_____ 1. Two opposed flat surfaces approximately equal in size; monoaxial movement is gliding or slightly twisting.

_____ 2. Saddle-shaped articulating surfaces at right angles; biaxial movement.

_____ 3. Convex cylinder applied to a concave bone; monoaxial movement.

_____ 4. Cylindrical process rotates within a ring; monoaxial rotation.

_____ 5. Round head fitting into a round depression; multiaxial.

_____ 6. Modified ball-and-socket joint; range of motion is biaxial.

D. Match the type of synovial joint with the correct example:

Ball-and-socket joint   Pivot joint
Ellipsoid joint   Plane joint
Hinge joint   Saddle joint

_____ 1. Articular process of adjacent vertebrae.

_____ 2. Carpal and metacarpal of thumb.

_____ 3. Between phalanges.

_____ 4. Between the atlas and axis.

_____ 5. Coxal (hip).

_____ 6. Atlas and occipital bone.

_____ 7. Metacarpals and phalanges.

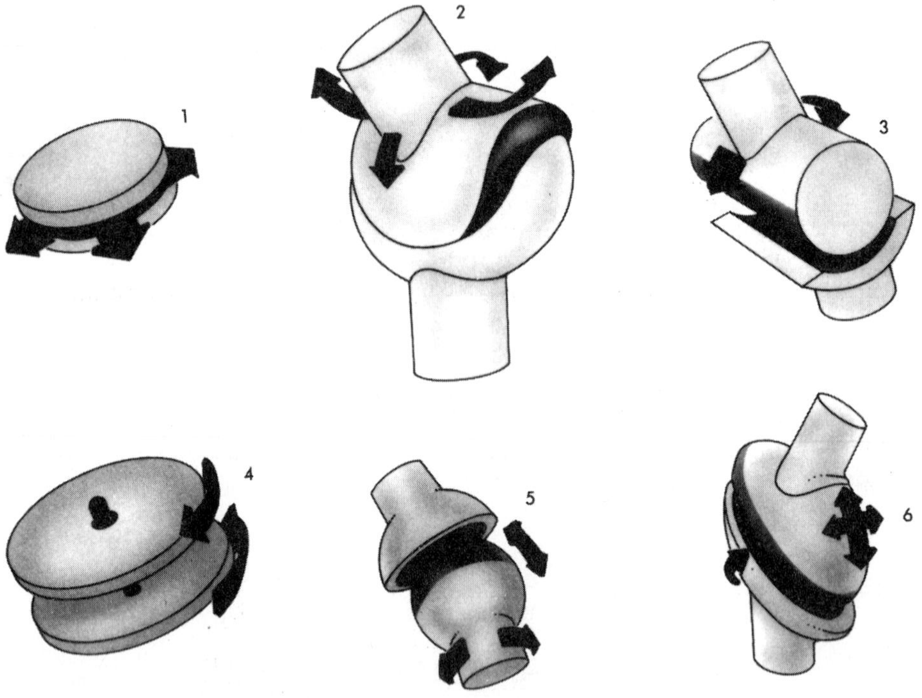

Figure 8-2

E. Match these terms with the correct type of synovial joint in Figure 8-2:

Ball-and-socket joint   Pivot joint
Ellipsoid joint   Plane joint
Hinge joint   Saddle joint

1. _____   3. _____   5. _____

2. _____   4. _____   6. _____

Angular Movements

Angular movements are those in which one part of a linear structure, such as the body or a limb, is bent to change the angle between the two parts.

Match these terms with the correct statement or definition:

Abduction	Extension
Adduction	Flexion
Dorsiflexion	Plantar flexion

_____ 1. Moving a part of the body in an anterior or ventral direction (except the knee).

_____ 2. Movement of the leg in an anterior direction.

_____ 3. Movement of the foot toward the plantar surface.

_____ 4. Movement toward the midline.

Circular Movements

Circular movements involve the rotation of a structure around an axis or movement in an arc.

Match these terms with the correct statement or definition:

Circumduction	Rotation
Pronation	Supination

_____ 1. Turning of a structure around its long axis.

_____ 2. Rotation of the palm so that it faces anteriorly. in relation to the anatomical position.

_____ 3. Combination of flexion, extension, abduction, and adduction.

Special Movements

Special movements are those movements that are unique to only one or two joints and do not fit neatly into one of the other categories.

Match these terms with the correct statement or definition:

Depression	Medial excursion
Elevation	Opposition
Eversion	Protraction
Inversion	Reposition
Lateral excursion	Retraction

_____ 1. Movement of a structure superiorly, e.g., the mandible.

_____ 2. Movement of a structure anteriorly, e.g., the scapula.

_____ 3. Movement of the mandible to the side.

_____ 4. Movement of the thumb and little finger together.

_____ 5. Turning the ankle so that the plantar surface faces laterally.

Movement Diagrams

Match these terms with the correct parts labeled in Figure 8-3:

Abduct
Adduct
Circumduct
Dorsiflex
Extend
Flex
Plantarflex
Pronate
Supinate

1. _____ 7. _____

2. _____ 8. _____

3. _____ 9. _____

4. _____ 10. _____

5. _____ 11. _____

6. _____

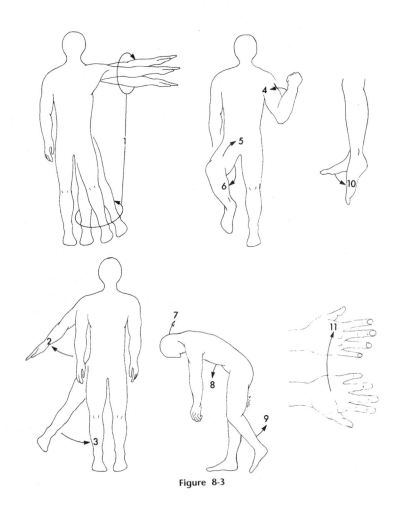

Figure 8-3

Temporomandibular Joint

The mandible articulates with the temporal bone to form the temporomandibular joint.

Using the terms provided, complete these statements:

Articular disk
Plane and ellipsoid
Ligaments

Mandibular
  condyle
Mandibular fossa

1. _____

2. _____

3. _____

4. _____

5. _____

In the temporomandibular joint the _(1)_ from the mandible articulates with the _(2)_ of the temporal bone. A fibrocartilage _(3)_ is interposed between the mandible and temporal bone, and the joint is surrounded by a fibrous capsule. The joint is also strengthened by lateral and accessory _(4)_ . The temporomandibular joint is a combination _(5)_ joint.

Shoulder Joint

The shoulder joint is a ball-and-socket joint in which stability is reduced and mobility is increased.

Match these terms with the
correct statement or definition:

Biceps brachii
Glenoid fossa
Glenoid labrum

Rotator cuff
Subscapular and subacromial
  bursae

_____

_____

_____

_____

1. Fibrocartilage ring that builds up the rim of the glenoid fossa.

2. Sacs with synovial fluid that open into the shoulder joint cavity.

3. Four muscles, collectively, that pull the humeral head superiorly and medially toward the glenoid fossa.

4. Tendon of this muscle passes through the articular capsule and helps to hold the humerus against the glenoid fossa.

Hip Joint

The femoral head articulates with the concave acetabulum of the coxa to form the coxal, or hip, joint.

Match these terms with the
correct statement or definition:

Acetabular labrum
Articular capsule
Iliofemoral ligament

Ligamentum teres
Transverse acetabular ligament

_____

_____

_____

_____

1. Lip of fibrocartilage; strengthens and deepens the acetabulum.

2. Ligament that passes inferior to, and strengthens, the acetabulum.

3. Strong connective tissue structure; extends from rim of the acetabulum to the neck of the femur, completely surrounding the femur head.

4. Ligament of the head of the femur; located inside the hip joint.

Knee Joint

"The knee joint, or genu, has traditionally been classified as a hinge joint, although it is actually a complex ellipsoid joint that allows flexion, extension, and a small amount of rotation of the leg."

Match these terms with the correct statement or definition:

Bursae
Collateral and popliteal
    ligaments

Cruciate ligaments
Menisci
Suprapatellar bursa

_____ 1. Thick articular disks at the margins of the tibia; deepen the articular surface.

_____ 2. Ligaments that extend between the fossa of the femur and the intercondylar eminence of the tibia; limit anterior and posterior displacement of the joint.

_____ 3. Ligaments that surround and strengthen the knee.

_____ 4. Superior extension of the joint capsule; allows movement of anterior thigh muscles over the distal end of the femur.

Ankle Joint and Arches of the Foot

"The ankle joint and arches of the foot provide support for the entire body."

Using the terms provided, complete these statements:

Arches
Dorsiflexion
Calcaneus

Hinge joint
Talocrural
Talus

The ankle, or __(1)__ joint, is a highly modified __(2)__ formed by the articulation of the tibia and fibula with the __(3)__. Plantar flexion and __(4)__, as well as limited inversion and eversion, can occur at this joint. There are three major __(5)__ in the foot that distribute the weight of the body between the __(6)__ and the ball of the foot.

1. _____
2. _____
3. _____
4. _____
5. _____
6. _____

QUICK RECALL

1. List the three major classes of joints.

2. Name the three types of fibrous joint and give an example of each.

3. Name the two types of cartilaginous joints; give an example of each.

4. Name the six types of synovial joints; give an example of each.

5. Name six types of angular movement.

6. List four types of circular movement.

7. Name ten types of special movement.

MASTERY LEARNING ACTIVITY

Place the letter corresponding to the correct answer in the space provided.

_____ 1. Which of these is used for classifying joints within the body?
   a. the material that connects the joints
   b. the way the joint moves
   c. the number of individual bones that articulate with one another
   d. a and b
   e. all of the above

_____ 2. Which of these types of joints contain fibrous connective tissue?
   a. syndesmosis
   b. suture
   c. gomphosis
   d. a and b
   e. all of the above

_____ 3. Which of these types of joints are held together by cartilage?
   a. symphysis
   b. synchondrosis
   c. synovial
   d. a and b
   e. all of the above

_____ 4. Which of these joints is NOT correctly matched with the type of joint?
   a. parietal bone to occipital bone; suture
   b. between the coxae; symphysis
   c. humerus and scapula; synovial
   d. sternum and ribs; syndesmosis

149

_____ 5. The epiphyseal plate can be described as a type of joint. Choose the term that describes the joint before growth in the length of the bone has terminated.
a. synchondrosis
b. synostosis
c. syndesmosis
d. symphysis
e. synovial

_____ 6. Which of these types of joints are often temporary?
a. syndesmoses
b. synovial
c. symphyses
d. synchondroses

_____ 7. Which of these joints are most movable?
a. sutures
b. syndesmoses
c. symphyses
d. synovial

_____ 8. The inability to produce the fluid that keeps most joints moist would likely be caused by a disorder of the
a. cruciate ligaments.
b. synovial membrane.
c. articular cartilage.
d. bursae.
e. none of the above

_____ 9. Which of these is NOT characteristic of a synovial joint?
a. articular surfaces covered with fibrous connective tissue
b. joint capsule
c. synovial membrane
d. synovial fluid
e. bursae

_____ 10. Synovial membranes form structures that, although not actually part of the synovial joint, are often associated with them. Which of these is an example of those structures?
a. bursae
b. tendon sheaths
c. covering of articular cartilage
d. a and b
e. all of the above

_____ 11. Assume that a sharp object penetrated a synovial joint. From these list of structures:
1. tendon or muscle
2. ligament
3. articular cartilage
4. fibrous capsule (of joint capsule)
5. skin
6. synovial membrane (of joint capsule)

Choose the order in which they would most likely be penetrated.
a. 5,1,2,6,4,3
b. 5,2,1,4,3,6
c. 5,1,2,6,3,4
d. 5,1,2,4,3,6
e. 5,1,2,4,6,3

_____ 12. Which of these do hinge joints and saddle joints have in common?
a. both are synovial joints
b. both have concave surfaces that articulate with a convex surface
c. both have movement in only one plane
d. a and b
e. all of the above

_____ 13. Which of these joints is correctly matched with the type of joint?
a. atlas to occipital; pivot
b. tarsals to metatarsals; saddle
c. femur to coxa; ellipsoid
d. tibia to talus; hinge

_____ 14. Once a door knob is grasped, what movement of the forearm is necessary to unlatch the door, i.e., turn the knob in a clockwise direction? (Assume you are right-handed.)
a. pronation
b. rotation
c. supination
d. flexion
e extension

15. After the door is unlatched, what movement of the forearm is necessary to open it? (Assume the door opens out, and you are on the outside.)
   a. protraction
   b. retraction
   c. flexion
   d. extension

16. After the door is unlatched, what movement of the arm is necessary to open it? (Still assuming the door opens out and you are on the outside.)
   a. flexion
   b. extension
   c. abduction
   d. adduction

17. When grasping a door knob, the thumb and little finger will undergo
   a. opposition.
   b. reposition.
   c. lateral excursion.
   d. medial excursion.

18. Tilting the head to the side is
   a. rotation.
   b. depression.
   c. abduction.
   d. lateral excursion.
   e. flexion.

19. A runner notices that the lateral side of her right shoe is wearing much more than the lateral side of her left shoe. This could mean that her right foot undergoes more _____ than her left foot.
   a. inversion
   b. eversion
   c. plantar flexion
   d. dorsiflexion

20. An articular disk is found in the
   a. temporomandibular joint.
   b. knee joint.
   c. ankle joint.
   d. a and b
   e. all of the above

21. A lip (labrum) of fibrocartilage deepens the joint cavity of the
   a. hip joint.
   b. shoulder joint.
   c. ankle joint.
   d. a and b
   e. all of the above

22. Which of these joints have a tendon inside the joint cavity?
   a. knee joint
   b. shoulder joint
   c. elbow joint
   d. ankle joint

23. Which of these joints have a ligament inside the articular capsule?
   a. knee joint
   b. hip joint
   c. shoulder joint
   d. a and b
   e. all of the above

24. Bursitis of the subacromial bursa could result from
   a. overuse of the shoulder joint.
   b. flexing the wrist.
   c. kneeling.
   d. running a long distance.

25. Which of these structures help to stabilize the shoulder joint?
   a. rotator cuff muscles
   b. cruciate ligaments
   c. medial and collateral ligaments
   d. articular disk
   e. all of the above

Use a separate sheet of paper to complete this section.

1. Using an articulated skeleton, examine the joints listed below. Describe the type of joint and the movement(s) possible.
   a. the joint between the nasal bones and the frontal bone
   b. the head of a rib and the body of a vertebra
   c. between the ribs and the sternum
   d. between the body of the sternum and the manubrium

2. For each of these joints, state what type of synovial joint it is and give the movements possible at the joint.
   a. between the phalanges
   b. between phalanx and metatarsal
   c. between the occipital condyles and the atlas
   d. between the atlas and axis (dens)
   e. between the tibia and the talus

3. For each muscle described below, describe the motion(s) produced when the muscle contracts. It may be helpful to use an articulated skeleton.
   a. The triceps muscle attaches to the infraglenoid tubercle of the scapula and the olecranon process of the ulna. How does contraction of this muscle affect the forearm? The arm?
   b. One of the muscles controlling movement of the index finger attaches to the distal third of the ulna. Its tendon runs over the dorsum (back) of the hand to the phalanx of the index finger. How does contraction of this muscle affect movement of the index finger? Of the wrist?
   c. The rectus abdominis attaches to the xiphoid process and the pubic crest. What movement of the vertebral column results when the rectus abdominis contracts?
   d. The semimembranosus muscle attaches to the ischial tuberosity and the medial condyle of the tibia. What movement of the thigh results when this muscle contracts? Of the leg?

4. What modifications to the skull and vertebrae would have to be made to make the joint between the skull and first vertebra into a ball-and-socket joint?

5. During a karate match a man was standing with his left leg forward, flexed, and supporting 60% of his weight. He was kicked just below the knee on the anterior tibia. What knee ligament was most likely damaged?

ANSWERS TO CHAPTER 8

Fibrous and Cartilaginous Joints
- A. 1. Suture; 2. Sutural ligament; 3. Fontanel;
  4. Synostosis; 5. Syndesmosis; 6. Gomphosis;
  7. Periodontal ligament; 8. Synchondrosis;
  9. Symphysis
- B. 1. Suture; 2. Synostosis; 3. Syndesmosis;
  4. Gomphosis; 5. Synchondrosis;
  6. Synchondrosis; 7. Symphysis

Synovial Joints
- A. 1. Articular cartilage; 2. Articular disk;
  3. Fibrous capsule; 4. Synovial membrane;
  5. Synovial fluid; 6. Bursa; 7. Tendon sheath
- B. 1. Bursa; 2. Articular cartilage; 3. Tendon
  sheath; 4. Joint cavity; 5. Joint capsule;
  6. Fibrous capsule; 7. Synovial membrane
- C. 1. Plane joint; 2. Saddle joint; 3. Hinge joint;
  4. Pivot joint; 5. Ball-and-socket joint;
  6. Ellipsoid joint;
- D. 1. Plane joint; 2. Saddle joint; 3. Hinge joint;
  4. Pivot joint; 5. Ball-and-socket joint;
  6. Ellipsoid joint; 7. Ellipsoid joint
- E. 1. Plane joint; 2. Saddle joint; 3. Hinge joint;
  4. Pivot joint; 5. Ball-and-socket joint;
  6. Ellipsoid joint

Angular Movements
- 1. Flexion; 2. Extension; 3. Plantar flexion;
  4. Adduction

Circular Movements
- 1. Rotation; 2. Supination; 3. Circumduction

Special Movements
- 1. Elevation; 2. Protraction; 3. Lateral excursion;
  4. Opposition; 5. Eversion

Movement Diagrams
- 1. Circumduct; 2. Abduct; 3. Adduct; 4. Flex;
  5. Flex; 6. Extend; 7. Extend; 8. Flex; 9. Flex;
  10. Plantar flex; 11. Pronate

Temporomandibular Joint
- 1. Mandibular condyle; 2. Mandibular fossa;
  3. Articular disk; 4. Ligaments; 5. Plane and
  ellipsoid

Shoulder Joint
- 1. Glenoid labrum; 2. Subscapular and
  subacromial bursae; 3. Rotator cuff; 4. Biceps
  brachii

Hip Joint
- 1. Acetabular labrum; 2. Transverse acetabular
  ligament; 3. Articular capsule; 4. Ligamentum
  teres

Knee Joint
- 1. Menisci; 2. Cruciate ligaments; 3. Collateral and
  popliteal ligaments; 4. Suprapatellar bursa

Ankle Joint and Arches of the Foot
- 1. Talocrural; 2. Hinge joint; 3. Talus;
  4. Dorsiflexion; 5. Arches; 6. Calcaneus

1. Fibrous, cartilaginous, and synovial joints
2. Sutures: any one of four examples; see Table 8-1 in textbook
   Syndesmosis: any one of four examples; see Table 8-1 in textbook
   Gomphosis: between the teeth and alveolar process in mandible or maxilla
3. Synchondrosis: any one of four examples; see Table 8-1 in textbook
   Symphysis: any one of four examples; see Table 8-1 in textbook
4. Plane joint: any one of nine examples; see Table 8-1 in textbook
   Saddle joint: between carpal and metacarpal of thumb

Hinge joint: any one of four examples; see Table 8-1 in textbook
Pivot joint: between atlas and axis, or between radius and ulna
Ball-and-socket joint: between coxa and femur, or between the scapula and humerus
Ellipsoid joint: any one of five examples; see Table 8-1 in textbook

5. Flexion, extension, plantar flexion, dorsiflexion, abduction, and adduction
6. Rotation, circumduction, pronation, and supination
7. Elevation, depression, protraction, retraction, lateral excursion, medial excursion, opposition, reposition, inversion, eversion

1. D.   One way of classifying joints is by the type of material that connects them (fibrous, cartilaginous, or synovial). Synovial joints are often grouped according to the movement allowed by the joint (plane, saddle, hinge, pivot, ball-and-socket, and ellipsoid).

2. E.   Sutures (between skull bones) and syndesmosis joints (interosseous membranes between the tibia and fibula) both contain fibers. Gomphoses are found between the teeth and jaws (periodontal ligaments).

3. D.   Symphyses are held together by fibrocartilage and synchondroses by hyaline cartilage. Synovial joints have articular cartilage, but it does not hold the bones together. Instead, it allows the bones to freely slide over each other.

4. D.   The sternum and ribs are joined by a synchondrosis joint.

5. A.   A joint that is held together by hyaline cartilage is classified as a synchondrosis. The epiphyseal plate is hyaline cartilage between the bone of the epiphysis and the bone of the diaphysis.

6. D.   Synchondrosis joints are held together by hyaline cartilage. The costal cartilages between the sternum and the ribs are examples of permanent synchondroses. Many synchondrosis joints are temporary, however, eventually being replaced by bone. For example, the epiphyseal plate is a synchondrosis joint where bone growth in length occurs. When growth stops, the epiphyseal plate is completely ossified and becomes a synostosis. It is then called the epiphyseal line.

7. D.   Synovial joints are freely movable, whereas the other joints listed allow limited movement. Sutures and syndesmosis joints are held together by fibrous connective tissue. Symphysis joints are connected by fibrocartilage.

8. B.   The synovial membrane, which lines the articular capsule of joints, produces the synovial fluid. This fluid acts as a lubricant between joint surfaces and provides nourishment to the articular cartilage.

9. A.   The articular surfaces of bones within a synovial joint are covered with hyaline cartilage, which provides a smooth surface where the bones meet. The joint capsule surrounds and supports the joint, and the synovial membrane lines the inside of the joint and produces synovial fluid, which fills the joint cavity. Bursae are synovial fluid sacs formed by the synovial membrane.

10. D.   Bursae and tendon sheaths are synovial sacs that are filled with synovial fluid. Bursae act as a cushion between the structures they separate. Tendon sheaths surround tendons, forming a double-walled cushion for the tendon to slide through. The articular cartilage, however, is not covered by synovial membrane.

11. E.   You must know the structure of a synovial joint, as well as the surrounding structures. List those structures from outside to inside in the correct order.

12. A.   Hinge joints and saddle joints are both synovial joints. In hinge joints a concave surface articulates with a convex surface (cylindrical surface), whereas in saddle joints both articular surfaces are concave. Hinge joints are capable of movement in one plane, and saddle joints are capable of movement in two planes.

13. D.   The tibia joins the talus as a hinge joint. The atlantooccipital joint is ellipsoid, the tarsometatarsal joint is plane, and the coxal joint is a ball-and-socket.

14. C.   Supination of the forearm would cause the knob to turn clockwise, unlatching the door.

15. C.   Flexion of the forearm would pull the door closer to the person, causing it to open.

16. B.   Extension of the arm would pull the door closer to the person, causing it to open.

17. A.   Opposition brings the thumb and little finger toward each other across the palm of the hand.

18. C.   Abduction is movement away from the midline.

19. A.   Inversion would turn the ankle so that more of the lateral surface of the foot would come into contact with the ground.

20. D.   Both the temporomandibular joint and knee joint have articular disks.

21. D. The acetabular labrum is found in the hip joint, and the glenoid labrum is part of the shoulder joint.

22. B. The tendon of the biceps brachii muscle passes through the shoulder joint cavity.

23. D. The cruciate ligaments are found within the knee joint, and the ligament of the head of the femur (ligamentum teres) is found within the hip joint.

24. A. The subacromial bursa is located in the shoulder.

25. A. Four rotator cuff muscles, the tendon of the biceps brachii muscle, three sets of ligaments, and the glenoid labrum all stabilize the shoulder joint. Cruciate ligaments, medial and collateral ligaments, and an articular disk are structures found in the knee.

 FINAL CHALLENGES

1. a. suture: little or no movement
   b. plane synovial: sliding movement
   c. synchondrosis: little or no movement
   d. symphysis: little or no movement

2. a. hinge: flexion and extension
   b. ellipsoid: flexion, extension, abduction, adduction, and circumduction
   c. ellipsoid: flexion, extension, abduction, adduction, and circumduction
   d. pivot: rotation
   e. hinge: plantar flexion, dorsiflexion, inversion, and eversion

3. a. extension of the forearm and arm
   b. extension of the finger and wrist
   c. flexion of the vertebral column
   d. extension of the thigh and flexion of the leg.

4. The occipital condyles could be converted into the "ball" by enlarging and extending anteriorly and posteriorly. This would form a ball structure with a hole (the foramen magnum) in its center. The superior articular facet of the axis could be converted into a "socket" by deepening and extending anteriorly and posteriorly. The dens of the axis would have to be eliminated, and the atlas would no longer rotate around the axis.

5. The posterior cruciate ligament prevents movement of the tibia in a posterior direction. The blow described would cause posterior movement of the tibia and likely damage the posterior cruciate ligament.

ANATOMY FREAK

I dig. I poke. I pry.
I agonize. I dentify.
All the body parts
That I have grown.

I know each muscle action,
From insertion to contraction,
The brain, the bowel,
The kidney and the bone.

All these parts within me
Used to seem so right.
Some worked on the day shift,
Some mostly at night.

Now my body parts are strangers,
With names I've never heard.
Terms with exotic syllables.
I can't spell a word!

What can stop this torment?
What can ease my plight?
What can make me dream of something
Besides A & P at night?

My organs all implore me
As they grumble and they creak,
NAME ME! NAME ME! NAME ME!

I am an anatomy freak.

Minyon Bond

9 Membrane Potentials

FOCUS: There is a higher concentration of K+ ions inside the cell than outside, and a higher concentration of Na+ ions outside the cell than inside. Slight leakage of K+ ions to the outside of the plasma membrane is responsible for establishing the resting membrane potential (RMP), a charge difference between the inside and the outside of the unstimulated plasma membrane. The outside of the plasma membrane is positively charged compared to the inside. An action potential is produced when the charge difference across the plasma membrane reverses and then returns to the resting condition. The charge reversal results from the movement of Na+ ions into the cell and the return to the resting condition occurs because of the movement of K+ ions out of the cell. Following the action potential, the sodium-potassium exchange pump restores ion concentrations by moving Na+ ions out of the cell and K+ ions into the cell. Action potentials are propagated along the plasma membrane of a cell and transferred across synapses to other cells. The strength of a stimulus determines the frequency of action potential generation, unless accommodation occurs.

CONTENT LEARNING ACTIVITY

Concentration Differences Across the Plasma membrane

66 *The plasma membrane is selectively permeable, allowing some, but not all,* 99 *substances to pass through the membrane.*

Match these terms with the correct statement or definition:

Cl- ions   Proteins
K+ ions   Sodium-potassium exchange
Na+ ions   pump

_____ 1. Large, negatively charged; doesn't cross the plasma membrane.

_____ 2. Repelled by negative charges; pass out through membrane channels.

_____ 3. The concentration of these ions increases inside the cell because of their attraction to negatively charged molecules.

_____ 4. Positively charged; pass through membrane channels easily.

_____ 5. Positively charged ions; greater concentration outside the cell.

_____ 6. Movement of K+ ions inward and Na+ ions outward across the plasma membrane; requires active transport (ATP molecules).

Resting Membrane Potential

"The RMP exists when cells are in an unstimulated, or resting, state.**"**

Using the terms provided, complete these statements:

Depolarization   Resting membrane
Hyperpolarization   potential
K^+ ions   Selectively permeable
Na^+ ions

1. _____

2. _____

3. _____

4. _____

5. _____

6. _____

7. _____

In nerve and skeletal muscle cells, the inside of the unstimulated plasma membrane is negative when compared to the outside of the plasma membrane. This potential difference is called the _(1)_. Unequal concentrations of charged molecules and ions that are separated by a _(2)_ plasma membrane are essential to resting membrane potential (RMP) development. Because the plasma membrane is somewhat permeable to _(3)_, these ions tend to diffuse from the inside to just outside of the plasma membrane. Their movement causes the outside of the plasma membrane to become positively charged. If the extracellular concentration of K^+ ions increases, fewer K^+ ions diffuse out of the cell, resulting in _(4)_ and a smaller RMP. Alternately, decreased K^+ ion concentration outside the cell causes more K^+ ions to diffuse outward and results in _(5)_. This effect can also be produced by an increase in permeability of the plasma membrane to _(6)_. Changes in the concentration of _(7)_ on either side of the plasma membrane do not markedly affect the RMP.

Movement of Sodium and Potassium Ions Through the Plasma membrane

"There are two separate systems for the movement of ions through the plasma membrane,**"** ion channels and the sodium-potassium exchange pump.

A. Match these terms with the correct statement or definition:

Ca^{2+} ions   Ligand-gated channels
Gating proteins   Voltage-sensitive channels
Ion channels

_____ 1. General term for pores in the plasma membrane that allow one type of ion to pass through it but not others.

_____ 2. Molecules that open and close ion channels.

_____ 3. Open or close in response to a change in the electrical charge across the plasma membrane, such as depolarization.

_____ 4. Open or close when a chemical such as acetylcholine binds to a receptor molecule.

_____ 5. Extracellular concentration of these cations influences the gating proteins for Na^+ ions.

B. Using the terms provided, complete the following statements:

ATP
Concentration gradients
Into

Out of
Resting membrane
   potential

1. _____

2. _____

3. _____

4. _____

5. _____

The sodium-potassium exchange pump transports approximately three Na$^+$ ions __(1)__ the cell for every two K$^+$ ions transported __(2)__ the cell. Very little of the __(3)__ is caused by the sodium-potassium exchange pump, whose major function is to maintain the normal __(4)__ across the plasma membrane. The movement of Na$^+$ and K$^+$ ions requires __(5)__ because it is accomplished by active transport, which moves the ions through the plasma membrane against their concentration gradients.

Electrically Excitable Cells

66 *Nerve and muscle cells are electrically excitable; thus the application of a stimulus causes* 99 *a series of characteristic changes in the RMP.*

A. Match these terms with the correct statement or definition:

Action potential
Local potential

_____

_____

_____

_____

_____

1. Confined to a small region of the plasma membrane.

2. Graded; magnitude of the response is proportional to the strength of the stimulus.

3. Can summate; one response can build on the previous response.

4. Occurs when the threshold potential level is reached.

5. Occurs in an all-or-none fashion.

B. Match these terms with the correct statement or definition:

Absolute refractory period
After potential
Depolarization

Relative refractory period
Repolarization

_____

_____

_____

_____

_____

1. Occurs as a result of Na$^+$ ion channels opening.

2. Occurs as a result of decreased membrane permeability to Na$^+$ ions, and increased membrane permeability to K$^+$ ions.

3. The membrane potential returns to RMP, becoming more negative.

4. Short period of hyperpolarization caused by increased K$^+$ ion permeability, observed after the action potential in many cells.

5. Period of time during the last part of the repolarization phase, during which a greater-than-threshold stimulus can initiate an action potential.

C. Match these terms with the parts labeled in Figure 9-1:

Absolute refractory period
Action potential
After potential
Depolarization

Relative refractory period
Repolarization
Threshold

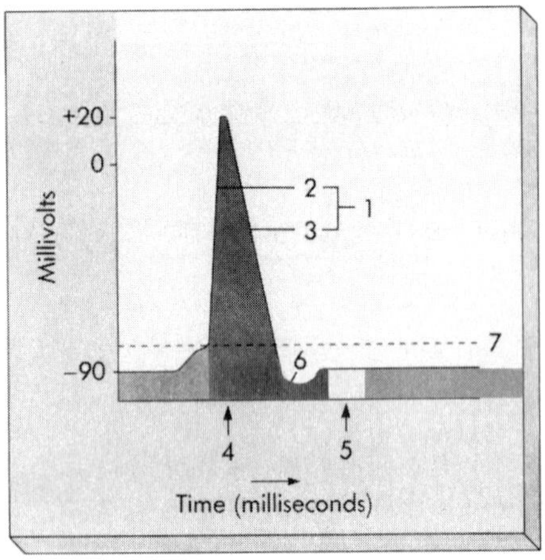

1. _____

2. _____

3. _____

4. _____

5. _____

6. _____

7. _____

Figure 9-1

Propagation of Action Potentials

66 *An action potential produced at any point on the plasma membrane acts as a stimulus* 99
to adjacent regions of the plasma membrane.

Match these terms with the correct statement or definition:

Absolute refractory period
Excitatory postsynaptic potential
Gap junction
Inhibitory postsynaptic potential

Na^+ ion channels
Neurotransmitter
Synapse

_____ 1. Action potentials open these structures, resulting in the production of additional action potentials.

_____ 2. Prevents an action potential from reversing its direction.

_____ 3. General term for a structure where action potentials in one cell can result in the production of action potentials in another cell.

_____ 4. Substance released from a nerve-cell process within a synapse.

_____ 5. Neurotransmitter binding to receptors can open these structures, resulting in the production of an action potential.

_____ 6. Electrical synapse where plasma membranes are joined, which allow depolarization in one cell to spread to the adjacent cell.

_____ 7. Local depolarization that causes the resting membrane potential to be closer to threshold.

_____ 8. Local depolarization that causes the resting membrane potential to be farther from threshold.

160

Action Potential Frequency

66*The action potential frequency is a measure of stimulus strength.*99

Match these terms with the correct statement or definition:

Accommodation   Supramaximal stimulus
Maximal stimulus   Threshold stimulus
Subthreshold stimulus

_____ 1. Stimulus resulting in a local potential so small that it does not reach threshold.

_____ 2. Stimulus just strong enough to cause a local potential to produce a single action potential.

_____ 3. Stimulus just strong enough to produce a maximum frequency of action potentials.

_____ 4. The local potential quickly returns to its RMP, even though the stimulus is applied for a long time.

☞ Between threshold and maximal stimulus strength, the action potential frequency increases in proportion to the strength of the stimulus.

Disorders

66*Several conditions illustrate the clinical consequences of abnormal membrane potential.*99

Match these terms with the correct description:

Hypocalcemia
Hypokalemia

_____ 1. Causes hyperpolarization of the RMP, resulting in muscular weakness and sluggish reflexes.

_____ 2. Causes action potentials to occur spontaneously, resulting in nervousness, muscular spasms, and tetany.

QUICK RECALL

1. List three causes of differences in intracellular and extracellular ion concentration.

2. List two factors of primary importance to the RMP.

3. List two separate systems for movement of ions through the plasma membrane.

4. List five characteristics of a local potential.

5. List four characteristics of an action potential.

6. Define the all-or-none law.

8. List five different strengths of stimuli.

MASTERY LEARNING ACTIVITY

Place the letter corresponding to the correct answer in the space provided.

_____ 1. Concerning concentration difference across the plasma membrane there are
   a. more K⁺ ions outside the cell than inside.
   b. more Na⁺ ions inside the cell than outside.
   c. more negatively charged proteins inside the cell than outside.
   d. a and b
   e. all of the above

_____ 2. Compared to the inside of the resting plasma membrane, the outside surface of the membrane is
   a. positively charged.
   b. electrically neutral.
   c. negatively charged.
   d. continuously reversing its electrical charge so that it is positive one second and negative the next.
   e. negatively charged whenever the sodium pump is in operation.

_____3. The resting membrane potential results when _____ accumulate on the outside of the plasma membrane.
   a. Na^+ ions
   b. K^+ ions
   c. Cl^- ions
   d. negatively charged proteins

_____4. If K^+ ion leakage increased, the resting membrane potential would _____. This is called _____.
   a. increase, hyperpolarization
   b. increase, depolarization
   c. decrease, hyperpolarization
   d. decrease, depolarization

_____5. Which of the following statements are correctly matched?
   a. depolarization: membrane potential becomes more negative
   b. hyperpolarization: membrane potential becomes more positive
   c. none of the above

_____6. Graphed below is a change in the resting membrane potential.

The indicated change represents a (an) _____ in the resting membrane potential and is called _____.
   a. increase, hyperpolarization
   b. increase, depolarization
   c. decrease, hyperpolarization
   d. decrease, depolarization

_____7. Which of the following statements about ion movement through the plasma membrane is true?
   a. Movement of Na^+ ions out of the cell requires energy (ATP).
   b. When Ca^{2+} ions bind to gating proteins the movement of Na^+ ions into the cell is inhibited.
   c. There is a specific ion channel for regulating the movement of Na^+ ions.
   d. all of the above

_____8. The MAJOR function of the sodium-potassium exchange pump is to
   a. pump Na^+ ions into and K^+ ions out of the cell.
   b. generate the resting membrane potential.
   c. maintain the concentration gradients of Na^+ ions and K^+ ions across the plasma membrane.
   d. oppose any tendency of the cell to undergo hyperpolarization.

_____9. A local potential is
   a. all-or-none.
   b. not propagated for long distances.
   c. variable in magnitude.
   d. a and b
   e. b and c

_____10. If one prevents Na^+ ions from diffusing into the cell,
   a. the membrane hyperpolarizes.
   b. there is no local potential.
   c. there still is a normal action potential.
   d. a and b

_____11. An action potential has the same intensity, whether it is triggered by a weak but threshold stimulus or by a strong, well above threshold stimulus. This is called
   a. an all-or-none response.
   b. a relative refractory response.
   c. an absolute refractory response.
   d. a latent period response.

_____12. During the depolarization phase of the action potential, the permeability of the membrane
   a. to K^+ ions is greatly increased.
   b. to Na^+ ions is greatly increased.
   c. to Ca^{2+} is decreased.
   d. is unchanged.

13. During depolarization the inside of the plasma membrane becomes more _____ than the outside of the plasma membrane.
    a. positive
    b. negative

14. Ms. Apeman went to the grocery store in time to take advantage of a sale on bananas. All morning and afternoon she ate bananas, which are high in potassium. That evening she didn't feel well. Which of the following symptoms is consistent with her overindulgence?
    a. increased heart rate
    b. hyperpolarization of her nerve cells
    c. action potentials that have an exaggerated amplitude in neurons
    d. all of the above

15. The repolarization phase of the action potential results from the
    a. increased permeability of the membrane to Na^+ ions.
    b. decreased permeability of the membrane to Na^+ ions.
    c. increased permeability of the membrane to K^+ ions.
    d. b and c
    e. a and c

16. During repolarization of a plasma membrane
    a. Na^+ ions rapidly move to the inside of the cell.
    b. Na^+ ions rapidly move to the outside of the cell.
    c. K^+ ions rapidly move to the inside of the cell.
    d. K^+ ions rapidly move to the outside of the cell.

17. The absolute refractory period
    a. limits how many action potentials can be produced in a given period of time.
    b. is the period of time when a strong stimulus can initiate a second action potential.
    c. is the period in which the permeability of the plasma membrane to K^+ ions is increasing.
    d. b and c
    e. none of the above

18. Decreasing the extracellular concentration of K^+ ions would affect the resting membrane potential by causing
    a. hyperpolarization.
    b. depolarization.
    c. no change.

19. Increasing the extracellular concentration of Ca^{2+} ions (hypercalcemia) would greatly _____ the ability of _____ to cross the plasma membrane.
    a. decrease, Na^+ ions
    b. increase, Na^+ ions
    c. increase, K^+ ions
    d. a and c

20. Graphed below is an action potential that resulted when a nerve cell was stimulated.

During the interval from A to B, mostly
    a. K^+ ions moved into the cell.
    b. K^+ ions moved out of the cell.
    c. Na^+ ions moved into the cell.
    d. Na^+ ions moved out of the cell.

21. What effect would decreasing the extracellular concentration of Ca^{2+} (hypocalcemia) have on the responsiveness of neurons to a stimulus?
    a. A greater than normal stimulus would be required to initiate an action potential.
    b. A smaller than normal stimulus would be required to initiate an action potential.
    c. The neuron may generate action potentials spontaneously.
    d. b and c

_____22. Mr. Twitchell has hyperpara-
thyroidism. The parathyroid gland
normally secretes a hormone that
causes bone resorption and the
release of bone minerals, including
calcium, into the circulatory system.
Mr. Twitchell's symptoms include
   a. increased heart rate.
   b. nervousness and tremors.
   c. tetany.
   d. none of the above

_____23. A patient suffering from
hypocalcemia exhibited
   1. tetany.
   2. increased permeability of cell
      membranes to Na^+ ions.
   3. increased action potential
      frequency in neurons.
   4. increased rate of movement of
      Na^+ ions across plasma
      membranes.
   5. increased release of chemicals at
      nerve-muscle synapses.

List the order in which the events
occur.
   a. 2, 3, 4, 5, 1
   b. 2, 4, 3, 1, 5
   c. 2, 4, 3, 5, 1
   d. 3, 2, 4, 5, 1
   e. 2, 4, 5, 3, 1

_____24. A subthreshold stimulus
   a. results in accommodation.
   b. produces a local potential.
   c. causes an all-or-none response.
   d. produces more action potentials
      than a submaximal stimulus.

_____25. Given the data below:

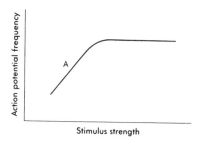

Choose the factor that is responsible
for determining the frequency of
action potentials at A. The duration
of the stimulus is constant and long
enough that more than one action
potential will be produced.
   a. the stimulus strength
   b. the absolute refractory period
   c. the relative refractory period
   d. the amplitude of the action
      potential

FINAL CHALLENGES

Use a separate sheet of paper to complete this section.

1. Predict the consequence of low extracellular
Na^+ ion concentration on the resting
membrane potential.

2. A man exhibited numbness, muscular
weakness, and occasional periods of
paralysis. His physician prescribed
potassium chloride tablets; their use
resulted in dramatic improvement in the
man's condition. Explain the symptoms and
the reason for their improvement.

3. Predict the consequence of introducing an
inhibitor of ATP synthesis into a nerve or
muscle cell on its response to a prolonged

threshold stimulus. Assume that
accommodation does not occur.

4. Fluorescent lights flicker very rapidly, but
they are perceived as producing a constant
light. Explain how this can happen.

5. A pregnant woman suffered from repeated
muscle cramps. Would you recommend she
take calcium, salt (NaCl), or potassium
tablets to relieve this condition? Explain.

6. One method of euthanasia involves
intravenous administration of potassium
chloride. Explain the rationale for using
potassium chloride.

ANSWERS TO CHAPTER 9

Concentration Differences Across the Plasma membrane
1. Proteins; 2. Cl⁻ ions; 3. K⁺ ions; 4. K⁺ ions; 5. Na⁺ ions; 6. Sodium-potassium exchange pump

Resting Membrane Potential
1. Resting membrane potential; 2. Selectively permeable; 3. K⁺ ions; 4. Depolarization; 5. Hyperpolarization; 6. K⁺ ions; 7. Na⁺ ions

Movement of Sodium and Potassium Ions Through the Plasma membrane
A. 1. Ion channels; 2. Gating proteins; 3. Voltage-sensitive channels; 4. Ligand-gated channels; 5. Ca²⁺ ions;
B. 1. Out of; 2. Into; 3. Resting membrane potential; 4. Concentration gradients; 5. ATP

Electrically Excitable Cells
A. 1. Local potential; 2. Local potential; 3. Local potential; 4. Action potential; 5. Action potential

B. 1. Depolarization; 2. Repolarization; 3. Repolarization; 4. After potential; 5. Relative refractory period
C. 1. Action potential; 2. Depolarization; 3. Repolarization; 4. Absolute refractory period; 5. Relative refractory period; 6. After potential; 7. Threshold

Propagation of Action Potentials
1. Na⁺ ion channels; 2. Absolute refractory period; 3. Synapse; 4. Neurotransmitter; 5. Na⁺ ion channels; 6. Gap junction; 7. Excitatory postsynaptic potential; 8. Inhibitory postsynaptic potential

Action Potential Frequency
1. Subthreshold stimulus; 2. Threshold stimulus; 3. Maximal stimulus; 4. Accommodation

Disorders
1. Hypokalemia; 2. Hypocalcemia

1. Permeability characteristics of the plasma membrane, presence of negatively charged proteins and other large anions within the cell, and the sodium-potassium exchange pump
2. Concentration difference of K⁺ ions across the membrane and the permeability of the membrane to K⁺ ions
3. Ion channels and the sodium-potassium exchange pump
4. Local potentials are graded, they can summate, they are propagated in a decremental fashion, they are depolarizations or hyperpolarizations, and they either remain constant as long as a stimulus is applied or they accommodate, .
5. Action potentials are produced when a local potential reaches threshold, are all-or-none, consist of a depolarization followed by a repolarization, and are propagated.
6. Once action potentials begin, they proceed without stopping and are constant in magnitude (the "all" part); and, if threshold is not reached, the action potentials do not occur (the "none" part).
7. Subthreshold, threshold, submaximal, maximal, and supramaximal stimuli

1. C. The inside of cells has more negatively charged proteins than the outside. The concentration of K⁺ ions is greatest inside the cell, and the concentration of Na⁺ ions is greatest outside the cell.

2. A. The outside of the resting plasma membrane is positively charged compared to the inside.

3. B. The membrane is somewhat permeable to K⁺ ions and they diffuse to the outside of the plasma membrane, producing a potential difference across the membrane.

4. A. When membrane permeability to K⁺ ions increases, K⁺ ions diffuse from the inside of the cell to the outside of the membrane, causing hyperpolarization.

5. C. During depolarization the membrane potential becomes more positive. During hyperpolarization the membrane potential becomes more negative.

6. D. The resting membrane potential decreases, and this is called depolarization.

7. D. Na^+ ions move out of the cell by active transport (sodium-potassium exchange pump) and diffuse into the cell through specific ion channels formed by proteins. Ca^{2+} ions binding to gating proteins in the ion channels inhibit the movement of Na^+ ions into the cell.

8. C. The major function of the sodium-potassium exchange pump is to maintain the sodium and potassium concentration gradients across the plasma membrane. It does not play a significant role in directly establishing the resting membrane potential. However, the concentration gradients it generates are important in maintaining the resting membrane potential and supporting the action potential.

9. E. A local potential is a depolarization or hyperpolarization that is variable in magnitude (not all-or-none). The magnitude of the local potential is graded (proportional to the strength of the stimulus that is responsible for its generation) and can summate. The local potential is not propagated for long distances.

10. B. Depolarization occurs when Na^+ ions move into the cell. Preventing Na^+ ion movement would prevent depolarization and there would be no local potential. Without a local potential there would be no action potential.

11. A. Once stimulated to produce an action potential the cell responds maximally (all). If the stimulus is below threshold, there is no action potential (none).

12. B. During depolarization, membrane permeability to Na^+ ions greatly increases, allowing an influx of positive charged Na^+ ions. Although K^+ ion channels begin to open during depolarization, they open more slowly than Na^+ ion channels. Consequently, much of the change in membrane permeability to K^+ ions does not occur until the repolarization phase.

13. A. As positively charged Na^+ ions move into the cell during depolarization the inside of the cell becomes more positive. Eventually enough Na^+ ions move into the cell to reverse the polarity across the plasma membrane, and the inside of the cell becomes positively charged compared to the outside of the plasma membrane.

14. A. Ms. Apeman's blood levels of K^+ ions are elevated because of the consumption of a large number of bananas. Therefore extracellular K^+ ion levels are elevated, resulting in depolarization of excitable tissues. Depolarization to threshold results in the generation of action potentials. Because the heart is an excitable tissue an increased rate of action potential generation and an increased heart rate would be expected.

15. D. During the repolarization phase of the action potential, at least two significant permeability changes occur. The permeability of the membrane to Na^+ ions decreases (it increased during the depolarization phase), and the permeability of the membrane to K^+ ions increases for a short time.

16. D. The concentration of K^+ ions is higher inside the cell than outside the cell. Because of an increase in membrane permeability to K^+ ions, K^+ ions move from the inside to the outside of the cell during repolarization.

17. A. During the absolute refractory period, no stimulus can evoke another action potential. A second action potential cannot be generated until AFTER the absolute refractory period. The relative refractory period is the time during which an above-threshold stimulus can initiate a second action potential and is the period during which membrane permeability to K^+ ions is increasing.

18. A. Decreasing the extracellular concentration of K^+ ions increases the concentration gradient for K^+ ions. Because the resting membrane potential is directly proportional to the potential for K^+ ions to diffuse out of the cell, the increased K^+ ions concentration gradient causes hyperpolarization.

19. A. When the extracellular concentration of Ca^{2+} ions increases, the Ca^{2+} ions bind to gating proteins in Na^+ ion channels, reducing the ability of Na^+ ions to diffuse into the cell.

20. C. From A to B is the depolarization phase, during which the membrane permeability to Na^+ ions increases, and the Na^+ ions diffuse down their concentration gradient into the cell.

21. D. When calcium levels are low, less Ca^{2+} ions bind to gating proteins in Na^+ ion channels, the channels are more open, and Na^+ ions can enter the cell with less difficulty, making it easier to initiate an action potential. If enough Na^+ ions enter, an action potential can be spontaneously generated.

22. D.   Chronically increased parathyroid hormone secretion can cause bone resorption at a rate great enough to elevate blood levels of calcium. Mr. Twitchell suffers from symptoms associated with hypercalcemia. Elevated Ca^{2+} ion levels decrease the permeability of the plasma membranes to Na^+ ions, and excitable tissues become less excitable (the probability that action potentials are generated is decreased because the depolarization that leads to an action potential is the result of an increase in the membrane permeability to Na^+ ions). One would expect a decreased electrical activity in excitable tissues. One would not expect to see an increased heart rate, nervousness and tremors, or tetany.

23. C.   Hypocalcemia causes an increased permeability of plasma membranes to Na^+ ions, which results in an increased rate of sodium movement across the plasma membrane. The result is an increased action potential frequency in neurons and an increased neurotransmitter release at the nerve-muscle synapse. Consequently, there would be an increased action potential frequency in skeletal muscle and tetany.

24. B   A subthreshold stimulus produces a local potential. A threshold stimulus produces an all-or-none action potential and a maximal stimulus produces a maximal rate of action potential production. A submaximal stimulus produces a higher frequency of action potentials than a threshold stimulus, but a lower frequency than a maximal stimulus. Accommodation results in no action potentials even though stimulation continues.

25. A.   The frequency of action potentials depends on the strength of the stimulus and the frequency of the stimulus. The stronger the strength of the stimulus, the greater the number of action potentials that will be generated. Recall that during the relative refractory period a stronger than normal stimulus causes another action potential. By increasing stimulus strength, action potentials are evoked sooner during the relative refractory period, resulting in an increased frequency of action potentials.

For stimuli of constant strength, the faster the frequency of stimulation, the greater the number of action potentials generated. The frequency of action potentials eventually is limited by the absolute refractory period.

★ FINAL CHALLENGES ★

1. The resting membrane is relatively impermeable to Na^+ ions and therefore a change in Na^+ ion concentration has little effect on the resting membrane potential.

2. Hyperpolarization of plasma membranes could account for all the symptoms. Increasing extracellular K^+ ion concentration (potassium chloride tablets) causes depolarization of plasma membranes and increases the ability of cells to have action potentials because the resting membrane potential is closer to threshold.

3. A prolonged threshold stimulus produces a continuous series of action potentials. Inhibiting ATP synthesis blocks the sodium-potassium exchange pump and prevents restoration of prestimulus ion concentrations across the plasma membrane. However, so few ions move in and out of the cell during each action potential that it takes 20,000 or more action potentials before ion concentrations are changed enough to affect the action potential.

4. When exposed to a flicker of light, the nerve cells in the eye produce local potentials that cause action potential generation. If the light flickers rapidly enough, the nerve cells receive another flicker of light at the end of their refractory period and generate another series of action potentials. The steady stream of action potentials is interpreted as a constant light.

5. The demands of the developing fetus for calcium for bone formation can result in low blood calcium levels in the mother. Consequently, gating proteins open more readily and Na^+ ions more easily enter nerve and muscle, making them more excitable. Calcium tablets could help reduce this condition by increasing the mother's blood calcium levels.

6. Intravenous administration of potassium chloride can greatly increase the extracellular concentration of potassium ions. As a result, nerve and muscle plasma membranes can depolarize and be unable to repolarize. Action potential production becomes impossible (because membranes are already depolarized) and death results.

Muscular System: Histology and Physiology

FOCUS: Muscle tissue is specialized to contract forcefully; it is responsible for the movement of body parts. The three types of muscle tissue are skeletal, smooth, and cardiac muscle. According to the sliding filament theory, the movement of actin myofilaments past myosin myofilaments results in the shortening of muscle fibers (cells) and therefore muscles. Muscle fibers respond to stimulation by contracting in an all-or-none fashion, but muscles show graded contractions caused by multiple motor unit summation and multiple wave summation. Muscles use aerobic respiration to obtain the energy (ATP) necessary for contraction but can also use anaerobic respiration for short periods of time. As a by-product of respiration muscles produce heat.

CONTENT LEARNING ACTIVITY

Characteristics of Muscle

❝*Muscle has four functional characteristics: contractility, excitability, extensibility, and elasticity.*❞

Match these terms with the correct statement or definition:

Cardiac muscle   Extensibility
Contractility   Skeletal muscle
Elasticity   Smooth muscle
Excitability

_____ 1. Ability to shorten forcefully.

_____ 2. Ability to be stretched to normal resting length or even beyond.

_____ 3. Ability to recoil to original resting length after being stretched.

_____ 4. Comprises about 40% of the body; under voluntary control; not capable of spontaneous contraction; responsible for locomotion, facial expressions, posture, and other body movements.

_____ 5. Found in the walls of hollow organs, the internal eye muscles, and the walls of blood vessels; under involuntary control and capable of spontaneous contractions; functions to regulate urine and blood flow and movement and mixing of food.

_____ 6. Found only in the heart; under involuntary control, capable of spontaneous contractions; functions to pump blood throughout the circulatory system.

Skeletal Muscle: Structure

"Skeletal muscles are composed of skeletal muscle fibers associated with smaller amounts of connective tissue, blood vessels, and nerves."

A. Using the terms provided, complete these statements:

Muscle cells   Nuclei
Myoblasts   Size
Number   Striated

Skeletal muscle fibers are skeletal __(1)__ . Each skeletal muscle fiber is a single cylindrical cell containing several __(2)__ located around the periphery of the fiber. Muscle fibers develop from less mature multinucleated cells called __(3)__ . The __(4)__ of skeletal muscle fibers remains relatively constant after birth; therefore enlargement of muscles is caused by an increase in __(5)__ , not number. As seen in longitudinal section, alternating light and dark bands give the muscle fiber a __(6)__ appearance.

1. _____
2. _____
3. _____
4. _____
5. _____
6. _____

B. Match these terms with the correct statement or definition:

Endomysium   Fascia
Epimysium   Perimysium
External lamina   Sarcolemma

_____ 1. Delicate layer of reticular fibers surrounding muscle fiber's cell membrane.

_____ 2. Muscle fiber cell membrane.

_____ 3. Surrounds each muscle fiber outside the external lamina.

_____ 4. Surrounds a bundle of muscle fibers (fasciculus).

_____ 5. Surrounds groups of fasciculi (a whole muscle).

_____ 6. Thick and fibrous; separates individual muscles and in some cases surrounds groups of muscles.

C. Match these terms with the correct parts labeled in Figure 10-1:

Endomysium
Epimysium
Fascia
Muscle fasciculus (bundle)
Muscle fibers (cells)
Perimysium

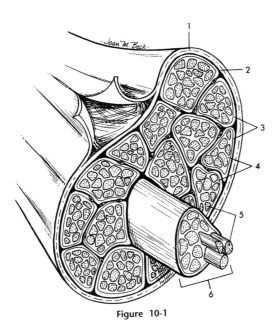

Figure 10-1

1. _____

2. _____

3. _____

4. _____

5. _____

6. _____

Muscle Fibers, Myofibrils, Sarcomeres, and Myofilaments

66_The interior of a skeletal muscle fiber is filled with myofibrils._**99**

A. Match these terms with the correct statement or definition:

Myofibril
Myofilaments
Sarcolemma

Sarcomere
Sarcoplasm

_____ 1. Cytoplasmic material of a muscle cell (without the myofibrils).

_____ 2. Threadlike structure that extends from one end of a muscle fiber to the other; consists of sarcomeres joined end-to-end.

_____ 3. Two proteins, actin and myosin, that are organized into sarcomeres.

_____ 4. Highly ordered unit of actin and myosin that extends from one Z-line to another.

 Myofibrils are composed of myofilaments.

B. Match these terms with the
   correct statement or definition:

A band   M line
Actin myofilaments   Myosin myofilaments
H zone   Z line
I band

_____ 1. Also called thin myofilaments.

_____ 2. Filamentous network of protein forming a disklike structure to
                              which actin myofilaments attach; the boundary of a sarcomere.

_____ 3. Area that extends from either side of a Z line to the ends of the
                              myosin myofilaments: the light band.

_____ 4. Area extending the length of myosin myofilaments within a
                              sarcomere; the dark band.

_____ 5. Area of a sarcomere composed of only myosin myofilaments;
                              no overlap of actin and myosin myofilaments.

_____ 6. Dark thin band in the middle of the H zone composed of
                              delicate filaments that attach to myosin myofilaments and hold
                              them in place.

The numerous myofibrils are oriented within each muscle fiber so that A bands and I
bands of parallel myofibrils are aligned and thus produce the striated pattern seen through
the microscope.

C. Match these terms with the
   correct statement or definition:

Cross bridge   Myosin head
F-actin   Tropomyosin
G-actin   Troponin

_____ 1. Polymer with two strands coiled to form a double helix.

_____ 2. Small globular units that combine to form a polymer of F-actin;
                              each unit has an active site to which myosin can bind.

_____ 3. Covers active sites on the actin molecule.

_____ 4. Composed of three subunits, each of which binds with either
                              actin, tropomyosin, or calcium ions.

_____ 5. Contains ATPase, an enzyme that splits ATP into ADP + P and
                              releases energy.

_____ 6. Structure that is formed when myosin binds with actin.

D. Match these terms with
the correct parts labeled
in Figure 10-2:

A band
Actin
F-actin
H zone
I band
Myosin
Myosin head
Sarcomere
Tropomyosin
Troponin
Z line

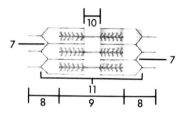

1 Type of myofilament

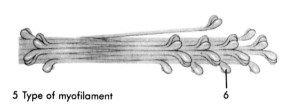

5 Type of myofilament

Figure 10-2

1. _____	5. _____	9. _____
2. _____	6. _____	10. _____
3. _____	7. _____	11. _____
4. _____	8. _____	

E. Match these terms with the correct statement or definition:

Sarcoplasmic reticulum   Terminal cisterna
T tubules   Triad

_____ 1. Tubelike invaginations of the sarcolemma that project into the sarcoplasm and wrap around sarcomeres near the ends of the A bands.

_____ 2. Highly specialized, smooth endoplasmic reticulum, the membrane of which actively transports calcium ions from the sarcoplasm into its lumen.

_____ 3. Enlargement of the sarcoplasmic reticulum next to a T tubule.

_____ 4. Grouping of a T tubule and its two adjacent terminal cisternae.

 Other organelles such as the numerous mitochondria and glycogen granules are packed between the myofibrils.

Sliding Filament Theory

66_Actin and myosin myofilaments slide past one another in a way that causes the sarcomeres to shorten._**99**

Using the terms provided, complete these statements:

A bands   H zones
Actin myofilaments   I bands
Gravity   Sarcomere

During contraction, cross bridges form between the heads of the myosin molecules and the actin molecules. Cross bridges forcefully cause the _(1)_ at each end of the sarcomere to slide past the myosin myofilaments toward the _(2)_. As a consequence, the _(3)_ and H zones become more narrow, but the _(4)_ remain constant in length. As the actin myofilaments slide over the myosin myofilaments, the Z lines are brought closer together, and the _(5)_ is shortened. During relaxation, cross bridges are released, and sarcomeres lengthen passively from the force of an antagonistic muscle or _(6)_.

1. _____
2. _____
3. _____
4. _____
5. _____
6. _____

Neuromuscular Junction

66_A branch of a motor neuron axon projects toward one muscle fiber and forms a neuromuscular junction,_**99** _or synapse, near the center of the muscle fiber._

A. Match these terms with the correct statement or definition:

Acetylcholinesterase   Presynaptic terminal
Motor neuron   Synaptic cleft
Neurotransmitter   Synaptic vesicles
Postsynaptic terminal
  (motor end plate)

_____ 1. Specialized nerve cells that propagate action potentials to skeletal muscle fibers.

_____ 2. The enlarged nerve terminal that rests in an invagination of the sarcolemma at the neuromuscular junction.

_____ 3. Space between the nerve fiber of the presynaptic terminal and the muscle fiber.

_____ 4. Muscle cell membrane in the area of the neuromuscular junction.

_____ 5. Spherical sacs located within each presynaptic terminal; contain acetylcholine.

_____ 6. Substance (such as acetylcholine) that is released from a presynaptic terminal, diffuses across the synaptic cleft, and stimulates (or inhibits) an action potential in the postsynaptic terminal.

_____ 7. Enzyme that breaks acetylcholine into acetic acid and choline; prevents accumulation of acetylcholine in the synaptic cleft.

B. Match these terms with the
   correct parts labeled
   in Figure 10-3:

Motor neuron
Muscle fiber

Synaptic cleft
Synaptic vesicles

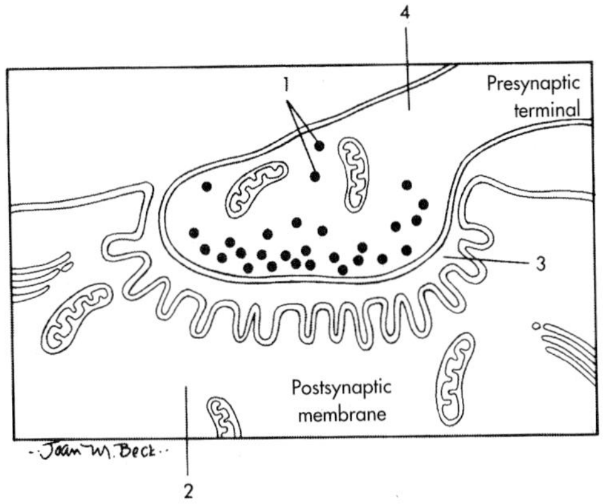

Figure 10-3

1. _____

2. _____

3. _____

4. _____

Excitation-Contraction Coupling

"Production of an action potential in a skeletal muscle fiber leads to contraction of the muscle fiber."

Using the terms provided, complete these statements:

ATP
Calcium ions
Cross bridge
Myosin
Sarcolemma

Sarcoplasmic
  reticulum
T tubules
Tropomyosin
Troponin

Action potentials are propagated along the _(1)_ and penetrate the _(2)_, which carry the depolarization to the muscle fiber's interior. As the action potential reaches the triads, calcium ions are released from the _(3)_. Calcium ions diffuse into the sarcoplasm surrounding the myofilaments and bind to _(4)_ of the actin myofilaments. This causes tropomyosin to move and expose actin active sites to _(5)_. Contraction occurs when actin and myosin bind, forming a _(6)_; myosin changes shape, and actin is pulled past myosin. Relaxation occurs when _(7)_ are taken up by the sarcoplasmic reticulum, _(8)_ binds to myosin, and troponin-_(9)_ covers the actin active sites.

1. _____

2. _____

3. _____

4. _____

5. _____

6. _____

7. _____

8. _____

9. _____

Energy Requirements for Contraction and Relaxation

"Energy from one ATP molecule is required for each cycle of cross bridge formation, movement, and release."

Using the terms provided, complete these statements:

Actin
Active site
ATP
ATPase
Calcium ions
Head

Myosin
Power
Recovery
Sarcoplasmic
reticulum

1. _____

2. _____

3. _____

4. _____

5. _____

6. _____

7. _____

8. _____

9. _____

10. _____

After a cross bridge has formed and movement has occurred, release of the myosin head from actin requires _(1)_ to bind to the head of the myosin molecule. ATP is broken down by _(2)_ in the head of the myosin myofilament and energy is stored in the _(3)_ of the myosin molecule. When the myosin molecule binds to _(4)_ to form another cross bridge, much of the energy is used for cross bridge formation and movement. Before the cross bridge can be released for another cycle, once again an ATP molecule must bind to the head of the _(5)_ molecule. Movement of the myosin molecule while the cross bridge is attached is a _(6)_ stroke, whereas return of the myosin head to its original position after cross bridge formation is a _(7)_ stroke. Relaxation occurs as a result of the active transport of _(8)_ back into the _(9)_, which allows the troponin-tropomyosin complex to block the _(10)_ on the actin molecules and cross bridges cannot reform.

Muscle Twitch

"A muscle twitch is a contraction of a whole muscle in response to a stimulus that causes an action potential."

Match these terms with the correct statement or definition:

Contraction phase
Lag phase

Relaxation phase

_____

_____

_____

_____

_____

1. An action potential causes the presynaptic terminal to release acetylcholine. Acetylcholine crosses the synaptic cleft and binds to postsynaptic receptors, causing an action potential.

2. An action potential propagates down the T tubules, causing the release of calcium ions from the sarcoplasmic reticulum.

3. Calcium ions bind with troponin, the troponin-tropomyosin complex changes position, and active sites on actin molecules are exposed to the heads of the myosin molecules.

4. Cross bridges between actin and myosin molecules form, move, release, and reform, causing sarcomeres to shorten.

5. Calcium ions are actively transported into the sarcoplasmic reticulum, troponin-tropomyosin complexes inhibit cross bridge formation, and muscle fibers lengthen passively.

Stimulus Strength and Muscle Contraction

❝*An isolated muscle fiber or motor unit follows the all-or-none law of muscle contraction.*❞
However, whole muscles exhibit characteristics that are more complex.

A. Match these terms with the
correct statement or definition:

Maximal stimulus
Multiple motor unit
  summation
Submaximal stimulus

Subthreshold stimulus
Supramaximal stimulus
Threshold stimulus

_____ 1. Stimulus just strong enough to produce an action potential in a single motor unit.

_____ 2. Stimulus strength between threshold and maximal values.

_____ 3. Stimulus that is stronger than necessary to activate all the motor units in a muscle.

_____ 4. Increasing stimulus strength, between threshold and maximum values, produces a graded increase in force of contraction of a muscle.

👉 A motor unit is a single neuron and all the muscle fibers it innervates.

B. Match these terms with the
correct parts of the graph
in Figure 10-4:

Maximal stimulus
Submaximal stimulus
Subthreshold stimulus

Supramaximal stimulus
Threshold stimulus

1. _____

2. _____

3. _____

4. _____

5. _____

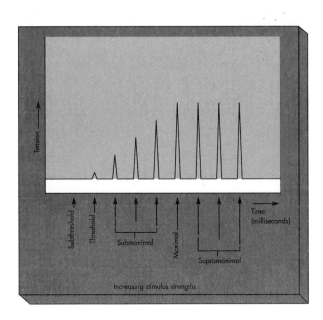

Figure10-4

Stimulus Frequency and Muscle Contraction

❝*The contractile mechanism does not exhibit a refractory period; muscle fibers do not have to completely relax before contracting again.*❞

A. Match these terms with the correct statement or definition:

Complete tetanus   Multiple wave summation
Incomplete tetanus   Treppe

_____

_____

_____

1. Increase in the force of a muscle contraction caused by an increased frequency of stimulation.

2. Stimuli occur so frequently that there is no muscle relaxation.

3. Graded response occurring in muscle that has rested for a prolonged period of time. If the muscle is stimulated with a maximal stimulus at a frequency that allows complete relaxation between stimuli, the second contraction is of a slightly greater magnitude than the first, and the third is greater than the second. After a few stimuli, all contractions are of equal magnitude.

B. Match these terms with the correct parts labeled in Figure 10-5:

Complete tetanus   Multiple wave summation
Incomplete tetanus   Treppe

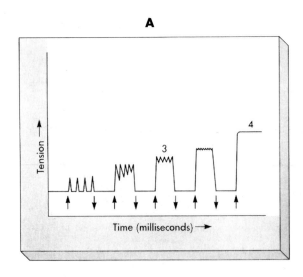

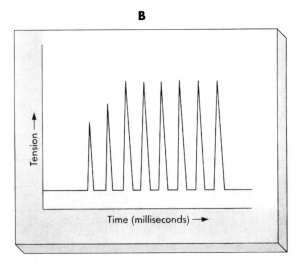

Figure 10-5

_____ 1. Demonstrated by graph A.

_____ 2. Demonstrated by graph B

_____ 3. Name part 3 of the graph.

_____ 4. Name part 4 of the graph.

Type of Muscle Contractions

66 *Muscle contractions are classified based on the type of contraction that predominates.* **99**

Match these terms with the
correct statement or definition:

Concentric   Isotonic
Eccentric   Muscle tone
Isometric

_____   1. Contractions that cause a change in muscle length but no
                              change in muscle tension.

_____   2. Contractions that cause a change in muscle tension but no
                              change in muscle length.

_____   3. Maintenance of steady tension for long periods of time.

_____   4. Contraction in which the muscle shortens and tension increases.

_____   5. Muscle tension is constant and the muscle increases in length.

Length vs. Tension

66 *Muscle contracts with decreased strength if its initial length is shorter or longer than optimum.* **99**

Match these terms with the
correct statement or definition:

Active tension   Total tension
Passive tension

_____   1. Produced when a muscle is stretched but is not stimulated.

_____   2. Produced when a muscle contracts.

_____   3. Sum of active and passive tension.

☞ Muscle length plotted against muscle tension in response to maximal stimuli is the active
     tension curve.

Fatigue, Physiological Contracture, and Rigor Mortis

66 *Fatigue is the decreased capacity to do work and the reduced efficiency of performance of a muscle.* **99**

A. Match these terms with the
   correct statement or definition:

Muscular fatigue   Synaptic fatigue
Psychological fatigue

_____   1. Involves the central nervous system; muscles are capable of
                              functioning, but the person "perceives" work is not possible.

_____   2. Result of ATP depletion; without adequate ATP levels in muscle
                              fibers, cross bridges cannot function normally.

_____   3. Occurs in the neuromuscular junction when the rate of
                              acetylcholine release is greater than the rate of acetylcholine
                              synthesis; rare but can occur after extreme exertion.

B. Match these terms with the
correct statement or definition:

Physiological contracture
Rigor mortis

_____ 1. Extreme muscular fatigue caused by a lack of ATP in which a muscle can neither contract nor relax.

_____ 2. Rigid muscles that occur after death; caused by calcium leakage from sarcoplasmic reticulum and no ATP to allow relaxation.

Energy Sources and Oxygen Debt

"*ATP is the immediate source of energy for muscle contraction.***"**

A. Match these terms with the
correct statement or definition:

Aerobic respiration
Anaerobic respiration

_____ 1. Occurs in the absence of oxygen and results in the breakdown of glucose to yield ATP and lactic acid.

_____ 2. Requires oxygen and breaks down glucose to produce ATP, carbon dioxide, and water.

_____ 3. More efficient (produces the most ATP for each molecule of glucose used) of the two types of respiration.

_____ 4. More suited to short periods of intense exercise.

B. Using the terms provided, complete these statements:

ADP   Creatine phosphate
Aerobic   Lactic acid
Anaerobic   Liver
ATP   Oxygen debt

The immediate source of energy for muscle contractions is
_(1)_ . During resting conditions only a small amount of
ATP is present in muscle cells. Energy is stored when ATP
transfers a high-energy phosphate to creatine to form _(2)_ .
During exercise, creatine phosphate releases a phosphate
that combines with _(3)_ to produce ATP. Resting muscles
or muscles undergoing long-term exercise depend primarily
upon _(4)_ respiration for ATP synthesis. On the other
hand, during short periods of intense exercise _(5)_ respiration combined with the breakdown of creatine phosphate
provides enough ATP for 1-3 minutes. These processes are
limited by the depletion of creatine phosphate and glucose
and the buildup of _(6)_ within muscle fibers. The extra
oxygen required after exercise above that required for resting metabolism is called the _(7)_ . During this time, ATP
and creatine phosphate levels are restored in muscle fibers,
and excess lactic acid is converted to glucose in the _(8)_ .

1. _____

2. _____

3. _____

4. _____

5. _____

6. _____

7. _____

8. _____

Slow and Fast Fibers

"Not all skeletal muscles have identical functional capabilities."

Match these terms with the
correct statement or definition:

Fast-twitch muscle fibers
Slow-twitch muscle fibers

_____ 1. Have a better developed blood supply.

_____ 2. Have very little myoglobin and fewer and smaller
                            mitochondria.

_____ 3. More fatigue-resistant.

_____ 4. Have large deposits of glycogen and are well adapted to
                            perform anaerobic respiration.

_____ 5. There is a greater concentration of this type of fiber in large
                            postural muscles.

☞ Fast-twitch muscle fibers are also called low oxidative, and slow-twitch muscle fibers are
also called high oxidative.

The Effects of Exercise

"Neither fast-twitch nor slow-twitch muscle fibers can be converted to muscle fibers of another type.
Nevertheless, training can increase the capacity of both types to perform more efficiently.

Using the terms provided, complete these statements:

Aerobic   Fatigue-resistant
Anaerobic   Hypertrophies
Atrophies   Motor units
Cardiovascular   Numbers
Fat   Size

Intense exercise resulting in _(1)_ respiration has the greatest effect on fast-twitch muscle fibers, causing them to
increase in strength and mass. The blood supply to both
fast-twitch and slow-twitch muscle fibers is increased by
endurance exercise requiring _(2)_ respiration, making both
types of fibers more _(3)_ . A muscle increases in size, or
_(4)_ , and increases in strength and endurance in response
to exercise. Conversely, a muscle that is not used _(5)_ . The
increase or decrease in size of individual muscles is caused
by a change in _(6)_ of muscle fibers. The increased strength
of a trained muscle also occurs because of the recruitment of
more _(7)_ , reduction of excess _(8)_ , greater ATP
production, and increased efficiency of the _(9)_ and
respiratory systems.

1. _____

2. _____

3. _____

4. _____

5. _____

6. _____

7. _____

8. _____

9. _____

Increased muscle contraction causes increased heat production, resulting in increased body temperature. The involuntary muscle contractions that occur during shivering increase heat production up to 18 times that of resting levels.

Structure of Smooth-Muscle Fibers

"Although smooth-muscle has characteristics common to skeletal muscle, it also has some**"** unique properties.

Using the terms provided, complete these statements:

Calcium ions
Calmodulin
Caveolae
Dense bodies
Intermediate filaments
Myosin

Myosin
 phosphatase
Single
Spindle-shaped
Striated

1. _____
2. _____
3. _____
4. _____
5. _____
6. _____
7. _____
8. _____
9. _____
10. _____

Skeletal muscle fibers are cylindrical cells, whereas smooth muscle fibers are __(1)__ , with a __(2)__ nucleus per cell. Skeletal muscle fibers appear to be __(3)__ , but smooth muscle fibers do not. Smooth muscle cells contain noncontractile __(4)__ , which attach to __(5)__ that are scattered throughout the cell. Although smooth muscle does not have a T tubule system, it does have __(6)__ , which are invaginations along the cell membrane. Smooth muscle lacks an extensive sarcoplasmic reticulum, so __(7)__ must enter the cell from the extracellular fluid to initiate contraction. Instead of binding to troponin, calcium binds to __(8)__ , and activates myosin kinase, an enzyme that transfers a phosphate group from ATP to __(9)__ and initiates cross bridge formation. Another enzyme, called __(10)__ removes the phosphate from myosin molecules, releasing the cross bridges.

Smooth-Muscle Types

"There are two types of smooth muscle: multiunit and visceral.**"**

Match the smooth-muscle type with the correct statement or definition:

Multiunit smooth-muscle
Visceral smooth-muscle

_____

_____

_____

_____

1. Often autorhythmic, has many gap junctions, and acts as a single unit.

2. Usually found in sheets (digestive, respiratory, and urinary tracts).

3. Has few gap junctions and normally contracts only when stimulated by nerves or hormones.

4. Occurs in sheets (blood vessels), small bundles (arrector pili), or single cells (spleen capsule).

Electrical and Functional Properties of Smooth-Muscle

66 *Smooth muscle has several properties not seen in skeletal muscle.* **99**

Using the terms provided, complete these statements:

Autorhythmic	Pacemaker cells
Gradual	Smooth muscle
Hormones	tone
Involuntary	Sudden

Spontaneous generation of action potentials in smooth muscle occurs because of the _(1)_ leakage of sodium and calcium ions into the cell. The action potential produces a contraction; therefore a smooth muscle is _(2)_ . Certain smooth muscle cells in the uterus, ureter, and digestive tract tend to develop action potentials more rapidly than other cells, and are called _(3)_ . Due to a _(4)_ stretch, some smooth muscle contracts. Despite a _(5)_ increase or decrease in length, smooth muscle maintains a constant tension and amplitude of contraction. This constant tension is called _(6)_ . Smooth muscle is innervated by the autonomic nervous system and is therefore under _(7)_ control. Some _(8)_ are also important in regulating smooth muscle, e.g., epinephrine and oxytocin.

1. _____

2. _____

3. _____

4. _____

5. _____

6. _____

7. _____

8. _____

Cardiac Muscle

66 *Cardiac muscle is found only in the heart.* **99**

Match these terms with the correct statement or definition:

Autorhythmic	Many
Intercalated disks	One
Involuntary	

_____ 1. Specialized cell-to-cell attachments between cardiac muscle cells.

_____ 2. Spontaneous, repetitive contraction of cardiac muscle cells.

_____ 3. Usual number of nuclei in a cardiac muscle cell.

1. List the four functional characteristics of muscle.

2. List the five connective tissue structures associated with skeletal muscle.

3. List the parts of a sarcomere found in the I band, A band, and H zone.

4. List the three substances with which troponin can combine.

5. List the events that result in the transfer of an action potential from a neuron to a skeletal muscle.

6. List three roles of ATP in muscle contraction and relaxation.

7. List the three phases of a muscle twitch.

8. State the all-or-none law of skeletal muscle contraction.

9. Explain how the force of contraction increases during multiple motor unit summation and multiple wave summation.

10. List four types of muscle contraction.

11. List three types of muscle fatigue.

12. Name two ways that ATP is produced in a muscle fiber during short periods of intense exercise.

13. List two types of skeletal muscle fibers.

14. List two types of smooth muscle.

MASTERY LEARNING ACTIVITY

Place the letter corresponding to the correct answer in the space provided.

_____ 1. Which of these is true of skeletal muscle?
   a. spindle-shaped cells
   b. under involuntary control
   c. many peripherally located nuclei per muscle cell
   d. forms the walls of hollow internal organs
   e. all of the above

_____ 2. The connective tissue sheath that surrounds a muscle fasciculus is the
   a. perimysium.
   b. endomysium.
   c. epimysium.
   d. hypomysium.

_____ 3. Given these structures:
1. whole muscle
2. muscle fiber (cell)
3. myofilament
4. myofibril
5. muscle fasciculus (bundle)

Choose the arrangement that lists the structures in the correct order from the largest to the smallest of a skeletal muscle.
a. 1,2,5,3,4
b. 1,2,5,4,3
c. 1,5,2,3,4
d. 1,5,2,4,3
e. 1,5,4,2,3

_____ 4. Each myofibril
a. is made up of many muscle fibers.
b. contains sarcoplasmic reticulum.
c. is made up of many sarcomeres.
d. contains T tubules.
e. is equivalent to a muscle fiber.

_____ 5. Myosin
a. is attached to the Z line.
b. is found primarily in the I band.
c. is thinner than actin.
d. none of the above

_____ 6. Which of these statements about the molecular structure of myofilaments is true?
a. Tropomyosin has a binding site for Ca^{2+} ions.
b. The head of the myosin molecule binds to an active site on G-actin.
c. ATPase is found on troponin.
d. Troponin binds to the rodlike portion of myosin.

_____ 7. The part of the sarcolemma that invaginates into the interior of a skeletal muscle cell is the
a. T tubule system.
b. sarcoplasmic reticulum.
c. myofibrils.
d. none of the above

_____ 8. Given these events:
1. acetylcholine broken down into acetic acid and choline
2. acetylcholine moves across the synaptic cleft
3. action potential reaches the terminal branch of motor neuron
4. acetylcholine combines with a receptor molecule
5. action potential produced on muscle cell membrane

Choose the arrangement that lists the events in the order they occur at a neuromuscular junction.
a. 2,3,4,1,5
b. 3,2,4,5,1
c. 3,4,2,1,5
d. 4,5,2,1,3

_____ 9. Given these events:
1. sarcoplasmic reticulum releases calcium
2. sarcoplasmic reticulum takes up calcium
3. calcium diffuses into myofibrils
4. action potential moves down the T tubule
5. sarcomere shortens
6. muscle relaxes

Choose the arrangement that lists the events in the order they occur following a single stimulation of a skeletal muscle cell.
a. 1,3,4,5,2,6
b. 2,3,5,4,6,2
c. 4,1,3,5,2,6
d. 4,2,3,5,1,6
e. 5,1,4,3,2,6

_____ 10. Given these events:
1. calcium combines with tropomyosin
2. calcium combines with troponin
3. tropomyosin pulls away from actin
4. troponin pulls away from actin
5. tropomyosin pulls away from myosin
6. troponin pulls away from myosin
7. myosin binds to actin

Choose the arrangement that lists the events in the order they occur during muscle contraction.
a. 1,4,7
b. 2,3,7
c. 1,3,7
d. 2,4,7
e. 2,5,6

_____ 11. Skeletal muscles
a. require energy in order to relax.
b. require energy in order to contract.
c. shorten with a force.
d. a and b
e. all of the above

_____ 12. Which of these events occurs during the lag (latent) phase of muscle contraction?
a. cross bridge formation
b. active transport of calcium ions into the sarcoplasmic reticulum
c. calcium ions combine with troponin
d. the sarcomere shortens

_____ 13. Which of these regions shortens during skeletal muscle contraction?
a. A band
b. I band
c. H zone
d. a and b
e. b and c

_____ 14. A motor unit is
a. a nerve fiber and all of the muscle fibers it innervates.
b. a nerve and the muscle it innervates.
c. a muscle fiber and the nerve fibers that innervate it.
d. none of the above

_____ 15. Under normal circumstances, at the level of a single skeletal muscle fiber, there are
a. graded contractions, depending on the strength of the stimulus input.
b. several rapid contractions resulting in the main contraction.
c. all-or-none responses.

_____ 16. Considering the force of contraction of a muscle cell, multiple wave summation occurs because of
a. increased frequency of action potentials on the cell membrane.
b. increased number of cross bridges formed.
c. increase in calcium concentration around the myofibrils.
d. all of the above

_____ 17. Graphed below is the response of an isolated skeletal muscle to a single stimulus of increasing strength.

Given the preceding information and what you know about skeletal muscle, choose the appropriate response(s) listed below.
a. Skeletal muscles respond in an all-or-none fashion.
b. Between stimulus strength A and B the number of sarcomeres that contract increases.
c. Between stimulus strength B and C the number of sarcomeres that contract increases.
d. a and b
e. a and c

_____ 18. Each muscle fiber responds in an all-or-none fashion. However, when maximal stimuli are applied to a skeletal muscle at an increasing frequency, the amplitude of contraction increases. Which of these statements is (are) consistent with this observation?
   a. Tetanic stimuli cause a greater degree of contraction of the sarcomeres.
   b. Early in a tetanic contraction the connective tissue components of muscle are stretched, allowing it to apply a greater tension to the load.
   c. A greater concentration of Ca^{2+} ions around myofilaments is achieved.
   d. a and c
   e. all of the above

_____ 19. A weight-lifter attempts to lift a weight from the floor, but the weight is so heavy he is unable to move it. The type of muscle contraction the weight-lifter used was mostly
   a. isometric.
   b. isotonic.
   c. isokinetic.

_____ 20. Given these graph:

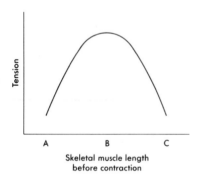

Choose the appropriate response.
   a. The maximum sarcomere length occurs at B.
   b. It is an advantage for one to greatly stretch one's muscles before trying to lift a heavy object.
   c. The overlap of actin and myosin is greater at B than at A.
   d. It would be an advantage if the resting length of one's muscles was near B.
   e. all of the above

_____ 21. Sam Macho was demonstrating to his friends the number of push-ups he could accomplish. After 30 push-ups he could do no more. After a short resting period he tried again. This time his friends vigorously yelled encouragement and he was able to do 35 push-ups. Which of these statements is most appropriate?
   a. Fatigue during the first attempt was primarily due to ATP depletion in muscle.
   b. Fatigue during the first attempt was primarily due to "fatigue" within the central nervous system.
   c. Following the depletion of creatine phosphate during the first attempt, the rate of anaerobic metabolism increased to provide a greater amount of ATP during the second attempt.
   d. After a vigorous "warm up" period one can nearly always perform better.

_____ 22. Marvin Amskray broke his school record of 19 seconds in the 220-yard dash. During his record-breaking run which of these provided the greatest amount of energy for muscular contraction?
   a. reactions that require oxygen
   b. reactions that do not use oxygen
   c. non-ATP compounds that have a phosphate group bound to them
   d. a and c
   e. b and c

_____ 23. In response to a homework assignment a group of students in an anatomy and physiology class had to run a quarter mile as fast as possible. Following the race, which of these conditions would be expected to exist in their bodies immediately after the run?
   a. elevated oxygen consumption
   b. increased formation of creatine phosphate
   c. increased conversion of lactic acid to glucose
   d. a and b
   e. all of the above

24. Which of these conditions would one expect to find within the leg muscles of an avid long-distance bicycle rider?
   a. myoglobin-poor
   b. contract very quickly
   c. creatine phosphate is a major supplier of energy for contraction
   d. primarily anaerobic
   e. none of the above

25. Aerobic exercise training can
   a. increase the fatigue resistance of slow-twitch muscle fibers.
   b. increase the number of myofibrils in slow-twitch fibers.
   c. convert fast-twitch muscle fibers into slow-twitch muscle fibers.
   d. a and b
   e. all of the above

26. Body temperature following exercise is higher than body temperature before exercise because
   a. the chemical reactions necessary for contraction release energy in the form of heat.
   b. the chemical reactions necessary to pay back the oxygen debt release energy in the form of heat.
   c. the rate of chemical reactions in contracting muscles increases.
   d. all of the above

27. After contraction has occurred in smooth muscle, the calcium
   a. is destroyed by acetylcholinesterase.
   b. is chemically bound to the filaments.
   c. is pumped out of the cell by active transport.
   d. is secreted by the Golgi apparatus.
   e. combines with ATP.

28. Which of these often have spontaneous contractions?
   a. multiunit smooth muscle
   b. visceral smooth muscle
   c. skeletal muscle
   d. a and b
   e. a and c

29. Compared to skeletal muscle, visceral smooth muscle
   a. has the same ability to be stretched.
   b. when stretched, loses the ability to contract forcefully.
   c. maintains about the same tension even when stretched.
   d. cannot maintain long, steady contractions.

30. Which of these applies to smooth muscle cells?
   a. Numerous sarcomeres give the cells a striated appearance.
   b. Calcium ions bind to calmodulin and cause cross bridge formation.
   c. Smooth muscle cells have extensive sarcoplasmic reticulum and T tubules.
   d. Slow waves in the RMP result from K^+ ions slowly diffusing into the cell.
   e. none of the above

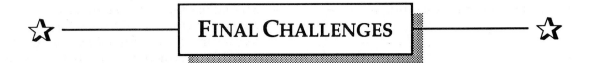

Use a separate sheet of paper to complete this section.

1. Given the characteristics listed below for a muscle tissue:
   1. multinucleated cells
   2. spindle-shaped cells
   3. striations
   4. under involuntary control
   5. cells branch

   List the minimum number of characteristics that would be sufficient to conclude that the type of muscle tissue was skeletal muscle.

2. A patient was poisoned with an unknown chemical. The symptoms and data listed below were exhibited:
   1. tetany of skeletal muscles
   2. elevated levels of acetylcholine in the circulatory system
   3. curare counteracted the tetany and caused flaccid paralysis
   4. the patient was then put on a respirator until he recovered

   The physician then concluded that the toxic substance inhibited acetylcholinesterase activity and therefore caused tetany. Do the data support that conclusion? Explain.

3. A single muscle cell was placed between two small clamps so that it was under slight tension. A very short stimulus was applied, and the force of contraction was measured as an increase in tension. After allowing the muscle cell to recover, the process was repeated. Each time the stimulus strength was increased slightly. The results are graphed below:

   Tension

   Stimulus strength

   What conclusion about the response of a muscle cell to stimulation can you make on the basis of this experiment?

4. Define the all-or-none law in skeletal muscle, and provide an explanation for it.

5. A very short stimulus was applied to the nerve supplying a whole muscle. A muscle twitch was produced, and the force of contraction was measured. After allowing the muscle to recover, the procedure was repeated. Each time the stimulus strength was increased slightly. The results are graphed below:

   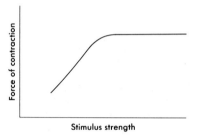

   Force of contraction

   Stimulus strength

   Given that the all-or-none law of skeletal muscle contraction is true, explain how the force of contraction gradually increases and then remains constant.

190

6. Given these graph of a muscle twitch:

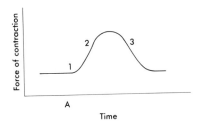

   If the muscle were stimulated at time A, during which part of the graph (1, 2, or 3) is there increased calcium uptake into the sarcoplasmic reticulum? What is the name for this part of a muscle twitch?

7. Predict the response of a whole muscle to a supramaximal stimulus applied to the nerve of the muscle at a very high frequency.

8. Two different muscle preparations, A and B, were stimulated. The stimuli were applied as a single stimulus to the nerve that supplies the muscle. The force of contraction that each muscle preparation generated in response to a single stimulus of increasing strength was determined, and the results are graphed below:

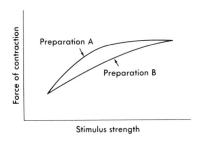

   Explain why the force of contraction of muscle preparation A increased more rapidly than the force of contraction of muscle preparation B.

9. Sally Gorgeous, an avid jogger, is running down the beach when she meets Sunny Beachbum, an avid weight-lifter. Sunny flirts with Sally, who decides he has more muscles than brains. She runs down the beach, but Sunny runs after her. After about a half mile, Sunny tires and gives up. Explain why Sally was able to outrun Sunny (i.e., do more muscular work) despite the fact that she obviously is less muscular.

10. These experiments were performed in an anatomy and physiology laboratory. The rate and depth of respiration for a resting student were determined. In experiment A the student ran in place for 30 seconds and then immediately sat down and relaxed, and respiration rate and depth were again determined. Experiment B was just like experiment A, except that the student held her breath while running in place. What differences in respiration would you expect for the two different experiments? Explain the basis for your predictions.

ANSWERS TO CHAPTER 10

Characteristics of Muscle
1. Contractility; 2. Extensibility; 3. Elasticity;
4. Skeletal muscle; 5. Smooth muscle; 6. Cardiac
muscle

Skeletal Muscle: Structure
A. 1. Muscle cells; 2. Nuclei; 3. Myoblasts;
4. Number; 5. Size; 6. Striated
B. 1. External lamina; 2. Sarcolemma;
3. Endomysium; 4. Perimysium;
5. Epimysium; 6. Fascia
C. 1. Fascia; 2. Epimysium; 3. Perimysium;
4. Endomysium; 5. Muscle fibers (cells);
6. Muscle fasciculus (bundle)

Muscle Fibers, Myofibrils, Sarcomeres, and
Myofilaments
A. 1. Sarcoplasm; 2. Myofibril; 3. Myofilaments;
4. Sarcomere
B. 1. Actin myofilaments; 2. Z line; 3. I band;
4. A band; 5. H zone; 6. M line
C. 1. F-actin; 2. G-actin; 3. Tropomyosin;
4. Troponin; 5. Myosin head; 6. Cross bridge
D. 1. Actin; 2. F-actin; 3. Tropomyosin;
4. Troponin; 5. Myosin; 6. Myosin head;
7. Z line; 8. I band; 9. A band; 10 H zone;
11. Sarcomere
E. 1. T tubules; 2. Sarcoplasmic reticulum;
3. Terminal cisterna; 4. Triad

Sliding Filament Theory
1. Actin myofilaments; 2. H zones; 3. I bands;
4. A bands; 5. Sarcomere; 6. Gravity

Neuromuscular Junction
A. 1. Motor neuron; 2. Presynaptic terminal;
3. Synaptic cleft; 4. Postsynaptic terminal
(motor end plate); 5. Synaptic vesicles;
6. Neurotransmitter; 7. Acetylcholinesterase
B. 1. Synaptic vesicles; 2. Muscle fiber;
3. Synaptic cleft; 4. Motor neuron

Excitation-Contraction Coupling
1. Sarcolemma; 2. T tubules; 3. Sarcoplasmic
reticulum; 4. Troponin; 5. Myosin; 6. Cross bridge;
7. Calcium ions; 8. ATP; 9. Tropomyosin

Energy Requirements for Contraction and
Relaxation
1. ATP; 2. ATPase; 3. Head; 4. Actin; 5. Myosin;
6. Power; 7. Recovery; 8. Calcium ions;
9. Sarcoplasmic reticulum; 10. Active site

Muscle Twitch
1. Lag phase; 2. Lag phase; 3. Lag phase;
4. Contraction phase; 5. Relaxation phase

Stimulus Strength and Muscle Contraction
A. 1. Threshold stimulus; 2. Submaximal
stimulus; 3. Supramaximal stimulus;
4. Multiple motor unit summation

B. 1. Subthreshold stimulus; 2. Threshold
stimulus; 3. Submaximal stimulus; 4. Maximal
stimulus; 5. Supramaximal stimulus

Stimulus Frequency and Muscle Contraction
A. 1. Multiple wave summation; 2. Complete
tetanus; 3. Treppe
B. 1. Multiple wave summation; 2. Treppe;
3. Incomplete tetanus; 4. Complete tetanus

Types of Muscle Contractions
1. Isotonic; 2. Isometric; 3. Muscle tone;
4. Concentric; 5. Eccentric

Length vs. Tension
1. Passive tension; 2. Active tension; 3. Total
tension

Fatigue, Physiological Contracture, and Rigor Mortis
A. 1. Psychological fatigue; 2. Muscular fatigue;
3. Synaptic fatigue
B. 1. Physiological contracture; 2. Rigor mortis

Energy Sources and Oxygen Debt
A. 1. Anaerobic respiration; 2. Aerobic
respiration; 3. Aerobic respiration;
4. Anaerobic respiration
B. 1. ATP; 2. Creatine phosphate; 3. ADP;
4. Aerobic; 5. Anaerobic; 6. Lactic acid;
7. Oxygen debt; 8. Liver

Slow and Fast Fibers
1. Slow-twitch muscle fibers; 2. Fast-twitch muscle
fibers; 3. Slow-twitch muscle fibers; 4. Fast-twitch
muscle fibers; 5. Slow-twitch muscle fibers

The Effects of Exercise
1. Anaerobic; 2. Aerobic; 3. Fatigue-resistant;
4. Hypertrophies; 5. Atrophies; 6. Size; 7. Motor
units; 8. Fat; 9. Cardiovascular

Structure of Smooth Muscle Fibers
1. Spindle-shaped; 2. Single; 3. Striated;
4. Intermediate filaments; 5. Dense bodies;
6. Caveolae; 7. Calcium ions; 8. Calmodulin;
9. Myosin; 10. Myosin phosphatase

Smooth Muscle Types
1. Visceral smooth muscle; 2. Visceral smooth
muscle; 3. Multiunit smooth muscle; 4. Multiunit
smooth muscle

Electrical and Functional Properties of Smooth
Muscle
1. Gradual; 2. Autorhythmic; 3. Pacemaker cells;
4. Sudden; 5. Gradual; 6. Smooth muscle tone;
7. Involuntary; 8. Hormones

Cardiac Muscle
1. Intercalated disks; 2. Autorhythmic; 3. One

1. Contractility, excitability, extensibility, and elasticity
2. External lamina, endomysium, perimysium, epimysium, and fascia
3. I band: a Z line and the actin myofilaments that extend from either side of the Z line to the ends of the myosin myofilaments; A band: extends the length of the myosin myofilaments in the sarcomere; H zone: only myosin myofilaments present; no overlapping actin myofilaments
4. Actin, tropomyosin, and calcium ions
5. a. An action potential arrives at the presynaptic terminal of the axon of a motor neuron.
   b. Ca^{2+} ion channels in the axon's cell membrane open, allowing calcium ions to diffuse into the cell.
   c. Ca^{2+} ions cause the acetylcholine in a few synaptic vesicles to be secreted into the synaptic cleft.
   d. Acetylcholine molecules diffuse across the cleft and bind to receptor molecules of the postsynaptic terminal.
   e. The membrane of the postsynaptic terminal (sarcolemma) becomes permeable to Na^+ ions.
   f. A local depolarization occurs that leads to an action potential in the muscle cell.

6. a. ATP must bind to myosin before cross bridges can form. The energy from ATP is stored in myosin and later released, causing cross bridge movement, i.e., muscle contraction.
   b. ATP is required for the cross bridges to be released, i.e., for muscle relaxation.
   c. ATP is required for Ca^{2+} ion uptake into the sarcoplasmic reticulum; Ca^{2+} ion uptake initiates relaxation.
7. Lag (latent), contraction, and relaxation phase
8. For a given condition, the skeletal muscle fiber contracts maximally or not at all.
9. Multiple motor unit summation: more motor units are recruited, i.e., more and more muscle fibers contract; multiple wave summation: each individual muscle fiber contracts more forcefully.
10. Isometric, isotonic, concentric, and eccentric contraction
11. Psychological, muscular, and synaptic fatigue
12. Breakdown of creatine phosphate and anaerobic respiration
13. Fast-twitch and slow-twitch muscle fibers
14. Multiunit and visceral smooth muscle

1. C. Skeletal muscle consists of long, cylindrical cells that are under voluntary control. They have many peripherally located nuclei. Smooth muscle or cardiac muscle makes up the walls of hollow internal organs.

2. A. By definition the connective tissue sheaths that surround the entire muscle, muscle fasciculus, and muscle fiber are the epimysium, perimysium, and endomysium, respectively.

3. D. Whole muscles are made up of muscle fasciculi (bundles). In each muscle fasciculus are many muscle fibers (cells). Each muscle fiber is composed of many myofibrils. The myofibrils consist of the myofilaments actin and myosin.

4. C. A myofibril consists of many sarcomeres linked end-to-end. There are many myofibrils within each muscle fiber. Each of the sarcomeres is made up of myofilaments (actin and myosin). The T tubule system is an extension of the sarcolemma (cell membrane of the muscle fiber) into the cell. The sarcoplasmic reticulum is endoplasmic reticulum associated with the myofibrils.

5. D. Myosin is thicker than actin. It is attached to the M line, not to the Z line. Myosin is found in the A band.

6. B. When the head of the myosin molecule binds to an active site on G-actin, a cross bridge is formed. Troponin has a binding site for Ca^{2+} ions, tropomyosin, and actin. ATPase is found on the myosin head.

7. A. The muscle cell membrane, the sarcolemma, dips into the cell to form the T tubule system. The sarcoplasmic reticulum is a specialized type of endoplasmic reticulum and is not part of the T tubule system.

8. B. At the neuromuscular junction the motor neuron produces acetylcholine that diffuses across the synaptic cleft. There is no direct contact between the neuron and the skeletal muscle cell. The acetylcholine combines with a receptor site in the skeletal muscle cell membrane. This leads to the production of an action potential in the skeletal muscle membrane. Acetylcholinesterase deactivates acetylcholine by splitting it into acetic acid and choline.

9. C. First, an action potential is formed in the muscle cell membrane at the neuromuscular junction. The action potential moves along the membrane and down into the T tubule system. This stimulates the sarcoplasmic reticulum to release calcium. The calcium diffuses into the myofibril and initiates contraction, and the sarcomere shortens. Finally, the sarcoplasmic reticulum takes up calcium by active transport and the muscle relaxes.

10. B. Calcium enters myofibrils and combines with troponin. This causes a reaction that results in tropomyosin pulling away from (exposing) actin. Myosin then binds to actin and there is cross bridge movement.

11. E. Energy (ATP) is required for active transport of calcium ions into the sarcoplasmic reticulum so relaxation can occur. Energy (ATP) is required for the movement of actin over myosin during contraction. Because of contraction, muscle is capable of shortening with force.

12. C. Calcium ions released from the sarcoplasmic reticulum combine with troponin during the latent phase, leading to contraction.

a is incorrect because cross bridge formation occurs during the contraction phase.

b is incorrect because active transport of calcium ions into the sarcoplasmic reticulum occurs during the relaxation phase.

d is incorrect because the sarcomere shortens during the contraction phase.

13. E. When sarcomeres shorten, the Z line is brought closer to the end of the myosin myofilament. Thus the I band must shorten. In addition, the ends of the actin myofilaments attached to a Z line approach the ends of the actin myofilaments of the other Z line in the same sarcomere. Thus the H zone must shorten.

14. A. By definition, a motor unit is a single nerve fiber and all of the muscle fibers it innervates.

15. C. A single skeletal muscle fiber responds in an all-or-none fashion. This means that, for a given condition, the fiber will respond maximally or not at all. Under favorable conditions a single skeletal muscle fiber may contract more forcibly than under less favorable circumstances.

16. D. Increased frequency of action potentials results in multiple wave summation, which leads to increased release of calcium from the sarcoplasmic reticulum. As calcium concentrations increase, the number of cross bridges that form is increased. If the frequency of stimulation is fast enough, tetanus results.

17. B. As the stimulus strength increases, the number of motor units and therefore sarcomeres responding to the stimulus increases. The increase in tension between A and B is consistent with that relationship. After all of the sarcomeres are responding (at point B), a further increase in the stimulus strength causes no more sarcomeres to respond. Skeletal muscle fibers and motor units respond in an all-or-none fashion, but skeletal muscles respond in a graded fashion.

18. E. At least two hypotheses explain the phenomenon. First, the connective tissue and cytoplasm of muscles have elastic characteristics. During a single muscle twitch, some of the tension generated by the muscle is used to stretch the elastic components. During tetanic contraction, the elastic component is stretched early, and then the tension generated by the muscle is applied to the load. The result is a greater measured tension. Second, the action potentials responsible for tetanic contraction may cause the release of enough calcium ions by the sarcoplasmic reticulum to result in a greater than normal concentration of calcium ions around the myofilaments. The greater concentration of the calcium ions may make the contraction of the sarcomeres more complete.

19. A. A contraction in which the length of the muscle does not change is isometric. Because the weight did not move, the weight-lifter's limbs did not change position, and therefore neither did the length of his muscles.

20. D. If the resting length of one's muscles was near B, one's muscles would be "adjusted" to achieve the maximum amount of tension nearly all the time.

a is incorrect. Maximum sarcomere length is at C.

b is incorrect. If one's muscles were greatly stretched (to C) the ability to generate tension would be reduced.

c is incorrect. The greatest amount of overlap for actin and myosin occurs at A. The optimal amount of overlap for contraction occurs at B.

21. B   In an intact animal, fatigue is often the result of the central nervous system. That encouragement increased the number of push-ups that same person was able to perform is consistent with central nervous system (psychological) fatigue.

22. E   For approximately a 20-second period, two anaerobic sources provide adequate energy to maintain muscular contraction. Creatine phosphate breaks down to ATP and creatine, and anaerobic metabolism produces ATP.

23. E   Following vigorous exercise the oxygen consumption remains elevated until the oxygen debt is "repaid." Events that occur during that process include the formation of creatine phosphate and the conversion of lactic acid into glucose.

24. E   An avid bicycle rider would certainly ride long distances. His leg muscles would probably be adapted for long-term exercise. Characteristics of such muscles are myoglobin-rich, contract relatively slowly, and depend primarily on aerobic metabolism for energy. Although creatine phosphate may be present, it does not, over long distances, supply a major proportion of the energy required for muscle contraction.

25. D   Aerobic exercise training increases the blood supply to muscle, making both slow-twitch and fast-twitch muscle fibers more resistant to fatigue. Aerobic training causes hypertrophy of slow-twitch fibers because of an increase in the number of myofibrils per cell. Exercise does not convert fast-twitch into slow-twitch fibers, or vice-versa.

26. D   During exercise chemical reactions in contracting muscles increase. Because the reactions release heat, body temperature increases and will remain elevated until the excess heat is lost following exercise. In addition, the chemical reactions involved with the oxygen debt also release heat and increase body temperature.

27. C   Smooth muscle has a very poorly developed sarcoplasmic reticulum. Calcium, which initiates the contractile process, comes from the extracellular fluid. After contraction, calcium is pumped out of the cell by active transport.

28. B   Visceral smooth muscle and cardiac muscle are capable of spontaneous contractions. Skeletal muscle and multiunit smooth muscle are stimulated by the somatic and autonomic nervous systems, respectively.

29. C   Even when stretched, smooth muscle can maintain the same tension. The force of contraction in skeletal muscle depends upon muscle length. Compared to skeletal muscle, smooth muscle can be stretched or shorten considerably and still maintain the ability to contract forcefully. This is a useful characteristic for muscle found in hollow organs that change shape. Also, smooth muscle can maintain a steady tension for long periods of time.

30. B   Smooth muscle cells have actin and myosin myofilaments; when calcium ions bind to the protein calmodulin, cross bridges form and contraction occurs. Smooth muscle cells have no T tubules or sarcomeres, and a sparse sarcoplasmic reticulum. Slow waves in the RMP are caused by an increase in permeability of the cell membrane to Na^+ and Ca^{2+} ions and subsequent depolarization of the cell.

 ☆ **FINAL CHALLENGES** ☆

1. Given that the tissue was muscle, the presence of multinucleated cells would be sufficient to identify the tissue as skeletal muscle, because smooth muscle and cardiac muscle have one nucleus per cell. As for the other characteristics listed:

    A. Smooth muscle has spindle-shaped cells, whereas cardiac and skeletal muscle cells are cylindrical.

    B. Striations are found in skeletal and cardiac muscle but not in smooth muscle.

    C. Skeletal muscle is under voluntary control, whereas cardiac and smooth muscle are under involuntary control.

    D. Branching cells are typical only of cardiac muscle.

2. The data do support the conclusion that the toxic substance inhibited acetylcholinesterase activity. However, there are other conclusions that are consistent with the data. For example, if the frequency of action potentials of motor neurons were increased by the toxic substance, one would expect to see a similar response.

3. In response to a single stimulus, either a single muscle cell does not contract, or it contracts with the same force. In other words, this experiment demonstrates the all-or-none law of skeletal muscle contraction.

4. The all-or-none law states that a skeletal muscle cell will either contract maximally or not contract at all.

   A. The all part of the all-or-none law can be explained as follows: once an action potential is generated in the muscle cell membrane, it travels down the T tubule system causing the release of calcium ions from the sarcoplasmic reticulum. The calcium ions initiate the contraction process. Because an action potential is of the same magnitude (for a given condition), no matter how strong the stimulus, the effect on the sarcoplasmic reticulum will be the same, and the force of contraction will be the same (maximal).

   B. The none part of the all-or-none law can also be explained on the basis of the action potential. A subthreshold stimulus produces no action potential and therefore no contraction. Once the stimulus is threshold or stronger, an action potential is generated, and contraction occurs.

5. As stimulus strength to the muscle nerve increases, the number of motor units stimulated increases (multiple motor unit summation), causing a gradual increase in the force of contraction. Once all the motor units are recruited, the force of contraction remains constant.

6. Part 1 of the graph is the lag (latent) phase, part 2 is the contraction phase, and part 3 is the relaxation phase of muscle contraction. During the relaxation phase there is increased calcium ion uptake by active transport into the sarcoplasmic reticulum. The removal of calcium ions from troponin allows tropomyosin to move between actin and myosin, and relaxation occurs.

7. A supramaximal stimulus (or a maximal stimulus) would stimulate all the motor units of a muscle causing a muscle twitch. A high frequency of supramaximal stimuli would cause multiple wave summation and result in tetanus.

8. Because the force of contraction is eventually the same for both muscle preparations it is reasonable to conclude that both preparations have the same number of muscle fibers. As stimulus strength increases, the number of motor units (and therefore the number of muscle fibers) recruited increases, resulting in a greater force of contraction. If preparation A has fewer motor units than preparation B, preparation A has more muscle fibers per motor unit than preparation B. For each motor unit recruited, the force of contraction will therefore be greater in preparation A than in preparation B, and the force of contraction will increase faster for preparation A.

9. Sally's aerobic exercise program of jogging has developed her slow-twitch muscle fibers and increased the fatigue resistance of her fast-twitch muscle fibers. Sunny's weight-lifting program consists of intense, but short, periods of exercise. This increases strength (hypertrophy of fast-twitch fibers); but, because it relies on anaerobic respiration, it does not develop aerobic, fatigue-resistant abilities. Sunny needs to do some aerobic exercises, which are included in most modern day weight-lifting programs. It is also possible Sally was less affected by psychological fatigue.

10. In experiment A the student used anaerobic respiration as she started to run in place, but aerobic respiration was also increased to meet most of her energy needs. When she stopped, her respiration rate would be increased over resting levels because of repayment of the oxygen debt due to anaerobic respiration. In experiment B almost all of her energy would come from anaerobic respiration because she is holding her breath while running in place. Consequently, she would have a much larger oxygen debt. One would predict that following running in place in experiment B her respiration rate would be greater than in experiment A, or that her respiration rate would be elevated for a longer period of time than in experiment A.

11 Muscular System: Gross Anatomy

FOCUS: Muscles cause movement by shortening. The movable end of the muscle is called the insertion, the fixed end is the origin. The force of contraction of a muscle is applied to a bone that acts as a lever across a movable joint (fulcrum). There are three classes of lever systems in the human body. A muscle that causes a particular movement such as flexing the forearm is called an agonist. Because muscles can only lengthen passively, the opposite movement (extending the forearm) is achieved by a different muscle, the antagonist. The interaction between agonist and antagonist allows precise control of movements. Often, several muscles work together to produce a movement; they are termed synergists. The study of muscle actions can be approached by examining groups of muscles first and then individual muscles within a group.

CONTENT LEARNING ACTIVITY

General Principles

❝Most muscles extend from one bone to another and cross at least one joint.**❞**

Match these terms with the correct statement or definition:

Aponeurosis
Antagonist
Belly
Fixators
Insertion

Origin (head)
Prime mover
Synergists
Tendon

_____ 1. General term for the connective tissue connection between muscles and bones.

_____ 2. A very broad tendon.

_____ 3. End of the muscle that moves the least.

_____ 4. Largest portion of a muscle, between its origin and insertion.

_____ 5. Muscles that work together to cause movement.

_____ 6. One muscle that plays the major role in accomplishing a desired movement.

_____ 7. Muscles that stabilize one or more joints crossed by the prime mover.

Muscle Shapes

"Muscles come in a wide variety of shapes."

Match the type of muscle with
the correct description:

Circular   Parallel
Convergent   Pennate

_____ 1. Fasciculi arranged like the barbs of a feather along a common tendon.

_____ 2. Fasciculi organized in direct line with the long axis of the muscle.

_____ 3. The base (origin) of the muscle is much wider than the insertion.

_____ 4. Best able to shorten, but shorten with less force than pennate muscles.

_____ 5. Acts as a sphincter to close an opening.

Nomenclature

"Muscles are named according to many different criteria."

Match the muscle nomenclature
with the correct meaning:

Brachialis   Maximus
Brevis   Pectoralis
Deltoid   Quadriceps
Gluteus   Rectus
Masseter   Teres

_____ 1. Chest

_____ 2. Buttock

_____ 3. Arm

_____ 4. Large

_____ 5. Short

_____ 6. Round

_____ 7. Triangular

_____ 8. Straight

_____ 9. Four heads

_____ 10. Chewer

Movements Accomplished by Muscles

❝When muscles contract, force is applied to levers (bones), resulting in movements❞ of the lever about a fulcrum (joint).

Match these terms with the correct statement or definition:

Class I lever system   Class III lever system
Class II lever system

_____ 1. The fulcrum is between the force and the weight; a child's seesaw is an example.

_____ 2. Weight is located between the fulcrum and the force; a wheelbarrow is an example.

_____ 3. The force is located between the fulcrum and the weight; a person operating a shovel is an example.

_____ 4. Most common type of lever system in the body.

Head Movement

❝The muscles of the neck that are attached to the skull provide❞ the force for a wide variety of movements.

Match the neck muscle group with the correct description or example:

Anterior neck muscles   Posterior neck muscles
Lateral neck muscles

_____ 1. Flexion of the head is largely caused by gravity but is assisted by this muscle group.

_____ 2. Mainly responsible for the extension of the head.

_____ 3. Involved in rotation and abduction of the head.

_____

_____ 4. Sternocleidomastoid is an example.

Facial Expression

" The skeletal muscles of the face are somewhat unique in that they " are not necessarily attached to the skeleton.

Match these muscles with the correct function:

Corrugator   Occipitofrontalis
Depressor anguli oris   Orbicularis oculi
Levator labii superioris   Orbicularis oris
Levator palpebrae superioris   Zygomaticus major and minor

_____ 1. Raises the eyebrows and furrows the skin of the forehead.

_____ 2. Closes the eyelids.

_____ 3. Raises the upper eyelid.

_____ 4. Furrows the skin between the eyebrows.

_____ 5. One of the kissing muscles.

_____ 6. Smiling.

_____ 7. This muscle makes sneering possible.

_____ 8. Muscle for frowning or pouting.

Mastication

" The muscles of mastication move the mandible. "

Match these muscles with the correct mandibular movements:

Lateral pterygoid muscle   Medial pterygoid muscle
Masseter muscle   Temporalis muscle

_____ 1. Three muscles that elevate the mandible.

_____

_____

_____ 2. Muscle of mastication that depresses the mandible.

☞   The tongue and buccinator muscles hold food in place between the teeth.

Tongue Movements

"The tongue is very important in mastication and speech."

Match the type of tongue muscle
with the correct description:

Extrinsic tongue muscles
Intrinsic tongue muscles

_____ 1. Muscles that make up the tongue.

_____ 2. Cause the shape of the tongue to change, as in rolling the
tongue into a tube.

_____ 3. Muscles that attach the tongue to the mandible, hyoid, and
skull.

_____ 4. Responsible for moving the tongue about, as in sticking out the
tongue.

Swallowing and the Larynx

*"The hyoid muscles and muscles of the soft palate, pharynx, and larynx
have several important functions."*

Match the type of hyoid muscle
with the correct description:

Infrahyoid muscles   Soft palate muscles
Laryngeal muscles   Suprahyoid muscles
Pharyngeal muscles

_____ 1. Depress the mandible.

_____ 2. Fix the hyoid bone when the mandible is depressed.

_____ 3. Elevate the larynx during swallowing.

_____ 4. Close the posterior opening to the nasal cavity during
swallowing.

_____ 5. Muscles from these two groups open the auditory tube to
equalize pressure between the middle ear and the atmosphere.

_____ 6. Constrict to force food into the esophagus.

_____ 7. Prevent food from entering the larynx; shorten vocal cords to
raise the pitch of the voice.

Movements of the Eyeball

66*The movements of the eyeball are made possible by six muscles.***99**

Match these terms with their correct location in the following statements:

Globe   Orbit
Oblique   Rectus

The eyeball muscles have their origin on the __(1)__ and insertion on the __(2)__. The four __(3)__ muscles cause the eyeball to move superiorly, inferiorly, medially, or laterally. The two __(4)__ muscles cause lateral-superior and lateral-inferior movements.

1. _____

2. _____

3. _____

4. _____

Muscles Moving the Vertebral Column

66*The muscles along the spine can be divided into a deep group and a superficial group.***99**

Match the muscle group of the back with the correct description:

Deep back muscles
Superficial back muscles

_____   1. Also called the erector spinae group.

_____   2. Produce the mass of muscles seen lateral to the midline of the back; in general, extend from the vertebrae to the ribs or from rib to rib.

_____   3. Extend from one vertebra to the next adjacent vertebra.

☞ The superficial and deep back muscles extend, laterally flex, and rotate the vertebral column.

Thoracic Muscles

66*The muscles of the thorax are involved almost entirely in the process of breathing.***99**

Match the muscle groups with their correct functions or descriptions:

Diaphragm   Internal intercostals
External intercostals   Scalenes

_____   1. Elevate the first two ribs during inspiration.

_____   2. Elevate the ribs and sternum, increasing the diameter of the thorax.

_____   3. Depress the ribs and sternum, forcing expiration.

_____   4. Responsible for the majority of volume change in the thoracic cavity during quiet breathing; dome-shaped when relaxed.

Abdominal Wall

"The muscles of the anterior abdominal wall flex and rotate the vertebral column."

Match these terms with the correct definition or description:

External abdominal oblique   Rectus abdominis
Internal abdominal oblique   Tendinous inscriptions
Linea alba   Transversus abdominis
Linea semilunaris

_____ 1. Tendinous area that produces a vertical line from the xiphoid process through the navel to the pubis.

_____ 2. Thick muscle, on either side of the midline, with fasciculi oriented vertically in a straight line.

_____ 3. Connective tissue that transects the rectus abdominis, causing it to appear segmented.

_____ 4. Most superficial lateral abdominal wall muscle.

_____ 5. Deepest lateral abdominal wall muscle.

Pelvic Floor and Perineum

"The inferior opening of the pelvis is closed by a muscular floor through which the anus and urogenital openings penetrate."

Match these terms with the correct description or definition :

Pelvic diaphragm   Urogenital diaphragm
Perineum

_____ 1. Mostly formed by the coccygeus and levator ani muscles; forms the pelvic floor; includes the sphincter ani externus.

_____ 2. Diamond-shaped area superficial to the pelvic floor.

_____ 3. Consists of the deep transverse perineus muscle and the sphincter urethrae muscle.

Scapular Movements

"The major scapular muscles fix or move the scapula."

Match these muscles with the correct function:

Levator scapulae   Serratus anterior
Pectoralis minor   Trapezius
Rhomboideus major and minor

_____ 1. Elevates, depresses, rotates, and fixes the scapula.

_____ 2. Retracts and fixes the scapula.

_____ 3. Rotates and protracts the scapula.

_____ 4. Depresses the scapula.

Arm Movements

"The arm is attached to the thorax and scapula by several muscles."

Match these muscles with the correct description:

Coracobrachialis   Subscapularis
Deltoid   Supraspinatus
Infraspinatus   Teres major
Latissimus dorsi   Teres minor
Pectoralis major

_____ 1. Two muscles that abduct the arm.

_____

_____ 2. Two muscles that can flex or extend the arm.

_____

_____ 3. Two muscles that attach to the trunk and adduct the arm.

_____

_____ 4. Three muscles that attach to the scapula and adduct the arm.

_____

_____

Forearm Movements

"Numerous muscles accomplish the movements of the forearm."

Match these muscles with
the correct description:

Anconeus	Pronator quadratus
Biceps brachii	Pronator teres
Brachialis	Supinator
Brachioradialis	Triceps brachii

_____

_____   1. Two muscles that extend the forearm.

_____

_____   2. Three muscles that flex the forearm.

_____

_____   3. Two muscles that supinate the forearm.

_____

_____   4. Muscle that flexes the arm.

_____   5. Muscle that extends the arm.

Wrist, Hand, and Finger Movements

"The forearm muscles can be divided into anterior and posterior groups."

A. Match the muscle group
with the correct statement:

Anterior forearm muscles
Posterior forearm muscles

_____   1. Most of these muscles originate on the medial epicondyle of
the humerus; responsible for flexion of the wrist and fingers.

_____   2. Most of these muscles originate on the lateral epicondyle of
the humerus; responsible for extension of the wrist and fingers.

B. Match these terms with the
correct description or definition:

Extrinsic hand muscles	Retinaculum
Hypothenar eminence	Thenar eminence
Intrinsic hand muscles	

_____   1. Located in the forearm with tendons that extend into the hand.

_____   2. Strong band of fibrous connective tissue that covers the flexor
and extensor tendons.

_____   3. General term for muscles located entirely with the hand.

_____   4. Fleshy prominence at the base of the thumb; responsible for
control of thumb movements.

Thigh Movements

"*Muscles located in the hip and thigh are responsible for thigh movements.*"

A. Match the hip muscles
with the correct function:

Deep thigh rotators   Iliacus
Gluteus maximus   Psoas major
Gluteus medius   Tensor fasciae latae
Gluteus minimus

_____ 1. Two anterior hip muscles that flex the thigh.

_____

_____ 2. Posterolateral hip muscle that extends, abducts, and laterally
rotates the thigh.

_____ 3. Two posterolateral hip muscles that abduct and medially
rotate the thigh.

_____

_____ 4. Muscle that inserts on the tibia through the iliotibial tract;
stabilizes the knee and abducts the thigh.

_____ 5. Group of hip muscles that laterally rotate the thigh.

B. Match the thigh muscles
with their function:

Anterior thigh muscles   Posterior thigh muscles
Medial thigh muscles

_____ 1. Thigh muscles that flex the thigh.

_____ 2. Thigh muscles that extend the thigh.

_____ 3. Thigh muscles that adduct the thigh.

Leg Movements

"*The muscles of the thigh that move the leg are contained in anterior, posterior,*
and medial compartments."

A. Match the muscle group
with the correct statement:

Anterior thigh muscles   Posterior thigh muscles
Medial thigh muscles

_____ 1. Muscle group responsible for extending the leg (except for the
sartorius muscle).

_____ 2. Muscle group mostly responsible for flexing the leg.

_____ 3. Muscle group that functions to adduct and flex the thigh.

_____ 4. Rectus femoris belongs to this group.

_____ 5. Biceps femoris belongs to this group.

B. Match the thigh muscle with its function:

Adductor brevis   Sartorius
Adductor longus   Semimembranosus
Adductor magnus   Semitendinosus
Biceps femoris   Vastus intermedius
Gracilis   Vastus lateralis
Pectineus   Vastus medialis
Rectus femoris

_____ 1. Four muscles that extend the leg; comprise most of the anterior thigh muscle.

_____

_____

_____

_____ 2. Five muscles that flex the leg.

_____

_____

_____

_____

_____ 3. Five muscles that adduct the thigh; comprise the medial compartment of the thigh.

_____

_____

_____

_____

_____ 4. Two muscles of the anterior thigh compartment that flex the thigh.

_____

_____ 5. Three muscles of the posterior thigh compartment that extend the thigh.

_____

_____

The quadriceps femoris consists of the rectus femoris, vastus lateralis, vastus intermedius, and vastus medialis.

Ankle, Foot, and Toe Movements

"Extrinsic foot muscles move the ankle, foot, and toes, whereas intrinsic foot muscles move the toes."

A. Match the muscle group with the correct statement:

Anterior leg muscles   Lateral leg muscles
Intrinsic foot muscles   Posterior leg muscles

_____   1. Acts to dorsiflex the foot and extend the toes.

_____   2. Acts to plantar flex and invert the foot and flex the toes.

_____   3. Acts to evert and plantar flex the foot.

_____   4. Muscles found within the foot that flex, extend, abduct, and adduct the toes.

B. Match the leg muscle with the correct description:

Gastrocnemius, soleus,   Tibialis anterior and posterior
  and plantaris
Peroneus brevis, longus, and tertius

_____   1. Muscles that invert the foot.

_____   2. Muscles that evert the foot.

_____   3. Muscles that join to form the calcaneal (Achilles) tendon; act to plantar flex the foot.

Location of Superficial Muscles

A. Match these muscles with
the correct parts labeled
in Figure 11-1:

Adductors of thigh
Biceps brachii
Deltoid
External abdominal oblique
Flexors of wrist and fingers
Gastrocnemius
Pectoralis major
Rectus abdominis
Rectus femoris
Sartorius
Serratus anterior
Sternocleidomastoid
Vastus lateralis
Vastus medialis

1. _____

2. _____

3. _____

4. _____

5. _____

6. _____

7. _____

8. _____

9. _____

10. _____

11. _____

12. _____

13. _____

14. _____

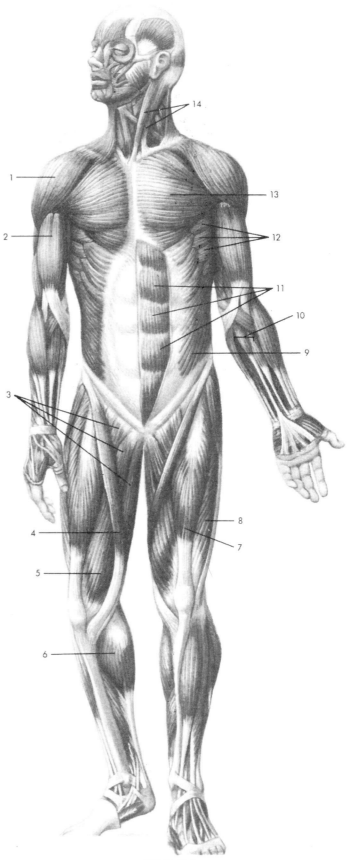

Figure 11-1

B. Match these muscles with
the correct parts labeled
in Figure 11-2:

Biceps femoris
Deltoid
Extensors of wrist and fingers
Gastrocnemius
Gluteus maximus
Infraspinatus
Latissimus dorsi
Semitendinosus
Teres major
Trapezius
Triceps brachii

1. _____

2. _____

3. _____

4. _____

5. _____

6. _____

7. _____

8. _____

9. _____

10. _____

11. _____

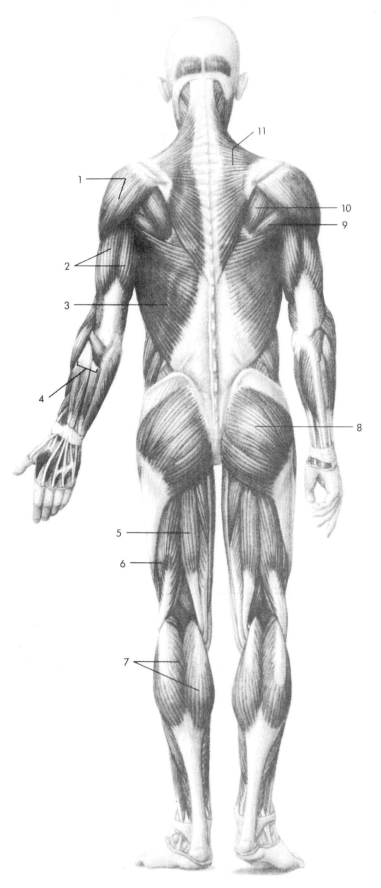

Figure 11-2

1. List the three basic parts of a muscle.

2. List four classes of muscle shape.

3. Name, and give an example of, the three classes of levers.

4. List the three groups of muscles that move the head.

5. Name the two muscles of facial expression that are circular muscles, and give their function.

6. List the muscles of mastication.

7. List the two major types of muscle groups that act on the vertebral column.

8. Name the three major thoracic muscles that are involved with inspiration.

9. List four muscles in the abdominal wall.

10. List two general functions for the muscles that attach the scapula to the thorax.

11. Define the rotator cuff.

12. List three muscles that flex and two muscles that extend the forearm.

13. List the two major muscle groups of the forearm, and give their major functions.

14. Define extrinsic and intrinsic hand muscles.

15. List the three groups of hip muscles and the three groups of thigh muscles that cause movement of the thigh.

16. List the three groups of thigh muscles that cause movement of the leg.

17. List the three muscle groups of the leg that act on the foot and toes.

MASTERY LEARNING ACTIVITY

Place the letter corresponding to the correct answer in the space provided.

_____ 1. Muscles that oppose one another are
a. synergists.
b. hateful.
c. levers.
d. antagonists.

_____ 2. The most movable attachment of a muscle is its
a. origin.
b. insertion.
c. fascia.
d. fulcrum.

_____ 3. Which of the following muscles is correctly matched with its type of fascicle orientation? (Hint: see diagrams of the muscles.)
a. pectoralis major - pennate
b. transversus abdominis - circular
c. temporalis - convergent
d. biceps femoris - parallel

_____ 4. The muscle whose name means the muscle is large and round?
a. gluteus maximus
b. vastus lateralis
c. teres major
d. latissimus dorsi
e. adductor magnus

_____ 5. In a class three lever system the
a. fulcrum is located between the force and the weight.
b. weight is located between the fulcrum and the force.
c. force is located between the fulcrum and the weight.

_____ 6. A prominent lateral muscle of the neck that can cause flexion of the neck or rotate the head is the
a. digastric.
b. mylohyoid.
c. sternocleidomastoid.
d. buccinator.
e. platysma.

_____ 7. Harry Wolf has just picked up his date for the evening. She is very attractive. Harry shows his appreciation by moving his eyebrows up and down, winking, smiling, and finally kissing her. Given the muscles listed below:
1. zygomaticus
2. levator labii superioris
3. occipitofrontalis
4. orbicularis oris
5. orbicularis oculi

In which order did he use the muscles:
a. 2,3,4,1
b. 2,5,3,1
c. 2,5,4,3
d. 3,5,1,4
e. 3,5,2,4

_____ 8. An aerial circus performer who supports herself only by her teeth while spinning around and around should have strong
a. temporalis muscles.
b. masseter muscles.
c. buccinator muscles.
d. a and b
e. all of the above

_____ 9. The tongue curls and folds PRIMARILY because of the action of the
a. intrinsic tongue muscles.
b. extrinsic tongue muscles.

_____ 10. The infrahyoid muscles
a. elevate the mandible.
b. move the mandible from side to side.
c. fix (prevent movement of) the hyoid.
d. a and b
e. all of the above

213

_____ 11. The soft palate muscles
   a. prevent food from entering the nasal cavity.
   b. force food into the esophagus.
   c. prevent food from entering the larynx.
   d. elevate the larynx.

_____ 12. The erector spinae muscles can cause
   a. extension of the vertebral column.
   b. lateral flexion of the vertebral column.
   c. rotation of the vertebral column.
   d. all of the above

_____ 13. Which of the following muscles are responsible for flexion of the vertebral column (below the neck)?
   a. deep back muscles
   b. superficial back muscles (erector spinae)
   c. rectus abdominis
   d. a and b
   e. all of the above

_____ 14. Which of the following muscles is involved with inspiration of air?
   a. diaphragm
   b. external intercostals
   c. scalenes
   d. a and b
   e. all of the above

_____ 15. Given the following muscles:
   1. external abdominal oblique
   2. internal abdominal oblique
   3. transversus abdominis

   Choose the arrangement that lists the muscles from most superficial to most deep.
   a. 1,2,3
   b. 1,3,2
   c. 2,1,3
   d. 2,3,1
   e. 3,1,2

_____ 16. The tendinous inscriptions
   a. attach the rectus abdominis muscles to the xiphoid process.
   b. divide the rectus abdominis muscles into subsections.
   c. separate the abdominal wall from the thigh.
   d. are the site of exit of blood vessels and nerves from the abdomen into the thigh.

_____ 17. Which of the following muscles is a synergist of the trapezius?
   a. rhomboids
   b. levator scapulae
   c. serratus anterior
   d. a and b
   e. all of the above

_____ 18. Which of the following muscles is a fixator (prevents protraction) of the scapula?
   a. trapezius
   b. levator scapulae
   c. pectoralis minor
   d. serratus anterior

_____ 19. Which of the following muscles adduct the arm (humerus)?
   a. latissimus dorsi
   b. teres major
   c. pectoralis major
   d. a and b
   e. all of the above

_____ 20. Which of the following muscles abduct the arm (humerus)?
   a. supraspinatus
   b. infraspinatus
   c. teres minor
   d. subscapularis
   e. all of the above

_____ 21. Which of the following muscles would one expect to be especially well developed in a boxer?
   a. biceps brachii
   b. brachialis
   c. deltoid
   d. triceps brachii
   e. supinator

_____22. Antagonist of the triceps brachii?
   a. biceps brachii
   b. brachialis
   c. anconeus
   d. a and b
   e. all of the above

_____23. The posterior group of forearm
   muscles is responsible for
   a. flexion of the wrist.
   b. flexion of the fingers.
   c. extension of the fingers.
   d. a and b

_____24. The intrinsic hand muscles that
   move the thumb?
   a. thenar muscles
   b. hypothenar muscles
   c. flexor pollicis longus
   d. extensor pollicis longus

_____25. Which of the following muscles can
   extend the femur?
   a. gluteus maximus
   b. gluteus minimus
   c. gluteus medius
   d. a and b
   e. all of the above

_____26. Given the following muscles:
   1. iliopsoas
   2. rectus femoris
   3 sartorius

   Which of the muscles act to flex the
   thigh?
   a. 1
   b. 1, 2
   c. 1, 3
   d. 2, 3
   e. 1, 2, 3

_____27. Which of the following muscles is
   found in the medial compartment of
   the thigh?
   a. rectus femoris
   b. sartorius
   c. gracilis
   d. vastus medialis
   e. semitendinosus

_____28. Which of the following muscles
   would be well developed in a
   football player whose specialty is
   kicking field goals?
   a. sartorius
   b. semitendinosus
   c. gastrocnemius
   d. quadriceps femoris

_____29. Which of the following muscles can
   flex the leg?
   a. biceps femoris
   b. sartorius
   c. gastrocnemius
   d. a and b
   e. all of the above

_____30. Which of the following muscles
   would one expect to be especially
   well developed in a ballerina?
   a. gastrocnemius
   b. gracilis
   c. gluteus maximus
   d. quadratus lumborum
   e. triangularis

_____31. Which of the following muscles are
   responsible for everting the foot?
   a. peroneus brevis
   b. peroneus longus
   c. peroneus tertius
   d. a and b
   e. all of the above

_____32. Which of the following muscles
   cause plantar flexion of the foot?
   a. tibialis anterior
   b. extensor digitorum longus
   c. peroneus tertius
   d. soleus

_____33. Which of the following muscles are
   attached to the coracoid process of
   the scapula?
   a. pectoralis minor
   b. coracobrachialis
   c. biceps brachii
   d. a and b
   e. all of the above

_____34. The anterior group of forearm muscles MOSTLY have their origin on the
a. medial epicondyle of the humerus.
b. lateral epicondyle of the humerus.
c. distal portion of the radius.
d. distal portion of the ulna.

_____35. Which of the following muscles spans two joints?
a. gastrocnemius
b. biceps femoris
c. sartorius
d. rectus femoris
e. all of the above

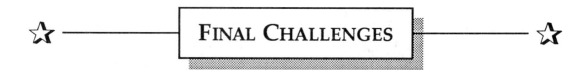

FINAL CHALLENGES

Use a separate sheet of paper to complete this section.

1. Describe the movement each of these muscles produces: quadratus lumborum, levator scapulae, latissimus dorsi, brachioradialis, gluteus minimus, semimembranosus, adductor magnus, and tibialis anterior. Name the muscle that acts as synergist and antagonist for each movement of the muscle.

2. What kind of lever system is in operation when the rectus femoris muscle moves the leg? When the semimembranosus muscle moves the leg?

3. In general, the origin of limb muscles is proximal, and the insertion is distal. What would happen if the opposite arrangement were true?

4. In Bell's palsy there is damage to the facial nerve. One of the symptoms is that the eye on the affected side remains open at all times. Explain why this happens.

5. Observe the actions of the abdominal muscles when you take a normal breath. Do they contract or relax? Why do they do this? What do the abdominal muscles do when you expel all possible air out of your lungs? Why do they do this?

6. Name the muscle that is involved in all the arm movements that occur when the arm swings back and forth during walking.

7. While trying to catch a baseball, the most distal digit of Bobby Bumblefingers's index finger was forcefully flexed, tearing a muscle tendon. What muscle was damaged, and what symptoms would Bumblefingers exhibit?

8. According to an advertisement, a special type of sandal makes the calf look better because the calf muscles are exercised when the toes curl and grip the sandal while walking. Explain why you believe or disbelieve this claim.

9. A patient suffered damage to a nerve that supplies the leg muscles. Consequently her left foot was plantar flexed and slightly inverted, creating a tendency for the toes to drag when she walked. To compensate she would raise her left leg higher than normal then slap the foot down. What muscles of the leg were no longer functional?

ANSWERS TO CHAPTER 11

General Principles
1. Tendon; 2. Aponeurosis; 3. Origin (head);
4. Belly; 5. Synergists; 6. Prime mover; 7. Fixators

Muscle Shapes
1. Pennate; 2. Parallel; 3. Convergent; 4. Parallel;
5. Circular

Nomenclature
1. Pectoralis; 2. Gluteus; 3. Brachialis;
4. Maximus; 5. Brevis; 6. Teres; 7. Deltoid;
8. Rectus; 9. Quadriceps; 10. Masseter

Movements Accomplished by Muscles
1. Class I lever system; 2. Class II lever system;
3. Class III lever system; 4. Class III lever system

Head Movement
1. Anterior neck muscles; 2. Posterior neck
muscles; 3. Posterior neck muscles; lateral neck
muscles; 4. Lateral neck muscles

Facial Expression
1. Occipitofrontalis; 2. Orbicularis oculi;
3. Levator palpebrae superioris; 4. Corrugator;
5. Orbicularis oris; 6. Zygomaticus major and
minor; 7. Levator labii superioris; 8. Depressor
anguli oris

Mastication
1. Temporalis muscle, masseter muscle, medial
pterygoid muscle; 2. Lateral pterygoid muscle

Tongue Movements
1. Intrinsic tongue muscles; 2. Intrinsic tongue
muscles; 3. Extrinsic tongue muscles; 4. Extrinsic
tongue muscles

Swallowing and the Larynx
1. Suprahyoid muscles; 2. Infrahyoid muscles;
3. Infrahyoid muscles; 4. Soft palate muscles;
5. Pharyngeal and soft palate muscles;
6. Pharyngeal muscles; 7. Laryngeal muscles.

Movements of the Eyeball
1. Orbit; 2. Globe; 3. Rectus; 4. Oblique

Muscles Moving the Vertebral Column
1. Superficial back muscles; 2. Superficial back
muscles; 3. Deep back muscles

Thoracic Muscles
1. Scalene; 2. External intercostals; 3. Internal
intercostals; 4. Diaphragm

Abdominal Wall
1. Linea alba; 2. Rectus abdominis; 3. Tendinous
inscriptions; 4. External abdominal oblique;
5. Transversus abdominis

Pelvic Floor and Perineum
1. Pelvic diaphragm; 2. Perineum; 3. Urogenital
diaphragm

Scapular Movements
1. Trapezius; 2. Rhomboideus major and minor;
3. Serratus anterior; 4. Pectoralis minor

Arm Movements
1. Deltoid, supraspinatus; 2. Deltoid, pectoralis
major; 3. Latissimus dorsi, pectoralis major;
4. Coracobrachialis, teres major, teres minor

Forearm Movements
1. Triceps brachii, anconeus; 2. Brachialis, biceps
brachii, brachioradialis; 3. Supinator, biceps
brachii; 4. Biceps brachii; 5. Triceps brachii

Wrist, Hand, and Finger Movements
A. 1. Anterior forearm muscles; 2. Posterior
forearm muscles
B. 1. Extrinsic hand muscles; 2. Retinaculum;
3. Intrinsic hand muscles; 4. Thenar
eminence

Thigh Movements
A. 1. Iliacus, psoas major; 2. Gluteus maximus;
3. Gluteus medius, gluteus minimus;
4. Tensor fasciae latae; 5. Deep thigh rotators
B. 1. Anterior thigh muscles; 2. Posterior thigh
muscles; 3. Medial thigh muscles

Leg Movements
A. 1. Anterior thigh muscles; 2. Posterior thigh
muscles; 3. Medial thigh muscles; 4. Anterior
thigh muscles; 5. Posterior thigh muscles
B. 1. Rectus femoris, vastus intermedius, vastus
lateralis, vastus medialis; 2. Biceps femoris,
semimembranosus, semitendinosus,
sartorius, gracilis; 3. Adductor brevis,
adductor longus, adductor magnus, gracilis,
pectineus; 4. Rectus femoris, sartorius;
5. Biceps femoris, semimembranosus,
semitendinosus

Ankle, Foot, and Toe Movements
A. 1. Anterior leg muscles; 2. Posterior leg
muscles; 3. Lateral leg muscles; 4. Intrinsic
foot muscles
B. 1. Tibialis anterior and posterior; 2. Peroneus
brevis, longus, and tertius; 3. Gastrocnemius,
soleus, and plantaris

Location of Superficial Muscles
A. 1. Deltoid; 2. Biceps brachii; 3. Adductors of
thigh; 4. Sartorius; 5. Vastus medialis;
6. Gastrocnemius; 7. Rectus femoris;
8. Vastus lateralis; 9. External abdominal
oblique; 10. Flexors of wrist and fingers;
11. Rectus abdominis; 12. Serratus anterior;
13. Pectoralis major; 14. Sternocleidomastoid
B. 1. Deltoid; 2. Triceps brachii; 3. Latissimus
dorsi; 4. Extensors of wrist and fingers;
5. Semitendinosus; 6. Biceps femoris;
7. Gastrocnemius; 8. Gluteus maximus;
9. Teres major; 10. Infraspinatus;
11. Trapezius

1. Origin (head), insertion, and belly
2. Pennate, parallel, convergent, and circular
3. Class I lever: a child's seesaw
   Class II lever: a wheelbarrow
   Class III lever: a shovel
4. Anterior, lateral, and posterior neck muscles
5. Orbicularis oculi: closing the eyelids; orbicularis oris: puckering the mouth
6. Temporalis, masseter, and medial and lateral pterygoids
7. Superficial and deep back muscles
8. Diaphragm, external intercostals, and scalenes
9. Rectus abdominis, external oblique, internal oblique, and transversus abdominis
10. Movement or fixation of the scapula
11. Four muscles that bind the humerus to the scapula and form a cuff or cap over the proximal humerus
12. Flex forearm: biceps brachii, brachialis, and brachioradialis
    Extend forearm: triceps brachii and anconeus
13. Anterior forearm: Flex fingers, thumb, and wrist; abduct or adduct wrist
    Posterior forearm: extend fingers, thumb, and wrist; abduct or adduct wrist
14. Extrinsic hand muscles are in the forearm, but have tendons that extend into the hand. Intrinsic hand muscles are entirely within the hand.
15. Hip muscles: Anterior; posterior and lateral; and deep thigh rotators
    Thigh muscles: Anterior, posterior, and medial
16. Anterior, posterior, and medial compartments
17. Anterior, posterior, and lateral compartments

MASTERY LEARNING ACTIVITY

1. D.   By definition two or more muscles that oppose one another are antagonists. Muscles that aid each other are synergists.

2. B.   By definition the most movable attachment of a muscle is the insertion. The origin is the "fixed" end of the muscle.

3. C.   The temporalis has a broad origin with fascicles that converge to insert on the mandible. The pectoralis major is convergent, the transversus abdominis is parallel, and the biceps femoris is pennate.

4. C.   Teres means round, and major means larger.

5. C.   The most common type of lever system in the body is a class III lever system in which the force is located between the fulcrum and the weight. In a class I lever system the fulcrum is located between the force and the weight. In a class II lever system the weight is located between the fulcrum and the force.

6. C.   The muscles and their functions are:

Muscle	Action
Sternocleidomastoid	Flex neck and rotate head
Digastric	Depress mandible, elevate hyoid
Mylohyoid	Depress mandible, elevate hyoid
Buccinator	Flatten cheek
Platysma	Depress lower lip, wrinkle skin of neck

7. D.   The muscles and their functions are:

Muscle	Function
Occipitofrontalis	Move eyebrows
Orbicularis oculi	Winking
Zygomaticus	Smiling
Orbicularis oris	Kissing
Levator labii superioris	Sneering

8. D.   The temporalis and masseter muscles close the jaw. The buccinator flattens the cheeks, acting to keep food between the teeth.

9. A   The intrinsic tongue muscles curl and fold the tongue, whereas the extrinsic tongue muscles move the tongue about as a unit.

10. C. The infrahyoid muscles fix the hyoid during mandibular depression. The suprahyoid muscles depress the mandible. The medial and lateral pterygoid muscles move the mandible from side to side.

11. A. The soft palate muscles prevent food from entering the nasal cavity. The pharyngeal muscles force food into the esophagus. The laryngeal muscles prevent food from entering the larynx.

12. D. The erector spinae and the deep back muscles can cause extension, lateral flexion, and rotation of the vertebral column.

13. C. The back muscles do not flex the vertebral column; that is accomplished by the abdominal muscles (rectus abdominis, external abdominal oblique, and internal abdominal oblique).

14. E. Contraction of the diaphragm, the major muscle of respiration, increases thoracic volume. Thoracic volume is also increased by the external intercostals, which elevate the ribs, and the scalenes, which elevate the first and second ribs.

15. A. From superficial to deep, the muscles of the lateral abdominal wall are the external abdominal oblique, internal abdominal oblique, and transversus abdominis.

16. B. The tendinous inscriptions subdivide the rectus abdominis, giving the abdomen a segmented appearance.

17. E. The trapezius and rhomboids retract (fix) the scapula, the trapezius and levator scapulae elevate the scapula, and the trapezius and serratus anterior rotate (upward) the scapula.

18. A. The trapezius and rhomboids fix the scapula.

19. E. The latissimus dorsi and pectoralis major attach to the trunk and cause adduction of the arm. The teres major attaches to the scapula and causes adduction of the arm.

20. A. The supraspinatus and deltoid abduct the arm.

21. D. Extension of the forearm during a punch is critical to a boxer: the triceps brachii performs that function.

22. D. The triceps brachii extends the forearm. Its antagonists (biceps brachii and brachialis) flex the forearm. The anconeus is a synergist of the triceps brachii and helps to extend the forearm.

23. C. The posterior group of forearm muscles is responsible for extension of the wrist and fingers, whereas the anterior group of forearm muscles is responsible for flexion of the wrist and fingers.

24. A. The thenar muscles are intrinsic (located within the hand) muscles that move the thumb. The hypo-thenar muscles are intrinsic hand muscles that move the little finger. The flexor pollicis longus and extensor pollicis longus both move the thumb, but they are extrinsic (located in the forearm) hand muscles.

25. A. The gluteus maximus extends, abducts, and laterally rotates the thigh. The gluteus minimus and medius abduct and medially rotate the thigh.

26. E. The iliopsoas is a hip muscle that flexes the thigh. The rectus femoris and sartorius are thigh muscles that flex the thigh.

27. C. The gracilis is found in the medial compartment; the semitendinosus in the posterior compartment; and the rectus femoris, vastus medialis, and sartorius in the anterior compartment.

28. D. The quadriceps femoris is composed of the rectus femoris, vastus intermedius, vastus medialis, and vastus lateralis. They extend the leg. The rectus femoris also flexes the thigh.

29. E. The biceps femoris (posterior thigh), sartorius (anterior thigh), and gastrocnemius (posterior leg) all flex the leg.

30. A. Because a ballerina often dances on her toes, plantar flexion of the foot is necessary. The gastrocnemius muscles perform that function.

31. E. The peroneal muscles evert the foot.

32. D. The soleus and gastrocnemius are the two muscles primarily responsible for plantar flexion of the foot. The other muscles listed cause dorsiflexion of the foot.

33. E. The pectoralis minor inserts on the coracoid process of the scapula, and the coracobrachialis and biceps brachii (short head) originate on the coracoid process of the scapula.

34. A. The anterior forearm muscles mostly originate on the medial humeral epicondyle, whereas the posterior forearm muscles mostly originate on the lateral humeral epicondyle.

35. E. The biceps femoris, sartorius, and rectus femoris extend from the coxa to the tibia (across the hip and knee joints). The gastrocnemius extends from the femur to calcaneus (across the knee and ankle joints).

1. Muscle	Action	Synergist	Antagonist
Quadratus lumborum	Abducts vertebral column	Erector spinae same side	Erector spinae opposite side; quadratus lumborum opposite side
Levator scapulae	Elevates scapula	Trapezius	Pectoralis minor
Latissimus dorsi	Adducts, extends, and medially rotates arm	Teres major; pectoralis major	Pectoralis major (flexes arm)
Brachioradialis	Flexes forearm	Brachialis; biceps brachii	Anconeus; triceps brachii
Gluteus minimus	Abducts and medially rotates thigh	Gluteus medius	Medial thigh muscles (adduct thigh); deep thigh rotators (laterally rotate thigh)
Semimembranosus	Flexes leg and extends thigh	Semitendinosus	Rectus femoris
Adductor magnus	Adducts, extends, and laterally rotates thigh	Gluteus maximus	Gluteus minimus
Tibialis anterior	Dorsiflexes foot	Peroneus tertius	Tibialis posterior; gastrocnemius; soleus

2. The foot is the weight of the lever system, and the knee joint is the fulcrum. The rectus femoris (the force) inserts on the tibial tuberosity between the weight and fulcrum; thus it is a class III lever system.

The semimembranosus inserts on the lateral condyle of the tibia; thus the force it generates is also between the weight and fulcrum, forming a class III lever system.

3. Limb movement would still be possible; but, when the muscles contracted, they would cause movement proximal to the muscles and the joint. This means the muscle would have to move (lift) itself, as well as part of the limb. In addition, we would all have limbs like Popeye the sailor!

4. Without stimulation from the facial nerve, the orbicularis oculi muscle is completely relaxed (flaccid paralysis), and the eye remains open.

5. During inspiration the abdominal muscles relax, allowing the abdominal organs to make room for the diaphragm as it descends inferiorly. This increases thoracic volume and the amount of air inspired. During forced expiration, the abdominal muscles push the abdominal organs and diaphragm superiorly, forcing more air out of the lungs.

6. The deltoid muscle abducts the arm, holding it away from the body. The deltoid also extends and flexes the arm as it swings back and forth.

7. Because the digit was forcefully flexed, the tendon of the muscle that opposes flexion would be stretched and torn (i.e., the extensor digitorum). Damage to this tendon would make him unable to extend the last phalanx of the index finger and unopposed contraction of the flexor digitorum profundus would keep the end of the index finger in a flexed position.

8. Disbelieve this advertisement. The toe flexors include the flexor hallucis longus and flexor digitorum longus. These muscles are small compared to the gastrocnemius and soleus muscles, which form the bulk of the leg. In addition, the toe flexors are deep to the gastrocnemius and soleus, and enlargement of the toe flexors caused by exercise does not have a noticeable effect on the appearance of the leg.

9. The muscles of the posterior compartment of the leg that cause plantar flexion and inversion of the foot are still functioning; but their antagonist, the muscles of the anterior and lateral compartments of the leg that cause dorsiflexion and eversion of the foot, are not functioning. Consequently the foot is plantar flexed and inverted.

Functional Organization of Nervous Tissue

FOCUS: The nervous system can be divided into the central nervous system (brain and spinal cord) and the peripheral nervous system (nerves and ganglia). Cells of the nervous system are neurons (multipolar, bipolar, or unipolar) and neuroglia cells (astrocytes, ependymal cells, microglia, oligodendrocytes, and neurolemmocytes). Neurons produce and conduct action potentials, whereas the neuroglia support and protect the neurons. Neurons connect with each other by means of synapses, and action potentials can be transferred in only one direction across the synapse. Neurotransmitters are chemicals released by presynaptic neurons that cross the synapse and excite or inhibit the postsynaptic neuron. The amount of neurotransmitter released can be increased through spatial or temporal summation. Neurons are arranged to form reflexes, convergent circuits, divergent circuits, and oscillating circuits.

CONTENT LEARNING ACTIVITY

Divisions of the Nervous System

66*Each division has structural and functional features that separate it from the other divisions.*99

A. Match these terms with the correct statement or definition:

Afferent division
Autonomic nervous system
Central nervous system
Efferent division

Parasympathetic division
Peripheral nervous system
Somatic nervous system
Sympathetic division

_____ 1. Consists of the brain and spinal cord.

_____ 2. Major site for processing information, initiating responses, and integrating mental processes.

_____ 3. Consists of nerves and ganglia; located outside the CNS.

_____ 4. PNS subdivision; impulses go from sensory organs to CNS.

_____ 5. Efferent subdivision; impulses go from CNS to skeletal muscle.

_____ 6. Efferent subdivision that transmits impulses from the CNS to smooth muscle, cardiac muscle, and glands.

_____ 7. Autonomic division; regulates resting or vegetative functions.

B. Match these terms with the
correct statement or definition:

Ganglia   12
Nerves   31

_____ 1. Bundles of axons and their sheaths that extend from the CNS to peripheral structures and from sensory organs to the CNS.

_____ 2. Collections of nerve cell bodies outside the CNS.

_____ 3. Number of cranial nerve pairs in the PNS.

_____ 4. Number of spinal nerve pairs in the PNS.

Neurons

66 *Neurons receive stimuli and transmit action potentials to other neurons or to effector organs.* 99

A. Match these terms with the
correct statement or definition:

Axon   Dendrite
Cell body (soma)

_____ 1. Portion of a neuron that contains the nucleus and other organelles such as rough endoplasmic reticulum.

_____ 2. Short, often highly branched cytoplasmic extension that is tapered from the neuron cell body to its tip.

_____ 3. Structure that is specialized to conduct action potentials to dendrites, cell bodies, or axons of other neurons.

B. Match these terms with the
correct statement or definition:

Axolemma   Chromatophilic substance
Axon hillock   (Nissl bodies)
Axoplasm   Telodendria
Dendritic spines   Terminal boutons
                              (presynaptic terminals)

_____ 1. Areas of rough endoplasmic reticulum found in the cytoplasm of the neuron cell body and base of dendrites; primary site of protein synthesis in a neuron.

_____ 2. Small extensions from dendrites; the location of synapses with axons of other neurons.

_____ 3. Enlarged area of each neuron cell body from which the axon arises.

_____ 4. Cytoplasm of the axon.

_____ 5. Branching ends of axons.

_____ 6. Enlarged ends of telodendria; contain vesicles with neurotransmitters.

☞ Dendrites respond to neurotransmitters from axons of other neurons by producing local potentials.

C. Match these terms with the correct parts labeled in Figure 12-1:

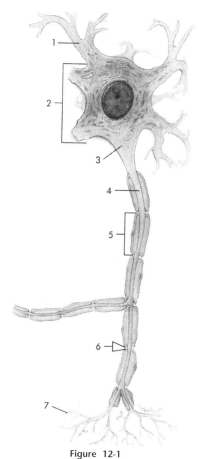

Axon
Axon hillock
Dendrite
Neuron cell body (soma)
Node of Ranvier
Neurolemmocyte
   (Schwann cell)
Telodendria

1. _____

2. _____

3. _____

4. _____

5. _____

6. _____

7. _____

Figure 12-1

Types of Neurons

❝Neurons can be classified by function or by structure.❞

A. Match these terms with the correct statement or definition:

Afferent (sensory) neurons   Efferent (motor) neurons
Association neurons (interneurons)

_____ 1. Conduct action potentials toward the CNS.

_____ 2. Conduct action potentials from the CNS to muscles or glands.

_____ 3. Conduct action potentials from one neuron to another.

B. Match these terms with the correct statement or definition:

Bipolar neuron   Unipolar neuron
Multipolar neuron

_____ 1. Neuron with several dendritic processes and a single axon.

_____ 2. Neuron with a single dendrite and a single axon.

_____ 3. Most CNS neurons, including motor neurons, are this type.

223

Neuroglia

66There are five types of neuroglia, each with unique structural and functional characteristics.99

Match these terms with the
correct statement or definition:

Astrocytes
Ependymal cells
Microglia

Neurolemmocytes
  (Schwann cells)
Oligodendrocytes
Satellite cells

_____ 1. Neuroglia that are star-shaped; help to regulate the
composition of extracellular fluid around neurons.

_____ 2. Neuroglia found in choroid plexuses and line the ventricles of
the brain; secrete and circulate cerebrospinal fluid.

_____ 3. Small cells; become mobile and phagocytic in response to
inflammation.

_____ 4. Neuroglia with cytoplasmic extensions that form myelin
sheaths around axons in the CNS.

_____ 5. Neuroglia cells; form myelin sheaths around axons in the PNS.

_____ 6. Specialized neurolemmocytes that surround, provide support,
and provide nutrients for neuron cell bodies in ganglia.

Axon Sheaths

66Axons, surrounded by sheaths, are structurally similar in both the PNS and the CNS.99

Using the terms provided, complete these statements:

Internodes
Myelin sheath
Myelinated axons
Nodes of Ranvier
Rapidly

Saltatory
  conduction
Slowly
Unmyelinated
  axons

 (1) rest in an invagination of the oligodendrocytes or the
neurolemmocytes, but the axons are not actually inside the
cell. Action potentials are propagated along the entire
axon membrane. In (2) of the PNS, each neurolemmocyte
repeatedly wraps around a segment of an axon to form the
 (3) , giving the axons a white appearance. Gaps between
neurolemmocytes are called (4) , whereas the areas
between the gaps are called (5) . Conduction of action
potentials from one node of Ranvier to another in
myelinated axons is called (6) . Action potentials are
conducted more (7) in myelinated than in unmyelinated
axons; small-diameter axons conduct action potentials more
 (8) than do large diameter axons.

1. _____

2. _____

3. _____

4. _____

5. _____

6. _____

7. _____

8. _____

Organization of Nervous Tissue

"Nervous tissue is organized so that axons form bundles.**"**

A. Match these terms with the correct statement or definition:

Cortex
Endoneurium
Epineurium
Ganglia
Gray matter
Nerve

Nerve fascicle
Nerve tract
Nuclei
Perineurium
White matter

_____ 1. Bundles of parallel axons with their associated sheaths that appear whitish in color.

_____ 2. Collection of nerve cell bodies that appears gray in color

_____ 3. Conduction pathway formed by white matter of the CNS.

_____ 4. Gray matter on the outer surface of the brain.

_____ 5. Collections of gray matter within the brain.

_____ 6. Collections of nerve cell bodies in the PNS.

B. Match these terms with the correct parts labeled in Figure 12-2:

Axon
Endoneurium
Epineurium
Fascicle

Nerve
Neurolemmocyte
Perineurium

1. _____

2. _____

3. _____

4. _____

5. _____

6. _____

7. _____

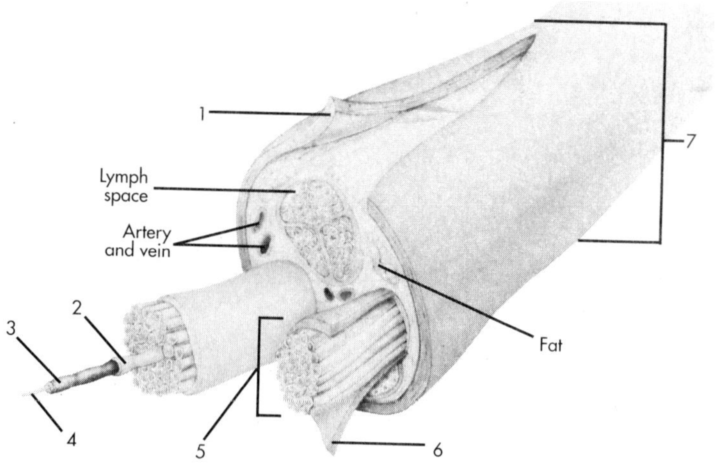

Figure 12-2

The Synapse

The synapse is a junction between two cells.

Match these terms with the correct statement or definition:

Acetylcholinesterase
Monoamine oxidase
Neuromodulator
Neurotransmitter

Postsynaptic membrane
Presynaptic terminal
Synaptic cleft
Synaptic vesicle

_____ 1. End of an axon that contains synaptic vesicles.

_____ 2. Space separating the axon ending and the cell with which it synapses.

_____ 3. Specific receptors that can bind to neurotransmitters are located here.

_____ 4. Chemical released from the presynaptic terminal.

_____ 5. Membrane-bound organelle that contains neurotransmitters.

_____ 6. Enzyme that breaks down acetylcholine in the synaptic cleft.

_____ 7. Nonneurotransmitter chemical that influences the likelihood that an action potential in the presynaptic terminal will result in the production of an action potential in the postsynaptic membrane.

☞ In the circulation, norepinephrine is taken up by the liver and kidney cells, where the enzymes monoamine oxidase and catechol-O-methyltransferase convert it to inactive metabolites.

EPSPs and IPSPs

The combination of neurotransmitters and their receptors causes either an EPSP or an IPSP.

A. Match these terms with the correct statement or definition:

Excitatory neuron
Excitatory postsynaptic potential (EPSP)

Inhibitory neuron
Inhibitory postsynaptic potential (IPSP)

_____ 1. Local depolarization of the postsynaptic membrane.

_____ 2. Caused by increase in permeability of cell membrane to Na^+ ions.

_____ 3. Neuron that causes EPSPs.

_____ 4. Local hyperpolarization of the postsynaptic membrane.

_____ 5. Caused by increase in permeability of cell membrane to K^+ or Cl^- ions.

B. Match these terms with
the correct parts labeled
in Figure 12-3:

EPSP
IPSP

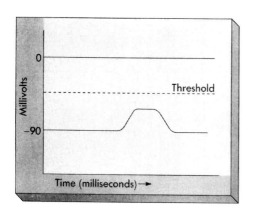

Graph A

Figure 12-3

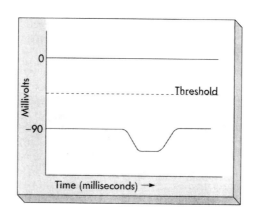

Graph B

_____ 1. The change in membrane potential seen in graph A.

_____ 2. The change in membrane potential seen in graph B.

C. Using the terms provided, complete these statements:

Axo-axonic   Presynaptic
Neuromodulators   inhibition
Presynaptic facilitation

1. _____

2. _____

3. _____

4. _____

Many of the synapses of the CNS are __(1)__ synapses, in which the axon of one neuron synapses with the presynaptic terminal (axon) of another neuron. When an action potential reaches the presynaptic terminal, __(2)__ released in the axo-axonic synapse can alter the amount of neurotransmitter released from the presynaptic terminal. In __(3)__, there is decreased neurotransmitter release from the presynaptic terminal, e. g., when endorphins and enkephalins are released in the brain and spinal cord. In __(4)__ there is increased neurotransmitter release from the presynaptic terminal, e. g., when nitric oxide from the postsynaptic membrane stimulates the release of glutamate from the presynaptic terminal.

227

Spatial and Temporal Summation

"_A single presynaptic action potential does not cause local depolarization in the postsynaptic_**"** _membrane sufficient to reach threshold and produce an action potential._

Using the terms provided, complete these statements:

IPSPs	Summation
Local potentials	Temporal
Spatial	Threshold

Presynaptic action potentials produce __(1)__ in the post-synaptic neuron. These combine in a process called __(2)__ at the axon hillock of the postsynaptic neuron. If this combination of local potentials exceeds __(3)__ at the axon hillock, an action potential is produced. __(4)__ summation results when two action potentials arrive simultaneously at two different presynaptic terminals that synapse with the same postsynaptic neuron. __(5)__ summation occurs when two action potentials arrive in very close succession at a single presynaptic terminal. Excitatory and inhibitory neurons may synapse with a single postsynaptic neuron. Summation occurs in the postsynaptic neuron, and the __(6)__ tend to cancel the EPSPs.

1. _____
2. _____
3. _____
4. _____
5. _____
6. _____

Reflexes

"_The reflex arc is the basic functional unit of the nervous system._**"**

A. Using the terms provided, complete these statements:

Afferent neurons	Efferent neurons
Association neurons	Reflex
Effector organs	Sensory receptors

Action potentials initiated in __(1)__ are propagated along __(2)__ within the PNS to the CNS where they usually synapse with __(3)__ . These neurons synapse with __(4)__ , which send axons out of the spinal cord. Action potentials of the efferent neurons cause __(5)__ to respond. This automatic response produced by a reflex arc is called a __(6)__ .

1. _____
2. _____
3. _____
4. _____
5. _____
6. _____

B. Match these terms with the correct parts labeled in Figure 12-4:

Afferent neuron
Association neuron
Effector organ
Efferent neuron
Sensory receptor

1. _____
2. _____
3. _____
4. _____
5. _____

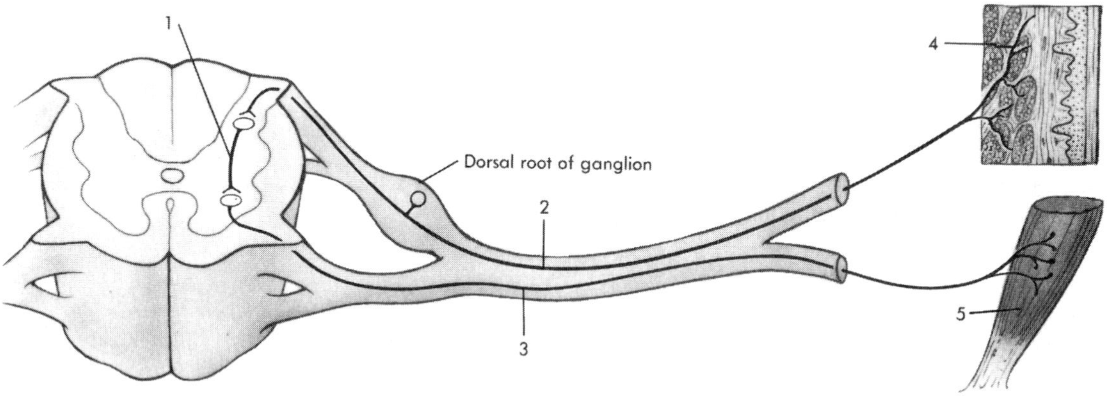

Dorsal root of ganglion

Figure 12-4

C. Using the terms provided, complete these statements:

Do
Do not

Reflexes _(1)_ require conscious thought and they _(2)_ vary in complexity. Reflexes _(3)_ function to maintain homeostasis. Because afferent neurons have many branches they _(4)_ function as isolated entities in the nervous system.

1. _____
2. _____
3. _____
4. _____

Neuronal Pathways and Circuits

66 *Neurons are organized within the CNS to form circuits.* 99

Match these terms with the correct statement or definition:

After discharge
Convergent circuit
Divergent circuit
Oscillating circuit

_____ 1. Circuit in which many neurons synapse with fewer neurons.

_____ 2. Innervation of motor neurons in the spinal cord is an example.

_____ 3. Afferent neurons carrying action potentials from pain receptors are an example.

_____ 4. Circuits with neurons arranged in a circular fashion.

_____ 5. A circular arrangement of neurons produces many action potentials.

_____ 6. Type of circuit responsible for periodic activity such as respiration.

1. Name two divisions of the efferent division of the nervous system. List two important characteristics of each.

2. List three types of neurons based on their shape. Give an example of each type.

3. List five types of neuroglia. Give a function for each type.

4. Contrast the structural and functional characteristics of myelinated and unmyelinated axons.

5. List two differences between white matter and gray matter. Give an example of each in the CNS and PNS.

6. Name three types of connective tissue coverings associated with nerves.

7. List three components of a synapse.

8. Distinguish between an EPSP and an IPSP.

9. List two types of summation and distinguish between them.

10. List the five components of a reflex arc.

11. List the three types of neuronal circuits.

MASTERY LEARNING ACTIVITY

Place the letter corresponding to the correct answer in the space provided.

_____ 1. The peripheral nervous system includes the
   a. somatic nervous system.
   b. brain.
   c. spinal cord.
   d. nuclei.

_____ 2. The part of the nervous system that controls smooth muscle, cardiac muscle, and glands is the
   a. somatic nervous system.
   b. autonomic nervous system.
   c. skeletal division.
   d. sensory nervous system.

_____ 3. Neurons have cytoplasmic extensions that connect one neuron to another neuron. Given these structures:
   1. axon
   2. dendrite
   3. dendritic spine
   4. terminal bouton

   Choose the arrangement that lists the structures in the order they would be found at the synapse between two neurons.
   a. 1,4,2,3
   b. 1,4,3,2
   c. 4,1,2,3
   d. 4,1,3,2

4. A neuron with many short dendrites and a single long axon is a
   a. multipolar neuron.
   b. unipolar neuron.
   c. bipolar neuron.
   d. none of the above

5. Most sensory neurons are _____ neurons.
   a. unipolar
   b. bipolar
   c. multipolar
   d. efferent
   e. a and b

6. Cells found in the choroid plexuses that secrete cerebrospinal fluid are
   a. astrocytes.
   b. microglia.
   c. ependymal cells.
   d. oligodendrocytes.
   e. Schwann cells.

7. Neuroglia that act as phagocytes (are phagocytic) within the central nervous system are
   a. oligodendrocytes.
   b. microglia.
   c. ependymal cells.
   d. astrocytes.
   e. somas.

8. Axons within nerves may have which of these associated with them?
   a. neurolemmocytes
   b. nodes of Ranvier
   c. oligodendrocytes
   d. a and b
   e. all of the above

9. Action potentials are conducted more rapidly
   a. in small-diameter axons than in large-diameter axons.
   b. in unmyelinated axons than in myelinated axons.
   c. along axons that have nodes of Ranvier.
   d. all of the above

10. Clusters of nerve cell bodies within the peripheral nervous system are
    a. ganglia.
    b. fascicles.
    c. nuclei.
    d. laminae.

11. Gray matter contains primarily
    a. myelinated fibers.
    b. neuron cell bodies.
    c. neurolemmocytes.
    d. all of the above

12. Given these connective tissue structures:
    1. endoneurium
    2. epineurium
    3. perineurium

    Choose the arrangement that lists the structures in order from the outside of a nerve to the inside.
    a. 1,2,3
    b. 1,3,2
    c. 2,1,3
    d. 2,3,1
    e. 3,2,1

13. Neurotransmitter substances are stored in vesicles that are located primarily in specialized portions of the
    a. soma.
    b. axon.
    c. dendrite.
    d. none of the above

14. Within a synapse,
    a. synaptic vesicles are found in both the presynaptic and the postsynaptic membranes.
    b. action potentials are conducted only from presynaptic to postsynaptic membranes.
    c. action potentials are conducted only from postsynaptic to presynaptic membranes.
    d. a and b

_____15. Given these list of information concerning synaptic transmission:
1. presynaptic action potentials
2. local depolarization
3. release of neurotransmitter
4. either metabolism or reabsorption of neurotransmitter
5. combination of neurotransmitter with receptors
6. postsynaptic action potentials

Choose the sequence listed below that best represents the order in which the events occur:
a. 1,3,2,4,5,6
b. 1,3,5,2,6,4
c. 4,3,5,2,1,6
d. 1,3,5,6,2,4
e. 1,3,2,5,6,4

_____16. An inhibitory neuron affects the neuron it synapses with by
a. producing an IPSP in the neuron.
b. hyperpolarizing the cell membrane of the neuron.
c. causing K^+ ions to diffuse out of the neuron.
d. causing Cl^- ions to diffuse into the neuron.
e. all of the above

_____17. Summation
a. is caused by a combining of several local potentials.
b. can occur when two action potentials arrive simultaneously at two different presynaptic terminals.
c. can occur when two action potentials arrive in very close succession at a single presynaptic terminal.
d. all of the above

_____18. Which of these structures is a component of the reflex arc?
a. afferent neuron
b. sensory receptor
c. efferent neuron
d. effector organ
e. all of the above

_____19. A convergent circuit
a. is a positive feedback system that produces many action potentials.
b. makes possible a reflex response and awareness of the reflex stimulus.
c. occurs when many neurons synapse with a smaller number of neurons.
d. all of the above

_____20. The output of a convergent circuit could be
a. an IPSP.
b. an EPSP.
c. an action potential.
d. no action potential.
e. all of the above

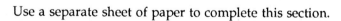

FINAL CHALLENGES

Use a separate sheet of paper to complete this section.

1. Describe the consequences of these situation in a neuronal circuit. Several neurons converge on a single neuron. One of the presynaptic neurons exhibits a constant frequency of action potentials and secretes a stimulatory neuromodulator. The other presynaptic neurons secrete stimulatory neurotransmitters. What effect does a drug that blocks the receptor molecules for the neuromodulator have on the response of the postsynaptic neuron to stimulation by the presynaptic neurons?

2. Given two series of neurons, explain why action potentials could be propagated along one series more rapidly than the other series.

3. In the diagram below, A and B represent different neuronal circuits. For each 10 times neuron 1 is stimulated, neuron 2 has 10 action potentials and neuron 3 has one action potential. Propose an explanation for the different number of action potentials produced in neurons 2 and 3.

4. The Hering-Breuer reflex limits overexpansion of the lungs during inspiration. Given that an oscillating circuit is responsible for the continual stimulation of respiratory muscles during inspiration and that stretch receptors in the lungs produce action potentials when the lungs are stretched, draw a simple neuronal circuit that would account for the Hering-Breuer reflex.

5. Nerve fibers that conduct the sensation of sharp, pricking, localized pain are myelinated fibers with a diameter of 2 to 6 micrometers. Nerve fibers that conduct the sensation of burning, aching, not well-localized pain are unmyelinated with a diameter of 0.5 to 2 micrometers. Describe the sequence of pain sensations produced when a painful stimulus is applied. Explain why there are two different systems for transmitting pain.

ANSWERS TO CHAPTER 12

Divisions of the Nervous System
A. 1. Central nervous system; 2. Central nervous system; 3. Peripheral nervous system; 4. Afferent division; 5. Somatic nervous system; 6. Autonomic nervous system; 7. Parasympathetic division
B. 1. Nerves; 2. Ganglia; 3. 12; 4. 31

Neurons
A. 1. Cell body (soma); 2. Dendrite; 3. Axon
B. 1. Chromatophilic substance (Nissl bodies); 2. Dendritic spines ; 3. Axon hillock; 4. Axoplasm; 5. Telodendria; 6. Terminal boutons (presynaptic terminals)
C. 1. Dendrite; 2. Neuron cell body (soma); 3. Axon hillock; 4. Axon; 5. Neurolemmocyte (Schwann cell); 6. Node of Ranvier; 7. Telodendria

Neuron Types
A. 1. Afferent (sensory) neurons; 2. Efferent (motor) neurons; 3. Association neurons (interneurons)
B. 1. Multipolar neuron; 2. Bipolar neuron; 3. Multipolar neuron

Neuroglia
1. Astrocytes; 2. Ependymal cells; 3. Microglia; 4. Oligodendrocytes; 5. Neurolemmocytes (Schwann cells); 6. Satellite cells

Axon Sheaths
1. Unmyelinated axons; 2. Myelinated axons; 3. Myelin sheath; 4. Nodes of Ranvier; 5. Internodes; 6. Saltatory conduction; 7. Rapidly; 8. Slowly

Organization of Nervous Tissue
A. 1. White matter; 2. Gray matter; 3. Nerve tract; 4. Cortex; 5. Nuclei; 6. Ganglia

B. 1. Epineurium; 2. Endoneurium; 3. Neurolemmocyte; 4. Axon; 5. Fascicle; 6. Perineurium; 7. Nerve

The Synapse
1. Presynaptic terminal; 2. Synaptic cleft; 3. Postsynaptic membrane; 4. Neurotransmitter; 5. Synaptic vesicle; 6. Acetylcholinesterase; 7. Neuromodulator

EPSPs and IPSPs
A. 1. Excitatory postsynaptic potential (EPSP); 2. Excitatory postsynaptic potential (EPSP); 3. Excitatory neuron; 4. Inhibitory postsynaptic potential (IPSP); 5. Inhibitory postsynaptic potential (IPSP)
B. 1. EPSP; 2. IPSP
C. 1. Axo-axonic; 2. Neuromodulators; 3. Presynaptic inhibition; 4. Presynaptic facilitation

Spatial and Temporal Summation
1. Local potentials; 2. Summation; 3. Threshold; 4. Spatial; 5. Temporal; 6. IPSPs

Reflexes
A. 1. Sensory receptors; 2. Afferent neurons; 3. Association neurons; 4. Efferent neurons; 5. Effector organs; 6. Reflex
B. 1. Association neuron; 2. Afferent neuron; 3. Efferent neuron; 4. Sensory receptor; 5. Effector organ
C. 1. Do not; 2. Do; 3. Do; 4. Do not

Neuronal Circuits
1. Convergent circuit; 2. Convergent circuit; 3. Divergent circuit; 4. Oscillating circuit; 5. After discharge; 6. Oscillating circuit

1. Somatic nervous system: innervates skeletal muscle, under voluntary control; autonomic nervous system: innervates cardiac muscle, smooth muscle, and glands, and is involuntary.
2. Multipolar neurons: most CNS neurons and motor neurons; bipolar neurons: rod and cone cells, olfactory receptors; unipolar neurons: most sensory neurons
3. Astrocytes: form a supporting matrix and help to regulate the composition of extracellular fluid around neurons in the CNS; microglia: macrophages that phagocytize foreign or necrotic tissue; ependymal cells: produce and circulate cerebrospinal fluid; neurolemmocytes: form myelin sheaths around axons in the PNS; oligodendrocytes: form myelin sheaths around axons in the CNS

4. Myelinated axon: wrapped by several layers of oligodendrocytes or neurolemmocytes, saltatory conduction (rapid); unmyelinated axon: rests in invagination of oligodendrocytes or neurolemmocytes, conducts action potentials more slowly

5. White matter: myelinated axons, propagates action potentials, forms nerve tracts in the CNS and nerves in the PNS; gray matter: neuron cell bodies and dendrites, site of integration (synapses), forms nuclei (in CNS) and ganglia (in PNS)
6. Endoneurium, perineurium, and epineurium
7. Presynaptic terminal, synaptic cleft, and postsynaptic membrane
8. EPSP: depolarization of postsynaptic membrane caused by increase in membrane permeability to sodium ions; IPSP: hyperpolarization of postsynaptic membrane caused by an increase in membrane permeability to chloride or potassium ions.
9. Spatial summation occurs when two or more presynaptic terminals simultaneously stimulate a postsynaptic membrane; temporal summation occurs when two or more action potentials arrive in succession at a single presynaptic neuron.
10. Sensory receptor, afferent neuron, association neuron, efferent neuron, and effector organ
11. Convergent, divergent, and oscillating circuits

MASTERY LEARNING ACTIVITY

1. A The nervous system is divided into the central nervous system (brain and spinal cord) and the peripheral nervous system, which includes the afferent (sensory) and efferent (motor) and autonomic divisions. The efferent division is divided into the somatic nervous system and the autonomic nervous system.

2. B The autonomic nervous system regulates the activities of smooth muscle, cardiac muscle, and glands. The somatic nervous system controls skeletal muscle.

3. B Axons extend toward other neurons, and the ends of the axons are the terminal boutons (presynaptic terminals). The terminal boutons are separated by the synaptic cleft from dendritic spines (postsynaptic membranes) on the dendrites of the neuron.

4. A Multipolar neurons have many dendrites and a single axon. Bipolar neurons have a single dendrite and a single axon. Unipolar neurons have a single axon and no dendrites.

5. E Efferent (away from central nervous system) neurons are multipolar motor neurons. Afferent (toward the central nervous system) neurons are sensory neurons. Most sensory neurons are unipolar, although a few are bipolar. Association neurons are multipolar and are found only within the central nervous system.

6. C Ependymal cells line the ventricles of the brain and central canal of the spinal cord. Within the choroid plexuses these cells are specialized to secrete cerebrospinal fluid. Astrocytes function as a supporting matrix and help to regulate the extracellular fluid around neurons in the CNS. Microglia are phagocytic cells in the CNS, and oligodendrocytes (CNS) and neurolemmocytes (PNS) form myelin sheaths around axons.

7. B Microglia are neuroglia that are phagocytic within the central nervous system.

8. D Nerves are part of the peripheral nervous system. Nerves contain axons that are surrounded by neurolemmocytes. In myelinated axons the neurolemmocytes cells give rise to the myelin sheath. The spaces between adjacent neurolemmocytes are called nodes of Ranvier. Oligodendrocytes are neuroglia cells found only within the central nervous system, where they form myelin sheaths around axons.

9. C Saltatory conduction from one node of Ranvier to the next node of Ranvier is the most rapid method of conducting action potentials. Myelinated axons are faster than unmyelinated axons, and large-diameter axons are faster than small-diameter axons.

10. A Clusters of nerve cell bodies within the peripheral nervous system are called ganglia. Clusters of nerve cell bodies within the central nervous system are called nuclei.

11. B Gray matter consists of unmyelinated neuron cell bodies, dendrites, and neuroglia cells. Myelinated neurons make up white matter. In the central nervous system myelin is produced by oligodendrocytes, and in the peripheral nervous system by neurolemmocytes.

12. D Nerves are surrounded by the epineurium. Nerve fascicles within the nerve are surrounded by the perineurium. Axons within the nerve fascicles are surrounded by the endoneurium.

13. B The presynaptic terminal that contains the neurotransmitter within synaptic vesicles is a specialized part of the axon. Release of neurotransmitter from only the axon ensures one-way transmission of nerve impulses.

236

14. B. Because only presynaptic terminals have synaptic vesicles with neurotransmitter, the action potential is always transferred from the presynaptic membrane to the postsynaptic membrane.

15. B. Synaptic transmission must involve (1) an action potential in the presynaptic neuron that (3) initiates the release of a neurotransmitter, which then diffuses across the synaptic cleft and (5) combines with receptors of the postsynaptic membrane. The combination of the receptor with the neurotransmitter initiates (2) a local depolarization, which (6) initiates postsynaptic action potentials if it is sufficiently large. Although metabolism or reabsorption of the neurotransmitter (4) probably begins as soon as the transmitter is released, it does not become a factor until after the postsynaptic response has occurred.

16. E. The inhibitory neuron releases a neurotransmitter that changes the permeability of the neuron with which it synapses to K^+ and Cl^- ions. Movement of K^+ ions out of the neuron and Cl^- ions into the neuron causes hyperpolarization of the neuron cell membrane. Because this increases the difference between the resting membrane potential and threshold, an IPSP (inhibitory postsynaptic potential) is produced.

17. D. Summation occurs when local potentials combine to reach threshold and produce an action potential. Spatial summation occurs when simultaneous stimulation from two or more presynaptic terminals cause the local potentials to combine. Temporal summation is caused by two or more successive stimulations from a single presynaptic terminal.

18. E. These components make up the reflex arc: sensory receptor, afferent neuron, association neuron (absent in some reflex arcs), efferent neuron, and effector organ.

19. C. A convergent circuit is the synapsing of many neurons with a smaller number of neurons. A divergent circuit occurs when a few neurons synapse with a larger number of neurons. Input to a reflex and the brain is an example. Oscillating circuits are positive-feedback systems that produce many action potentials.

20. E. The simplest example of a convergent circuit is two neurons synapsing with a third neuron. If one of the two were an inhibitory neuron, an IPSP would be produced. If one of the two were an excitatory neuron, an EPSP would be produced. If both neurons were simultaneously stimulated, there could be two possible outcomes. If the IPSP effects dominate, the third neuron's membrane would hyperpolarize and there would be no action potential. If the EPSP effects dominate, the third neuron's membrane would depolarize to threshold level and an action potential would result.

☆ **FINAL CHALLENGES** ☆

1. The continual release of the stimulatory neuromodulator makes the postsynaptic neuron more sensitive to the neurotransmitter. Blocking the receptors for the neuromodulator would result in less sensitivity to the neurotransmitter, making it more difficult to produce an action potential.

2. If one series of neurons had more neurons, it would have more synapses, which would slow down the rate of action potential propagation. Or, if one series were unmyelinated and the other myelinated, or if one series had smaller diameter axons, the rate would also be slower.

3. Suppose that the neurons in circuit B require temporal summation. Thus, for an action potential to be passed from neuron to neuron in circuit B, several action potentials would be necessary to cause the release of adequate amounts of neurotransmitter. The requirement for temporal summation could result from small

amounts of neurotransmitter release, reduced numbers of neurotransmitter receptors, or neuromodulators in circuit B.

4. The oscillating circuit diagram below provides continual stimulation of the respiratory muscles until it is inhibited by an inhibitory neuron activated when the lungs overstretch.

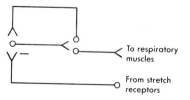

5. The large-diameter, myelinated axons transmit action potentials more rapidly (15 to 120 m/second) than the small-diameter, unmyelinated fibers (2 m/second or less) Therefore a painful stimulus produces two kinds of sensation: a sharp, localized, pricking pain followed by a burning, aching pain that is not as well localized. The fast-conducting pain fibers allow us to be quickly aware of pain and its source and to respond quickly to new pain stimuli. For example, the hand can be rapidly removed from a hot object. However, the fast-conducting pain fibers take up extra space because of their large diameter and associated myelin sheath. The slow-conducting fibers take up less space and allow us to be aware of chronic pain for which a quick response is not necessary.

13 Central Nervous System: Brain and Spinal Cord

FOCUS: The central nervous system consists of the brain and spinal cord. The brain can be subdivided into the brainstem (medulla oblongata, pons, and midbrain), diencephalon (thalamus, hypothalamus, and others), cerebrum, and cerebellum. The brainstem regulates circulatory and respiratory functions, contains the nuclei for most of the cranial nerves, and functions as a tract system to connect the spinal cord to different areas of the brain. The diencephalon functions as a relay center between the brainstem and cerebrum, controls the activities of the pituitary gland, and regulates autonomic functions. The cerebrum is the site of conscious sensation, thoughts, and control of body movements, whereas the cerebellum is involved with balance and coordinated movements. The brain contains hollow spaces, the ventricles, which are filled with cerebrospinal fluid. The cerebrospinal fluid circulates and enters the meninges, membranes that surround and protect the brain. The spinal cord consists of neuron cell bodies (gray matter organized into horns) and ascending and descending nerve tracts formed by nerve cell axons (white matter). Each tract carries specific information such as fine touch, pain and temperature, and voluntary motor activity. Reflexes are stereotyped responses to stimuli that are integrated within the spinal cord and brain. Important reflexes include the stretch reflex (stretching a muscle results in muscle contraction), the Golgi tendon reflex (excessive tension in a muscle causes the muscle to relax), and the withdrawal reflex (a limb is removed from a painful stimulus).

CONTENT LEARNING ACTIVITY

Development

"The CNS develops on the upper surface of the embryo.**"**

A. Match these terms with the correct statement or definition:

Central canal   Neural tube
Neural crest cells   Ventricles
Neural plate

_____ 1. Flat plate of tissue on the upper surface of the embryo.

_____ 2. Structure formed when lateral sides of the neural plate are elevated as waves, move toward each other, and fuse.

_____ 3. Separate from neural waves and give rise to part of the PNS.

_____ 4. Center of the spinal cord continuous with the ventricles of the brain.

B. Match these terms with the correct statement or definition:

Diencephalon   Myelencephalon
Mesencephalon   Telencephalon
Metencephalon

_____ 1. Develops into the cerebrum in the adult.

_____ 2. Develops, in part, into the thalamus in the adult.

_____ 3. Becomes the pons and cerebellum in the adult.

_____ 4. Becomes the medulla oblongata in the adult.

Brainstem

" The brainstem connects the spinal cord to the remainder of the brain "
and is responsible for many essential functions.

A. Match the structure with the part of the brainstem in which it is located:

Medulla oblongata
Midbrain
Pons

_____ 1. Pyramids.

_____ 2. Heart rate centers.

_____ 3. Swallowing center.

_____ 4. Respiratory centers.

_____

_____ 5. Colliculi.

_____ 6. Cerebral peduncles.

The reticular formation is a diffuse network of neurons scattered throughout the brainstem.

B. Match these parts of the brain with the correct function:

Cerebral peduncles   Red nuclei
Colliculi   Reticular formation
Olives   Substantia nigra
Pyramids

_____ 1. Skeletal muscle nerve tracts that descend and decussate.

_____ 2. Balance, coordination, and modulation of sound.

_____ 3. Involved with visual reflexes and auditory pathways.

_____ 4. Relay information from the cerebrum to the spinal cord; a major CNS motor pathway.

_____ 5. Maintaining consciousness and arousal.

C. Match these terms with the correct parts of the diagram labeled in Figure 13-1:

Cerebral peduncles
Inferior colliculus
Medulla oblongata
Midbrain
Pineal body
Pons
Spinal cord
Superior colliculus
Thalamus

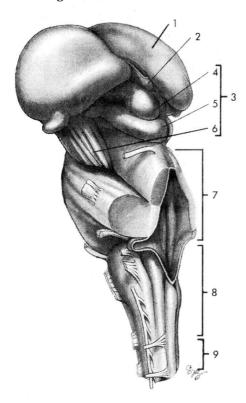

Figure 13-1

1. _____

2. _____

3. _____

4. _____

5. _____

6. _____

7. _____

8. _____

9. _____

Diencephalon

66*The diencephalon is the part of the brain between the brainstem and the cerebrum.*99

A. Match these parts of the diencephalon with the correct descriptions or functions:

Epithalamus   Subthalamus
Hypothalamus   Thalamus

_____   1. Most sensory input projects to this point.

_____   2. Involved in controlling motor functions.

_____   3. Contains the pineal body, which plays a role in the onset of puberty.

_____   4. Contains the mamillary bodies, which are involved in olfactory reflexes.

_____   5. Regulates the secretion of hormones by the pituitary gland.

_____   6. Involved in control of many autonomic functions such as movement of food through the digestive tract.

_____   7. Contains hunger and thirst control centers.

B. Match these terms with the correct parts of the diagram labeled in Figure 13-2:

Corpus callosum   Pineal body
Hypothalamus   Pituitary gland
Intermediate mass   Pons
Mamillary body   Subthalamus
Midbrain   Thalamus
Optic chiasma

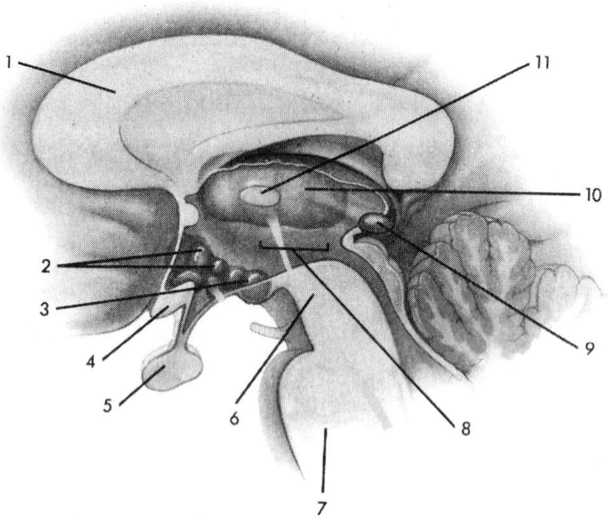

1. _____

2. _____

3. _____

4. _____

5. _____

6. _____

7. _____

8. _____

9. _____

10. _____

11. _____

Figure 13-2

Cerebrum

66*The cerebrum is the largest portion of the brain.*99

A. Match these terms with the correct statement or definition:

Central sulcus   Longitudinal fissure
Gyri   Sulci
Lateral fissure

_____ 1. Deep groove that separates the right and left cerebral hemispheres.

_____ 2. Raised folds or ridges on the surface of the cerebrum.

_____ 3. Groove that separates the frontal and parietal lobes.

_____ 4. Deep groove that separates the temporal lobe from the rest of the cerebrum.

B. Match these lobes of the cerebrum with the correct primary function:

Frontal lobe   Parietal lobe
Occipital lobe   Temporal lobe

_____ 1. Voluntary motor function, motivation, aggression, and mood.

_____ 2. Evaluation of most sensory input (excluding smell, hearing and vision).

_____ 3. Reception and integration of visual input.

_____ 4. Reception and integration of olfactory and auditory input.

_____ 5. Abstract thought and judgment ("psychic cortex").

C. Match these terms with the correct statement or definition:

Association fibers   Projection fibers
Commissural fibers   White matter
Gray matter

_____ 1. Nuclei and the cortex of the cerebrum are composed of this.

_____ 2. Nerve tracts of the cerebrum between the cortex and nuclei (the cerebral medulla) are composed of this.

_____ 3. Nerve tracts that connect areas of the cerebral cortex within the same hemisphere.

_____ 4. Nerve tracts that connect one cerebral hemisphere to the other cerebral hemisphere.

_____ 5. Nerve tracts that connect the cerebrum with other parts of the brain and spinal cord.

Cerebral Cortex

"The gray matter on the outer surface of the cerebrum is the cortex."

A. Match these terms with the correct statement or definition:

Association areas   Primary somesthetic area
Primary sensory areas   Projection

_____ 1. General term for regions of the cerebral cortex where sensory sensations are perceived.

_____ 2. Area that receives pain, pressure, and temperature input; located in the postcentral gyrus.

_____ 3. Conscious awareness of sensation on some part of the body.

_____ 4. Cortical area that is involved in recognition of sensations and is immediately adjacent to the primary sensory area.

_____ 5. Present sensory input is compared to past sensory input in this area.

The primary somesthetic area is organized topographically with the superior part of the body represented on the inferior part of the cortex. Further, areas of the body with many sensory receptors are represented by larger areas of the cortex than are areas of the body with fewer sensory receptors.

B. Match these areas of the cerebral cortex with the correct description:

Prefrontal area   Primary motor cortex
Premotor area

_____ 1. Located in the precentral gyrus.

_____ 2. Organized topographically similar to the primary somesthetic area.

_____ 3. Staging area where motor functions are organized before they are actually initiated.

_____ 4. Motivation to plan and initiate movements occur here.

Speech

"Many separate processes are necessary to produce speech.**"**

Match these areas of the cerebral cortex with their correct role in speech production.

Broca's area   Primary motor cortex
Premotor area   Wernicke's area

_____ 1. Association area necessary for understanding and formulating coherent speech.

_____ 2. Receives input from Wernicke's area and initiates the movements necessary for speech; also known as the motor speech area.

_____ 3. Broca's area sends action potentials here.

_____ 4. Activation of this area is the last step in speech production.

Brain Waves

"Simultaneous action potentials in large numbers of neurons in the brain can be recorded**"** from electrodes placed on the scalp.

Match these brain waves with the activity that produces them:

Alpha wave   Delta wave
Beta wave   Theta wave

_____ 1. Quiet resting state with eyes closed.

_____ 2. Intense mental activity.

_____ 3. Deep sleep.

_____ 4. In children or frustrated adults.

Memory

66*Memory can be classified in several ways.*99

Using the terms provided, complete these statements:

Calpain
Cerebellum and premotor area
Declarative memory
Hippocampus and amygdala
Long-term memory

Long-term potentiation
Memory engrams
Procedural memory
Sensory memory
Short-term memory

1. _____
2. _____
3. _____
4. _____
5. _____
6. _____
7. _____
8. _____
9. _____
10. _____

_(1)_ is the very short-term retention of sensory input while something is scanned, evaluated, and acted upon. However, data may be considered valuable enough to move into _(2)_, where information is retained for a few seconds to a few minutes. Certain pieces of information are transferred into _(3)_, which may involve a physical change in neuron shape. This process is called _(4)_. During this process calcium influx into the cell activates an enzyme called _(5)_, which aids in the change of dendrite shape. A whole series of neurons, called _(6)_, are probably involved in the long-term retention of a given piece of information. There are two types of long-term memory. _(7)_ involves the retention of names, dates, and places, whereas _(8)_ involves the learning of a skill such as riding a bicycle. Declarative memory is stored in the _(9)_, whereas procedural memory is stored in the _(10)_.

Right and Left Cortex

66*Language and perhaps some other functions are not shared equally between the two hemispheres.*99

Match these terms with the correct statement or definition:

Corpus callosum
Left cerebral hemisphere

Right cerebral hemisphere

_____

_____

_____

_____

1. Controls muscular activity in and receives sensory input from the right half of the body.

2. In most people the analytical hemisphere involved in mathematics and speech.

3. In most people the hemisphere involved in spatial perception and musical ability.

4. The largest commissure between the right and left hemispheres.

Basal Ganglia

❝_The basal ganglia are a group of functionally related nuclei in the inferior cerebrum,_**❞** _diencephalon, and midbrain._

Match these terms with the correct statement or definition:

Caudate nucleus	Substantia nigra
Corpus striatum	Subthalamic nucleus
Lentiform nucleus	

_____ 1. Located in the diencephalon.

_____ 2. Located in the midbrain.

_____ 3. Collectively, the nuclei in the cerebrum.

_____ 4. Two nuclei that are part of the corpus striatum.

_____

☞ The basal ganglia are involved in coordinating motor movements and posture.

Limbic System

❝_Portions of the cerebrum and diencephalon are grouped together under the title limbic system._**❞**

Using the terms provided, complete these statements:

Basal ganglia   Hypothalamus
Fornix   Olfactory cortex
Habenular nuclei   Olfactory nerves
Hippocampus

Structurally, the limbic system consists of many parts. These parts consist of certain cerebral cortical areas, including the cingulate gyrus and _(1)_ ; various nuclei, including the _(2)_ in the epithalamus; parts of the _(3)_ ; the _(4)_ , especially the mamillary bodies; the _(5)_ , associated with the sense of smell; and tracts such as the _(6)_ connecting these areas. One of the major sources of sensory input into the limbic system is the _(7)_ .

1. _____

2. _____

3. _____

4. _____

5. _____

6. _____

7. _____

☞ Emotions, the visceral responses to these emotions, motivation, mood, and sensations of pain and pleasure are all influenced by the limbic system.

Cerebellum

❝ *The major function of the cerebellum is motor coordination.* **❞**

A. Match these terms with the correct statement or definition:

Cerebellar peduncles   Lateral hemispheres
Flocculonodular lobe   Vermis

_____ 1. Nerve tracts that communicate with other areas of the CNS.

_____ 2. Small, inferior portion of the cerebellum involved in balance.

_____ 3. Narrow central portion of the cerebellum; involved with gross motor coordination.

_____ 4. Paired structures involved in fine motor coordination.

☞ The cerebellum is organized like the cerebrum with gray matter on the inside as nuclei and on the outside as cortex.

B. Using the terms provided, complete these statements:

Cerebellum   Motor cortex
Comparator   Proprioceptors

Action potentials that initiate voluntary movements are sent from the motor cortex to skeletal muscles and to the __(1)__ . At the same time, __(2)__ send information about body position to the cerebellum. If a difference is detected between the intended movement and the actual movement, action potentials are sent from the cerebellum to the __(3)__ and spinal cord to correct the difference. Thus a major function of the cerebellum is that of a __(4)__ , resulting in smooth, coordinated movements.

1. _____

2. _____

3. _____

4. _____

Spinal Cord

❝ *The spinal cord is extremely important to the overall function of the nervous system.* **❞**

A. Match these terms with the correct statement or definition:

Cauda equina   Lumbar enlargement
Cervical enlargement Filum terminale
Conus medullaris

_____ 1. Region where nerves that supply the upper limbs enter and exit the spinal cord.

_____ 2. Inferior end of the spinal cord (extends to L2).

_____ 3. Connective tissue filament that anchors the inferior end of the spinal cord to the coccyx.

_____ 4. Conus medullaris and numerous nerves extending inferiorly from it.

247

B. Match these terms with the correct statement or definition:

Anterior (ventral) horns   Gray and white commissures
Dorsal root   Lateral horn
Dorsal root ganglion   Posterior (dorsal) horn
Funiculi   Ventral root

_____ 1. Columns of white matter; consist of fasciculi, or tracts.

_____ 2. Many sensory neurons synapse with neurons in this structure.

_____ 3. Contains the cell bodies of autonomic neurons.

_____ 4. Allows communication between the halves of the spinal cord.

_____ 5. Conveys efferent nerve processes away from the cord.

_____ 6. Contains the cell bodies of sensory neurons.

C. Match these terms with the correct parts in Figure 13-3:

Anterior (ventral) horn   Lateral horn
Dorsal root   Posterior (dorsal) horn
Dorsal root ganglion   Ventral root
Funiculi

1. _____

2. _____

3. _____

4. _____

5. _____

6. _____

7. _____

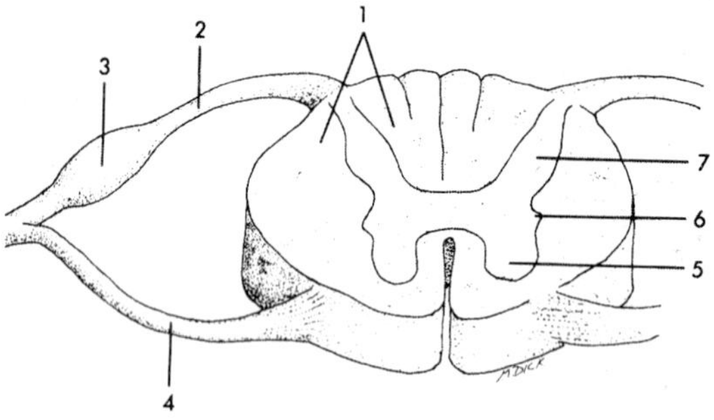

Figure 13-3

Spinal Reflexes

" Automatic reactions to stimuli that occur without conscious thought are reflexes. "

A. Match these terms with the correct description or function:

Crossed extensor reflex   Stretch reflex
Golgi tendon reflex   Withdrawal reflex
Reciprocal innervation

_____ 1. Stretch of a muscle results in contraction of the muscle.

_____ 2. Excess tension causes the muscle to relax.

_____ 3. Removes a limb from a painful stimulus.

_____ 4. Causes the antagonist of a muscle to relax.

_____ 5. Prevents falling when the withdrawal reflex occurs in one leg.

B. Match these terms with the
   correct statement or definition:

Alpha motor neuron   Golgi tendon organ
Gamma motor neuron   Muscle spindle

_____ 1. Sensory receptor for the stretch reflex.

_____ 2. These innervate the muscle in which a muscle spindle is
                            embedded.

_____ 3. Action potentials carried by these cause muscle spindles to
                            contract.

_____ 4. Stimulation of these causes inhibition of alpha motor neurons.

Ascending Spinal Pathways

"*There are several major ascending tracts in the spinal cord.*"

A. Match these spinal pathways
   with the correct function:

Anterior spinothalamic   Spinocerebellar
Dorsal column-medial lemniscal   Trigeminothalamic
Lateral spinothalamic

_____ 1. Pain and temperature from most of the body.

_____ 2. Pain and temperature from the face and teeth.

_____ 3. Light touch, pressure, tickle, and itch.

_____ 4. Two-point discrimination (fine touch), proprioception,
                            pressure, and vibration.

_____ 5. Proprioception to the cerebellum (two).

_____

☞ The dorsal column-medial lemniscal system can be divided into the fasciculus gracilis
   (inferior to midthorax) and the fasciculus cuneatus (superior to midthorax).

B. Match the nerve tract with the
   location where it crosses over
   (decussates) from one side of
   the body to the other:

Cerebellum
Medulla oblongata
Spinal cord
Uncrossed

_____ 1. Lateral spinothalamic tract.

_____ 2. Anterior spinothalamic tract.

_____ 3. Dorsal column-medial lemniscal system.

_____ 4. Posterior spinocerebellar tract.

_____ 5. Anterior spinocerebellar tract (crosses twice).

_____

C. Match the type of neuron with
   the correct location of the
   neuron cell body:

Primary neuron
Secondary neuron
Tertiary neuron

_____   1. Dorsal root ganglion.

_____   2. Spinal cord.

_____   3. Medulla oblongata.

_____   4. Thalamus.

Descending Spinal Pathways

66*Most of the descending pathways are involved in the control of motor functions.***99**

A. Match these terms with the
   correct statement or definition:

Extrapyramidal system
Pyramidal system

_____   1. Involved in muscle tone and fine movements such as dexterity.

_____   2. Involved in less precise control of motor functions and with
                                 overall body coordination and cerebellar function.

_____   3. Includes the corticospinal and corticobulbar tracts.

_____   4. Includes the rubrospinal, vestibulospinal, and reticulospinal
                                 tracts.

B. Match the nerve tract with
   the correct function:

Corticobulbar tract   Rubrospinal tract
Corticospinal tract   Vestibulospinal tract
Reticulospinal tract

_____   1. Primarily involved with control of upper limb movement.

_____   2. Control of head and neck movement.

_____   3. Cerebellar-like function in control of distal hand and arm
                                 movements.

_____   4. Primarily involved in maintenance of upright posture.

☞ The descending pathways reduce the transmission of pain action potentials by secreting
   endorphins (natural analgesics).

C. Match the nerve tract with the
   location where it decussates
   from one side to the other:

Medulla oblongata
Spinal cord

_____   1. Lateral corticospinal tract.

_____   2. Anterior corticospinal tract.

D. Match the type of neuron with the correct location of the neuron cell body:

Lower motor neuron
Upper motor neuron

_____ 1. Cerebral cortex, cerebellum, and brainstem.

_____ 2. Spinal cord.

_____ 3. Cranial nerve nuclei.

Mentinges

66_Three connective tissue layers surround and protect the spinal cord._**99**

A. Match these terms with the correct statement or definition:

Arachnoid layer   Pia mater
Denticulate ligaments   Subarachnoid space
Dura mater   Subdural space
Epidural space

_____ 1. Most superficial and thickest meningeal layer; continuous with the periosteum of the cranial vault.

_____ 2. Divides to form the dural sinuses.

_____ 3. Space that separates the dura mater from the periosteum of the vertebral canal.

_____ 4. Middle, wispy meningeal layer.

_____ 5. Space between the dura mater and the arachnoid layer.

_____ 6. Space between the arachnoid layer and the pia mater.

_____ 7. Connective tissue strands that hold the spinal cord in place.

B. Match these terms with the correct part of the diagram labeled in Figure 13-4:

Arachnoid granulation
Arachnoid layer
Dura mater

Pia mater
Subarachnoid space
Superior sagittal sinus
(dural sinus)

1. _____

2. _____

3. _____

4. _____

5. _____

6. _____

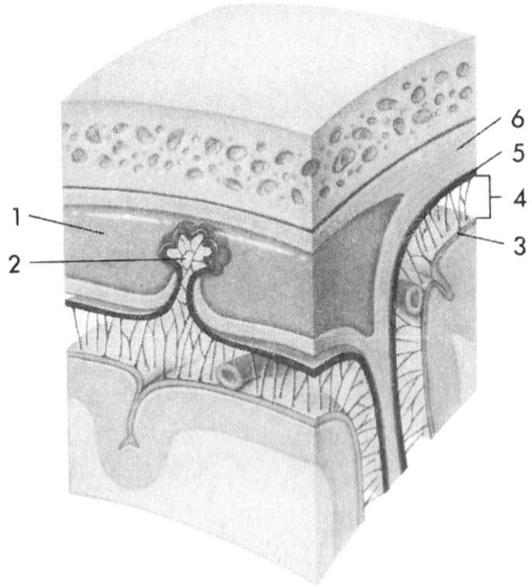

Figure 13-4

Ventricles

“The CNS is a hollow tube that may be quite reduced in some areas and expanded in others.**”**

A. Match these terms with the correct statement or definition:

Central canal
Cerebral aqueduct
Fourth ventricle
Interventricular foramina

Lateral ventricle
Septa pellucida
Third ventricle

_____ 1. Separates the lateral ventricles from each other.

_____ 2. Midline cavity located between the lobes of the thalamus.

_____ 3. Connect the lateral ventricles to the third ventricle.

_____ 4. Cavity located in the inferior pontine and superior medullary regions.

_____ 5. Connects the third and fourth ventricles.

_____ 6. Continuation of the fourth ventricle into the spinal cord.

B. Match these terms with the correct parts of the diagram labeled in Figure 13-5:

Arachnoid granulation
Central canal
Cerebral aqueduct
Choroid plexus
Fourth ventricle

Interventricular foramen
Lateral ventricles
Subarachnoid space
Superior sagittal sinus
Third ventricle

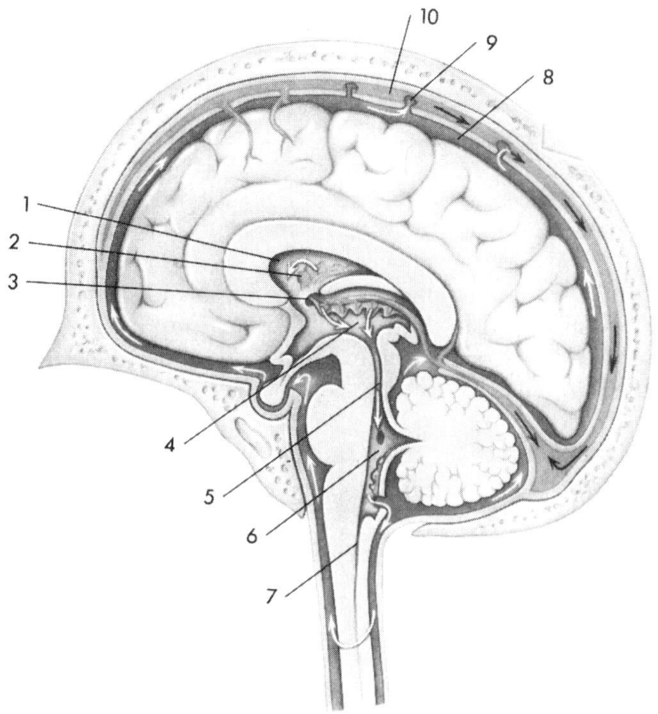

Figure 13-5

1. _____
2. _____
3. _____
4. _____
5. _____
6. _____
7. _____
8. _____
9. _____
10. _____

Cerebrospinal Fluid

66*Cerebrospinal fluid provides a protective cushion around the CNS*99
and provides some nutrients to CNS tissue.

Match these terms with the correct statement or definition:

Arachnoid granulations
Cerebrospinal fluid
Choroid plexuses

Dural sinuses
Lateral and median foramina
Subarachnoid space

_____ 1. Fills ventricles and subarachnoid spaces.

_____ 2. Site of production of cerebrospinal fluid.

_____ 3. Cerebrospinal fluid from the fourth ventricle enters this.

_____ 4. Passageways from the fourth ventricle into the subarachnoid space.

_____ 5. Point where cerebrospinal fluid reenters the bloodstream.

_____ 6. Cerebrospinal fluid enters the dural sinuses through these.

Clinical Applications

"There are many different causes of central nervous system disorders."

Match these terms with the correct statement or definition:

Dyskinesias   Multiple sclerosis
Encephalitis   Rabies
Hydrocephalus   Reye's syndrome
Meningitis   Stroke

_____ 1. An accumulation of cerebrospinal fluid within the ventricles or subarachnoid space.

_____ 2. Includes cerebral palsy, Huntington's chorea, and Parkinson's disease.

_____ 3. Results from demyelination of the brain and spinal cord.

_____ 4. May develop in children following an influenza or chickenpox infection.

_____ 5. Inflammation of the brain caused by a virus, or sometimes bacteria.

_____ 6. Hemorrhage, thrombosis, embolism, or spasm of cerebral blood vessels that result in an infarct.

QUICK RECALL

1. Complete the following table:

STRUCTURE	FUNCTION
Cerebral cortex	
Cerebral medulla	
Basal ganglia	
Thalamus	
Hypothalamus	
Limbic system	
Midbrain	
Pons	
Medulla oblongata	
Cerebellum	

2. Name the three parts of the brainstem.

3. List four major parts of the diencephalon.

4. List the four largest lobes of the cerebrum, and list their major functions.

5. List the three types of nerve tracts found in the cerebral medulla.

6. List five major reflexes.

7. Complete the following table for Ascending Spinal Pathways:

SPINAL PATHWAY	FUNCTION
Spinothalamic system	
Dorsal column-medial lemniscal system	
Spinocerebellar system	

8. Distinguish between the pyramidal and extrapyramidal systems.

9. List the three meninges and three spaces associated with them.

10. List the four ventricles of the brain and the openings that connect them to each other.

MASTERY LEARNING ACTIVITY

Place the letter corresponding to the correct answer in the space provided.

_____ 1. If a section was made that separated the brainstem from the rest of the brain, the cut would be made between the
   a. medulla and pons.
   b. pons and midbrain.
   c. midbrain and diencephalon.
   d. thalamus and cerebrum.

_____ 2. Important centers for vasoconstriction and cardiac control are located in the
   a. cerebrum
   b. medulla.
   c. cerebellum.
   d. basal ganglia.

_____ 3. Important respiratory centers are located in the
   a. cerebrum.
   b. cerebellum.
   c. pons and medulla.
   d. sacral plexus.

_____ 4. Our conscious state is maintained by activity generated in the
   a. cerebellum.
   b. reticular formation.
   c. limbic system.
   d. medulla.

_____ 5. The major relay station for action potentials going to and from the cerebral cortex is the
   a. hypothalamus.
   b. thalamus.
   c. pons.
   d. cerebellum.
   e. cranial nerves.

_____ 6. The furrows on the surface of the cerebrum are called the
   a. gyri.
   b. sulci.
   c. commissures.
   d. tracts.
   e. none of the above

_____ 7. A cutaneous nerve to the hand is severed at the elbow. The proximal end of the nerve at the elbow is then stimulated. The subject reports
   a. no sensation because the receptors are gone.
   b. a sensation limited to the region of the elbow.
   c. a sensation localized in region of the hand.
   d. a vague sensation on the side of the body containing the cut nerve.

_____ 8. General sensory inputs (pain, touch, temperature) to the cerebrum end in the
   a. precentral gyrus.
   b. postcentral gyrus.
   c. central sulcus.
   d. arachnoid.

_____ 9. Concerning the visual centers of the brain,
   a. they are located in the parietal lobes of the cerebrum.
   b. the primary visual cortices relate past to present visual experiences and evaluate what is being seen.
   c. the visual association areas are the areas where the image is "seen."
   d. none of the above

_____ 10. Given the following areas of the cerebrum:
   1. superior part of the primary motor cortex
   2. inferior part of the primary motor cortex
   3. prefrontal area
   4. premotor area

Suppose Harry Wolf wanted to flirt with a girl by winking and smiling. In what order would the areas be used?
a. 4,1
b. 3,4,1
c. 3,4,2
d. 4,1,2
e. 4,1,2,3

_____11. Given the following areas of the cerebral cortex:
1. Broca's area
2. premotor area
3. primary motor cortex
4. Wernicke's area

If a person hears and understands a word and then says the word out loud, in what order would the areas be used?
a. 1,4,2,3
b. 1,4,3,2
c. 4,1,2,3
d. 4,1,3,2

_____12. During meditation an adult would generate
a. alpha waves.
b. gamma waves.
c. delta waves.
d. theta waves.

_____13. Long-term memory involves
a. a change in the shape of neurons.
b. movement of calcium into the neuron.
c. activation of the enzyme calpain.
d. all of the above

_____14. The main connection between the left and right hemispheres of the cerebrum is the
a. intermediate mass.
b. corpus callosum.
c. vermis.
d. unmyelinated nuclei.

_____15. Concerning the basal ganglia,
a. the corpus striatum is an example.
b. they are gray matter located in the cerebrum, diencephalon, and midbrain.
c. they can act to inhibit muscle movement.
d. all of the above

_____16. A football player exhibited the following symptoms after a third-quarter play in which he was injured:
1. uncontrolled rhythmic contraction of skeletal muscles
2. abnormally great tension in muscles

The injury probably involved what area of the brain?
a. medulla oblongata
b. basal ganglia
c. cerebral cortex
d. cerebellum
e. hypothalamus

_____17. The limbic system is involved in the control of
a. sleep and wakefulness.
b. maintaining posture.
c. higher intellectual processes.
d. emotion.

_____18. A nurse is caring for a patient who exhibits the following symptoms:
1. inability to stand and walk normally
2. normal intelligence
3. capable of performing voluntary movements although the movements are not smooth and precise
4. normal tension in skeletal muscles
5. no obvious palsy

The patient is probably suffering from a condition that affected the
a. cerebral ganglia.
b. cerebellum.
c. cerebral cortex.
d. medulla oblongata.
e. pons.

_____19. The spinal cord extends from the
a. medulla to coccyx.
b. level of the third cervical vertebra to coccyx.
c. level of the axis to the lowest lumbar vertebra.
d. medulla to level of the second lumbar vertebra.
e. axis to sacral vertebra 5.

20. All sensory neurons entering the spinal cord
   a. enter through the dorsal horn.
   b. have their cell bodies in the dorsal root ganglia.
   c. are part of a spinal nerve.
   d. all of the above

21. Which of the following events occur when a person steps on a tack with their right foot?
   a. The right foot is pulled away from the tack because of the Golgi tendon reflex.
   b. The left leg is extended to support the body because of the stretch reflex.
   c. The flexor muscles of the thigh contract, and the extensor muscles of the thigh relax because of reciprocal innervation.
   d. all of the above

22. Several of the events that occurred between the time that a physician struck a patient's patellar ligament with a rubber hammer and the time his quadriceps femoris contracted (knee jerk reflex) are listed below:
   1. increased frequency of action potentials in the afferent neurons
   2. stretch of the muscle spindles
   3. increased frequency of action potentials in the alpha motor neurons
   4. stretch of the quadriceps femoris
   5. contraction of the quadriceps femoris

   Which of the following lists most closely describes the sequence of events as they normally occur?
   a. 4,1,2,3,5
   b. 4,1,3,2,5
   c. 1,4,3,2,5
   d. 4,2,1,3,5
   e. 4,2,3,1,5

23. The extrapyramidal system is mainly concerned with
   a. skilled, learned movements.
   b. water balance.
   c. gross movements and posture.
   d. transmission of motor signals from the medulla to the cerebrum.

24. Most fibers of the pyramidal system
   a. decussate in the medulla.
   b. synapse in the pons.
   c. descend in the rubrospinal tract.
   d. do not decussate in the brain.
   e. a and c

25. If one severed the lateral spinothalamic tract on the right side of the spinal cord,
   a. pain sensations below the damaged area on the right side would be eliminated.
   b. pain sensations below the damaged area on the left side would be eliminated.
   c. temperature sensation would be unaffected.
   d. none of the above

26. A person with a spinal cord injury is suffering from paresis (partial paralysis) in her right lower limb. Which of the following spinal pathways is probably involved?
   a. left lateral corticospinal tract
   b. right lateral corticospinal tract
   c. left dorsal column-medial lemniscal system
   d. right dorsal column-medial lemniscal system

27. Given the following symptoms:
   1. loss of sensations from the left upper limb
   2. loss of sensations from the right upper limb
   3. ataxia (failure of motor coordination) in the left upper limb
   4. ataxia (failure of motor coordination) in the right upper limb

   Which of the symptoms is (are) consistent with damage to the thalamus on the right side?
   a. 1
   b. 4
   c. 1,3
   d. 1,4
   e. 2,4

_____28. The most superficial of the meninges is a thick, tough membrane called the
a. pia mater.
b. arachnoid.
c. dura mater.
d. epidural mater.

_____29. The ventricles of the brain are interconnected. Which of the following ventricles are NOT correctly matched with the structures that connect them?
a. lateral ventricle to the third ventricle through the interventricular foramina
b. left lateral ventricle to right lateral ventricle through the central canal
c. third ventricle to fourth ventricle through the cerebral aqueduct
d. fourth ventricle to subarachnoid space through the median and lateral foramina

_____30. Cerebrospinal fluid is produced by the _____, circulates through the ventricles, and enters the subarachnoid space. The cerebrospinal fluid leaves the subarachnoid space through the _____.
a. choroid plexus, arachnoid granulations
b. arachnoid granulations, choroid plexus
c. dural sinuses, dura mater
d. dura mater, dural sinuses

FINAL CHALLENGES

Use a separate sheet of paper to complete this section.

1. How could one distinguish between damage to the hypothalamus and damage to the medulla?

2. Although alcohol has effects on other areas of the brain, it has a considerable effect on cerebellar function. What kinds of motor tests would reveal the drunken condition?

3. Would a patient with Parkinson's disease be expected to have reduced (hyporeflexive) or exaggerated (hyperreflexive) reflexes? Explain.

4. In pneumoencephalography the ventricles are partially filled with air and X-rays are taken. Abnormalities in the shape of the ventricles can be used to locate tumors or areas of atrophy. One way this procedure is accomplished is by lumbar puncture. If the patient is sitting up with the back and neck flexed when the air is injected, trace the route taken by an air bubble into the ventricles. If the patient lies supine after the air is injected, what part of the ventricles would fill with air?

5. A patient suffered a small lesion in the center of the spinal cord at the level of the sixth thoracic vertebra. What symptoms would it be possible to observe?

6. A patient suffered a loss of fine touch and pressure on the right side of the body. Voluntary movement of muscles was not affected, and pain and thermal sensations were normal. Is it possible to conclude that the right side of the spinal cord was damaged?

ANSWERS TO CHAPTER 13

Development
- A. 1. Neural plate; 2. Neural tube; 3. Neural crest cells; 4. Central canal
- B. 1. Telencephalon; 2. Diencephalon; 3. Metencephalon; 4. Myelencephalon

Brainstem
- A. 1. Medulla oblongata; 2. Medulla oblongata; 3. Medulla oblongata; 4. Pons and medulla oblongata; 5. Midbrain; 6. Midbrain
- B. 1. Pyramids; 2. Olives; 3. Colliculi; 4. Cerebral peduncles; 5. Reticular formation
- C. 1. Thalamus; 2. Pineal body; 3. Midbrain; 4. Superior colliculus; 5. Inferior colliculus; 6. Cerebral peduncles; 7. Pons; 8. Medulla oblongata; 9. Spinal cord

Diencephalon
- A. 1. Thalamus; 2. Subthalamus; 3. Epithalamus; 4. Hypothalamus; 5. Hypothalamus; 6. Hypothalamus; 7. Hypothalamus
- B. 1. Corpus callosum; 2. Hypothalamus; 3. Mamillary body; 4. Optic chiasma; 5. Pituitary gland; 6. Midbrain; 7. Pons; 8. Subthalamus; 9. Pineal body; 10. Thalamus; 11. Intermediate mass

Cerebrum
- A. 1. Longitudinal fissure; 2. Gyri; 3. Central sulcus; 4. Lateral fissure
- B. 1. Frontal lobe; 2. Parietal lobe; 3. Occipital lobe; 4. Temporal lobe; 5. Temporal lobe
- C. 1. Gray matter; 2. White matter; 3. Association fibers; 4. Commissural fibers; 5. Projection fibers

Cerebral Cortex
- A. 1. Primary sensory areas; 2. Primary somesthetic area; 3. Projection; 4. Association areas; 5. Association areas
- B. 1. Primary motor cortex; 2. Primary motor cortex; 3. Premotor area; 4. Prefrontal area

Speech
- 1 Wernicke's area; 2. Broca's area; 3. Premotor area; 4. Primary motor cortex

Brain Waves
- 1. Alpha wave; 2. Beta wave; 3. Delta wave; 4. Theta wave

Memory
- 1. Sensory memory; 2. Short-term memory; 3. Long-term memory; 4. Long-term potentiation; 5. Calpain; 6. Memory engrams; 7. Declarative memory; 8. Procedural memory; 9. Hippocampus and amygdala; 10. Cerebellum and premotor area

Right and Left Cortex
- 1. Left cerebral hemisphere; 2. Left cerebral hemisphere; 3. Right cerebral hemisphere; 4. Corpus callosum

Basal Ganglia
- 1. Subthalamic nucleus; 2. Substantia nigra; 3. Corpus striatum; 4. Caudate nucleus, lentiform nucleus

Limbic System
- 1. Hippocampus; 2. Habenular nuclei; 3. Basal ganglia; 4. Hypothalamus; 5. Olfactory cortex; 6. Fornix; 7. Olfactory nerves

Cerebellum
- A. 1. Cerebellar peduncles; 2. Flocculonodular lobe; 3. Vermis; 4. Lateral hemispheres
- B. 1. Cerebellum; 2. Proprioceptors; 3. Motor cortex; 4. Comparator

Spinal Cord
- A. 1. Cervical enlargement; 2. Conus medullaris; 3. Filum terminale; 4. Cauda equina
- B. 1. Funiculi; 2. Posterior (dorsal) horn; 3. Lateral horn; 4. Gray and white commissures; 5. Ventral root; 6. Dorsal root ganglion
- C. 1. Funiculi; 2. Dorsal root; 3. Dorsal root ganglion; 4. Ventral root; 5. Anterior (ventral) horn; 6. Lateral horn; 7. Posterior (dorsal) horn

Spinal Reflexes
- A. 1. Stretch reflex; 2. Golgi tendon reflex; 3. Withdrawal reflex; 4. Reciprocal innervation; 5. Crossed extensor reflex
- B. 1. Muscle spindle; 2. Alpha motor neuron; 3. Gamma motor neuron; 4. Golgi tendon organ

Ascending Spinal Pathways
- A. 1. Lateral spinothalamic; 2. Trigemino-thalamic; 3. Anterior spinothalamic; 4. Dorsal column-medial lemniscal; 5. Spinocerebellar and dorsal column-medial lemniscal
- B. 1. Spinal cord; 2. Spinal cord; 3. Medulla oblongata; 4. Uncrossed; 5. Spinal cord and cerebellum
- C. 1. Primary neuron; 2. Secondary neuron; 3. Secondary neuron; 4. Tertiary neuron

Descending Spinal Pathways
- A. 1. Pyramidal system; 2. Extrapyramidal system; 3. Pyramidal system; 4. Extrapyramidal system
- B. 1. Corticospinal tract; 2. Corticobulbar tract; 3. Rubrospinal tract; 4. Vestibulospinal tract
- C. 1. Medulla oblongata; 2. Spinal cord
- D. 1. Upper motor neuron; 2. Lower motor neuron; 3. Lower motor neuron

Meninges
    A. 1. Dura mater; 2. Dura mater; 3. Epidural
        space; 4. Arachnoid layer; 5. Subdural space;
        6. Subarachnoid space; 7. Denticulate
        ligaments
    B. 1. Superior sagittal sinus (dural sinus);
        2. Arachnoid granulations; 3. Pia mater;
        4. Subarachnoid space; 5. Arachnoid layer;
        6. Dura mater

Ventricles
    A. 1. Septa pellucida; 2. Third ventricle;
        3. Interventricular foramina; 4. Fourth
        ventricle; 5. Cerebral aqueduct; 6. Central
        canal

    B. 1. Lateral ventricle; 2. Choroid plexus;
        3. Interventricular foramen; 4. Third
        ventricle; 5. Cerebral aqueduct; 6. Fourth
        ventricle; 7. Central canal; 8. Subarachnoid
        space; 9. Arachnoid granulation; 10. Superior
        sagittal sinus (dural sinus);

Cerebrospinal Fluid
    1. Cerebrospinal fluid; 2. Choroid plexuses;
    3. Subarachnoid space; 4. Lateral and median
    foramina; 5. Dural sinuses; 6. Arachnoid
    granulations

Clinical Applications
    1. Hydrocephalus; 2. Dyskinesias; 3. Multiple
    sclerosis; 4. Reye's syndrome; 5. Encephalitis;
    6. Stroke

QUICK RECALL

1. Cerebral cortex: interprets sensations, initiates motor activities, thoughts, reasoning, speech, and other higher brain functions; cerebral medulla: tract system that connects parts of a cerebral hemisphere with other parts of the same hemisphere, with the opposite cerebral hemisphere, or with other parts of the brain and spinal cord; basal ganglia: coordinate motor movement and posture, inhibit unwanted muscular activity; thalamus: relay center for sensory and motor functions; hypothalamus: endocrine control, autonomic control, temperature regulation, hunger, thirst, and emotions; limbic system: emotions; midbrain: visual and auditory reflexes (colliculi), motor pathways (cerebral peduncles), connects forebrain with hindbrain; pons: connects cerebellum to brain (cerebellar peduncles), regulates respiration, and contains ascending and descending tracts; medulla oblongata: motor pathways (pyramids), balance (olives), autonomic reflexes (e.g., heart rate, breathing, swallowing), ascending and descending tracts, consciousness (reticular activating system); cerebellum: muscle coordination and balance

2. Medulla oblongata, pons, and midbrain

3. Thalamus, subthalamus, hypothalamus, and epithalamus

4. Frontal lobe: voluntary motor function, motivation, aggression, and mood; parietal lobe: major center for reception and evaluation of most sensory information; occipital lobe: reception and integration of visual input; temporal lobe: reception and evaluation of olfactory and auditory input; memory

5. Association, commissural, and projection tracts

6. Stretch reflex, Golgi tendon reflex, withdrawal reflex, reciprocal innervation, and crossed extensor reflex

7. Spinothalamic system: ascending pathway carrying pain, temperature, light touch, pressure, tickle, and itch sensations; dorsal column-medial lemniscal system: ascending pathways carrying senses of two-point discrimination, proprioception, vibration, and pressure; spinocerebellar system: ascending pathways carrying unconscious proprioception

8. Pyramidal system: conscious control of skeletal muscles, especially concerned with fine movements and speed, increases muscle tone; extrapyramidal system: conscious and unconscious control of skeletal muscle, involved with error-correcting function of cerebellum, maintenance of posture and balance, control of large movements (e.g., in the trunk and limbs), inhibits muscle activity

9. Dura mater, arachnoid layer, and pia mater. The subdural space separates the dura mater and arachnoid layer and the subarachnoid space separates the arachnoid layer and the pia mater. In the spinal cord the dura mater is separated from the periosteum of the vertebral canal by the epidural space.

10. Lateral ventricles are connected to the third ventricle by the interventricular foramina; the third ventricle is connected to the fourth ventricle by the cerebral aqueduct.

261

1. C. The brainstem consists of the medulla, pons, and midbrain. The midbrain connects to the diencephalon (thalamus, hypothalamus, subthalamus, and epithalamus).

2. B. The medulla contains many nuclei that are responsible for many basic life functions. Severe damage to the medulla usually results in death. Other medullary functions include swallowing, vomiting, sneezing, and coughing.

3. C. Respiratory centers are located in the pons and medulla.

4. B. The reticular formation is a loose network of nerve fibers and nuclei scattered throughout the medulla and extending up through the brainstem to the thalamus. When the reticular formation is active, it stimulates other areas of the brain, resulting in a conscious or awake state. Barbiturates and general anesthetics depress the activity of the reticular formation.

5. B. The thalamus is the major relay center for action potentials going to and from the cerebral cortex. Most sensory tracts (olfaction is an exception) synapse in the thalamus.

6. B. There are many rounded ridges or convolutions on the surface of the cerebrum called gyri. Separating the ridges are furrows called sulci. Fissures divide the brain into major parts. The longitudinal fissure divides the cerebrum into left and right halves. The transverse fissure separates the cerebrum from the cerebellum. The white matter of the brain makes up nerve tracts. There are three basic kinds of tracts in the cerebrum:
1. Association tracts: limited to one cerebral hemisphere.
2. Commissural tracts: cross from one cerebral hemisphere to another (corpus callosum)
3. Projection tracts: from the cerebral hemispheres to other parts of the nervous system.

7. C. Normally, when a specific body part such as the hand is stimulated, action potentials are generated that travel to the brain. There, through projection, awareness that the hand was stimulated occurs. When the severed nerve was stimulated, action potentials were generated in the axon supplying the hand. When these action potentials reached the brain, they were projected to the hand, and it felt as though the hand was stimulated.

8. B. General sensory input terminates in the postcentral gyrus (primary somesthetic area). The precentral gyrus (primary motor cortex) is concerned with the control of skeletal muscles. The central sulcus is the furrow that separates the precentral gyrus from the postcentral gyrus. The arachnoid is one of the meninges.

9. D. The visual centers are located in the occipital lobes. The primary visual cortices are locations where the image is "seen", whereas the association visual areas relate past to present visual experiences.

10. C. The motivation and foresight to initiate movement of the facial muscle occur in the prefrontal area. Then the premotor area determines which muscles contract and the order in which they contract. Finally the primary motor cortex sends action potentials to each specific muscle. The inferior part of the primary motor cortex controls the face.

11. C. Wernicke's area is necessary for understanding and formulating speech. Broca's area initiates the series of movements necessary for speech. Then the premotor area programs the movements and finally muscle movement is initiated in the primary motor cortex.

12. A. Alpha waves are produced in an awake, but quietly resting person. Theta waves usually occur in children, and delta waves occur in deep sleep. There are no gamma waves.

13. D. Calcium movement into the neuron activates calpain, which partially degrades the neuron cytoskeleton, causing a change in shape.

14. B. The left and right cerebral hemispheres are connected by the corpus callosum. The intermediate mass connects the left and right thalamus. The vermis connects the left and right cerebellar hemispheres.

15. D. The basal ganglia are nuclei (gray matter) located in the cerebrum, diencephalon, and midbrain. The corpus striatum in the cerebrum is an example. The basal ganglia function to plan and coordinate motor movements and posture. One major effect of the basal ganglia is to inhibit unwanted muscular activity.

16. B. One function of the basal ganglia is to inhibit unwanted muscular activity. Loss of this inhibitory effect could result in uncontrolled rhythmic contractions of skeletal muscle and greater than normal tension in skeletal muscle.

17. D. The limbic system includes those structures that affect emotional responses. This includes some hypothalamic and thalamic nuclei, parts of the basal ganglia, the olfactory bulbs, parts of the cerebral cortex, the fornix, and the mamillary bodies.

18. B. The symptoms listed suggest that the patient has the ability to perform voluntary movements and has normal tension in the skeletal muscles. Therefore the cerebrum does not appear to be involved. The patient's ability to perform coordinated movements is affected, which suggests that the comparator function of the cerebellum is involved. None of the data suggest that the medulla or pons is involved.

19. D. During fetal development the vertebral column grows faster than the spinal cord. As a result, the spinal cord reaches only to the level of the first or second lumbar vertebra. Below that region is a mass of descending nerves called the cauda equina. Lumbar punctures and spinal anesthetics are usually administered between the third and fourth lumbar vertebrae.

20. D. Sensory neuron cell bodies are located in the dorsal root ganglia, which are in the dorsal roots of spinal nerves. The sensory neuron axons enter the spinal cord through the dorsal (posterior) horn. In contrast, motor neuron cell bodies are located in the ventral (anterior) horn of the spinal cord. They exit through the ventral root of the spinal nerve.

21. C. The right foot is removed from the tack as the flexor muscles contract (withdrawal reflex) and the extensor muscles relax (reciprocal innervation). The left leg is extended because of the crossed extensor reflex. The Golgi tendon reflex causes muscles under excessive strain to relax.

22. D. The rubber hammer striking the patellar ligament causes the quadriceps femoris to stretch (4), which also results in the stretch of the muscle spindles (2). The stretch of the muscle spindles increases the frequency of afferent neuron action potentials (1), stimulating alpha motor neurons (3), that stimulate the quadriceps femoris to contract (5).

23. C. The extrapyramidal system is involved with gross movement and posture. The extrapyramidal system is complex, including parts of the cerebrum, cerebellum, basal ganglia, thalamus, and brainstem. The pyramidal system consists of motor areas in the cerebral cortex and the nerve tracts leading to skeletal muscles. It is concerned with voluntary, skilled, learned movements.

24. A. Most (75%-85%) pyramidal fibers decussate in the medulla and descend in the lateral corticospinal tracts. Some (15%-25%) pyramidal fibers do not decussate in the brain and descend in the anterior corticospinal tract. The upper motor neurons of the pyramidal system synapse with lower motor neurons in the spinal cord.

25. B. The lateral spinothalamic tract carries pain and thermal sensations, and it decussates (crosses) within the spinal cord. Interruption of the lateral spinothalamic tract on the right side results in the loss of pain and thermal sensations on the left side below the damaged area.

26. B. Disruption of the lateral corticospinal pathway results in paresis. Because it decussates within the medulla, disruption of the right lateral corticospinal pathway affects the right side of the body. The patient is probably suffering from damage to her right lateral corticospinal tract.

27. C. Because most sensory information passes through the thalamus on the side opposite to the side where the sensory receptors are located, 1 and 3 are correct. Ataxia on the left side would result from the loss of proprioception sensations from that side, because the dorsal column-medial lemniscal system carries most of the proprioceptive information from the upper limb to the cerebrum.

28. C. Three layers of connective tissue called meninges cover the entire central nervous system. From outside to inside, they are the dura mater, arachnoid, and pia mater.

29. B. The lateral ventricles are separated by the septa pellucida and are not interconnected. The central canal is a continuation of the fourth ventricle into the spinal cord.

30. A. The pia mater and the arachnoid are modified to form the choroid plexus, which produces cerebrospinal fluid. The arachnoid granulations reabsorb the cerebrospinal fluid from the subarachnoid space. The fluid then enters the dural sinuses.

1. One strategy is to look for easily observed abnormalities associated with the hypothalamus but not the medulla: e.g., abnormal emotional behavior (fear, rage), hunger, thirst, sweating, shivering, or urine production. Observation of functions that could be affected by either the hypothalamus or the medulla, such as heart rate, sleep (consciousness), or blood vessel diameter, would not be useful.

2. Because the cerebellum acts to match intended movements with actual movements, reduced cerebellar function results in an inability to point precisely to an object (such as one's nose). It also results in poor balance.

3. Parkinson's disease results in decreased activity of inhibitory neurons. Therefore the neurons to muscles become overstimulated, resulting in tremors. The same hyperexcited state also results in exaggerated reflexes.

4. In a lumbar puncture air is injected into the subarachnoid space around the spinal cord. The air travels superiorly and enter the fourth ventricle through the lateral foramina. From the fourth ventricle air can pass through the cerebral aqueduct to the third ventricle and from there into the lateral ventricles through the interventricular foramina. In the supine position, the anterior portions of the lateral ventricles are filled with air.

5. Because the lesion is in the center of the spinal cord, only those tracts that cross over in the spinal cord are affected. Therefore one expects to see loss of pain and temperature sensation (lateral spinothalamic tract). The loss occurs bilaterally at the level of the injury, but not below the injury. Because the dorsal column-medial lemniscal system (fine touch) and lateral corticospinal tract (voluntary motor activity) cross over in the medulla, they are unaffected by a lesion in the center of the spinal cord.

6. It is possible that the dorsal column-medial lemniscal system within the right side of the spinal cord is damaged. However, it is also possible that this tract system could be damaged within the medulla, where neurons synapse and cross over to the left side of the brain, or within the tracts on the left side that ascend from the medulla to the thalamus. Another possibility is damage to the cerebral cortex on the left side. Additional information is needed to decide exactly where the injury is located.

Peripheral Nervous System: Cranial Nerves and Spinal Nerves

FOCUS: The peripheral nervous system consists of 31 pairs of spinal nerves and 12 pairs of cranial nerves. The spinal nerves branch to form dorsal rami, which supply the dorsal trunk, and ventral rami. The ventral rami give rise to the intercostal nerves or join with each other to form plexuses. The major plexuses are the cervical, brachial, lumbar, sacral, and coccygeal plexuses. Within a plexus, the fibers from different spinal nerves join together to form nerves that leave the plexus. The cranial nerves all originate within nuclei of the brain and supply the structures of the head and neck, with the exception of the vagus nerve, which supplies the visceral organs of the thorax and abdomen. Cranial nerves can have sensory functions (e.g., optic nerve for vision), somatic motor and proprioceptive functions (e.g., oculomotor nerve which controls movements of the eyeball), or parasympathetic functions (e.g., oculomotor nerve which controls the size of the pupil).

CONTENT LEARNING ACTIVITY

Cranial Nerves

66 *The 12 pairs of cranial nerves have several different combinations of function.* **99**

A. Match these terms with the correct statement or definition:

Parasympathetic   Sensory
Proprioception   Somatic motor

_____ 1. Includes vision, touch, and pain.

_____ 2. Involves control of skeletal muscles through motor neurons.

_____ 3. Informs the brain about the position of various body parts, including joints and muscles.

_____ 4. Involves the regulation of glands, smooth muscle, and cardiac muscle.

_____ 5 This category is included with somatic motor function because the nerves supplying the motor input to muscles also convey impulses from those muscles to the CNS.

_____ 6. Two types of ganglia associated with cranial nerves.

_____

265

B. Match the name of the cranial nerve with its number:

Abducens   Olfactory
Accessory   Optic
Facial   Trigeminal
Glossopharyngeal   Trochlear
Hypoglossal   Vagus
Oculomotor   Vestibulocochlear

_____ I.

_____ II.

_____ III.

_____ IV.

_____ V.

_____ VI.

_____ VII.

_____ VIII.

_____ IX.

_____ X.

_____ XI.

_____ XII.

👉 Two mnemonics: On Old Olympus' Towering Top, A Finn And* German Viewed A Hop
Oh, Oh, Oh, To Touch And Feel Very Good Velvet, AH!

*Note that VIII was formerly called the auditory rather than the vestibulocochlear.

C. Match these cranial nerves with their correct sensory function:

Facial (VII)   Trigeminal (V)
Glossopharyngeal (IX)   Vagus (X)
Olfactory (I)   Vestibulocochlear (VIII)
Optic (II)

_____ 1. Sensory from face, teeth, upper and lower jaw, and oral cavity.

_____ 2. Alveolar nerves arise from maxillary and mandibular branches.

_____ 3. Sense of taste from anterior two thirds of tongue.

_____ 4. Sense of hearing and balance.

_____ 5. Sense of taste from posterior one third of tongue; sensory from receptors in the carotid arteries and aortic arch.

_____ 6. Sensory from the thoracic and abdominal organs; sense of taste from posterior tongue.

D. Match these cranial nerves with their correct motor function:

Abducens (VI)
Accessory (XI)
Facial (VII)
Glossopharyngeal (IX)
Hypoglossal (XII)

Oculomotor (III)
Trigeminal (V)
Trochlear (IV)
Vagus (X)

_____ 1. Motor to four extrinsic eye muscles.

_____ 2. Two nerves, each motor to an extrinsic eye muscle.

_____ 3. Motor to muscles of mastication.

_____ 4. Motor to muscles of facial expression.

_____ 5. Motor to soft palate, pharynx, and laryngeal muscles (voice).

_____ 6. Joins vagus (X); motor to sternocleidomastoid and trapezius.

_____ 7. Motor to intrinsic and extrinsic tongue muscles.

E. Match these terms with the correct statement or definition:

Facial (VII)
Glossopharyngeal (IX)

Oculomotor (III)
Vagus (X)

_____ 1. Parasympathetic to pupil of eye and ciliary muscle of lens.

_____ 2. Parasympathetic to two salivary glands and lacrimal glands.

_____ 3. Parasympathetic to the parotid salivary gland.

_____ 4. Parasympathetic to the thoracic and abdominal viscera.

Spinal Nerves

66Spinal nerves arise from many rootlets along the dorsal and ventral surfaces of the spinal cord.99

A. Match these terms with the correct statement or definition:

Dermatome
Dorsal rami
Dorsal root ganglion

Plexuses
Ventral rami
Ventral root

_____ 1. Joins with the dorsal root to form the spinal nerve.

_____ 2. Collection of nerve cell bodies in the dorsal root of the spinal nerve.

_____ 3. Skin with sensory innervation by a pair of spinal nerves.

_____ 4. Innervate most of the deep muscles of the dorsal trunk.

_____ 5. Spinal nerve branches that become intercostal nerves or plexuses.

_____ 6. Organization produced by intermingling of nerves.

B. Match these terms with the correct parts labeled in Figure 14-1:

Dorsal ramus
Dorsal root
Dorsal root ganglion
Spinal nerve
Ventral ramus
Ventral root

Figure 14-1

1. _____

2. _____

3. _____

4. _____

5. _____

6. _____

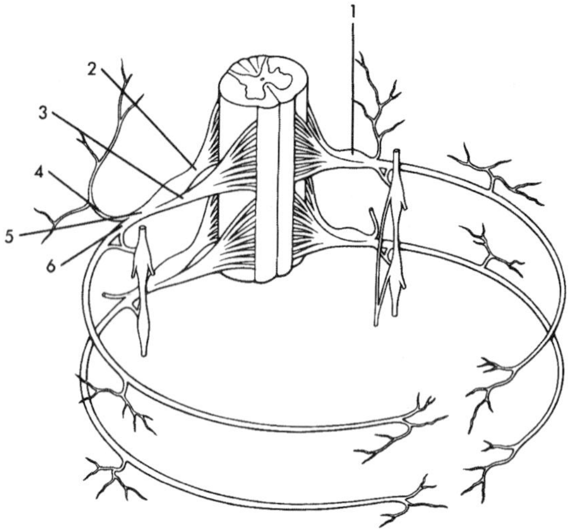

C. Match the spinal nerve region with the number of pairs of that spinal nerve:

Cervical   Sacral
Coccygeal   Thoracic
Lumbar

_____   1. One pair.

_____   2. Five pairs (two regions)

_____

_____   3. Eight pairs.

_____   4. Twelve pairs.

Cervical Plexus

66 *The cervical plexus is a relatively small plexus originating from spinal nerves C1 to C4.* **99**

Match these terms with the correct statement or definition:

Diaphragm   Neck and posterior head
Hyoid muscles

_____   1. Cutaneous innervation of the cervical plexus.

_____   2. Motor innervation from the cervical plexus (except phrenic nerve).

_____   3. Phrenic nerve innervates this structure.

Brachial Plexus

"The brachial plexus originates from spinal nerves C5 to T1."

A. Match these nerves with their correct motor function or innervation:

    Axillary nerve   Radial nerve
    Median nerve   Ulnar nerve
    Musculocutaneous nerve

_____ 1. Innervates the deltoid; abduction of the arm.

_____ 2. Innervates all the extensor muscles of the upper limb.

_____ 3. Innervates the anterior muscles of the arm; flexes forearm.

_____ 4. Innervates most of the intrinsic hand muscles.

_____ 5. Two nerves that innervate the flexor muscles of the forearm.

_____

B. Match these terms with the correct statement or definition:

    Axillary nerve   Radial nerve
    Median nerve   Ulnar nerve
    Musculocutaneous nerve

_____ 1. Sensory innervation to the shoulder joint and skin over the shoulder.

_____ 2. Sensory innervation to the posterior arm, forearm, and hand.

_____ 3. Sensory innervation to the lateral surface of the forearm.

_____ 4. Sensory innervation to the ulnar side of the hand.

☞ Additional brachial plexus nerves supply most of the muscles acting on the scapula and arm, and supply the cutaneous innervation of the medial arm and forearm.

Lumbar and Sacral Plexuses

"Because of their overlapping relationship, the lumbar and sacral plexuses are often considered together as a single lumbosacral plexus."

A. Match these nerves with the correct description:

    Common fibular (peroneal) nerve   Tibial nerve
    Ischiadic (sciatic) nerve

_____ 1. Tibial and common fibular (peroneal) nerves bound together in the same connective tissue sheath.

_____ 2. Branches to form medial plantar, lateral plantar, and sural nerves.

_____ 3. Branches to form deep and superficial fibular (peroneal) nerves.

B. Match these nerves with their correct motor function or innervation:

Common fibular (peroneal) nerve
Femoral nerve
Medial and lateral plantar nerves

Obturator nerve
Tibial nerve

_____ 1. Innervates the muscles that adduct the thigh.

_____ 2. Innervates the iliopsoas, sartorius, and quadriceps femoris muscles.

_____ 3. Innervates most of the posterior thigh and leg muscles.

_____ 4. Innervates the anterior and lateral muscles of the leg and foot.

☞ Several other nerves from the lumbosacral plexus innervate the lower abdominal, hip, and abdominal floor muscles.

C. Match these nerves with their correct sensory innervation:

Common fibular (peroneal) nerve
Femoral nerve

Obturator nerve
Tibial nerve

_____ 1. Sensory innervation to the medial side of the thigh.

_____ 2. Sensory innervation to the anterior and lateral thigh and medial leg.

_____ 3. Sensory innervation to the lateral and posterior one third of the leg and the sole of the foot.

☞ Several other nerves from the lumbosacral plexus innervate the skin of the external genitalia, the superior medial thigh, and the posterior thigh. The coccygeal plexus is a very small plexus that supplies motor innervation to the muscles of the pelvic floor and cutaneous innervation over the coccyx.

Clinical Applications

66 *There are many PNS disorders.* **99**

A. Match these terms with the correct statement or definition:

Anesthesia
Hyperesthesia
Neuralgia

Neuritis
Paresthesia

_____ 1. Increased sensitivity to stimuli.

_____ 2. General term for inflammation of a nerve.

_____ 3. Severe spasms of throbbing pain along a nerve pathway; an example is sciatica.

_____ 4. Abnormal spontaneous sensations such as tingling, prickling, or burning.

B. Match these terms with the
correct statement or definition:

Herpes simplex I   Myasthenia gravis
Herpes simplex II   Poliomyelitis
Herpes zoster (varicella)

_____ 1. Viral disease that causes fever blisters or cold sores.

_____ 2. Viral disease that causes chicken pox and shingles.

_____ 3. An autoimmune disorder resulting in a decrease of
acetylcholine receptors at neuromuscular junctions.

_____ 4. Viral disease that affects motor neurons in the anterior horn of
gray matter of the spinal cord.

QUICK RECALL

1. List the 12 cranial nerves.

2. List the three basic functions of the cranial nerves.

3. List three types of rami that branch from spinal nerves, and state what type of nerve
fiber (sensory, motor, or sympathetic) is found in each.

4. Name five major plexuses formed by the spinal nerves, and list the level of the spinal
cord from which each plexus arises.

5. List the five major nerves originating from the brachial plexus.

6. List the four major nerves originating from the lumbosacral plexus.

Place the letter corresponding to the correct answer in the space provided.

_____ 1. The cranial nerve involved in
chewing food?
a. trochlear (IV)
b. trigeminal (V)
c. abducens (VI)
d. facial (VII)
e. vestibulocochlear (VIII)

_____ 2. The cranial nerve responsible for
focusing the eye (innervates the
ciliary muscle of the eye)?
a. optic (II)
b. oculomotor (III)
c. trochlear (IV)
d. abducens (VI)
e. facial (VII)

_____ 3. The cranial nerve involved in
feeling a toothache?
a. trochlear (IV)
b. trigeminal (V)
c. abducens (VI)
d. facial (VII)
e. vestibulocochlear (VIII)

_____ 4. From these list of cranial nerves:
1. olfactory (I)
2. optic (II)
3. oculomotor (III)
4. abducens (VI)
5. vestibulocochlear (VIII)

Select the nerves that are sensory
only.
a. 1,2,3
b. 1,2,5
c. 2,3,4
d. 2,3,5
e. 3,4,5

_____ 5. From these list of cranial nerves:
1. optic (II)
2. oculomotor (III)
3. trochlear (IV)
4. trigeminal (V)
5. abducens (VI)

Select the nerves that are involved
in "rolling the eyes."
a. 1,2,3
b. 1,2,4
c. 2,3,4
d. 2,4,5
e. 2,3,5

_____ 6. From these list of cranial nerves:
1. trigeminal (V)
2. facial (VII)
3. glossopharyngeal (IX)
4. vagus (X)
5. hypoglossal (XII)

Select the nerves that are involved
in the sense of taste.
a. 1,2,3
b. 1,4,5
c. 2,3,4
d. 2,3,5
e. 3,4,5

_____ 7. From these list of cranial nerves:
1. trigeminal (V)
2. facial (VII)
3. glossopharyngeal (IX)
4. vagus (X)
5. hypoglossal (XII)

Select the nerves that innervate the
salivary glands.
a. 1,2
b. 2,3
c. 3,4
d. 4,5
e. 3,5

_____ 8. From these list of cranial nerves:
1. oculomotor (III)
2. trigeminal (V)
3. facial (VII)

4. vestibulocochlear (VIII)
5. glossopharyngeal (IX)
6. vagus (X)

Select the nerves that are part of the parasympathetic nervous system.
a. 1,2,4,5
b. 1,3,5,6
c. 1,4,5,6
d. 2,3,4,5
e. 2,3,5,6

_____ 9. After examining a patient, the patient's physician concluded that these symptoms were indicative of a lesion in the trigeminal nerve on the right side:
1. Opened jaw deviated to the right.
2. The right temporalis and right masseter muscles were atrophied.
3. There were no sensations on the right side of the scalp.
4. There was no corneal sensation on the right side.
5. There was no sensation to pin pricks in the right side of the oral cavity.

Which of the above symptoms suggest that the maxillary and mandibular branches of the trigeminal nerve are involved?
a. 1,2,4
b. 1,2,5
c. 1,2,3,4
d. 1,2,3,5
e. 2,3,4,5

_____ 10. Which of these would be a correct count of the spinal nerves?
a. 9 cervical, 12 thoracic, 5 lumbar, 5 sacral, 1 coccygeal
b. 8 cervical, 12 thoracic, 5 lumbar, 5 sacral, 1 coccygeal
c. 7 cervical, 12 thoracic, 5 lumbar, 5 sacral, 1 coccygeal
d. 8 cervical, 11 thoracic, 4 lumbar, 6 sacral, 1 coccygeal
e. none of the above

_____ 11. Given these structures:
1. dorsal ramus
2. dorsal root
3. plexus
4. ventral ramus

5. ventral root

Choose the arrangement that lists the structures in the order that an action potential would pass through them, given that the action potential originated in the spinal cord and propagated to a peripheral nerve.
a. 2,1,3
b. 2,3,1
c. 3,4,5
d. 5,3,4
e. 5,4,3

_____ 12. Damage to the dorsal ramus of a spinal nerve would result in
a. loss of sensation.
b. loss of motor control.
c. a and b

_____ 13. A collection of spinal nerves that join together after leaving the spinal cord is called a
a. ganglion.
b. nucleus.
c. projection nerve.
d. plexus.

_____ 14. A dermatome
a. is the area of skin supplied by a pair of spinal nerves.
b. may be supplied by more than one nerve from a plexus.
c. can be used to locate the site of spinal cord injury.
d. all of the above

_____ 15. Anesthetic injected into the cervical plexus would
a. prevent pain in the leg.
b. prevent pain in the thigh and leg.
c. prevent pain in the arm.
d. interfere with the ability of the patient to breathe.

16. The skin of the index and middle fingers is supplied by the
    a. median nerve.
    b. radial nerve.
    c. ulnar nerve.
    d. a and b
    e. all of the above

17. The extensor muscles of the upper limb are supplied by the
    a. musculocutaneous nerve.
    b. radial nerve.
    c. median nerve.
    d. ulnar nerve.

18. The intrinsic hand muscles, other than those that move the thumb, are supplied by the
    a. musculocutaneous nerve.
    b. axillary nerve.
    c. popliteal nerve.
    d. intrinsic nerve.
    e. ulnar nerve.

19. The ischiadic (sciatic) nerve is actually two nerves combined within the same sheath. The two nerves are the
    a. femoral and obturator.
    b. femoral and gluteal.
    c. common fibular (peroneal) and tibial.
    d. common fibular (peroneal) and obturator.
    e. tibial and gluteal.

20. The muscles of the anterior thigh compartment are supplied by the
    a. obturator nerve.
    b. gluteal nerve.
    c. ischiadic (sciatic) nerve.
    d. femoral nerve.
    e. ilioinguinal nerve.

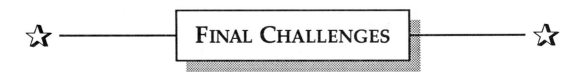

FINAL CHALLENGES

Use a separate sheet of paper to complete this section.

1. A patient has a severe case of hiccoughs (spasmodic contractions of the diaphragm). The physician injects an anesthetic solution into the neck about an inch above the clavicle. What nerve was injected?

2. A woman slips on a wet kitchen floor. As she falls, she forcefully strikes the edge of the kitchen table with her elbow. Name the nerve she is most likely to damage. What symptoms (motor and/or sensory) would you expect her to develop?

3. Jock Player was tackled and slammed into the ground while playing football, resulting in a dislocated shoulder. After the humerus was reset into the glenoid fossa of the scapula, it was discovered that Jock could not abduct his arm. Explain. Where would you expect there to be a loss of sensation in the skin?

4. A stab wound in the proximal, medial part of the arm produced impairment of flexion and supination of the forearm. There was a loss of sensation from the lateral surface of the forearm. What nerve was damaged? Explain why there was only impairment and not complete loss of flexion and supination.

5. A skier breaks his ankle. As part of his treatment, the ankle and leg are placed in a plaster cast. Unfortunately, the cast is too tight around the proximal portion of the leg and presses in against the neck of the fibula. Where would you predict the patient would experience tingling or numbness in the leg? Explain.

6. One day you and a friend observe a man walking down the street in a peculiar fashion. He seems to raise his left foot much higher than normal to keep his toes from dragging on the ground. Apparently he is unable to dorsiflex his left foot. When he places the foot on the ground, it makes a clopping or flapping sound. You recognize this as foot drop, and, turning to your friend, you comment that the man must have tibial nerve damage. Your friend disagrees. Who is right?

7. A patient was diagnosed as having tic douloureux, a condition in which a light touch to the upper lip, cheek, or under the eye, or chewing, produces a sudden sensation of severe pain in and around the area touched. Although the cause of tic douloureux is unknown, what nerve must be involved? Can you determine which branch of the nerve is most likely affected in this patient?

8. While playing hockey, Skip Puck received a deep cut on the left side of his neck just lateral to the larynx. After the wound was sutured, these observations were made: the tongue, when protruded, deviated to the left, the left shoulder was drooping, and Skip had difficulty in turning his head to the right side. Name the nerves that were damaged, and explain why this damage produced the observed symptoms.

ANSWERS TO CHAPTER 14

Cranial Nerves

A. 1. Sensory; 2. Somatic motor;
3. Proprioception; 4. Parasympathetic;
5. Proprioception; 6. Parasympathetic,
sensory

B. I. Olfactory; II. Optic;
III. Oculomotor; IV. Trochlear;
V. Trigeminal; VI. Abducens;
VII. Facial; VIII. Vestibulocochlear;
IX. Glossopharyngeal; X. Vagus;
XI. Accessory; XII. Hypoglossal

C. 1. Trigeminal (V); 2. Trigeminal (V);
3. Facial (VII); 4. Vestibulocochlear
(VIII); 5. Glossopharyngeal (IX); 6. Vagus (X)

D. 1. Oculomotor (III); 2. Trochlear (IV),
abducens (VI); 3. Trigeminal (V);
4. Facial (VII); 5. Vagus (X); 6. Accessory
(XI); 7. Hypoglossal (XII)

E. 1. Oculomotor (III); 2. Facial (VII);
3. Glossopharyngeal (IX); 4. Vagus (X)

Spinal Nerves

A. 1. Ventral root; 2. Dorsal root ganglion;
3. Dermatome; 4. Dorsal rami; 5. Ventral
rami; 6. Plexuses

B. 1. Dorsal root ganglion; 2. Dorsal root;
3. Ventral root; 4. Dorsal ramus; 5. Spinal
nerve; 6. Ventral ramus

C. 1. Coccygeal; 2. Lumbar, Sacral; 3. Cervical;
4. Thoracic

Cervical Plexus

1. Neck and posterior head; 2. Hyoid muscles;
3. Diaphragm

Brachial Plexus

A. 1. Axillary nerve; 2. Radial nerve;
3. Musculocutaneous nerve; 4. Ulnar nerve;
5. Median nerve, ulnar nerve

B. 1. Axillary nerve; 2. Radial nerve;
3. Musculocutaneous nerve; 4. Ulnar nerve

Lumbar and Sacral Plexuses

A. 1. Ishciadic (sciatic) nerve; 2. Tibial nerve;
3. Common fibular (peroneal) nerve

B. 1. Obturator nerve; 2. Femoral nerve; 3. Tibial
nerve; 4. Common fibular (peroneal) nerve

C. 1. Obturator nerve; 2. Femoral nerve; 3. Tibial
nerve

Clinical Applications

A. 1. Hyperesthesia; 2. Neuritis; 3. Neuralgia;
4. Paresthesia

B. 1. Herpes simplex I; 2. Herpes zoster;
3. Myasthenia gravis; 4. Poliomyelitis

1. Olfactory (I), Optic (II), Oculomotor (III),
Trochlear (IV), Trigeminal (V), Abducens (VI),
Facial (VII), Vestibulocochlear (VIII),
Glossopharyngeal (IX), Vagus (X), Accessory (XI),
and Hypoglossal (XII)

2. Sensory, somatic motor and proprioceptive, and
parasympathetic

3. Dorsal rami: sensory and motor nerves fibers
Ventral rami: sensory and motor nerve fibers
Sympathetic rami: sympathetic nerve fibers

4. Cervical plexus: C1 to C4
Brachial plexus: C5 to T1
Lumbar plexus: L1 to L4
Sacral plexus: L4 to S4
Coccygeal plexus: S4, S5, and coccygeal nerve

5. Axillary, radial, musculocutaneous, ulnar, and
median nerves

6. Obturator, femoral, tibial, and common fibular
(peroneal)

1. B. The trigeminal nerves supply the muscles of
mastication. The facial nerves supply the
muscles of facial expression.

2. B. Parasympathetic innervation through the
oculomotor nerve causes the eye to focus and the
pupil to constrict. Sympathetic stimulation
causes the pupil to open.

3. B. The trigeminal nerve is the major sensory
nerve of the head. The maxillary branch supplies
the teeth of the upper jaw, and the mandibular
branch the teeth of the lower jaw.

4. B. The olfactory (sense of smell), optic (sense
of vision), and vestibulocochlear (sense of
hearing and balance) are sensory only.

5. E. The oculomotor supplies the inferior oblique and the superior, medial, and inferior rectus muscles. The trochlear supplies the superior oblique, and the abducens supplies the lateral rectus.

6. C. The sense of taste from the anterior two thirds of the tongue is carried by the facial nerve and from the posterior one third by the glossopharyngeal nerve. The vagus nerve also is involved with the sense of taste from the root of the tongue.

7. B. The facial nerves supply the submandibular and sublingual glands, and the glossopharyngeal nerves supply the parotid glands.

8. B. The oculomotor, facial, glossopharyngeal, and vagus nerves have parasympathetic fibers.

9. B. The temporalis and mandibular muscles are innervated by the maxillary and mandibular branches of the trigeminal nerve. A deviated jaw with atrophied masseter and temporalis muscles suggests the involvement of these branches. Additionally, the maxillary and mandibular branches carry pain sensations from the oral cavity to the central nervous system. Therefore 1, 2, and 5 are correct.

   Sensations from the scalp and cornea are carried by the ophthalmic branch of the trigeminal nerve. Therefore, 3 and 4 are not correct.

10. B. One must know the correct count.

11. E. The efferent axon from the spinal cord passes through the ventral root into a spinal nerve. The ventral rami of spinal nerves join to form plexuses from which the peripheral nerves arise.

12. C. The dorsal ramus is a branch of a spinal nerve. Spinal nerves contain both afferent (sensory) and efferent (motor) fibers.

13. D. A plexus is a collection of spinal nerves. Ganglia are aggregates of cell bodies in the peripheral nervous system. Nuclei are aggregates of cell bodies in the central nervous system.

14. D. The distribution of a pair of spinal nerves in the skin is a dermatome. Spinal nerves pass through a plexus, and their fibers can exit the plexus through two or more peripheral nerves. Therefore a dermatome can be supplied by more than one spinal nerve. Loss of sensation in a dermatome could indicate damage to the segment of the spinal cord (or more superior segments) associated with a spinal nerve.

15. D. The phrenic nerves, which supply the diaphragm, arise from the cervical plexus.

16. D. The median nerve supplies the lateral two thirds of the hand, mostly on the anterior surface, whereas the radial nerve supplies the lateral two thirds of the hand, mostly on the posterior surface.

17. B. The radial nerve supplies the muscles that extend the forearm, wrist, and digits.

18. E. The intrinsic hand muscles are found only within the hand, and they are mostly supplied by the ulnar nerve. The median nerve supplies most of the intrinsic hand muscles that move the thumb.

19. C. The common fibular (peroneal) and tibial nerves form the ischiadic (sciatic) nerve within the posterior thigh.

20. D. The femoral nerve supplies the anterior thigh compartment, the obturator nerve the medial thigh compartment, and the ischiadic (sciatic) nerve (mostly the tibial nerve) supplies the posterior thigh compartment.

 FINAL CHALLENGES

1. The phrenic nerve, which supplies the diaphragm, was injected.

2. The ulnar nerve is the nerve most likely to be damaged. One might expect a loss of sensation from the medial one third of the hand. If motor function is affected, one might expect an inability to abduct/adduct the medial four fingers, adduct the wrist, or flex the distal interphalangeal joints of the fourth and fifth digits. There could be impairment of wrist flexion resulting from loss of function of the flexor carpi ulnaris muscle.

3. When the shoulder was dislocated, the axillary nerve was stretched and damaged, resulting in loss of function of the deltoid muscle and therefore the ability to abduct the arm. Loss of sensation in the skin of the shoulder, over the deltoid, is expected.

4. The musculocutaneous nerve was damaged, causing the loss of function of the biceps brachii and part of the brachialis muscle. There is only impairment because the radial nerve supplies part of the brachialis and the brachioradialis muscles, which cause flexion. Although the supinator function of the biceps brachii is lost, the supinator muscle, supplied by the radial nerve, still functions.

5. The plaster cast is pressing against the common fibular (peroneal) nerve at the neck of the fibula. Tingling is expected along the lateral and anterior leg and the dorsum of the foot.

6. Your friend is right. Injury to the deep fibular (peroneal) nerve (a branch of the common fibular [peroneal] nerve) results in the loss of dorsiflexion. Tibial nerve damage produces an inability to plantar flex the foot and toes.

7. The maxillary branch of the trigeminal nerve.

8. Damage to the hypoglossal nerve to the tongue produces the deviation of the tongue. Damage to the accessory nerves could cause the shoulder to droop (trapezius muscle) and result in an inability to turn the head (sternocleidomastoid muscle).

15

The Senses

FOCUS: The special senses include olfaction, taste, vision, hearing, and balance. Odors are detected by olfactory neurons in the nasal cavity. Taste bud cells on the tongue detect the four basic tastes: sour, salty, bitter, and sweet. The eye is responsible for the sense of vision. Light entering the eye is refracted by the cornea and the lens so that an image is focused on the retina. The retina has cones, which respond best to bright light and are responsible for visual acuity and color vision, and rods, which respond best to low levels of light. The ear is divided into three parts. The auricle and external auditory meatus of the external ear funnel sound waves to the tympanic membrane. In the middle ear the ear ossicles transmit vibrations of the tympanic membrane to the perilymph of the inner ear. Sound waves in the perilymph cause the vestibular membrane to vibrate, which produces waves in the endolymph of the cochlear duct. These waves cause the basilar membrane to vibrate, producing action potentials in the spiral organ. The saccule and the utricle of the inner ear are responsible for detecting the position of the head relative to the ground, whereas the semicircular canals detect the rate of movement of the head in different planes.

CONTENT LEARNING ACTIVITY

Classification of the Senses

66*The senses are the means by which the brain receives information about the outside world.*99

Match these terms with the correct statement or definition:

Chemoreceptors
Mechanoreceptors
Nociceptors
Photoreceptors

Somatic
Special
Thermoreceptors
Visceral

_____ 1. The modality that includes touch, pressure, temperature, proprioception, and pain.

_____ 2. The modality that provides primarily pain and pressure information about internal organs.

_____ 3. The modalities of smell, taste, sight, sound, and balance.

_____ 4. Receptors that respond to compression, bending, or stretching.

_____ 5. Receptors that respond to temperature.

_____ 6. Receptors that respond to painful mechanical, chemical, or thermal stimuli

Sensation

66 *Sensation, or perception, is the conscious awareness of stimuli received by sensory receptors.* 99

Match these terms with the correct statement or definition:

Adaptation   Projection
Phasic   Tonic

_____ 1. The sensation of a specific stimulus in a particular body part.

_____ 2. A decreased sensitivity to a continued stimulus.

_____ 3. Receptors that respond only to changes in stimuli.

Types of Afferent Nerve Endings

66 *There are at least eight major types of sensory nerve endings.* 99

Match these types of nerve ending with their correct function:

Exteroceptors   Muscle spindle
Free nerve endings   Pacinian corpuscles
Golgi tendon apparatus Proprioceptors
Hair follicle receptors Ruffini's end organs
Meissner's corpuscles   Visceroceptors
Merkel's disks

_____ 1. General term for receptors providing positional information.

_____ 2. General term for receptors on the surface of the body; provide information about the environment.

_____ 3. Simplest nerve ending; pain, temperature, itch, and movement.

_____ 4. Tactile disks; light touch and superficial pressure.

_____ 5. Complex nerve endings resembling an onion; respond to deep cutaneous pressure and vibration.

_____ 6. Distributed through the dermal papillae and involved in fine discriminative touch.

_____ 7. Located in the dermis; respond to skin displacement in continuous touch or pressure.

_____ 8. Proprioception associated with tendon movement.

_____ 9. Helps control muscle tone.

Olfaction

66 *Olfaction occurs in response to odors that enter the extreme superior portion of the nasal cavity.* 99

A. Match these terms with the correct statement or definition:

Basal cells   Olfactory hairs
Olfactory bulbs   Olfactory tracts
Olfactory epithelium   Olfactory vesicles

_____ 1. Specialized cells lining the olfactory recess.

_____ 2. Axons of olfactory neurons project through the cribriform plate to these structures.

_____ 3. These structures project from the olfactory bulbs to the cerebral cortex.

_____ 4. Bulbous enlargements of the dendrites of olfactory neurons.

_____ 5. Have chemoreceptors that bind to molecules, resulting in action potential production.

_____ 6. Lost olfactory neurons in the olfactory epithelium are replaced by these cells.

B. Using the terms provided, complete these statements:

Association neurons   Medial olfactory area
Intermediate olfactory area Mitral cells
Lateral olfactory area   Olfactory nerves

Axons from olfactory neurons form the _(1)_, which enter the olfactory bulbs, where the axons synapse with _(2)_. The latter cells relay olfactory information to the brain through olfactory tracts and synapse with _(3)_ in the olfactory bulb. Each olfactory tract terminates in the olfactory cortex of the brain, which has three distinct areas. The _(4)_ is involved in the conscious perception of smell, whereas the _(5)_ is responsible for visceral and emotional reactions to odors. The _(6)_ has axons that extend to the olfactory bulbs where incoming sensory information is modified.

1. _____

2. _____

3. _____

4. _____

5. _____

6. _____

Taste

66 _Taste buds are specialized sensory structures that detect gustatory, or taste, stimuli._ 99

A. Match these terms with the correct statement or definition:

Circumvallate papillae   Foliate papillae
Filiform papillae   Fungiform papillae

_____ 1. Largest but least numerous papillae; surrounded by a groove or valley.

_____ 2. Mushroom-shaped papillae; appear as small red dots scattered irregularly over the tongue.

_____ 3. Leaf-shaped papillae; distributed over the sides of the tongue and containing the most sensitive taste buds.

_____ 4. Filament-shaped papillae; most numerous papillae, but with no taste buds.

B. Match these terms with the correct statement or definition:

Epithelial cells   Gustatory (taste) pores
Gustatory (taste) cells Taste buds
Gustatory hairs

_____ 1. Oval structures, consisting of two types of cells, embedded in the epithelium of the tongue and mouth.

_____ 2. About 50 of these cells are found internally in each taste bud.

_____ 3. Microvilli found on a gustatory cell.

_____ 4. Opening in the epithelium containing gustatory hairs.

☞ The taste buds detect four basic tastes: sour, salty, bitter, and sweet.

C. Match these nerves with their function or description:

Facial nerve (VII)   Vagus nerve (X)
Glossopharyngeal nerve (IX)

_____ 1. Branch of this nerve (chorda tympani) crosses the tympanic membrane.

_____ 2. Carries taste from the anterior two thirds of the tongue.

_____ 3. Carries taste from the posterior one third of the tongue .

Accessory Structures

66 *Accessory structures aid the function of the eye in many ways.* 99

A. Match these terms with the correct statement or definition:

Bulbar conjunctiva   Meibomian glands
Canthi   Palpebrae
Caruncle   Palpebral conjunctiva
Chalazion   Sty
Ciliary glands   Tarsal plate

_____ 1. Eyelids.

_____ 2. Angles where the eyelids join .

_____ 3. Reddish-pink mound in the medial canthus.

_____ 4. Dense connective tissue; maintains the shape of the eyelid.

_____ 5. Modified sweat glands that open into the eyelash follicles.

_____ 6. Inflamed ciliary gland.

_____ 7. Eyelid glands that produce sebum.

_____ 8. Infection or blockage of a meibomian gland; a meibomian cyst.

_____ 9. Mucous membrane covering the inner surface of the eyelids.

B. Match these terms with the
   correct statement or definition:

Lacrimal apparatus   Lacrimal sac
Lacrimal canaliculus   Nasolacrimal duct
Lacrimal gland   Punctum
Lacrimal papilla

_____ 1. Gland that produces tears.

_____ 2. Passageway in the medial corner of the eye into which excess
                              tears flow.

_____ 3. Opening into the lacrimal canaliculi.

_____ 4. Small lump upon which the punctum is located.

_____ 5. Structure located between the lacrimal canaliculus and the
                              nasolacrimal duct.

_____ 6. Passageway that opens into the nasal cavity.

☞ Tears moisten the surface of the eye, lubricate the eyelids, wash away foreign objects,
  and contain lysozyme, which kills certain bacteria.

C. Match these terms with the correct parts of the diagram   Lacrimal canaliculi
   labeled in Figure 15-1:   Lacrimal gland
                                                                   Lacrimal sac
                                                                   Nasolacrimal duct
                                                                   Puncta

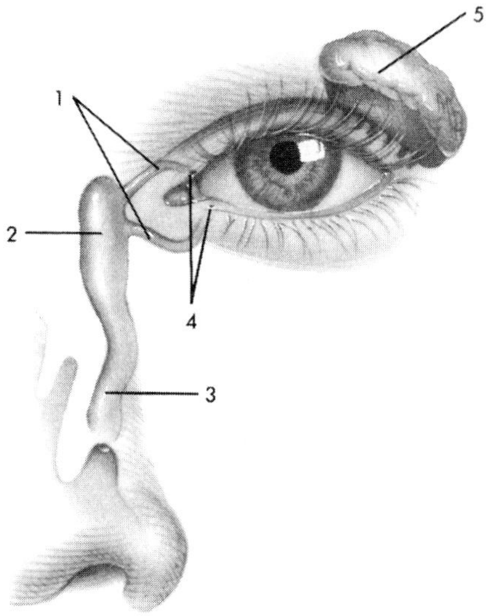

Figure 15-1

1. _____

2. _____

3. _____

4. _____

5. _____

D. Match these extrinsic eye muscles   Oblique muscles
   with the correct description: Rectus muscles

_____ 1. Extrinsic eye muscles that run more or less straight
                              anteroposteriorly.

_____ 2. Extrinsic eye muscles that are placed at an angle to the eye.

_____ 3. There are four of these muscles.

Anatomy of the Eye

"The eye is composed of three coats or tunics."

A. Match these terms with the correct statement or definition:

Cornea	Sclera
Fibrous tunic	Vascular tunic
Nervous tunic	

_____ 1. Outer layer of the eye, consisting of the sclera and cornea.

_____ 2. Middle eye layer; the choroid, ciliary body, and iris.

_____ 3. Inner layer of the eye, consisting of the retina.

_____ 4. Firm, opaque, white outer layer of the posterior five sixths of the eye.

_____ 5. Avascular, transparent structure; permits light entry.

B. Match these terms with the correct statement or definition:

Choroid	Dilator pupillae
Ciliary body	Iris
Ciliary processes	Pupil
Ciliary ring	Sphincter pupillae

_____ 1. Vascular tunic associated with the scleral portion of the eye.

_____ 2. Attached to the anterior margin of the choroid and the lateral margin of the iris.

_____ 3. Part of the ciliary body that contains smooth muscles (intrinsic eye muscles); attaches to the lens by the suspensory ligaments and changes the shape of the lens.

_____ 4. Capillaries and epithelium that produce aqueous humor.

_____ 5. Smooth muscle structure surrounding the pupil.

_____ 6. Radial group of iris muscles; increases the size of the pupil.

C. Match these terms with the correct statement or definition:

Fovea centralis	Pigmented retina
Macula lutea	Rods and cones
Optic disc	Sensory retina

_____ 1. Outer part of the retina consisting of simple cuboidal epithelium.

_____ 2. Photoreceptor cells in the retina.

_____ 3. Small yellow spot near the center of the posterior retina.

_____ 4. Small pit in the retina with the greatest visual acuity.

_____ 5. Blind spot containing no photoreceptor cells; where blood vessels enter and nerve processes exit the eye.

D. Match these terms with the correct parts of the diagram labeled in Figure 15-2:

Choroid   Optic nerve
Ciliary body   Pupil
Cornea   Retina
Iris   Sclera
Lens   Suspensory ligaments

1. _____

2. _____

3. _____

4. _____

5. _____

6. _____

7. _____

8. _____

9. _____

10. _____

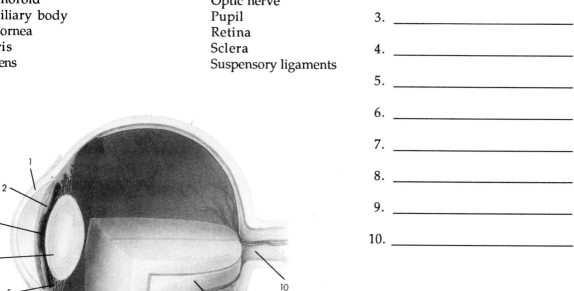

Figure 15-2

E. Match these terms with the correct statement or definition:

Anterior chamber   Posterior chamber
Aqueous humor   Vitreous humor
Canal of Schlemm

_____   1. Part of the anterior eye cavity that lies between the iris and lens.

_____   2. Fluid that fills the anterior and posterior chambers of the eye; maintains intraocular pressure.

_____   3. Venous ring that returns aqueous humor to the circulatory system.

_____   4. Transparent jellylike substance that fills the posterior cavity of the eye.

F. Using the terms provided, complete these statements:

Capsule
Crystallines
ligaments
Cuboidal epithelial

Lens fibers
Suspensory

1. _____

2. _____

3. _____

4. _____

5. _____

The lens consists of a layer of __(1)__ cells on its anterior surface and a posterior portion of very long columnar epithelial cells called __(2)__. Lens fibers lose their nuclei and other cellular organelles and accumulate a special set of proteins called __(3)__. The lens is covered by a highly elastic transparent __(4)__. The lens is suspended between the two eye compartments by the __(5)__, which are connected to the lens capsule and to the ciliary body.

Function of the Complete Eye

"*The eye functions much like a camera.***"**

A. Match these terms with the correct statement or definition:

Concave lens surface
Convex lens surface
Focal point
Focusing

Reflection
Refraction
Visible light (spectrum)

_____ 1. Portion of the electromagnetic spectrum that can be detected by the human eye.

_____ 2. Bending of light rays as they pass into a new medium, such as light passing from air into water.

_____ 3. Causes light rays to converge.

_____ 4. Point where convergent light rays cross.

_____ 5. Act of causing light rays to converge.

_____ 6. Light rays that bounce off a nontransparent object.

B. Match these terms with the correct statement or definition:

Accommodation
Ciliary muscles
Convergence
Depth of focus

Emmetropia
Far point of vision
Near point of vision
Suspensory ligaments

_____ 1. Structures that maintain elastic pressure on the lens.

_____ 2. Normal resting, flattened condition of the lens.

_____ 3. Structures that contract and reduce the tension on the lens.

_____ 4. Point beyond which accommodation is not required; usually 20 feet or more from the eye.

_____ 5. Process of allowing the lens to assume a more spherical (convex) shape.

_____ 6. The greatest distance through which an object can be moved and still remain in focus.

_____ 7. Medial movement of the eyeballs that keeps objects in view as they move closer to the eye.

☞ When focusing on a nearby object the lens becomes more spherical (accommodation), the pupil constricts (increasing depth of field), and the eyeballs turn medially (convergence).

Structure and Function of the Retina

"_The retina consists of several distinct layers._**"**

A. Match these parts of the retina with the correct description:

Pigmented retina
Sensory retina

_____ 1. This portion of the retina has three layers of neurons.

_____ 2. This portion of the retina is a layer of cells filled with melanin.

☞ Melanin pigment enhances visual acuity by isolating individual photoreceptors in a black matrix.

B. Match these terms with the correct statement or definition:

Bleaching   Retinal
Cone cells   Rhodopsin
Fovea centralis   Rod cells
Iodopsin

_____ 1. Bipolar photoreceptor cells that cannot detect color, most important for vision under conditions of reduced light.

_____ 2. Molecule containing opsin in loose chemical combination with retinal; found in rod cells.

_____ 3. Pigment molecule derived from vitamin A.

_____ 4. Process that occurs when rhodopsin is exposed to light, resulting in the separation of retinal and opsin.

_____ 5. Bipolar photoreceptor cells that are sensitive to blue, red, or green light, most important in visual acuity.

_____ 6. Visual pigment in cones.

_____ 7. Part of the retina that has only cone cells.

Away from the fovea centralis, the concentration of cones decreases, and the concentration of rods increases.

C. Match the type of adaptation to changing light conditions with the correct statement:

Dark adaptation
Light adaptation

_____ 1. Amount of rhodopsin increases.

_____ 2. Pupil constricts.

_____ 3. Rod function increases, cone function decreases.

_____ 4. More rapid of the two processes.

D. Match these terms with the correct statement or definition:

Association neurons   Optic nerve
Bipolar cells   Photoreceptor cells
Ganglion cells

_____ 1. Neurons in the outermost layer of the sensory retina; next to the pigmented retina.

_____ 2. Neurons in the middle layer of the sensory retina.

_____ 3. Innermost layer of neurons in the sensory retina; closest to the vitreous humor.

_____ 4. Collection of all the ganglion cells as they exit the eye.

_____ 5. Neurons that modify the signal from photoreceptor cells before they leave the retina.

E. Using the terms provided, complete these statements:

Bipolar   Increases
Decreases   Spatial
summation
Ganglion

In the retina, numerous rods usually synapse with one __(1)__ cell. Further, many of these cells then synapse with one __(2)__ cell. This arrangement results in __(3)__ that __(4)__ light sensitivity and __(5)__ visual acuity. In comparison, cones exhibit little convergence. This decreases light sensitivity and increases visual acuity.

1. _____

2. _____

3. _____

4. _____

5. _____

Neuronal Pathways

"*There are several neuronal pathways from the eyes through the brain.***"**

A. Using the terms provided, complete these statements:

Optic chiasma   Superior colliculi
Optic nerve   Thalamus
Optic radiations   Visual cortex
Optic tract

Ganglion cells from the sensory retina converge and exit the
eye in the __(1)__ , then pass into the cranial vault through
the optic foramen. Just inside the vault the optic nerves
are connected at the __(2)__ , where some of the axons cross to
the opposite side of the brain. Beyond the optic chiasma,
the route of the ganglionic axons is called the __(3)__ . Most of
the optic tract axons end in the __(4)__ , although some
separate from the optic tract and terminate in the
__(5)__ , the center for reflexes initiated by visual stimuli.
Neurons of the thalamus form the __(6)__ , projecting to the
__(7)__ in the occipital lobe.

1. _____

2. _____

3. _____

4. _____

5. _____

6. _____

7. _____

B. Match these terms with the
   correct statement or definition:

Binocular vision   Nasal half
Depth perception   Temporal portion of retina
Nasal portion of retina   Temporal half

_____ 1. Images from this part of a visual field falls on the temporal
                           portion of the retina.

_____ 2. Images from this part of the retina do not cross to the opposite
                           side of the brain.

_____ 3. Area of overlap of the visual fields.

_____ 4. Ability to distinguish between near and far objects.

☞ Images from the right half of both visual fields (e.g., the nasal half of the left visual
field and the temporal half of the right visual field) project to the left side of the brain,
and vice versa.

Clinical Applications

*"*There are many common disorders of the eye.*"*

A. Match these terms with the correct statement or definition:

Astigmatism   Myopia
Hyperopia   Presbyopia

_____ 1. Occurs when the lens and cornea are optically too strong, or when the eyeball is too long.

_____ 2. The eye becomes less able to accommodate; a result of aging.

_____ 3. Occurs when the cornea or lens is not uniformly curved.

_____ 4. Corrected by a concave lens.

_____ 5. Corrected by reading glasses or bifocals.

_____ 6. Corrected by lenses with curvature opposite to the defect.

B. Match these terms with the correct statement or definition:

Cataract   Glaucoma
Color blindness   Strabismus
Diabetes   Trachoma

_____ 1. Lack of parallelism of light paths through the eye.

_____ 2. Disease resulting from increased intraocular pressure.

_____ 3. Recessive X-linked trait producing a deficiency of cone pigments.

_____ 4. Clouding of the lens resulting from a buildup of proteins.

_____ 5. Results in defective circulation to the eye; a leading cause of blindness.

Auditory Structures and Their Functions

*"*The external, middle, and inner ear are all involved in hearing.*"*

A. Match these terms with the correct description or definition:

Auricle (pinna)   External auditory meatus
Cerumen   Tympanic membrane

_____ 1. Fleshy part of the external ear on the outside of the head.

_____ 2. Passageway from the outside to the eardrum.

_____ 3. Modified sebum, commonly called earwax, that helps to prevent foreign objects from reaching the tympanic membrane.

_____ 4. Thin, semitransparent, nearly oval, three-layered membrane that vibrates in response to sound waves.

B. Match these terms with the correct description or definition:

Auditory tube   Oval window
Incus   Round window
Malleus   Stapes
Mastoid air cells

_____ 1. Opening with the foot plate of the stapes attached.

_____ 2. Membrane-covered opening on the medial side of the middle ear with nothing attached.

_____ 3. Spaces in the mastoid processes of the temporal bones.

_____ 4. Structure that allows air pressure to equalize between the middle ear and the outside air.

_____ 5. Auditory ossicle that is attached to the tympanic membrane.

_____ 6. Middle auditory ossicle.

C. Match these terms with the correct parts of the diagram labeled in Figure 15-3:

Auditory tube
Auricle (pinna)
Cochlea
External auditory meatus
External ear
Incus
Inner ear
Malleus
Middle ear
Oval window
Round window
Semicircular canals
Stapes
Tympanic membrane
Vestibule

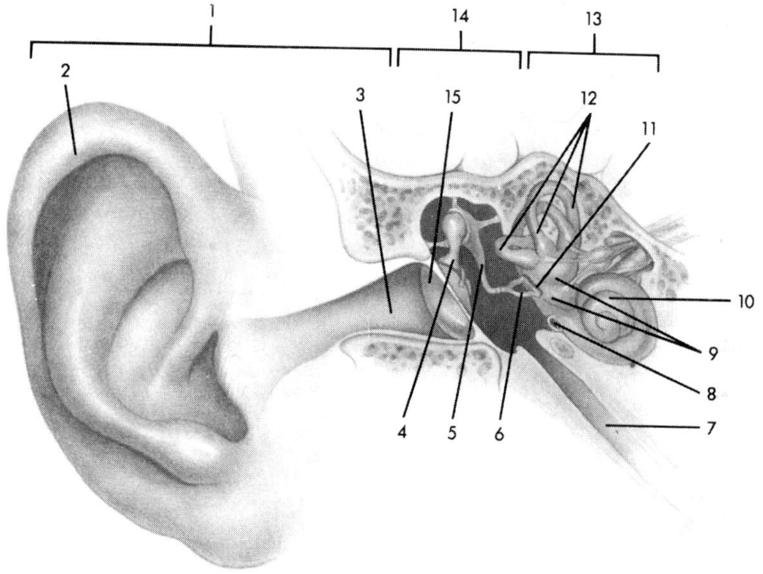

Figure 15-3

1. _____   6. _____   11. _____

2. _____   7. _____   12. _____

3. _____   8. _____   13. _____

4. _____   9. _____   14. _____

5. _____   10. _____   15. _____

D. Match these terms with the correct statement or definition:

Basilar membrane
Bony labyrinth
Cochlea
Cochlear duct
Helicotrema

Membranous labyrinth
Modiolus
Scala tympani
Scala vestibuli
Vestibular membrane

_____ 1. Interconnecting bony tunnels and chambers in the petrous portion of the temporal bone; subdivided into the cochlea, vestibule, and semicircular canals.

_____ 2. Part of the inner ear that can be divided into scala tympani, scala vestibuli, and cochlear duct.

_____ 3. Cochlear chamber that contains perilymph and extends from the oval window to the helicotrema.

_____ 4. Cochlear chamber that contains perilymph and extends from the helicotrema to the round window.

_____ 5. Opening between the scala vestibuli and the scala tympani.

_____ 6. Wall of the membranous labyrinth bordering the scala vestibuli.

_____ 7. Wall of the membranous labyrinth bordering the scala tympani.

_____ 8. Interior of the membranous labyrinth containing endolymph; space between the vestibular and basilar membranes.

_____ 9. Bony core of the cochlea, which has a projection called the spiral lamina.

_____ 10. Attached to the spiral lamina and the spiral ligament.

E. Using the terms provided, complete these statements:

Cochlear ganglion
Cochlear nerve
Hair cells

Spiral organ
Tectorial membrane
Vestibulocochlear nerve

The cells inside the cochlear duct are highly modified to form a structure called the _(1)_, or organ of Corti. This structure contains supporting epithelial cells and specialized sensory cells called _(2)_, which have specialized hairlike projections at their ends. The apical ends of the hairs are embedded within an acellular gelatinous shelf called the _(3)_, which is attached to the spiral lamina. Hair cells have no axons, but the basilar region of each hair cell is covered by synaptic terminals of sensory neurons, the cell bodies of which are grouped into a _(4)_ within the cochlear modiolus. The proximal, afferent fibers of these neurons join to form the _(5)_, which then joins the vestibular nerve to become the _(6)_.

1. _____

2. _____

3. _____

4. _____

5. _____

6. _____

F. Using the terms provided, complete these statements:

Amplitude
Attenuation reflex
Auditory ossicles
Basilar membrane
Endolymph
Frequency

Round window
Scala vestibuli
Tectorial membrane
Timbre
Tympanic membrane

1. _____

2. _____

3. _____

4. _____

5. _____

6. _____

7. _____

8. _____

9. _____

10. _____

11. _____

Vibrations are propagated through the air as sound waves. Volume (loudness) of sound is a function of the _(1)_ (height) of the waves, and pitch is a function of _(2)_ (how far the waves are apart). The resonance quality, or overtones, of a sound is _(3)_. Sound waves are collected by the auricle and are conducted through the external auditory meatus toward the _(4)_. Sound waves strike the tympanic membrane and cause it to vibrate, which in turn causes the _(5)_ to vibrate. Two small skeletal muscles attached to the ossicles reflexively dampen excessively loud sounds. The contraction of these muscles, called the _(6)_, protects the delicate ear structures from being damaged by loud noises. The vibration of the auditory ossicles is transferred to the oval window and causes vibration in the perilymph of the _(7)_. This produces waves in the perilymph, which causes the vestibular membrane and _(8)_ of the cochlear duct to vibrate. Consequently, the _(9)_ vibrates, causing the hairs embedded in the _(10)_ to bend, inducing action potentials in cochlear neurons. Vibration of the basilar membrane also causes vibration of the perilymph within the scala tympani. These vibrations are dissipated by vibration of the membrane of the _(11)_.

G. Match these terms with the correct parts of the diagram labeled in Figure 15-4:

Basilar membrane
Cochlear duct
Helicotrema
Oval window
Round window
Scala tympani
Scala vestibuli
Spiral organ
Tectorial membrane
Vestibular membrane

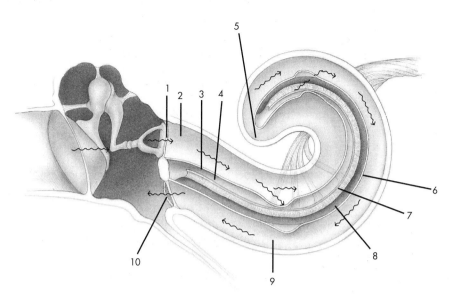

Figure 15-4

1. _____   5. _____   8. _____

2. _____   6. _____   9. _____

3. _____   7. _____   10. _____

4. _____

Neuronal Pathways for Hearing

“The special senses of hearing and balance are both transmitted by the vestibulocochlear nerve.”

Using the terms provided, complete these statements:

Auditory cortex   Superior colliculus
Inferior colliculi   Superior olivary nucleus
Medulla   Thalamus

The neurons from the cochlear ganglion synapse with the
central nervous system neurons in the dorsal or ventral
cochlear nucleus in the _(1)_ . These neurons in turn either
synapse in or pass through the _(2)_ . Ascending neurons
from this point travel in the lateral lemniscus. All
ascending fibers synapse in the _(3)_ , and neurons from
there project to the medial geniculate nucleus of the _(4)_ ,
where they synapse with neurons that terminate in the
_(5)_ in the dorsal portion of the temporal lobe. Neurons
from the inferior colliculus also project to the _(6)_ and
cause reflexive turning of the head and eyes toward a loud
sound.

1. _____

2. _____

3. _____

4. _____

5. _____

6. _____

Balance

"The organs of balance are found within the inner ear."

Match these terms with the
correct statement or definition:

Ampulla	Macula
Crista ampullaris	Otoliths
Cupula	Semicircular canals
Kinetic labyrinth	Static labyrinth

_____ 1. Consists of the utricle and saccule; involved in evaluating the position of the head relative to gravity or linear acceleration.

_____ 2. Specialized epithelium in the utricle and saccule.

_____ 3. Add weight to the gelatinous mass that embeds hair cells.

_____ 4. Arranged in three planes; they enable a person to sense movement in all directions.

_____ 5. Specialized sensory epithelium found in each ampulla.

_____ 6. Curved gelatinous mass suspended over the crista ampullaris.

Neuronal Pathways for Balance

"Balance is a complex process not simply confined to one kind of input."

Using the terms provided, complete these statements:

Nystagmus   Vestibular ganglion
Proprioceptive nerves   Vestibular nucleus

Neurons synapsing on the hair cells of the maculae and cristae ampullares converge into the __(1)__, where their cell bodies are located. Afferent fibers from these neurons terminate in the __(2)__ within the medulla. The vestibular nuclear complex also receives input from __(3)__ throughout the body. Reflex pathways exist between the kinetic portion of the vestibular system and the nuclei controlling the extrinsic eye muscles. Spinning the head causes the eyes to track slowly in the direction of motion and return with a rapid recovery movement. This is called __(4)__.

1. _____

2. _____

3. _____

4. _____

Clinical Applications

"There are several disorders that affect the ear."

Match these terms with the
correct statement or definition:

Otitis media	Tinnitus
Otosclerosis	

_____ 1. Caused by spongy bone growth immobilizing the stapes.

_____ 2. Disorder characterized by noises in the ears.

1. Name the three functional areas of the olfactory cortex and the function of each.

2. List the four basic tastes detected by the taste buds.

3. List the three coats (tunics) of the eye.

4. Name the two types of photoreceptor cells in the retina, and list two important differences between them.

5. List the two major compartments (cavities) of the eye and the substances that fill each.

6. List the three layers of neurons in the sensory retina.

7. List the visual pigments found in rod and cone cells.

8. List the three types of cone cells.

9. Name the three auditory ossicles found in the middle ear.

10. List the three subdivisions of the bony labyrinth, and give their function.

11. List the three cochlear chambers and the fluid found in each.

12. List the two functional parts of the organs of balance.

13. Name the structures that relieve pressure in the middle ear and the inner ear.

MASTERY LEARNING ACTIVITY

Place the letter corresponding to the correct answer in the space provided.

_____ 1. Olfactory neurons
   a. have projections called cilia.
   b. have axons that combine to form the olfactory nerves.
   c. connect to the olfactory bulb.
   d. have receptors that react with molecules dissolved in fluid.
   e. all of the above

_____ 2. The olfactory cortex can be divided into three areas.
   a. The lateral olfactory area is involved with the emotional reactions to smell.
   b. The medial olfactory area is involved in the conscious perception of smell.
   c. The intermediate olfactory area modifies olfactory bulb input.
   d. all of the above

_____ 3. Gustatory (taste) cells
   a. are found only on the tongue.
   b. extend through tiny openings called taste buds.
   c. release neurotransmitter when stimulated.
   d. have axons that extend to the taste area of the cerebral cortex.
   e. all of the above

_____ 4. Which of the following is NOT one of the basic tastes?
   a. spicy
   b. sweet
   c. sour
   d. salt
   e. bitter

5. Tears
   a. are released onto the surface of the eye near the medial corner of the eye.
   b. in excess are removed by the canal of Schlemm.
   c. in excess can cause a sty.
   d. can eventually end up in the nasal cavity.

6. The fibrous tunic of the eye includes the
   a. conjunctiva.
   b. sclera.
   c. choroid.
   d. iris.
   e. retina.

7. The ciliary body
   a. contains smooth muscles that attach to the lens by suspensory ligaments.
   b. produces the vitreous humor.
   c. is part of the iris of the eye.
   d. is part of the sclera.

8. The lens normally focuses light onto the
   a. optic disc.
   b. iris.
   c. macula lutea.
   d. cornea.

9. Given the following structures:
   1. lens
   2. aqueous humor
   3. vitreous humor
   4. cornea

   Choose the arrangement that lists the structures in the order that light entering the eye would encounter them.
   a. 1,2,3,4
   b. 1,4,2,3
   c. 4,1,2,3
   d. 4,2,1,3
   e. 4,3,1,2

10. Aqueous humor
   a. is the pigment responsible for the black color of the choroid.
   b. exits the eye through the canal of Schlemm.
   c. is produced by the iris.
   d. can cause cataracts if overproduced.

11. Contraction of the smooth muscle in the ciliary body causes the
   a. lens to flatten.
   b. lens to become more spherical.
   c. pupil to constrict.
   d. pupil to dilate.

12. Given the following events:
   1. medial rectus contracts
   2. lateral rectus contracts
   3. pupils dilate
   4. pupils constrict
   5. lens of the eye flattens
   6. lens of the eye becomes more spherical

   Assume that you were looking at an object that was 30 feet away. If you suddenly looked at an object that was 1 foot away, which of the above events would occur?
   a. 1,3,6
   b. 1,4,5
   c. 1,4,6
   d. 2,3,6
   e. 2,4,5

13. Concerning the arrangement of neurons within the retina,
   a. photoreceptor cells synapse with bipolar cells.
   b. axons of ganglionic cells leave the retina to form the optic nerve.
   c. light must pass through a layer of ganglionic cells and a layer of bipolar cells to reach the photoreceptor cells.
   d. all of the above

14. Which of the following photoreceptor cells is correctly matched with its function?
   a. rods - vision in low light
   b. cones - visual acuity
   c. cones - color vision
   d. all of the above

15. Concerning dark adaptation,
   a. the amount of rhodopsin increases.
   b. the pupils constrict.
   c. it occurs more rapidly than light adaptation.
   d. all of the above

_____ 16. In the retina there are cones that are most sensitive to a particular color. Given the following list of colors:
1. red
2. yellow
3. green
4. blue

Indicate which colors correspond to specific types of cones.
a. 2,3
b. 3,4
c. 1,2,3
d. 1,3,4
e. 1,2,3,4

_____ 17. Given the following areas of the retina:
1. macula lutea
2. fovea centralis
3. optic disc
4. periphery of the eye

Choose the arrangement that lists the areas according to their cone density, starting with the area that has the highest density of cones.
a. 1,2,3,4
b. 1,3,2,4
c. 2,1,4,3
d. 2,4,1,3
e. 3,4,1,2

_____ 18. Concerning axons in the optic nerve from the right eye,
a. they all go to the right occipital lobe.
b. they all go to the right lobe of the thalamus.
c. some go to the right occipital lobe, and some go to the left occipital lobe.
d. some go to the right thalamic lobe, and some go to the left thalamic lobe.

_____ 19. A lesion that destroyed the left optic tract of a boy would eliminate vision in his
a. left nasal visual field.
b. left temporal visual field.
c. right temporal visual field.
d. a and b
e. a and c

_____ 20. A person with an abnormally long eyeball (anterior to posterior) would be _____ and would wear a _____ to correct his or her vision.
a. nearsighted, concave lens
b. nearsighted, convex lens
c. farsighted, concave lens
d. farsighted, convex lens

_____ 21. Which of the following structures is found within or is a part of the external ear?
a. oval window
b. auditory tube
c. ossicles
d. auricle
e. cochlear duct

_____ 22. Given the following ear bones:
1. incus
2. malleus
3. stapes

Choose the arrangement that lists the ear bones in order from the tympanic membrane to the inner ear.
a. 1,2,3
b. 1,3,2
c. 2,1,3
d. 2,3,1
e. 3,2,1

_____ 23. Given the following structures:
1. perilymph
2. endolymph
3. vestibular membrane
4. basilar membrane

Choose the arrangement that lists the structures in the order sound waves coming from the outside would encounter them in producing sound.
a. 1,3,2,4
b. 1,4,2,3
c. 2,3,1,4
d. 2,4,1,3

_____ 24. The spiral organ is found within the
a. cochlear duct.
b. scala vestibuli
c. scala tympani.
d. vestibule

25. The _____ allows for release of excess pressure within the INNER ear.
    a. tectorial membrane
    b. basilar membrane
    c. vestibular membrane
    d. round window
    e. auditory tube

26. An increase in the loudness of sound occurs as a result of an increase in the
    a. frequency of the sound wave.
    b. amplitude of the sound wave.
    c. resonance of the sound wave.
    d. a and b

27. If a person's attenuation reflex were nonfunctional, how would the perception of sound be affected when that person was exposed to a loud noise?
    a. The noise would seem louder than normal.
    b. The noise would seem quieter than normal.
    c. The noise would seem normal.

28. Interpretation of different sounds is possible because of the ability of the _____ to vibrate at different frequencies and stimulate the _____.
    a. vestibular membrane, vestibular nerve
    b. vestibular membrane, spiral organ
    c. basilar membrane, vestibular nerve
    d. basilar membrane, spiral organ

29. Which structure is a specialized receptor found within the utricle?
    a. macula
    b. crista ampullaris
    c. spiral organ
    d. cupula

30. Damage to the semicircular canals would affect the ability to detect
    a. linear acceleration.
    b. the position of the head relative to the ground.
    c. movement of the head in all directions.
    d. none of the above

FINAL CHALLENGES

Use a separate sheet of paper to complete this section.

1. At the battle of Bunker Hill, it could have been said, "Don't shoot until you see the _____ of their eyes!"

2. Crash McBang, a 20-year-old student, wears glasses when he reads. However, he never wears glasses when he drives. What eye disorder does Crash probably have? What kind of lens is in his reading glasses?

3. A woman has lost the ability to see "out of the corner of both of her eyes" (i.e., loss of peripheral vision). What visual fields are affected? What parts of her retina are affected? If the loss is a result of nerve tract damage, where is the most likely place for that damage to have occurred?

4. Retinitis pigmentosa is a hereditary disorder that typically begins in childhood with the development of night blindness. As the disease progresses, rods and cones degenerate and the visual fields become concentrically smaller until little useful vision is left. Explain the symptoms produced as the disease progresses.

5. Otto Harp notices that he can't hear music when it is played softly, but he can hear perfectly well when it is played loudly. Following his doctor's advice, he gets a hearing aid that solves his hearing problems. Do you think Otto's problem is in the middle or the inner ear? How did the hearing aid help? Explain.

6. As part of a game you are spun around in a clockwise direction. When you suddenly stop spinning, it feels like you are spinning in a counterclockwise direction. Explain.

ANSWERS TO CHAPTER 15

Classification of the Senses
1. Somatic; 2. Visceral; 3. Special;
4. Mechanoreceptors; 5. Thermoreceptors;
6. Nociceptors

Sensation
1. Projection; 2 Adaptation; 3. Phasic

Types of Afferent Nerve Endings
1. Proprioceptors; 2. Exteroceptors; 3. Free nerve endings; 4. Merkel's disks; 5. Pacinian corpuscles; 6. Meissner's corpuscles; 7. Ruffini's end organs; 8. Golgi tendon apparatus; 9. Muscle spindle

Olfaction
A. 1. Olfactory epithelium; 2. Olfactory bulbs;
   3. Olfactory tracts; 4. Olfactory vesicles;
   5. Olfactory hairs; 6. Basal cells
B. 1. Olfactory nerves; 2. Mitral cells;
   3. Association neurons; 4. Lateral olfactory area; 5. Medial olfactory area;
   6. Intermediate olfactory area

Taste
A. 1. Circumvallate papillae; 2. Fungiform papillae; 3. Foliate papillae; 4. Filiform papillae
B. 1. Taste buds; 2. Gustatory (taste) cells;
   3. Gustatory hairs; 4. Gustatory (taste) pores
C. 1. Facial nerve (VII); 2. Facial nerve (VII);
   3. Glossopharyngeal nerve (IX)

Accessory Structures
A. 1. Palpebrae; 2. Canthi; 3. Caruncle; 4. Tarsal plate; 5. Ciliary glands; 6. Sty; 7. Meibomian glands; 8. Chalazion; 9. Palpebral conjunctiva
B. 1. Lacrimal gland; 2. Lacrimal canaliculus
   3. Punctum; 4. Lacrimal papilla; 5. Lacrimal sac; 6. Nasolacrimal duct
C. 1. Lacrimal canaliculi; 2. Lacrimal sac;
   3. Nasolacrimal duct; 4. Puncta; 5. Lacrimal gland
D. 1. Rectus muscles; 2. Oblique muscles;
   3. Rectus muscles

Anatomy of the Eye
A. 1. Fibrous tunic; 2. Vascular tunic; 3. Nervous tunic; 4. Sclera; 5. Cornea
B. 1. Choroid; 2. Ciliary body; 3. Ciliary ring;
   4. Ciliary processes; 5. Iris; 6. Dilator pupillae
C. 1. Pigmented retina; 2. Rods and cones;
   3. Macula lutea; 4. Fovea centralis; 5. Optic disc
D. 1. Cornea; 2. Iris; 3. Pupil; 4. Lens;
   5. Suspensory ligaments; 6. Ciliary body;
   7. Sclera; 8. Choroid; 9. Retina; 10. Optic nerve
E. 1. Posterior chamber; 2. Aqueous humor;
   3. Canal of Schlemm; 4. Vitreous humor
F. 1. Cuboidal epithelial; 2. Lens fibers;

3. Crystallines; 4. Capsule; 5. Suspensory ligaments

Function of the Complete Eye
A. 1. Visible light (spectrum); 2. Refraction;
   3. Convex lens surface; 4. Focal point;
   5. Focusing; 6. Reflection
B. 1. Suspensory ligaments; 2. Emmetropia;
   3. Ciliary muscles; 4. Far point of vision;
   5. Accommodation; 6. Depth of focus;
   7. Convergence

Structure and Function of the Retina
A. 1. Sensory retina; 2. Pigmented retina
B. 1. Rod cells; 2. Rhodopsin; 3. Retinal;
   4. Bleaching; 5. Cone cells; 6. Iodopsin;
   7. Fovea centralis
C. 1. Dark adaptation; 2. Light adaptation;
   3. Dark adaptation; 4. Light adaptation
D. 1. Photoreceptor cells; 2. Bipolar cells;
   3. Ganglion cells; 4. Optic nerve;
   5. Association neurons
E. 1. Bipolar; 2. Ganglion; 3. Spatial summation;
   4. Increases; 5. Decreases

Neuronal Pathways
A. 1. Optic nerve; 2. Optic chiasma; 3. Optic tract; 4. Thalamus; 5. Superior colliculi;
   6. Optic radiations; 7. Visual cortex
B. 1. Nasal half; 2. Temporal portion of retina;
   3. Binocular vision; 4. Depth perception

Clinical Applications
A. 1. Myopia; 2. Presbyopia; 3. Astigmatism;
   4. Myopia; 5. Presbyopia; 6. Astigmatism
B. 1. Strabismus; 2. Glaucoma; 3. Color blindness; 4. Cataract; 5. Diabetes

Auditory Structures and Their Functions
A. 1. Auricle; 2. External auditory meatus;
   3. Cerumen; 4. Tympanic membrane
B. 1. Oval window; 2. Round window; 3. Mastoid air cells; 4. Auditory tube; 5. Malleus; 6. Incus
C. 1. External ear; 2. Auricle; 3. External auditory meatus; 4. Malleus; 5. Incus 6. Stapes;
   7. Auditory tube; 8. Round window;
   9. Vestibule; 10. Cochlea; 11. Oval window;
   12. Semicircular canals; 13. Inner ear;
   14. Middle ear; 15. Tympanic membrane
D. 1. Bony labyrinth; 2. Cochlea; 3. Scala vestibuli; 4. Scala tympani; 5. Helicotrema;
   6. Vestibular membrane; 7. Basilar membrane; 8. Cochlear duct; 9. Modiolus;
   10. Basilar membrane
E. 1. Spiral organ; 2. Hair cells; 3. Tectorial membrane; 4. Cochlear ganglion; 5. Cochlear nerve; 6. Vestibulocochlear nerve
F. 1. Amplitude; 2. Frequency; 3. Timbre;
   4. Tympanic membrane; 5. Auditory ossicles;
   6. Attenuation reflex; 7. Scala vestibuli;

8. Endolymph; 9. Basilar membrane;
10. Tectorial membrane; 11. Round window
G. 1. Oval window; 2. Scala vestibuli;
3. Vestibular membrane; 4. Cochlear duct;
5. Helicotrema; 6. Basilar membrane;
7. Tectorial membrane; 8. Spiral organ;
9. Scala tympani; 10. Round window

Neuronal Pathways for Hearing
1. Medulla; 2. Superior olivary nucleus; 3. Inferior
colliculi; 4. Thalamus; 5. Auditory cortex;

6. Superior colliculus

Balance
1. Static labyrinth; 2. Macula; 3. Otoliths;
4. Semicircular canals; 5. Crista ampullaris;
6. Cupula

Neuronal Pathways for Balance
1. Vestibular ganglion; 2. Vestibular nucleus;
3. Proprioceptive nerves; 4. Nystagmus

Clinical Applications
1. Otosclerosis; 2. Tinnitus

QUICK RECALL

1. Lateral olfactory area: conscious perception of smell; Intermediate olfactory area: modulating smell: Medial olfactory area: visceral and emotional responses to smell
2. Sour, salty, bitter, and sweet
3. Fibrous, vascular, and nervous tunics
4. Rods: high sensitivity to light, but lower visual acuity; shades of gray
Cones: lower sensitivity to light, but greater visual acuity; color
5. Anterior compartment filled with aqueous humor and posterior compartment filled with vitreous humor
6. Photoreceptors (rods and cones), bipolar cells, and ganglion cells

7. Rods: rhodopsin; Cones: iodopsin
8. Red-sensitive, green-sensitive, and blue-sensitive cone cells
9. Malleus, incus, and stapes
10. Vestibule, cochlea, and semicircular canals. The vestibule and semicircular canals are involved primarily in balance, and the cochlea is involved in hearing.
11. Scala vestibuli: perilymph; Scala tympani: perilymph; and the cochlear duct, filled with endolymph
12. Static labyrinth and kinetic labyrinth
13. Middle ear: auditory tube
Inner ear: round window

MASTERY LEARNING ACTIVITY

1. E   The olfactory neurons are bipolar neurons with projections called cilia. The cilia have receptors that react to airborne molecules that have dissolved in the fluid covering the olfactory epithelium. Action potentials generated in the olfactory neurons are propagated through the olfactory neuron axons (which form the olfactory nerve) to the olfactory bulb.

2. C   The axons from the intermediate area synapse with association neurons in the olfactory bulb and function to modify olfactory bulb input to the olfactory cortex. The lateral olfactory area is involved with conscious sensation of odors and the medial olfactory area with the emotional reactions to smell.

3. C   Taste cells are found within taste buds, which are found in the tongue and mouth. The gustatory hairs of taste cells extend through openings called taste pores. When receptors react with substances dissolved in fluid, the taste cells release a neurotransmitter that stimulates action potentials in the neurons of cranial nerves (facial, glossopharyngeal, and vagus). The neurons synapse with neurons that extend to the

thalamus. Thalamic neurons then extend to the taste area of the cerebral cortex.

4. A   The four basic tastes are sour, salty, bitter, and sweet.

5. D   The lacrimal glands are located in the superolateral corner of the orbit. Released tears move across the anterior surface of the eye because of gravity and blinking. The tears exit through the lacrimal canaliculi in the medial corner of the eye and eventually reach the nasal cavity.

6. B   The fibrous tunic includes the sclera and cornea. The vascular tunic includes the choroid, ciliary body, and iris. The nervous tunic consists of the retina. The conjunctiva is a mucous membrane that covers the anterior sclera and inner surface of the eyelids.

7. A   The ciliary body, which is part of the choroid, consists of an outer ciliary ring (smooth muscles that attach to the lens by suspensory ligaments) and an inner group of ciliary processes that produce aqueous humor.

302

8. C.   The macula lutea is located almost exactly at the posterior pole of the eye. A depression, the fovea centralis, is located in the center of the macula lutea. The fovea centralis is the most sensitive portion of the retina. The optic disc contains no photoreceptor cells and is also called the blind spot. The iris regulates the amount of light reaching the lens.

9. D.   Light passes through the cornea, aqueous humor, lens, and vitreous humor.

10. B.   The ciliary processes produce a fluid, aqueous humor, which circulates through the anterior cavity of the eye and exits through the canal of Schlemm. The aqueous humor maintains intraocular pressure, provides nutrients, and refracts light. Overproduction of aqueous humor results in glaucoma.

11. B.   When the ciliary muscles contract, tension on the suspensory ligaments is reduced, and the lens becomes more spherical.

12. C.   The medial rectus would contract, directing the gaze more medially. The pupil would constrict, increasing the depth of focus. The lens would become more spherical (accommodation).

13. D.   Photoreceptor cells synapse with bipolar cells that synapse with ganglionic cells. The axons of the ganglionic cells form the optic nerve. Light passes through the ganglionic and bipolar layers to reach the photoreceptor cells, which are located next to the pigmented retina.

14. D.   Rods are responsible for vision in low light. Cones are responsible for vision in bright light, color vision, and visual acuity (the fovea centralis has only cones).

15. A.   During dark adaptation the production of rhodopsin occurs more rapidly than the breakdown of rhodopsin, and the amount of rhodopsin increases, making the rods more sensitive to low levels of light. This process occurs more slowly than light adaptation, which is a decrease in rhodopsin that occurs when the rate of breakdown of rhodopsin exceeds its rate of production. During dark adaptation the pupils dilate, and during light adaptation they constrict.

16. D.   The three kinds of cones are blue sensitive, red sensitive, and green sensitive.

17. C.   The fovea centralis has the highest density of cones (and has no rods). The macula lutea, in which the fovea centralis is located, has the next highest density of cones. From the macula lutea toward the periphery the number of cones decreases (and the number of rods increases). The optic disc has no cones or rods.

18. D.   Some of the axons in the optic nerve go to the right thalamic lobe, and others cross to the left thalamic lobe through the optic chiasma. In the thalamus the axons synapse with thalamic neurons that extend to the visual cortex of the occipital lobes through the optic radiations.

19. E.   Light from the nasal half of the left visual field falls on the left temporal retina from which axons extend through the left optic nerve to the left optic tract. Light from the temporal half of the right visual field falls on the right nasal retina, from which axons extend through the right optic nerve and cross over in the optic chiasma to the left optic tract.

20. A.   An abnormally long eyeball would result in images being focused anterior to the retina. This condition is nearsightedness and is corrected with a concave lens.

21. D.   The external ear includes the auricle (pinna), external auditory meatus, and tympanic membrane. The middle ear includes the ossicles, oval window, round window, and auditory tube. The cochlear duct is part of the inner ear.

22. C.   The malleus is attached to the tympanic membrane. The incus connects the malleus to the stapes, which is attached to the oval window.

23. A.   Movement of the stapes produces sound waves in the perilymph of the scala vestibuli. The waves cause the vestibular membrane to vibrate, which produces waves in the endolymph of the cochlear duct. Waves in the endolymph cause the basilar membrane to vibrate.

24. A.   The cochlea is subdivided into three parts: the cochlear duct, the scala vestibuli, and the scala tympani. The spiral organ is found within the cochlear duct. The vestibule is the entryway of the inner ear that connects to the middle ear through the oval window.

25. D.   The round window is covered by a membrane. When pressure in the inner ear increases, the membrane bulges outward, relieving the pressure. The auditory tube equalizes pressure between the middle ear and the external environment.

26. B.   The loudness of a sound is directly proportional to the amplitude of the sound wave, whereas the pitch of the sound is a function of the wave frequency. Timbre is the resonance quality of sound.

27. A.   When one is exposed to a loud noise, the acoustic reflex initiates contraction of the tensor tympani and stapedius muscles. This causes the ossicles of the middle ear to become more rigid and dampen the amplitude of sound vibrations. The result is that loud sounds are perceived as being more quiet. The reflex may function to protect the inner ear from loud noises. Without the reflex a noise would be louder than normal.

28. D.   High-pitched sounds cause the base of the basilar membrane to vibrate the most, whereas low-pitched sounds cause the apex to vibrate the most. Vibration of the basilar membrane causes hair cells in the spiral organ to be bent. The hair cells induce action potentials in the cochlear nerve (the vestibular nerve is involved with balance).

29. A.   The macula is found within the utricle. The cupula is a gelatinous mass that is part of the crista ampullaris. The crista ampullaris is part of the semicircular canals. The spiral organ is found within the cochlea.

30. C.   The semicircular canals detect the rate of movement of the head in all directions. The utricle and the saccule detect linear acceleration and the position of the head relative to the ground.

 FINAL CHALLENGES

1. Sclera. The sclera forms the "whites of the eyes."

2. Farsightedness is characterized by normal vision at a distance but requires corrective convex lens to see close up because the near point of vision is greater than normal. Because of the student's youth, presbyopia is not likely.

3. Loss of peripheral vision "out of the corners of the eyes" results from an inability to see the temporal halves of the left and right visual fields. Light from the temporal half of the left visual field falls on the left nasal retina, and light from the temporal half of the right visual field falls on the right temporal retina. The most likely place for damage is in the optic chiasma, because the optic nerve fibers from the nasal retinas cross from one side of the body to the other within the optic chiasma.

4. Early in its development the disease affects rods, causing night blindness. Also, the disease must start at the periphery of the retina and gradually extend toward the macula lutea, thus accounting for the concentric narrowing of vision.

5. Otto's hearing problem is in the middle ear and is caused by reduced transfer of sound from the tympanic membrane to the perilymph of the inner ear. The hearing aid amplifies sound and increases the movement of the tympanic membrane and thus the ear ossicles. The increased movement of the ossicles better transfers sound waves to the perilymph of the inner ear. A problem with the spiral organ or nerve pathways is not indicated, because Otto hears all frequencies of sound when they are loud enough.

6. When a person is spinning in a clockwise direction, at first the endolymph in the semicircular canals lags behind the movement of the skull bones. Consequently, the hair cells embedded in the cupula are pushed in the opposite direction of the spin; this is interpreted by the brain as clockwise spin. Note that the perceived motion of spin (clockwise) is the opposite of the movement of the cupula (counterclockwise). As spinning continues, the endolymph catches up with the skull bones, and the cupula is no longer pushed over. When spinning stops, the skull bones immediately stop moving, but the endolymph in the semicircular canals continues to move, causing the hair cells embedded in the cupula to bend in the direction of the spin. The clockwise movement of the cupula is interpreted as counterclockwise movement, even though no such movement of the head is actually occurring.

16 Autonomic Nervous System

FOCUS: The autonomic nervous system regulates the activities of smooth muscle, cardiac muscle, and glands and can be divided into sympathetic and parasympathetic divisions. The cell bodies of the preganglionic neurons of the sympathetic division are located in the lateral horns of the spinal cord. Their axons pass through spinal nerves T1 to L2 into white rami communicantes to the sympathetic chain ganglia to synapse with postganglionic neurons. The axons of the postganglionic neurons reenter the spinal nerves through gray rami communicantes or exit the sympathetic chain ganglia as sympathetic nerves. In some cases the preganglionic neurons pass through the sympathetic chain ganglia (forming splanchnic nerves) and synapse in collateral ganglia or in the adrenal medulla. The parasympathetic division arises from the brain as part of the cranial nerves (III, VII, IX, or X) or in the lateral horns of the spinal cord from S2 to S4 (giving rise to the pelvic nerves). Preganglionic neurons synapse with postganglionic neurons within terminal ganglia that are near or on their effector organs. The preganglionic neurons of both divisions secrete acetylcholine which binds to nicotinic receptors of the postganglionic neurons. All the postganglionic neurons of the parasympathetic division and some postganglionic neurons of the sympathetic division secrete acetylcholine which binds to muscarinic receptors of the effector organs. Most postganglionic neurons of the sympathetic division secrete norepinephrine which binds to adrenergic receptors (alpha and beta). Generally the parasympathetic division maintains homeostasis on a day-to-day basis, whereas the sympathetic division prepares the body for activity. Both divisions are capable of producing excitatory or inhibitory effects. If they innervate the same organ, the two divisions usually produce opposite effects.

CONTENT LEARNING ACTIVITY

Contrasting the Somatic Motor and Autonomic Nervous Systems

❝*All efferent neurons belong to the somatic motor nervous system or the autonomic nervous system.*❞

A. Match the type of neuron with the correct description:

Afferent neurons   Efferent neurons
Autonomic neurons   Somatic motor neurons

_____ 1. Neurons that propagate action potentials from sensory receptors to the CNS.

_____ 2. Extend from the CNS to skeletal muscle.

_____ 3. Two neurons present between the CNS and effector organs.

_____ 4. Effect on target tissues can be excitatory or inhibitory.

305

B. Match these terms from the autonomic division with the correct statement or definition:

Autonomic ganglion   Postganglionic neuron
Effector organ   Preganglionic neuron

_____ 1. First neuron of ANS between the CNS and the organs innervated.

_____ 2. Cell bodies are located in the brainstem or spinal cord.

_____ 3. Second neuron of ANS between the CNS and the organs innervated.

_____ 4. Location of cell bodies of the postganglionic neurons.

C. Match the terms from part B with the parts labeled on the diagram in Figure 16-1:

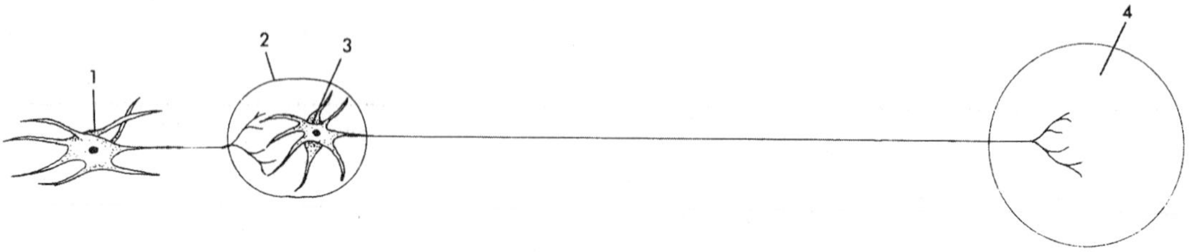

Figure 16-1

1. _____   3. _____

2. _____   4. _____

Sympathetic Division

"_Stimulation of sympathetic neurons generally prepares the individual for physical activity._**"**

A. Using the terms provided, complete these statements:

Adrenal medulla
Collateral (prevertebral)
  ganglion
Convergence
Divergence
Epinephrine
Gray ramus communicans
Lateral horns

Postganglionic
Splanchnic nerve
Sympathetic
  chain ganglia
Sympathetic
  nerve
Thoracolumbar
White ramus
  communicans

The _(1)_ of the spinal cord gray matter from T1 to L2 contain the cell bodies of the sympathetic preganglionic neurons. Because of the location of the preganglionic cell bodies, this division is sometimes called the _(2)_ division. Axons of sympathetic preganglionic neurons run through ventral roots of spinal nerves and project to _(3)_ on either side of the vertebral column. The short connection between the spinal nerve and the sympathetic chain ganglion, through which preganglionic axons pass, is called a _(4)_ because it is myelinated. After entering the sympathetic chain ganglion, preganglionic neurons may synapse with _(5)_ neurons at the same level as, inferior to, or superior to, the ganglion that the preganglionic neuron first entered. The postganglionic neurons then pass through a _(6)_ (unmyelinated axons) to enter a spinal nerve or project through a _(7)_ to the organs they innervate. An alternate course occurs for preganglionic axons that originate between T5 and T12 of the spinal cord. These axons exit the chain ganglion without synapsing, and pass through a _(8)_ to a _(9)_, where they synapse with postganglionic neurons. The celiac, superior mesenteric, and inferior mesenteric are examples of collateral ganglia. Yet another course occurs for the preganglionic axons of splanchnic nerves. Instead of synapsing in collateral ganglia, the preganglionic axons extend to the _(10)_, where they synapse with highly specialized postganglionic neurons that contain _(11)_ and norepinephrine, which can be released into the blood. A single preganglionic neuron synapses with many postganglionic neurons; thus _(12)_ is a characteristic of the sympathetic division.

1. _____
2. _____
3. _____
4. _____
5. _____
6. _____
7. _____
8. _____
9. _____
10. _____
11. _____
12. _____

B. Match these terms with the correct parts labeled in Figure 16-2:

Collateral ganglion
Gray ramus communicans
Splanchnic nerve
Sympathetic chain ganglion
White ramus communicans

1.  _____
2. _____
3. _____
4. _____
5. _____

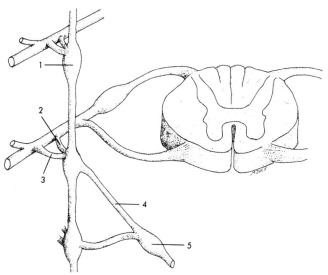

Figure 16-2

Parasympathetic Division

"Increased activity of the parasympathetic division is consistent with resting conditions, during**"** which vegetative functions are emphasized.

Using the terms provided, complete these statements:

Cranial nerve nuclei	Postganglionic
Craniosacral	Preganglionic
Lateral horns	Terminal ganglia
Pelvic nerves	Vagus (X)

1. _____

2. _____

3. _____

4. _____

5. _____

6. _____

7. _____

8. _____

Preganglionic cell bodies of the parasympathetic division are either within __(1)__ or within the __(2)__ of gray matter from S2 to S4 in the spinal cord. Therefore, this division is sometimes called the __(3)__ division. Axons of the __(4)__ neurons pass through cranial and pelvic nerves to __(5)__ either near or embedded in the wall of the organ innervated. The axons of the __(6)__ neurons extend the relatively short distance from the terminal ganglia to the target organ. The cranial nerves that carry parasympathetic fibers include the oculomotor (III), facial (VII), glossopharyngeal (IX), and vagus (X) nerves. Approximately 75% of all parasympathetic neurons pass through the __(7)__ nerves. Parasympathetic preganglionic axons whose cell bodies are in the sacral region of the spinal cord course through __(8)__ to innervate the bladder, lower colon, rectum, and organs of the reproductive system.

Contrasting the Sympathetic and Parasympathetic Divisions

"There are a number of structural differences between the sympathetic and parasympathetic**"** divisions of the ANS.

Match these terms with the correct statement or definition:

Parasympathetic division
Sympathetic division

_____ 1. The ratio of postganglionic to preganglionic neurons is much less for this division.

_____ 2. Some axons from this division exit through cranial nerves.

_____ 3. Cell bodies of the preganglionic neurons are located in the thoracic and lumbar regions of the spinal cord.

_____ 4. Cell bodies of the postganglionic neurons are located in chain ganglia and collateral ganglia.

_____ 5. Axons of the postganglionic neurons travel only a short distance to the effector organ.

Neurotransmitter Substances and Receptors

❝*The sympathetic and parasympathetic nerve endings secrete one of two transmitters.***❞**

Match these terms with the
correct statement or definition:

Adrenergic neuron   Cholinergic neuron
Adrenergic receptor   Muscarinic receptors
  (alpha or beta)   Nicotinic receptors

_____ 1. Neuron that secretes acetylcholine.

_____ 2. All preganglionic neurons of the autonomic division are of this type.

_____ 3. All postganglionic neurons of the parasympathetic division are this type.

_____ 4. Most postganglionic neurons of the sympathetic division are of this type.

_____ 5. Found in postganglionic neurons, these cholinergic receptors produce an excitatory response to acetylcholine.

_____ 6. Found in effector organs, these cholinergic receptors may produce either an excitatory or inhibitory response to acetylcholine.

_____ 7. Norepinephrine binds to and activates these receptors, which are found in effector organs; response can be excitatory or inhibitory.

👉 Autonomic reflexes allow an individual to maintain homeostasis by mediating responses to stressful conditions and changes in physical and mental activity. However, higher centers of the brain can also affect autonomic functions.

Functional Generalizations About the Autonomic Nervous System

❝*Most generalizations about the influence of the ANS on effector organs have exceptions.***❞**

Match these terms with the
correct statement or definition:

Parasympathetic division
Sympathetic division

_____ 1. Sweat glands and blood vessels are innervated almost exclusively by this division.

_____ 2. Stimulates digestive glands and contracts the gallbladder.

_____ 3. Has the greatest activity during physical activity or stress.

_____ 4. Increases metabolism and causes release of glucose from liver.

_____ 5. Causes an increase in heart rate and force of contraction and causes dilation of the pupils of the eye.

_____ 6. This division has a more localized effect.

1. Complete these table:

CHARACTERISTIC	SOMATIC MOTOR	AUTONOMIC
Number of neurons		
Effector organs		
Neurotransmitter		
Conscious vs. unconscious control		

2. Complete these table by indicating which ANS division is involved, and what kind of cell body is found in each location.

LOCATION	PARASYMPATHETIC OR SYMPATHETIC DIVISION	PREGANGLIONIC OR POSTGANGLIONIC CELL BODY
Cranial nuclei		
Lateral horns T1-L2		
Lateral horns S2-S4		
Chain ganglia		
Collateral ganglia		
Terminal ganglia		

3. Name two types of neurotransmitters released by the ANS and indicate where each is released in both the sympathetic and parasympathetic neurons.

4. List four types of receptors found in the ANS and give their location.

5. List six functional generalizations concerning the ANS.

Place the letter corresponding to the correct answer in the space provided.

_____ 1. Given these statements:
1. neuron cell bodies in the motor nuclei of cranial nerves
2. neuron cell bodies in the lateral horn of the spinal cord
3. two synapses between the CNS and effector organs
4. effector organs include smooth muscle

Which of the statements are true for the autonomic division, but not the somatic motor division, of the PNS?
a. 1,3
b. 2,4
c. 1,2,3
d. 2,3,4
e. 1,2,3,4

_____ 2. Given these structures:
1. gray ramus communicans
2. white ramus communicans
3. spinal nerve
4. sympathetic chain ganglion

Choose the arrangement that lists the structures in the order action potentials pass through them from a spinal nerve to an effector organ.
a. 1,4,2,3
b. 2,4,1,3
c. 3,1,4,2
d. 3,4,1,2

_____ 3. Given these structures:
1. collateral ganglion
2. sympathetic chain ganglion
3. white ramus communicans
4. splanchnic nerve

Choose the arrangement that lists the structures in the order action potentials pass through them on the way from a spinal nerve to an effector organ.
a. 1,3,2,4
b. 1,4,2,3
c. 3,1,4,2
d. 3,2,4,1

_____ 4. The white ramus communicans contains
a. preganglionic sympathetic fibers.
b. postganglionic sympathetic fibers.
c. preganglionic parasympathetic fibers.
d. postganglionic parasympathetic fibers.

_____ 5. The cell bodies of the postganglionic neurons of the sympathetic division are located in the
a. sympathetic chain ganglia.
b. collateral ganglia.
c. lateral horns of the thoracic and lumbar regions of the spinal cord.
d. dorsal root ganglia
e. a and b

_____ 6. Splanchnic nerves
a. are part of the parasympathetic division.
b. have preganglionic neurons that synapse in the collateral ganglia.
c. exit from the cervical region of the spinal cord.
d. all of the above

_____ 7. Which of these are part of the parasympathetic division?
a. cranial nerves III, VII, IX, and X
b. pelvic nerves
c. thoracic spinal nerves
d. lumbar spinal nerves one and two
e. a and b

_____ 8. Concerning the preganglionic neurons of the autonomic nervous system,
a. in the parasympathetic division they secrete acetylcholine.
b. in the sympathetic division they secrete acetylcholine.
c. in the sympathetic division they secrete norepinephrine.
d. a and b
e. a and c

9. Generally speaking, the parasympathetic division
   a. has more postganglionic neurons per preganglionic neuron than the sympathetic division.
   b. has fewer postganglionic neurons per preganglionic neuron than the sympathetic division.
   c. has longer preganglionic neurons than the sympathetic division.
   d. a and c
   e. b and c

10. A cholinergic neuron
    a. secretes acetylcholine.
    b. reacts with acetylcholine.
    c. secretes norepinephrine.
    d. reacts with norepinephrine.

11. Which of the neurotransmitters is correctly matched with its receptor?
    a. acetylcholine - nicotinic receptor
    b. acetylcholine - muscarinic receptor
    c. norepinephrine - alpha adrenergic receptor
    d. norepinephrine - beta adrenergic receptor
    e. all of the above

12. Nicotinic receptors are located in
    a. postganglionic neurons of the parasympathetic division.
    b. postganglionic neurons of the sympathetic division.
    c. effector organs of the parasympathetic division.
    d. a and b
    e. all of the above

13. Alpha adrenergic receptors
    a. can be excitatory or inhibitory.
    b. are found on preganglionic neurons.
    c. are found on postganglionic neurons.
    d. are part of the parasympathetic division.

14. The sympathetic division
    a. is always stimulatory.
    b. is always inhibitory.
    c. is usually under conscious control.
    d. generally opposes the actions of the parasympathetic division.
    e. a and c

15. Which of these is expected if the sympathetic division is stimulated?
    a. Blood flow to the visceral organs increases.
    b. Blood flow to skeletal muscles increases.
    c. Heart rate decreases.
    d. Glucose release from the liver decreases.
    e. a and b

☆ ——————— FINAL CHALLENGES ——————— ☆

Use a separate sheet of paper to complete this section.

1. When trying to remember whether the sympathetic or parasympathetic division has an excitatory or inhibitory effect on a particular organ, what generalization is useful?

2. A patient with Horner's syndrome exhibits these symptoms on the left side of the face: pupillary constriction, flushing of the skin, and absence of sweating. Injury (loss of function) of what part of the autonomic nervous system could produce these symptoms?

3. Given these parts of the autonomic nervous system:
   1. cranial nerve
   2. pelvic nerve
   3. spinal nerve
   4. splanchnic nerve
   5. sympathetic nerve
   6. vagus nerve

   Match the part of the autonomic nervous system that, <u>if damaged</u>, produces these symptoms:
   A. inability to produce "goose flesh"
   B. flushed skin
   C. increased heart rate
   D. inability to defecate (move feces through the inferior end of the large intestine)
   E. dry eyes

4. A man is accidentally poisoned with mushrooms that contain muscarine. Which of these symptoms, a dry mouth, diarrhea or even involuntary defecation, or contracted pupils would you expect to observe?

5. For which of these conditions, nasal congestion, asthma attack, high blood pressure, or tachycardia, would epinephrine be effective? Explain.

ANSWERS TO CHAPTER 16

Contrasting the Somatic motor and Autonomic Nervous Systems
- A. 1. Afferent neurons; 2. Somatic motor neurons; 3. Autonomic neurons; 4. Autonomic neurons
- B. 1. Preganglionic neuron; 2. Preganglionic neuron; 3. Postganglionic neuron; 4. Autonomic ganglion
- C. 1. Preganglionic neuron; 2. Autonomic ganglion; 3. Postganglionic neuron; 4. Effector organ

Sympathetic Division
- A. 1. Lateral horns; 2. Thoracolumbar; 3. Sympathetic chain ganglia; 4. White ramus communicans; 5. Postganglionic; 6. Gray ramus communicans; 7. Sympathetic nerve; 8. Splanchnic nerve; 9. Collateral (prevertebral) ganglion; 10. Adrenal medulla; 11. Epinephrine; 12. Divergence
- B. 1. Sympathetic chain ganglion; 2. Gray ramus communicans; 3. White ramus communicans; 4. Splanchnic nerve; 5. Collateral ganglion

Parasympathetic Division
- 1. Cranial nerve nuclei; 2. Lateral horns; 3. Craniosacral; 4. Preganglionic; 5. Terminal ganglia; 6. Postganglionic; 7. Vagus (X); 8. Pelvic nerves

Contrasting the Sympathetic and Parasympathetic Divisions
- 1. Parasympathetic division; 2. Parasympathetic division; 3. Sympathetic division; 4. Sympathetic division; 5. Parasympathetic division

Neurotransmitter Substances and Receptors
- 1. Cholinergic neuron; 2. Cholinergic neuron; 3. Cholinergic neuron; 4. Adrenergic neuron; 5. Nicotinic receptors; 6. Muscarinic receptors; 7. Adrenergic receptors (alpha or beta)

Functional Generalizations About the Autonomic Nervous System
- 1. Sympathetic division; 2. Parasympathetic division; 3. Sympathetic division; 4. Sympathetic division; 5. Sympathetic division; 6. Parasympathetic division

1. Number of neurons: somatic motor, one; autonomic, two
   Effector organs: somatic motor, skeletal muscle; autonomic, smooth muscle, cardiac muscle, and glands
   Neurotransmitter: somatic motor, acetylcholine; autonomic, acetylcholine or norepinephrine
   Conscious vs. unconscious control: somatic motor, conscious; autonomic, unconscious

2. Cranial nuclei: parasympathetic division, preganglionic cell body; lateral horns T1-L2: sympathetic division, preganglionic cell body; lateral horns S2-S4: parasympathetic, preganglionic cell body; chain ganglia: sympathetic, postganglionic cell body; collateral ganglia: sympathetic, postganglionic cell body; terminal ganglia: parasympathetic, postganglionic cell body

3. All preganglionic fibers: acetylcholine; postganglionic fibers of parasympathetic: acetylcholine; postganglionic fibers of sympathetic: mostly norepinephrine, some release acetylcholine

4. Alpha and beta adrenergic receptors: effector organs of sympathetic division; nicotinic cholinergic receptors: postganglionic receptors for both sympathetic and parasympathetic divisions; muscarinic cholinergic receptors: effector organ receptors of parasympathetic and some sympathetic divisions

5. In most cases the influence of the two autonomic divisions is opposite on structures that receive dual innervation; each division can produce inhibitory or excitatory effects; and the parasympathetic division is consistent with resting conditions, whereas the sympathetic division is consistent with physical activity or stress; most organs that receive autonomic neurons are innervated by both sympathetic and parasympathetic divisions; one division alone or both divisions acting together can coordinate the activities of different structures; and the sympathetic division has a more general effect when activated.

1. D. Motor neurons of the somatic motor division are located in cranial nerve nuclei and the ventral horns of the spinal cord. The preganglionic neurons of the autonomic division are located in cranial nerve nuclei and lateral horns of the spinal cord. Because the preganglionic neurons synapse with the postganglionic neurons and the postganglionic neurons synapse with the effector organs, there are two synapses. The effector organs for the autonomic division include smooth muscle, cardiac muscle, and glands, whereas the effector organs for the somatic motor division are skeletal muscles.

2. B. From the spinal nerve, preganglionic axons pass through the white ramus communicans into a sympathetic chain ganglion. At the same or a different level the preganglionic axons synapse with a postganglionic neuron, the axons of which pass through a gray ramus communicans into a spinal nerve.

3. D. Preganglionic fibers pass through the white ramus communicans, the sympathetic chain ganglion, and splanchnic nerve to synapse with a postganglionic neuron in the collateral ganglion.

4. A. The white ramus communicans contains presynaptic sympathetic fibers, which are generally myelinated (thus the white color). The gray ramus communicans contains postsynaptic sympathetic fibers that are unmyelinated (hence the gray color).

5. E. Sympathetic postganglionic neuron cell bodies are found in sympathetic chain ganglia and in collateral ganglia. Sympathetic preganglionic neuron cell bodies are found in the lateral horns of the spinal cord. Sensory neuron cell bodies are found in the dorsal root ganglia.

6. B. Splanchnic nerves have preganglionic fibers that synapse in collateral ganglia. They are part of the sympathetic nervous system and originate between T5 and T12 of the spinal cord.

7. E. The parasympathetic nerves include four cranial nerves (III, VII, IX, and X), and the pelvic nerves. The thoracic spinal nerves and the first two lumbar spinal nerves are part of the sympathetic nervous system.

8. D. All preganglionic neurons of the autonomic nervous system secrete acetylcholine.

9. E. The parasympathetic division has fewer preganglionic neurons per postganglionic neuron than does the sympathetic division. However, the preganglionic neurons of the parasympathetic division are longer because they extend to terminal ganglia on the effector organ, whereas sympathetic preganglionic neurons extend to the sympathetic chain ganglia or collateral ganglia.

10. A. By definition, cholinergic neurons secrete acetylcholine, and adrenergic neurons secrete norepinephrine.

11. E. Acetylcholine binds to nicotinic or muscarinic receptors. Norepinephrine binds to both alpha and beta adrenergic receptors, although it has a greater affinity for alpha receptors.

12. D. Postganglionic neurons of both divisions of the autonomic nervous system have nicotinic receptors. The effector organs of the parasympathetic division have muscarinic receptors.

13. A. Alpha and beta receptors can be either excitatory or inhibitory. They are found on effector organs as part of the sympathetic division.

14. D. When innervating the same organ, the sympathetic division generally produces the opposite effect of the parasympathetic division. The sympathetic division or the parasympathetic division can produce stimulatory or inhibitory effects (i.e., they have a stimulatory effect in one organ and an inhibitory effect in another organ). The sympathetic and parasympathetic divisions are under involuntary control.

15. B. The sympathetic division prepares the body for activity. Blood flow to muscle increases, heart rate increases, and blood sugar levels rise. Meanwhile, activities not immediately necessary for activity are inhibited, such as decreasing blood flow to visceral organs.

1. In general, the parasympathetic division has a greater effect under resting conditions, whereas the sympathetic division has a major role under conditions of physical activity or stress. The sympathetic division increases heart rate, causes dilation of skeletal muscle blood vessels and constriction of visceral blood vessels, promotes glucose release from the liver, and decreases intestinal tract activities.

2. Horner's syndrome results from interruption of sympathetic nerves. Consequently there is an inability to sweat in the area affected, and blood vessels dilate, producing flushing of the skin. Without sympathetic stimulation of the pupil, the parasympathetic division dominates and the pupil constricts.

3. A. Primarily spinal nerves, some innervation through sympathetic nerves to head and neck
   B. Primarily spinal nerves, some innervation through sympathetic nerves to head and neck
   C. Damage to the vagus nerve could eliminate its parasympathetic inhibitory effect on the heart, resulting in an increased heart rate (sympathetic effect mediated through sympathetic nerves)
   D. Pelvic nerve
   E. Cranial nerve (facial nerve VII)

4. Muscarine poisoning should result in the activation of muscarinic receptors on effector organs of the parasympathetic nervous system, increasing salivation, increasing motility of the intestine (diarrhea or involuntary defecation), and constricted pupils.

5. Epinephrine would produce similar effects to sympathetic stimulation. Epinephrine would cause constriction of blood vessels, and by reducing blood flow, help relieve nasal congestion (reduce fluid loss into the nasal cavity). Epinephrine could also cause dilation of the airway system of the lungs and help relieve asthma symptoms. Epinephrine would increase blood pressure by causing vasoconstriction and would increase heart rate (tachycardia).

Functional Organization of the Endocrine System

FOCUS: Endocrine tissues internally produce hormones that are released into the blood, where they are carried to target tissues and produce a response. Hormones are either proteins, polypeptides, amino acid derivatives, or lipids. The release of hormones can be regulated by nonhormone substances (e.g., glucose, calcium, sodium), by other hormones, or by the nervous system. Some hormones are released at relatively constant rates, others are released suddenly in response to certain stimuli, and a few increase or decrease in a cyclic fashion. Once in the blood, hormones are transported dissolved in plasma or bound to plasma proteins. Hormones are eliminated from the blood by excretion into urine or bile and by enzymatic degradation. Hormones bind to protein or glycoprotein receptors in their target tissues. Each hormone has a specific receptor, and target tissues have more than one kind of receptor. The target tissue response to the hormone can be constant; it may decrease because of a decrease in the number of receptors (down regulation); or increase because of an increase in the number of receptors (up regulation). When a hormone binds to a membrane-bound receptor, membrane permeability can change, or a second messenger (e.g., cyclic AMP) can be produced. The change in membrane permeability or the second messenger is responsible for activating the response of the target tissue to the hormone. When a hormone binds to an intracellular receptor, genes are activated and enzymes are synthesized, producing the response of the target tissue.

CONTENT LEARNING ACTIVITY

General Characteristics of the Endocrine System

66*Endocrine glands secrete their products internally and influence tissues that are separated by* **99** *some distance from the endocrine glands.*

A. Match these terms with the correct statement or definition:

Amplitude-modulated   Frequency-modulated
Endocrine system   Hormones

_____ 1. Composed of glands that secrete their products into the circulatory system.

_____ 2. The secretions of endocrine glands; acts on target tissues.

_____ 3. Communication between cells that involves increases or decreases in the concentration of hormones in the body fluids.

_____ 4. The method of communication between neurons and their effectors; an all-or-none response.

B. Match these chemical messengers with their correct location in the following table:

Hormone
Neurohormone
Neuromodulator

Neurotransmitter
Parahormone
Pheromone

INTERCELLULAR CHEMICAL MESSENGER	PRODUCED BY	FUNCTION	EXAMPLE
1. _____	Specialized cells	Travels in blood; influences specific activities	Thyroxine
2. _____	Neurons	Like hormones	Oxytocin, ADH
3. _____	Neurons	Released from presynaptic terminals; affects postsynaptic terminals	Acetylcholine, norepinephrine
4. _____	Neurons	Released from presynaptic terminals; alters sensitivity of postsynaptic neurons to neurotransmitters	Prostaglandins, endorphins
5. _____	Wide variety of tissue	Secreted into tissue space; localized effect	Prostaglandins, histamine

☞ Hormones and neurohormones are proteins, polypeptides, amino acid derivatives, or lipids.

Control of Secretion Rate

66 *Most hormones are not secreted at a constant rate; most endocrine glands increase and decrease* 99 *their secretory activity dramatically over time.*

Using the terms provided, complete these statements:

Hormone   Nonhormone
Neural

1. _____

2. _____

3. _____

There are three major patterns of regulation for hormones. An example is the effect glucose has on the secretion of insulin. This is a case of a (1) influencing an endocrine gland. The release of antidiuretic hormone from the pituitary is a different situation. In this pattern the release of a hormone is regulated by (2) control. A third pattern occurs when thyroid-stimulating hormone causes the thyroid to release thyroid hormones. In this case a (3) controls the secretory activity of an endocrine gland.

Negative-feedback mechanisms play essential roles in maintaining hormone levels within normal concentration ranges. There are a few examples of positive-feedback mechanisms (e.g., oxytocin during delivery).

Transport and Distribution in the Body

❝*Because hormones circulate in the blood, they are distributed throughout the body.***❞**

Using the terms provided, complete these statements:

Equilibrium   Plasma proteins
Free hormone   Target tissue

Hormones are dissolved in blood plasma and are transported either in a free form or bound to _(1)_ . An _(2)_ exists between the unbound hormone and those bound to plasma proteins. A large increase or decrease in plasma protein concentration can influence the concentration of _(3)_ in the blood. In general, the amount of free hormone that reaches the _(4)_ is directly correlated with the concentration of hormone in the blood.

1. _____
2. _____
3. _____
4. _____

Metabolism and Excretion

❝*The destruction and elimination of hormones limits the length of time they are active.***❞**

A. The length of time it takes for half a dose of a substance to be eliminated from the circulatory system is its half-life. Match the length of the half-life with the correct hormone characteristic:

Long half-life
Short half-life

_____

_____

_____

1. Typical of water-soluble hormones (proteins, glycoproteins, and epinephrine and norepinephrine).

2. These hormones generally regulate activities that have a rapid onset and a short duration.

3. The concentration of these hormones is maintained at a relatively constant level through time.

319

B. Match these factors with the correct statement or definition:

Active transport   Metabolism
Conjugation   Reversible binding
Excretion   Structural protection

_____ 1. Elimination of hormones from the blood into the urine or bile.

_____ 2. In this process hormones are enzymatically degraded.

_____ 3. Hormones are made less active and eliminated by attaching compounds, such as sulfates or glucuronic acid, to them.

_____ 4. Method of prolonging the half-life of hormones by reacting with plasma proteins.

_____ 5. Half-life of hormones is prolonged by the carbohydrate components of glycoprotein hormones.

Interaction of Hormones with Their Target Tissue

"*Hormones bind to receptors in their target tissues and alter the rate at which activities occur.***"**

Match these terms with the correct statement or definition:

Down regulation   Up regulation
Hormone receptors

_____ 1. Protein or glycoprotein molecules of specific three-dimensional shape that bind to a single type of hormone.

_____ 2. Decrease in the number of hormone receptors after exposure to certain hormones.

_____ 3. Increase in the number of hormone receptors after exposure to certain hormones.

☞ A hormone may bind to a number of different types of receptors. Different cell types can have different types of receptors.

Classes of Hormone Receptors

"*There are two major classes of hormone receptors.***"**

A. Match these terms with the correct statement or definition:

Intracellular receptors
Membrane-bound receptors

_____ 1. Binds to water-soluble or large molecular weight hormones.

_____ 2. Binds to lipid-soluble hormones (e.g., steroids, thyroid hormones).

_____ 3. When activated, a change in membrane permeability can result, or the activity of already existing enzymes is changed.

_____ 4. When activated, new enzymes (proteins) are produced.

B. Match these terms with the correct statement or definition:

Cascade effect   Protein kinase
First messenger   Second messenger

_____ 1. Hormone binding to a receptor molecule.

_____ 2. Molecule produced as a result of a first messenger binding to a receptor molecule; carries a signal from the membrane into the cell.

_____ 3. Cyclic AMP and cyclic GMP are examples; function to activate protein kinases.

_____ 4. Attach phosphate groups to enzymes, altering their activity.

_____ 5. A few second messengers activate several enzymes that each activate many more enzymes, and so on.

C. Using the terms provided, complete these statements:

Binds to channels   Membrane
Enzymes   Splits apart
G proteins

1. _____
2. _____
3. _____
4. _____
5. _____
6. _____

Many receptor molecules have regulatory proteins, called _(1)_ , which are associated with the intracellular part of the plasma membrane. The G proteins have alpha, beta, and gamma subunits. When a hormone _(2)_ the receptor molecule, the G protein splits apart and the alpha subunit _(3)_ guanine triphosphate (GTP). The GTP-bound alpha subunit activates _(4)_ or binds to _(5)_ . Consequently, other activities result that produce a response from the cell. The activity of the alpha subunit is terminated when the GTP _(6)_ to form guanine diphosphate. The alpha subunit then recombines with the beta and gamma subunits.

D. Using the terms provided, complete these statements:

DNA   mRNA
Enzymes   Protein synthesis

1. _____
2. _____
3. _____
4. _____

Hormones may bind to intracellular receptors in the cytoplasm or the nucleus of the cell. Once in the nucleus, the hormone-receptor combination has fingerlike projections that interact with _(1)_ to regulate the synthesis of _(2)_ , which moves into the cytoplasm and initiates _(3)_ . The _(4)_ that result produce the response of the cell.

1. List six types of chemical messengers produced by cells.

2. Explain the three major patterns for regulation of hormone secretion, and give an example of each type.

3. List three important patterns of secretion of hormones.

4. List four means by which hormones are eliminated from the circulatory system.

5. List two means by which the half-life of hormones is prolonged.

6. List the two major classes of hormone receptors, and list the types of hormone molecules that bind to each.

7. Name the two major effects of G proteins.

Mastery Learning Activity

Place the letter corresponding to the correct answer in the space provided.

_____ 1. An endocrine gland
   a. lacks a duct.
   b. releases its hormone into the surrounding interstitial fluid.
   c. depends on the blood for transport of its secretion.
   d. all of the above

_____ 2. When comparing the endocrine system with the nervous system, generally speaking, the endocrine system
   a. is faster acting than the nervous system.
   b. produces effects that are of a shorter duration.
   c. is regulated by less elaborate feedback mechanisms.
   d. uses amplitude-modulated signals.

_____ 3. Given the following list of molecule types:
   1. amino acid derivatives
   2. fatty acid derivatives
   3. polypeptides
   4. proteins
   5. steroids

   Of the molecule types listed, which include hormones?
   a. 1,2,3
   b. 2,3,4
   c. 1,2,3,4
   d. 2,3,4,5
   e. 1,2,3,4,5

_____ 4. The secretion of a hormone from an endocrine tissue is regulated by
   a. other hormones
   b. nonhormone substances in the blood
   c. the nervous system
   d. all of the above

_____ 5. Hormones are released into the blood
   a. at relatively constant levels.
   b. in large amounts in response to a stimulus.
   c. in a cyclic fashion (increasing and decreasing in amounts in a regular fashion).
   d. all of the above

_____ 6. Concerning the half-life of hormones,
   a. lipid-soluble hormones generally have a longer half-life
   b. hormones with shorter half-lives regulate body action more precisely than hormones with longer half-lives.
   c. hormones with a longer half-life are maintained at more constant levels in the blood.
   d. all of the above

_____ 7. Given the following observations:
   1. A hormone will affect only a specific tissue (i.e., a hormone will not affect all tissues).
   2. A tissue can respond to more than one hormone.
   3. Some tissues respond rapidly to a hormone, whereas others take many hours to respond.

   Which of the observations can be explained by hormone receptors?
   a. 1
   b. 1,2
   c. 2,3
   d. 1,3
   e. 1,2,3

8. Down regulation
   a. results in a decrease in the number of receptors in the target tissue.
   b. produces an increase in the sensitivity of the target tissue to the hormone.
   c. is found in target tissues that respond to hormones that are maintained at constant levels.
   d. all of the above

9. When a hormone binds to a membrane-bound receptor,
   a. the receptor may change shape.
   b. the permeability of the membrane may change.
   c. enzyme activity may be altered.
   d. all of the above

10. Given the following events:
    1. activation of cyclic AMP
    2. activation of genes
    3. increase of enzyme activity

    Which of the events occurs when a hormone binds to an intracellular hormone receptor?
    a. 1
    b. 1,2
    c. 2,3
    d. 1,2,3

FINAL CHALLENGES

Use a separate sheet of paper to complete this section.

1. A diabetic hears about a method of birth control that uses a skin patch containing estrogen. The estrogen diffuses from the patch through the skin, ensuring a steady release of small amounts of estrogen into the blood. The estrogen then inhibits the development of the ovum (see Chapter 28). The diabetic wonders if such a system could be used to administer insulin. What would you tell him?

2. Assume that negative feedback is the major means by which a hormone's secretion rate is controlled. Assume also that the hormone causes the concentration of a substance called X to decrease in the blood. Predict the effect on the rate of hormone secretion if the levels of X in the blood are caused to remain higher than normal.

3. Suppose the following experimental data were collected: (1) a few minutes after exposure to a hormone the target tissue

began producing a secretion, and (2) after several hours of continual exposure to the hormone the production of secretion by the target tissue decreased. What conclusions can you come to about the regulatory mechanisms that control the target tissue's secretion? Would the response of the target tissue be mediated by a second messenger mechanism or by an intracellular receptor mechanism? Explain.

4. Given that you could measure any two intracellular substances, which two would you choose if you wanted to determine if the response of a cell to a hormone was mediated by the second messenger mechanism or by the intracellular receptor mechanism?

5. Antidiuretic hormone decreases the amount of urine produced by the kidneys. What would happen to urine production if the metabolism of antidiuretic hormone in the liver were impaired?

ANSWERS TO CHAPTER 17

General Characteristics of the Endocrine System
- A. 1. Endocrine system; 2. Hormones; 3. Amplitude-modulated; 4. Frequency-modulated
- B. 1. Hormone; 2. Neurohormone; 3. Neurotransmitter; 4. Neuromodulator; 5. Parahormone

Control of Secretion Rate
1. Nonhormone; 2. Neural; 3. Hormone

Transport and Distribution in the Body
1. Plasma proteins; 2. Equilibrium; 3. Free hormone; 4. Target tissue

Metabolism and Excretion
- A. 1. Short half-life; 2. Short half-life; 3. Long half-life
- B. 1. Excretion; 2. Metabolism; 3 Conjugation; 4. Reversible binding; 5. Structural protection

Interaction of Hormones with Their Target Organs
1. Hormone receptors; 2. Down regulation; 3. Up regulation

Classes of Hormone Receptors
- A. 1. Membrane-bound receptors; 2. Intracellular receptors; 3. Membrane-bound receptors; 4. Intracellular receptors
- B. 1. First messenger; 2. Second messenger; 3. Second messenger; 4. Protein kinase; 5. Cascade effect
- C. 1. G proteins; 2. binds to; 3. binds to; 4. Enzymes; 5. Membrane channels; 6. Splits apart
- D. 1. DNA; 2. mRNA; 3. Protein synthesis; 4. Enzymes

1. Hormones, neurohormones, neurotransmitters, neuromodulators, parahormones, and pheromones
2. Regulation by a nonhormone (glucose); by neural control (ADH); or by a hormone (thyroid hormone)
3. Relatively constant; sudden change in response to stimuli; and change occurring in cycles
4. Excretion, metabolism (enzymatic degradation), conjugation, and active transport
5. Reversible binding and structural protection
6. Membrane-bound receptors: water-soluble or large molecular weight hormones
   Intracellular receptors: lipid-soluble hormones
7. Open membrane channels and activate second messengers.

1. D. An endocrine gland has no ducts, releases its secretions into the surrounding interstitial fluid, and depends on the circulatory system to transport the secreted hormone to its target tissue.

2. D. The endocrine system uses amplitude-modulated signals (increase or decrease in hormone concentration), whereas the nervous system uses frequency-modulated signals (the number of action potentials produced). The endocrine system also tends to be slower acting than the nervous system but produces longer lasting effects. The endocrine system also has more elaborate feedback mechanisms.

3. E. Hormones can be classified into a "protein" group (i.e., amino acid derivatives, polypeptides, proteins, and glycoproteins) and a lipid group (i.e., steroids and fatty acid derivatives).

4. D. Other hormones, nonhormone substances, and the nervous system can regulate hormone secretion.

5. D. Some hormones are released at a constant rate (i.e., hormones involved with long-term maintenance), some are released in large amounts in response to a stimulus (allowing rapid adjustment to changing conditions), and some are released in a cyclic fashion (allowing periodic changes).

6. D. Lipid-soluble hormones (lipid hormones, thyroid hormones) bind to plasma proteins, increasing their half-life and contributing to the maintenance of constant levels of lipid-soluble hormones in the blood. Water-soluble hormones (e.g., protein hormones and epinephrine) are rapidly degraded by enzymes and have a short half-life. Hormones with a long half-life are maintained at constant levels in the blood, whereas hormones with a short half-life regulate activities that have a rapid onset and a short duration. Hormones with a short half-life allow precise regulation of body actions because the level of the hormones can rapidly be adjusted.

7. E. A hormone receptor is specific for a given hormone, and only tissues with that receptor can respond to the hormone. Because a tissue can have more than one kind of receptor, a tissue can respond to more than one kind of hormone. Membrane-bound receptors produce rapid responses in tissues, whereas intracellular receptors may take hours to respond.

8. A. Down regulation results in a decrease in the number of receptors in the target tissue, because of decreased receptor synthesis and/or increased receptor degradation. The decrease in receptors decreases the sensitivity of the target tissue to the hormone. Down regulation is found in target tissues that respond to short-term increases in hormone concentration.

9. D. Membrane-bound receptors operate in different ways. In some cases the receptor changes shape, producing a permeability change in the membrane. Other receptors activate a second messenger, which leads to an alteration of enzyme activity.

10. C. The combination of a hormone with an intracellular receptor leads to activation of genes. The gene produces mRNA, which leaves the nucleus; at the ribosome the mRNA is involved in the production of enzymes (proteins). The enzyme activity that results is responsible for the response of the cell to the hormone.

 ## FINAL CHALLENGES

1. Insulin levels normally change in order to maintain normal blood sugar levels, despite periodic fluctuations in sugar intake. A constant supply of insulin from a skin patch would result in insulin levels that would be too low when blood sugar levels were high (after a meal) and would be too high when blood sugar levels were low (between meals). In addition, insulin is a protein hormone that would diffuse through the lipid barrier of the skin (see Chapter 5) with much more difficulty than estrogen, which is a steroid that is lipid soluble.

2. Negative feedback mechanisms operate by returning values to normal levels (homeostasis). If substance X levels are higher than normal, the rate of hormone secretion should increase, because this would normally cause the level of substance X to decrease back toward a normal value.

3. One possibility is down regulation, in which exposure to the hormone resulted in a decrease in the number of hormone receptors and therefore a decreased responsiveness to the hormone. Another possibility is that the secretion of the target tissue produced an effect elsewhere in the body such as the production of another hormone, and the other hormone then inhibited the target tissue. The response of the target tissue is most likely mediated by a second messenger mechanism. Intracellular receptor mechanisms normally require several hours to produce a response.

4. An increase in cyclic AMP or other second messengers would indicate the second messenger mechanism, and an increase in mRNA would indicate the intracellular receptor mechanism.

5. Impairment of antidiuretic hormone conjugation in the liver increases the half-life of the hormone, increasing the concentration of the hormone in the blood. Consequently, one would expect a lower-than-normal production of urine.

18 Endocrine Glands

FOCUS: The major endocrine structures are in the brain (i.e., hypothalamus, pituitary gland, and pineal body) or are specialized glands (i.e., thyroid gland, parathyroid glands, thymus gland, pancreas, adrenal glands, testes, or ovaries). However, other tissues (e.g., the stomach, small intestine, liver, kidneys, and placenta) also produce hormones. The hypothalamus plays a major role in regulating the secretions of the pituitary gland. Through the hypothalamohypophyseal portal system the anterior pituitary receives regulatory hormones from hypothalamic neurosecretory cells. In response to these regulatory hormones, the anterior pituitary releases its own hormones. The posterior pituitary receives and stores hormones produced by the hypothalamic neurosecretory cells. These hormones are released in response to action potentials from the hypothalamus. The pituitary hormones affect a number of different target tissues, including other endocrine glands, and they regulate a variety of functions. Growth hormone (GH), thyroid-stimulating hormone (TSH), and adrenocorticotropic hormone (ACTH) all affect metabolism: GH directly promotes growth and stimulates the production of somatomedins, which also affect growth; TSH stimulates the thyroid gland to produce thyroid hormones, which increase metabolic rate; and ACTH stimulates the adrenal cortex to release cortisol, which promotes fat and protein breakdown. A number of hormones from the pituitary are involved in reproduction: oxytocin increases uterine contractions during delivery and is responsible for milk release during lactation; prolactin stimulates milk production; follicle-stimulating hormone (FSH) and luteinizing hormone (LH) are necessary for the development of sperm cells or oocytes and they also regulate the production of male (testosterone) and female (estrogen and progesterone) sex hormones. Other pituitary hormones include antidiuretic hormone (ADH), which increases water reabsorption in the kidneys, and melanocyte-stimulating hormone (MSH), which increases skin pigmentation. Many important body functions are regulated by other endocrine glands. For example, calcium levels are regulated by parathyroid hormone from the parathyroid glands and by calcitonin from the parafollicular cells of the thyroid gland; blood sugar levels are maintained by insulin and glucagon from the pancreas; sodium, potassium, and hydrogen ion levels are regulated by aldosterone from the adrenal cortex; epinephrine from the adrenal medulla assists the sympathetic nervous system during times of physical activity; thymosin from the thymus gland is necessary for the immune system to function properly; and melatonin from the pineal body may be involved with the onset of puberty.

327

Introduction

Match these terms with the correct parts of the diagram labeled in Figure 18-1:

Adrenals
Hypothalamus
Ovaries
Pancreas
Parathyroids

Pineal body
Pituitary
Testes
Thymus
Thyroid

1. _____

2. _____

3. _____

4. _____

5. _____

6. _____

7. _____

8. _____

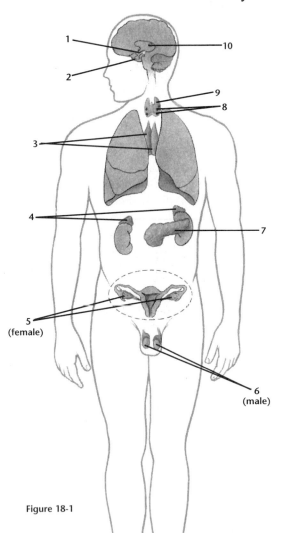

Figure 18-1

Pituitary Gland and Hypothalamus

66 *The hypothalamus and pituitary are where the body's two major regulatory systems interact.* **99**

A. Match these terms with the correct statement or definition:

Action potentials
Adenohypophysis
Hypothalamohypophyseal
  portal system
Hypothalamohypophyseal
  tract

Hypothalamus
Infundibulum
Neurohormone
Neurohypophysis

_____ 1. Connection between the pituitary gland and hypothalamus.

_____ 2. Also called the posterior pituitary.

_____ 3. Develops from the roof of the embryonic oral cavity.

_____ 4. Vessels from the hypothalamus to the adenohypophysis.

_____ 5. Produced in the hypothalamus; travel to adenohypophysis and act as releasing hormones or inhibiting hormones.

_____ 6. Source of neurohormones secreted from the neurohypophysis.

_____ 7. Source of hormones secreted from the adenohypophysis.

_____ 8. Regulate the release of hormones from the neurohypophysis.

B. Match these terms with the correct parts of the diagram labeled in Figure 18-2:

Anterior pituitary
Hypothalamohypophyseal
  portal system
Hypothalamohypophyseal
  tract

Hypothalamus
Neurosecretory cell
Posterior pituitary

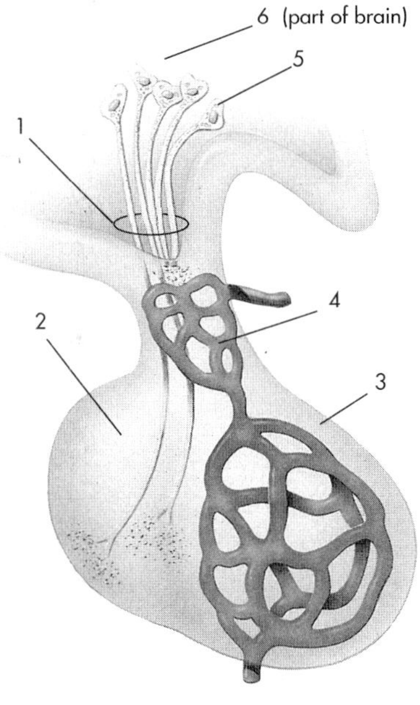

1. _____

2. _____

3. _____

4. _____

5. _____

6. _____

Figure 18-2

Neurohypophyseal Hormones

"The neurohypophysis stores and secretes two polypeptide neurohormones."

A. Match these terms with the correct statement as it pertains to the effect of increased ADH:

Decreases
Increases

_____ 1. Effect of ADH on urine volume.

_____ 2. Effect of ADH on blood osmolality.

_____ 3. Effect of ADH on blood volume.

_____ 4. Effect of ADH on blood vessel constriction and blood pressure.

B. Match these terms with the correct statement:

Blood osmolality
Blood volume

_____ 1. Osmoreceptors in the hypothalamus respond to this factor.

_____ 2. Sensory receptors for blood pressure indirectly measure this factor.

_____ 3. Increase in this factor results in an increase of ADH secretion.

_____ 4. Increase in this factor results in a decrease of ADH secretion.

C. Match these terms with the correct statement:

Hypersecretion of ADH
Hyposecretion of ADH

_____ 1. Production of a large amount of dilute urine; diabetes insipidus.

_____ 2. Production of small quantities of very concentrated urine.

D. Match these terms with the correct statement:

Decreases
Increases

_____ 1. Effect of oxytocin on smooth muscle contraction in the uterus.

_____ 2. Effect of oxytocin on milk ejection in lactating females.

E. Match these terms with the correct statement:

Decreases oxytocin secretion
Increases oxytocin secretion

_____ 1. Effect of stretching of uterus.

_____ 2. Effect of stimulation of nipples by nursing.

Adenohypophyseal Hormones

"Adenohypophyseal secretions are influenced by neurohormones from the hypothalamus.**"**

A. Match these terms with the correct statement as it pertains to the effect of increased growth hormone:

Decreases
Increases

_____ 1. Effect on growth and metabolic rate.

_____ 2. Effect on amino acid uptake and protein synthesis by cells.

_____ 3. Effect on rate of lipid breakdown and blood glucose levels.

_____ 4. Effect on somatomedin production by liver and skeletal muscle.

_____ 5. Effect of insulin-like growth factor I and II on cartilage and bone growth.

B. Match these terms with the correct statement:

Decreases GH secretion
Increases GH secretion

_____ 1. Effect of low blood glucose levels.

_____ 2. Effect of stress.

_____ 3. Effect of GH-RH from the hypothalamus.

_____ 4. Effect of GH-IH (somatostatin) from the hypothalamus.

C. Match these terms with the correct statement:

Hypersecretion of GH
Hyposecretion of GH

_____ 1. In children this produces dwarfism.

_____ 2. In children this produces giantism.

_____ 3. In adults this produces acromegaly.

D. Match these terms with the correct statement or definition:

ACTH   MSH
Beta-endorphins   Proopiomelanocortin
Lipotropins

_____ 1. Large precursor molecule that produces ACTH and others.

_____ 2. Cause fat breakdown and release of fatty acids.

_____ 3. Have the same effect as opiate drugs.

_____ 4. Two hormones that stimulate melanocytes in the skin to produce melanin.

E. Match these terms with the correct statement or definition:

FSH and LH
GnRH
PIH
PRH
Prolactin

_____ 1. Secreted from the adenohypophysis; regulate gamete and reproductive hormone production.

_____ 2. Neurohormone that stimulates the secretion of LH and FSH.

_____ 3. Responsible for milk production.

_____ 4. Neurohormone that stimulates prolactin production.

The Thyroid Gland

66 *The thyroid gland is composed of two lobes connected by a narrow band of thyroid tissue* **99** *called the isthmus.*

A. Match these terms with the correct statement or definition:

Calcitonin
Follicle
Parafollicular cells
Thyroid hormones

_____ 1. Small sphere of cuboidal epithelium in the thyroid that is filled with thyroglobulin.

_____ 2. Scattered cells between follicles in the thyroid.

_____ 3. Product of parafollicular cells.

B. Match these terms with the correct statement or definition:

I⁻ ions
Thyroglobulins
Tetraiodothyronine (T$_4$)
Thyroxin-binding globulin
Triiodothyronine (T$_3$)
Tyrosine

_____ 1. Large proteins synthesized in thyroid follicles.

_____ 2. Amino acid used in the synthesis of thyroid hormones.

_____ 3. Actively absorbed into thyroid follicles; oxidized and bound to tyrosines.

_____ 4. Also called thyroxine; the major secretory product of the thyroid gland.

_____ 5. The major thyroid hormone that interacts with target cells.

_____ 6. Transports most thyroid hormones; increases the half-life of thyroid hormones.

C. Match these terms with the correct statement as it pertains to the effects of increased thyroid hormones:

Decreases
Increases

_____ 1. Effect on glucose, fat, and protein metabolism.

_____ 2. Effect on body temperature.

D. Match these terms with the correct statement:

Hypersecretion of thyroid hormones
Hyposecretion of thyroid hormones

_____ 1. Increased metabolic rate, weight loss, sweating.

_____ 2. Hyperactivity, rapid heart rate, exophthalmos.

_____ 3. Weight gain; reduced appetite; dry, cold skin.

_____ 4. Myxedema, decreased iodide uptake, cold intolerance.

_____ 5. Decreased iodide uptake resulting from iodine deficiency in the diet; goiter.

_____ 6. Cretinism.

_____ 7. Grave's disease.

_____ 8. Thyroid storm.

E. Match these terms with the correct statement:

Decreases
Increases

_____ 1. Effect of stress or exposure to cold on TRH secretion.

_____ 2. Effect of prolonged fasting on TRH secretion.

_____ 3. Effect of increase of TRH on TSH secretion.

_____ 4. Effect of TSH increase on synthesis and secretion of T_3 and T_4.

_____ 5. Effect of increases in T_3 and T_4 on TRH secretion.

_____ 6. Effect of increases of T_3 and T_4 on TSH secretion.

F. Match these terms with the correct statement:

Decreases
Increases

_____ 1. Effect of calcitonin on breakdown of bone by osteoclasts.

_____ 2. Effect of calcitonin on calcium and phosphate in blood.

_____ 3. Effect of increased blood calcium levels on calcitonin secretion.

The Parathyroid Glands

66 *The parathyroid glands are usually embedded in the posterior portion of each lobe of the* 99 *thyroid gland.*

A. Match these terms with the correct statement as it pertains to the effects of increased parathyroid hormone:

Decreases
Increases

_____ 1. Effect on osteoclast activity in bone.

_____ 2. Effect on calcium reabsorption in the kidneys.

_____ 3. Effect on the formation of active vitamin D synthesis, which increases the rate of calcium and phosphate absorption in the intestine.

_____ 4. Effect on blood calcium levels.

_____ 5. Effect of PTH on blood phosphate levels.

B. Match the correct term with this statement:

Decreases PTH secretion
Increases PTH secretion

_____ 1. Effect of low blood calcium levels on PTH secretion.

C. Match these terms with the correct symptom:

Hypersecretion of PTH
Hyposecretion of PTH

_____ 1. Kidney stones, osteomalacia, muscular weakness, constipation.

_____ 2. Increased muscular excitability, tachycardia, muscle tetany, diarrhea.

_____ 3. Increased cell permeability to sodium ions, depolarization of cell membrane.

Adrenal Glands

66 *The adrenal glands are near the superior pole of each kidney.* 99

Match these terms with the correct statement or definition:

Adrenal cortex
Adrenal medulla

_____ 1. Inner portion of adrenal glands.

_____ 2. Derived from neural crest cells; part of the sympathetic nervous system.

_____ 3. Contains the zona glomerulosa, zona fasciculata, and zona reticularis layers.

Adrenal Medulla

"The adrenal medulla secretes two major hormones."

A. Match these terms with the correct statement:

Epinephrine   Norepinephrine
Neurohormone

_____   1. Adrenal medullary hormone secreted in larger quantity.

_____   2. General category of hormones produced by the adrenal medulla.

B. Match these terms with the correct statement as it pertains to the effects of increased adrenal medullary hormones:

Decreases
Increases

_____   1. Effect on heart rate and the force of contraction of the heart.

_____   2. Effect on blood glucose level.

_____   3. Effect on blood flow to the skin, kidneys, and digestive system.

_____   4. Effect on blood flow to heart and skeletal muscle.

C. Match these terms with the correct statement:

Decreases adrenal medullary hormone secretion
Increases adrenal medullary hormone secretion

_____   1. Effect of emotional excitement, stress, exercise, or injury.

_____   2. Effect of stimulation of sympathetic neurons.

_____   3. Effect of low blood glucose levels.

D. Match these terms with the correct statement:

Hypersecretion of adrenal medullary hormones
Hyposecretion of adrenal medullary hormones

_____   1. A result of pheochromocytoma.

_____   2. Hypertension, pallor, and sweating.

Adrenal Cortex

"The adrenal cortex secretes three types of steroid hormones."

A. Match these terms with the correct statement as it pertains to the adrenal cortex:

Androstenedione
Glucocorticoids
Mineralocorticoids

_____ 1. Secreted by the zona glomerulosa; example is aldosterone.

_____ 2. Secreted by the zona fasciculata; example is cortisol.

_____ 3. Secreted by the zona reticularis; converted into testosterone.

B. Match these terms with the correct statement as it pertains to the effects of increased aldosterone:

Decreases
Increases

_____ 1. Effect on sodium ion concentration in the blood.

_____ 2. Effect on potassium ion concentration in the blood.

_____ 3. Effect on hydrogen ion concentration in the blood.

C. Match these terms with the correct statement:

Hypersecretion of aldosterone
Hyposecretion of aldosterone

_____ 1. Hypernatremia and high blood pressure.

_____ 2. Alkalosis.

_____ 3. Hyperkalemia and skeletal muscle tetany.

D. Match these terms with the correct statement as it pertains to the effects of increased cortisol:

Decreases
Increases

_____ 1. Effect on fat and protein metabolism.

_____ 2. Effect on glucose and amino acid uptake by skeletal muscle.

_____ 3. Effect on the synthesis of glucose from amino acids (gluconeogenesis).

_____ 4. Effect on blood glucose levels and glycogen deposits in cells.

_____ 5. Effect on the intensity of the inflammatory response.

E. Match these terms with the
    correct statement or definition:

ACTH   Hypoglycemia or stress
Cortisol   Hypothalamus
CRH

_____   1. Location of CRH production.

_____   2. Production of ACTH is stimulated by this neurohormone.

_____   3. Hormone that stimulates cortisol production.

_____   4. Two hormones that inhibit CRH secretion.

_____

_____   5. Hormone that inhibits ACTH production.

_____   6. External factors that stimulate CRH production.

F. Match these terms with the
    correct statement:

Hypersecretion of cortisol
Hyposecretion of cortisol

_____   1. Hyperglycemia leading to diabetes mellitus.

_____   2. Osteoporosis, muscle atrophy, and weakness.

_____   3. Fat redistributed, resulting in moon face and buffalo hump.

_____   4. Increased skin pigmentation.

☞   The immune system is depressed in both hyposecretion and hypersecretion of cortisol.

G. Match these diseases with the
    correct conditions:

Addison's disease
Aldosteronism
Cushing's syndrome

_____   1. Hypersecretion of aldosterone.

_____   2. Hypersecretion of cortisol and androgens.

_____   3. Hyposecretion of aldosterone and cortisol.

H. Match these terms with the
    correct statement as it pertains
    to the effects of increased
    adrenal androgens:

Decreases
Increases

_____   1. Effect on amount of pubic and axillary hair in women.

_____   2. Effect on sex drive in women.

In males the effects of adrenal androgens are negligible in comparison with the testosterone produced by the testes.

The Pancreas and Pancreatic Hormones
"*The pancreas is both an exocrine gland and an endocrine gland.***"**

A. Match these terms with the
correct statement as it pertains
to the pancreas:

Alpha cells   Ducts and acini
Beta cells   Pancreatic islets
Delta cells

_____ 1. Constitute the exocrine portion of the pancreas.

_____ 2. Islet cells that secrete glucagon.

_____ 3. Islet cells that secrete insulin.

B. Match these terms with the
correct statement as it pertains
to the effects of increased insulin
or glucagon:

Decreases
Increases

_____ 1. Effect of insulin on uptake and use of glucose and amino acids in muscle cells.

_____ 2. Effect of insulin on glycogen and fat synthesis.

_____ 3. Effect of insulin on blood sugar levels.

_____ 4. Effect of glucagon on breakdown of liver glycogen to glucose.

_____ 5. Effect of glucagon on glucose synthesis from amino acids and fats.

_____ 6. Effect of glucagon on blood sugar levels.

_____ 7. Effect of glucagon on fat breakdown and production of ketones.

 Insulin has little direct effect on the nervous system, except to increase glucose uptake by the satiety (hunger) center.

C. Match these terms with the
correct statement:

Decreases insulin secretion
Increases insulin secretion

_____ 1. Hyperglycemia, certain amino acids.

_____ 2. Parasympathetic stimulation or gastrointestinal hormones.

D. Match these terms with the
   correct statement:

Decreases glucagon secretion
Increases glucagon secretion

_____ 1. Hypoglycemia, certain amino acids.

_____ 2. Sympathetic stimulation.

E. Match these terms with the
   correct symptoms:

Diabetes mellitus
Insulin shock

_____ 1. Polyuria, polydipsia, and polyphagia.

_____ 2. Glucosuria and hyperglycemia.

_____ 3. Headache, drowsiness, fatigue, convulsions, and death.

_____ 4. Sweating, pale skin, and tachycardia.

_____ 5. Ketonuria, acetone breath, and acidosis.

_____ 6. Atherosclerosis and peripheral vascular disease.

Hormonal Regulation of Nutrients

"Several hormones function together to regulate blood nutrient levels."

A. Match these terms with the
   correct statement for conditions
   immediately after a meal:

Decreased
Increased

_____ 1. Levels of glucagon, cortisol, growth hormone, and epinephrine.

_____ 2. Insulin secretion.

_____ 3. Uptake of glucose, amino acids, and fats.

_____ 4. Glucose converted to glycogen.

B. Match these terms with the
   correct statement for conditions
   two hours after a meal:

Decreased
Increased

_____ 1. Levels of glucagon, cortisol, growth hormone, and epinephrine.

_____ 2. Insulin secretion.

_____ 3. Uptake of glucose.

_____ 4. Glycogen converted to glucose.

_____ 5. Fat and protein metabolism for most tissues.

339

C. Match these terms with the
correct statement for conditions
during exercise:

Decreased
Increased

_____ 1. Sympathetic nervous system stimulation.

_____ 2. Release of epinephrine and glucagon.

_____ 3. Release of insulin.

_____ 4. Fatty acids, triacylglycerols, and ketones in blood.

_____ 5. Fat and glycogen metabolism in skeletal muscle.

Hormones of the Pineal Body, Thymus Gland, and Others

66*Hormones are produced at several other locations in the body.*99

Match these terms with the
correct statement or definition:

Melatonin   Thymosin
Pineal body

_____ 1. Endocrine gland in the epithalamus which secretes hormones that inhibit reproductive function.

_____ 2. One secretion of the pineal body; production of this hormone decreases as day length increases.

_____ 3. Hormone produced by the thymus; affects the immune system.

Hormonelike Substances

66*Paracrine regulatory substances are released from cells near the target cells and reach the target*99 *cells by diffusion.*

Match these terms with the
correct statement or definition:

Endorphins and enkephalins
Prostaglandins

_____ 1. Substances that initiate some symptoms of inflammation and pain.

_____ 2. Substances involved in regulation of uterine contractions.

_____ 3. Small polypeptides that bind to the same receptors as morphine.

QUICK RECALL

A. Match these endocrine glands with the correct hormone each secretes:

Adenohypophysis (anterior pituitary)
Adrenal cortex
Adrenal medulla
Hypothalamus
Neurohypophysis (posterior pituitary)
Ovaries
Pancreas
Parathyroid glands
Pineal body
Testes
Thymus gland
Thyroid gland (follicle cells)
Thyroid gland (parafollicular cells)

_____ 1. ACTH.

_____ 2. ADH.

_____ 3. Adrenal androgens.

_____ 4. Aldosterone.

_____ 5. Calcitonin.

_____ 6. Cortisol.

_____ 7. CRH.

_____ 8. Epinephrine.

_____ 9. Estrogen.

_____ 10. FSH.

_____ 11. GH.

_____ 12. GH-RH and GH-IH.

_____ 13. Glucagon.

_____ 14. GnRH.

_____ 15. Insulin.

_____ 16. LH.

_____ 17. Melatonin.

_____ 18. MSH.

_____ 19. Norepinephrine.

_____ 20. Oxytocin.

_____ 21. Parathyroid hormone.

_____ 22. Progesterone.

_____ 23. Prolactin.

_____ 24. PRH and PIH.

_____ 25. Testosterone.

_____ 26. Tetraiodothyronine and triiodothyronine.

_____ 27. Thymosin.

_____ 28. TRH.

_____ 29. TSH.

B. Match these target tissues with the correct hormone that affects them:

Adipose tissue, liver, and skeletal muscle
Adrenal cortex
Blood vessels and heart
Bone
Immune tissues
Kidneys
Liver
Mammary glands
Most tissues of the body
Ovaries or testes
Skin
Thyroid gland
Uterus

_____ 1. ADH.

_____ 2. Oxytocin (two).

_____

_____ 3. GH.

_____ 4. TSH.

_____ 5. ACTH.

_____ 6. MSH.

_____ 7. LH and FSH.

_____ 8. Prolactin.

_____ 9. Thyroid hormones.

_____ 10. Calcitonin.

_____ 11. PTH (two).

_____

_____ 12. Epinephrine (two).

_____

_____ 13. Aldosterone.

_____ 14. Glucocorticoids.

_____ 15. Insulin.

_____ 16. Glucagon.

_____ 17. Testosterone, progesterone, and estrogen.

_____ 18. Thymosin.

C. Simmonds' disease results from a tumor in the pituitary gland or from a lack of blood supply to the pituitary. Symptoms are similar to those that occur when the pituitary gland is removed. What effects (increase, decrease, or no effect) would Simmonds' disease have on these variables?

_____ 1. Metabolic rate.

_____ 2. Height, if condition developed in a child.

_____ 3. Ability to deal with stress.

_____ 4. Blood sugar levels.

_____ 5. Blood calcium levels.

_____ 6. Blood sodium levels.

_____ 7. Blood pH.

_____ 8. Milk production.

_____ 9. Ability to produce sperm or egg cells.

MASTERY LEARNING ACTIVITY

Place the letter corresponding to the correct answer in the space provided.

_____ 1. The pituitary gland
a. is derived from the brain.
b. is derived from the roof of the mouth.
c. is divided into two parts.
d. all of the above

_____ 2. The hypothalamohypophyseal portal system
a. contains one capillary bed.
b. carries hormones from the adenohypophysis to the body.
c. carries hormones from the neurohypophysis to the body.
d. carries hormones from the hypothalamus to the adenohypophysis.
e. carries hormones from the hypothalamus to the neurohypophysis.

_____ 3. Hormones secreted from the neurohypophysis
a. are produced in the hypothalamus.
b. are transported to the neurohypophysis within axons.
c. include ADH and oxytocin.
d. all of the above

_____ 4. Which of these is a consequence of an elevated blood osmolality?
a. decreased prolactin secretion
b. increased prolactin secretion
c. decreased ADH secretion
d. increased ADH secretion

_____ 5. Oxytocin is responsible for
a. preventing milk release from the mammary glands.
b. preventing goiter.
c. causing contraction of the uterus.
d. maintaining normal calcium levels.

_____ 6. Growth hormone
a. increases the usage of glucose.
b. increases the breakdown of lipids.
c. decreases the synthesis of proteins.
d. all of the above

_____ 7. Hypersecretion of growth hormone
a. results in giantism if it occurs in children.
b. causes acromegaly in adults.
c. increases the probability that one will develop diabetes mellitus.
d. a and b
e. all of the above

_____ 8. LH and FSH
a. are produced in the hypothalamus.
b. production is increased by TSH.
c. promote the production of gametes and reproductive hormones.
d. inhibit the production of prolactin.

_____ 9. Thyroid hormones
a. require iodine for their production.
b. are made from the amino acid tyrosine.
c. are transported in the blood bound to thyroxin-binding globulin.
d. all of the above

_____ 10. Which of these symptoms are associated with hyposecretion of the thyroid gland?
a. hypertension
b. nervousness
c. diarrhea
d. weight loss with a normal or increased food intake
e. decreased metabolic rate

344

11. Which of these would probably occur if a normal person received an injection of thyroid hormone?
    a. The secretion rate of TSH would decline.
    b. The person would develop symptoms of hypothyroidism.
    c. The person would develop hypercalcemia.
    d. The person would secrete more thyroid-releasing hormone.

12. Which of these would occur as a response to a thyroidectomy (removal of the thyroid gland)?
    a. increased calcitonin secretion
    b. increased T_3 and T_4 secretion
    c. decreased TRH secretion
    d. increased TSH secretion

13. Choose the statement below that most accurately predicts the long-term effect of a substance that prevents active transport of iodide by the thyroid gland.
    a. Large amounts of thyroid hormone would accumulate within the thyroid follicles, and little would be released.
    b. The person would exhibit symptoms of hypothyroidism.
    c. The anterior pituitary would secrete smaller amounts of TSH.
    d. The circulating levels of T_3 and T_4 would be increased.

14. A patient exhibited symptoms of hyperthyroidism. She had elevated blood levels of T_3 and T_4 and reduced TSH levels. There was no evidence of a tumor in her thyroid gland. Which of these explanations agrees with the data?
    a. She experienced hypersecretion of TRH.
    b. Neither the hypothalamus nor the pituitary was sensitive to the negative-feedback effect of T_3 and T_4.
    c. She had an elevated antibody in her blood that is similar in structure to TSH, but which could not be measured by the test for TSH.
    d. She suffered from anterior pituitary hypofunction.

15. Calcitonin
    a. is produced by the parathyroid glands.
    b. levels increase when blood calcium levels decrease.
    c. causes blood calcium levels to decrease.
    d. insufficiency results in weak bones and tetany.

16. Parathyroid hormone secretion increases in response to
    a. a decrease in blood calcium levels.
    b. increased production of parathyroid-stimulating hormone from the anterior pituitary.
    c. increased secretion of parathyroid-releasing hormone from the hypothalamus.
    d. all of the above

17. If parathyroid hormone levels increase, which of these would be expected?
    a. Osteoclast activity is increased.
    b. Calcium absorption from the small intestine is inhibited.
    c. Calcium reabsorption from urine is inhibited.
    d. Less active vitamin D would be formed in the kidneys.

18. Choose the response below that is consistent with the development of a tumor that destroys the parathyroid glands.
    a. The person may develop tetany.
    b. The person will be highly lethargic.
    c. The person's bones will break more easily than normal.
    d. Increased blood levels of calcium will result.

19. The adrenal medulla
    a. is formed from a modified portion of the sympathetic nervous system.
    b. has epinephrine as its major secretory product.
    c. increases its secretions during exercise.
    d. all of the above

20. Pheochromocytoma is a condition in which a benign tumor results in hypersecretion of the adrenal medulla. The symptoms that one would expect include
    a. hypotension.
    b. bradycardia.
    c. pallor.
    d. lethargy.
    e. a and b

21. Hormones secreted from the adrenal cortex include
    a. aldosterone.
    b. cortisol.
    c. androgen.
    d. a and b
    e. all of the above

22. If aldosterone secretions decrease,
    a. blood potassium levels decrease, and cells become more excitable.
    b. blood potassium levels increase, and cells become less excitable.
    c. blood hydrogen levels decrease, and acidosis results.
    d. blood hydrogen levels increase, and alkalosis results.
    e. blood sodium levels decrease, and blood volume decreases.

23. Glucocorticoids (cortisol)
    a. increase breakdown of fats.
    b. increase breakdown of proteins.
    c. increase the blood sugar levels.
    d. decrease inflammation.
    e. all of the above

24. The release of cortisol from the adrenal cortex is regulated by other hormones. Which of these hormones is correctly matched with its origin and function?
    a. CRH: secreted by the hypothalamus; stimulates the adrenal cortex to secrete cortisol
    b. CRH: secreted by the anterior pituitary; stimulates the adrenal cortex to secrete cortisol
    c. ACTH: secreted by the hypothalamus; stimulates the adrenal cortex to secrete cortisol
    d. ACTH: secreted by the anterior pituitary; stimulates the adrenal cortex to secrete cortisol

25. Which of these would be expected in Cushing's syndrome?
    a. loss of hair in women
    b. deposition of fat in the face, neck, and abdomen
    c. low blood glucose
    d. low blood pressure

26. Within the pancreas, the islets of Langerhans produce
    a. insulin.
    b. glucagon.
    c. digestive enzymes.
    d. a and b
    e. all of the above

27. Insulin
    a. increases the uptake of glucose by its target tissues.
    b. increases the uptake of amino acids by its target tissues.
    c. increases glycogen synthesis in the liver and in skeletal muscle.
    d. a and b
    e. all of the above

28. Which of these tissues is least affected by insulin?
    a. adipose tissue
    b. heart
    c. skeletal muscle
    d. brain

29. Glucagon
    a. primarily affects the liver.
    b. causes blood sugar levels to decrease.
    c. decreases fat metabolism.
    d. all of the above

30. Concerning ketones,
    a. they are a byproduct of the breakdown of fats.
    b. excess production of ketones can produce acidosis.
    c. ketones can be used as a source of energy.
    d. all of the above

31. When blood sugar levels increase,
    a. insulin and glucagon secretion increase.
    b. insulin and glucagon secretion decrease.
    c. insulin secretion increases, and glucagon secretion decreases.
    d. insulin secretion decreases, and glucagon secretion increases.

32. If a person who has diabetes mellitus forgot to take an insulin injection, symptoms that may soon appear include
    a. acidosis.
    b. hyperglycemia.
    c. increased urine production.
    d. all of the above

33. Melatonin
    a. is produced by the posterior pituitary.
    b. production increases as day length increases.
    c. inhibits development of the reproductive system.
    d. all of the above

34. A researcher noticed that following a heavy meal in the evening, rich in carbohydrates, he would be hungrier than usual in the morning. He decided to do an experiment to determine the cause for the condition. In the first experiment, several volunteers fasted for 24 hours. Then they consumed 1 g of glucose per kilogram of body weight (time 0 hours in the graph below). For the next 4 hours blood glucose levels were measured.

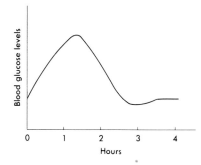

In the second experiment, three different groups were used. Each group fasted for 24 hours before the experiment. Group A then consumed 1 g of glucose per kilogram of body weight. Group B consumed a heavy meal rich in carbohydrates. Group C continued to fast. Measurement of blood insulin levels was made immediately following each treatment (time 0 in the graph below) and at 1 and 3 hours.

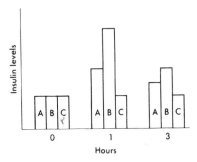

On the basis of the data, the investigator concluded that
a. large amounts of insulin are secreted in response to food consumption, and insulin levels remain high in the blood long enough for blood levels of glucose to decrease below normal, creating the hunger sensation.
b. blood glucose levels become elevated following food consumption and remain elevated for several hours because of the large amounts of insulin secreted, creating the hunger sensation.
c. the ability of the pancreas to produce insulin is limited to a 2- to 3-hour period. Thereafter only small amounts are secreted, creating the sensation of hunger.
d. the hunger sensation cannot be related to either blood glucose levels or insulin levels.

35. Which of these hormones, produced by most tissues of the body, can promote inflammation?
    a. endorphin
    b. enkephalin
    c. thymosin
    d. prostaglandin

Use a separate sheet of paper to complete this section.

1. If there is insufficient dietary intake of iodide, goiter can develop. Would the levels of T_3, T_4, TSH, and TRH be higher or lower than normal? Explain.

2. During pregnancy and/or lactation, parathyroid hormone levels may be elevated. Explain why this occurs and give a reason why it might be harmful.

3. A patient has pheochromocytoma. Would you expect her pupils to be dilated or constricted? Explain.

4. One of the symptoms of Addison's disease is increased skin pigmentation resulting from high levels of ACTH. Explain why ACTH levels are high.

5. A young boy (6 years old) exhibited marked and rapid development of sexual characteristics. On examination his testicles were not found to be larger than normal, but his plasma testosterone levels were elevated. As a mental exercise a student nurse decided that she would propose a cure. She considered the symptoms and decided on surgery to remove an adrenal tumor. Explain why you agree or disagree with her diagnosis.

6. The glucose tolerance test determines the ability of a person to dispose of a standard dose of glucose. If the rate of removal of glucose from the blood is significantly slower than normal, a diagnosis of diabetes mellitus is made. However, the diagnosis can be wrong. One explanation for false test results is patient anxiety. Explain how anxiety could produce a false test result.

7. Would you expect a person who was blind from birth to enter puberty at an earlier or later age than a normally sighted person? Explain.

ANSWERS TO CHAPTER 18

Introduction
1. Hypothalamus; 2. Pituitary; 3. Thymus;
4. Adrenals; 5. Ovaries; 6. Testes; 7. Pancreas;
8. Parathyroids; 9. Thyroid; 10. Pineal body

Pituitary Gland and Hypothalamus
A. 1. Infundibulum; 2. Neurohypophysis;
3. Adenohypophysis;
4. Hypothalamohypophyseal portal system;
5. Neurohormones; 6. Hypothalamus;
7. Adenohypophysis; 8. Action potentials
B. 1. Hypothalamohypophyseal tract;
2. Posterior pituitary; 3. Anterior pituitary;
4. Hypothalamohypophyseal portal system;
5. Neurosecretory cell; 6. Hypothalamus

Neurohypophyseal Hormones
A. 1. Decreases; 2. Decreases; 3. Increases;
4. Increases
B. 1. Blood osmolality; 2. Blood volume; 3. Blood
osmolality; 4. Blood volume
C. 1. Hyposecretion; 2. Hypersecretion
D. 1. Increases; 2. Increases
E. 1. Increases; 2. Increases

Adenohypophyseal Hormones
A. 1. Increases; 2. Increases; 3. Increases;
4. Increases; 5. Increases
B. 1. Increases; 2. Increases; 3. Increases;
4. Decreases
C. 1. Hyposecretion; 2. Hypersecretion;
3. Hypersecretion
D. 1. Proopiomelanocortin; 2. Lipotropins;
3. Beta-endorphin; 4. MSH and ACTH
E. 1. FSH and LH; 2. GnRH; 3. Prolactin; 4. PRH

The Thyroid Gland
A. 1. Follicle; 2. Parafollicular cells; 3. Calcitonin
B. 1. Thyroglobulins; 2. Tyrosine; 3. I⁻ ions;
4. Tetraiodothyronine (T_4);
5. Triiodothyronine (T_3); 6. Thyroxine-binding
globulin
C. 1. Increases; 2. Increases
D. 1. Hypersecretion; 2. Hypersecretion;
3. Hyposecretion; 4. Hyposecretion;
5. Hyposecretion; 6. Hyposecretion;
7. Hypersecretion; 8. Hypersecretion
E. 1. Increases; 2. Decreases; 3. Increases;
4. Increases; 5. Decreases; 6. Decreases
F. 1. Decreases; 2. Decreases; 3. Increases

The Parathyroid Glands
A. 1. Increases; 2. Increases; 3. Increases;
4. Increases; 5. Decreases
B. 1. Increases
C. 1. Hypersecretion; 2. Hyposecretion;
3. Hyposecretion

Adrenal Glands
1. Adrenal medulla; 2. Adrenal medulla;
3. Adrenal cortex

Adrenal Medulla
A. 1. Epinephrine; 2. Neurohormone
B. 1. Increases; 2. Increases; 3. Decreases;
4. Increases
C. 1. Increases; 2. Increases; 3. Increases
D. 1. Hypersecretion; 2. Hypersecretion

Adrenal Cortex
A. 1. Mineralocorticoids; 2. Glucocorticoids;
3. Androstenedione
B. 1. Increases; 2. Decreases; 3. Decreases
C. 1. Hypersecretion; 2. Hypersecretion;
3. Hyposecretion
D. 1. Increases; 2. Decreases; 3. Increases;
4. Increases; 5. Decreases
E. 1. Hypothalamus; 2. CRH; 3. ACTH; 4. ACTH
and cortisol; 5. Cortisol; 6. Hypoglycemia or
stress
F. 1. Hypersecretion of cortisol;
2. Hypersecretion of cortisol;
3. Hypersecretion of cortisol;
4. Hyposecretion of cortisol
G. 1. Aldosteronism; 2. Cushing's syndrome;
3. Addison's disease
H. 1. Increases; 2. Increases

The Pancreas and Pancreatic Hormones
A. 1. Ducts and acini; 2. Alpha cells; 3. Beta cells
B. 1. Increases; 2. Increases; 3. Decreases;
4. Increases; 5. Increases; 6. Increases;
7. Increases
C. 1. Increases; 2. Increases
D. 1. Increases; 2. Increases
E. 1. Diabetes mellitus; 2. Diabetes mellitus;
3. Insulin shock; 4. Insulin shock; 5. Diabetes
mellitus; 6. Diabetes mellitus

Hormonal Regulation of Nutrients
A. 1. Decreased 2. Increased 3. Increased
4. Increased
B. 1. Increased 2. Decreased 3. Decreased
4. Increased; 5. Increased
C. 1. Increased; 2. Increased; 3. Decreased
4. Increased; 5. Increased

Hormones of the Pineal Body, Thymus Gland, and Others
1. Pineal body; 2. Melatonin; 3. Thymosin

Hormonelike Substances
1. Prostaglandins; 2. Prostaglandins;
3. Endorphins and enkephalins

A.
1. Adenohypophysis
2. Neurohypophysis
3. Adrenal cortex
4. Adrenal cortex
5. Thyroid gland (parafollicular cells)
6. Adrenal cortex
7. Hypothalamus
8. Adrenal medulla
9. Ovaries
10. Adenohypophysis
11. Adenohypophysis
12. Hypothalamus
13. Pancreas (alpha cells)
14. Hypothalamus
15. Pancreas (beta cells)
16. Adenohypophysis
17. Pineal body
18. Adenohypophysis
19. Adrenal medulla
20. Neurohypophysis
21. Parathyroid glands
22. Ovaries
23. Adenohypophysis
24. Hypothalamus
25. Testes
26. Thyroid gland (follicular cells)
27. Thymus gland
28. Hypothalamus
29. Adenohypophysis

B.
1. Kidney
2. Uterus; mammary gland

3. Most tissues of the body
4. Thyroid gland
5. Adrenal cortex
6. Skin
7. Ovaries or testes
8. Mammary glands
9. Most tissues of the body
10. Bone
11. Bone; kidney
12. Blood vessels and heart; adipose tissue, liver, and skeletal muscle
13. Kidneys
14. Most tissues
15. Adipose tissue, liver, and skeletal muscle
16. Liver
17. Most tissues of the body
18. Immune cells

C.
1. Decrease (TSH, T_4, T_3)
2. Decrease (GH)
3. Decrease (ACTH, cortisol)
4. Possible decrease (ACTH, cortisol), but GH, glucagon, and epinephrine might stabilize.
5. No effect (PTH and calcitonin)
6. No effect; controlled by renin-angiotensin-aldosterone
7. No effect; controlled by renin-angiotensin-aldosterone
8. Decrease (prolactin)
9. Decrease (FSH, LH)

MASTERY LEARNING ACTIVITY

1. D. The pituitary gland can be divided into the anterior pituitary (adenohypophysis) and the posterior pituitary (neurohypophysis). The anterior pituitary is derived from the roof of the mouth, and the posterior pituitary is derived from the brain.

2. D. The hypothalamohypophyseal portal system originates as a capillary bed within the hypothalamus and extends to the adenohypophysis, where it terminates as a capillary bed. It is responsible for carrying hormones from the hypothalamus to the adenohypophysis.

3. D. Hormones such as ADH and oxytocin are produced by neurosecretory cells in the hypothalamus. The hormones are transported within the axons of the neurosecretory cells (through the hypothalamohypophyseal tract) to the neurohypophysis. The hormones are stored in the ends of the axons, and they are secreted in response to action potentials.

4. D. Increased blood osmolality increases the frequency of action potentials produced by osmoreceptors within the hypothalamus. In response, an increased amount of ADH is secreted to initiate water conservation in the kidneys. Increased prolactin results in production of milk.

5. C. Oxytocin causes uterine contractions. It is released during parturition (delivery) and helps stimulate the uterus to contract. Oxytocin also causes the myoepithelial cells that surround the alveoli of the mammary glands to contract. Consequently, in the lactating female oxytocin assists in "ejecting" milk from the mammary gland.

6. B. Growth hormone spares glucose usage, increases the breakdown of lipids, and increases the synthesis of proteins.

7. E. Hypersecretion of growth hormone increases bone growth in length, if it occurs before the epiphyseal plates are ossified (in children), and in width. In adults the epiphyseal plates are ossified, so the bone growth in width is most obvious (acromegaly). Because growth hormone increases blood glucose levels, those who suffer from hypersecretion of growth hormone are likely to develop diabetes mellitus.

8. C. Produced in the adenohypophysis, LH and FSH promote the production of gametes (oocytes and sperm cells) and reproductive hormones (testosterone in the male, estrogen and progesterone in the female). GnRH from the hypothalamus stimulates the release of LH and FSH. Prolactin is regulated by PRH and PIH from the hypothalamus.

9. D. Iodine binds with the amino acid tyrosine to form the thyroid hormones T_3 and T_4. The thyroid hormones are stored within the thyroid follicles as components of thyroglobulin. When released from the follicles, thyroid hormones are transported in the blood bound to thyroxin-binding globulin (TBG) or albumin.

10. E. Hypotension, lethargy, sluggishness, constipation, weight gain, and decreased metabolic rate are symptoms of hypothyroidism (hyposecretion of the thyroid gland).

11. A. An injection of thyroid hormone (T_3 and T_4) would result in an elevated blood level of these hormones. It would mimic hyperthyroidism. The elevated T_3 and T_4 levels in the blood would act on both the hypothalamus and the anterior pituitary to decrease TRH and TSH secretion, respectively.

12. D. Removal of the thyroid gland would eliminate T_3 and T_4 secretion, because they are the thyroid hormones. In response to the decreased T_3 and T_4 secretion, both TRH and TSH would be released in larger amounts. Calcitonin is secreted by the parafollicular cells of the thyroid gland. Its removal would eliminate calcitonin secretion.

13. B. The reduced iodine uptake results in decreased T_3 and T_4 synthesis. The T_3 and T_4 stored within the thyroid gland will be depleted, and the T_3 and T_4 levels in the blood will decrease. The reduced blood levels of T_3 and T_4 will result in the development of hypothyroidism and also remove a negative feedback effect on TSH secretion, resulting in increased TSH secretion.

14. C. Elevated T_3 and T_4 levels are normally the result of elevated TSH levels in the blood. However, in this case the T_3 and T_4 levels were elevated, whereas the TSH levels were suppressed. The only response that is consistent with the data is that there is another substance (an antibody) that has a TSH-like effect on the thyroid gland (as in Graves' disease).

15. C. Calcitonin causes blood calcium levels to decrease. It is produced by the parafollicular cells of the thyroid gland, it increases in response to increased blood calcium levels, and there is no known pathology associated with lack of calcitonin secretion.

16. A. Parathyroid hormone secretion is mediated through changes in blood calcium levels.

17. A. Increased parathyroid hormone levels result in increased blood calcium levels. Parathyroid hormone acts in several ways to increase blood calcium levels: it increases osteoclast activity, which increases bone breakdown and release of calcium into the blood; it promotes the production of active vitamin D, which increases calcium absorption from the small intestine; and it increases calcium reabsorption from urine.

18. A. Degeneration of the parathyroid gland results in decreased parathyroid hormone (PTH) secretion. PTH increases calcium absorption in the gut, reabsorption in the kidney, and resorption from bone. Decreased PTH results in hypocalcemia. The only symptom listed that is consistent with reduced PTH secretion and reduced blood levels of calcium is tetany.

19. D. The medulla is a modified portion of the sympathetic nervous system. Its major secretory products are epinephrine and small amounts of norepinephrine, which are released during exercise, stress, injury, or emotional excitement or in response to low blood glucose levels.

20. C. Hypersecretion of the adrenal medulla would result in large amounts of epinephrine released into the circulatory system, causing cutaneous vasoconstriction and pallor. Epinephrine would increase heart rate, stroke volume, and vasoconstriction in the viscera and skin. Hypertension is therefore consistent with pheochromocytoma. Epinephrine also tends to induce hyperexcitability rather than lethargy.

21. E. The zona glomerulosa of the adrenal cortex secretes aldosterone, the zona fasciculata secretes cortisol, and the zona reticularis secretes androgens.

22. E   A decrease in aldosterone results in a decrease in blood sodium levels and therefore blood volume. Aldosterone decrease also causes blood potassium levels to increase, resulting in more excitable cells (hypopolarization of cell membranes), and an increase in blood hydrogen ion levels, resulting in acidosis.

23. E   Glucocorticoids increase the breakdown of fats and proteins and increase blood sugar levels. They also decrease the intensity of the inflammatory response.

24. D   ACTH is secreted by the anterior pituitary and stimulates the adrenal cortex to secrete cortisol. CRH is secreted by the hypothalamus and stimulates the secretion of ACTH from the anterior pituitary.

25. B   Cushing's syndrome results from hypersecretion of cortisol and androgens. The cortisol causes fat to be redistributed to the face (moon face), neck (buffalo hump), and abdomen. It also causes hyperglycemia and depresses the immune system. The increased level of androgen causes hirsutism (excessive facial and body hair). Low blood pressure and low blood glucose are associated with decreased aldosterone levels (Addison's disease).

26. D   Cells within the islets of Langerhans produce insulin and glucagon. The beta cells produce insulin, and the alpha cells produce glucagon. The exocrine portion of the pancreas produces digestive enzymes.

27. E   Insulin promotes the uptake of glucose and amino acids, both of which can be used as an energy source. In the liver and in skeletal muscle the glucose is stored as glycogen. In adipose tissue glucose is converted into fat. The amino acids can be used to synthesize proteins or glucose.

28. D   Insulin affects brain tissue less than adipose, heart, or skeletal muscle. Glucose enters the brain cells without the presence of insulin. An exception is the satiety (hunger) center within the hypothalamus of the brain.

29. A   Glucagon primarily affects the liver, causing the breakdown of glycogen to glucose and the synthesis of glucose from amino acids and fats. Consequently, blood sugar levels increase. Glucagon also promotes fat metabolism.

30. D   When blood levels decrease, insulin levels drop, and glucagon levels increase. One result is a switch to the use of fats as a source of energy. One product produced during fat metabolism is ketones, which can be used by most tissues for energy. Ketones are also acidic and can produce acidosis.

31. C   Both insulin and glucagon secretions are regulated by blood sugar levels, but in the opposite direction.

32. D   A lack of insulin would be the result of a diabetic forgetting to take an insulin injection. Symptoms consistent with hyposecretion of insulin would develop, such as acidosis, hyperglycemia, and increased urine production.

33. C   Melatonin inhibits the development of the reproductive system in some animals and may be involved with the onset of puberty in humans. Melatonin is produced by the pineal body, and its production increases as day length decreases.

34. A   The data in the first experiment indicate that following glucose (or food) intake, the blood glucose levels increase and remain elevated for roughly 2 hours. Then blood glucose falls and remains less than normal for several hours. From the second experiment, it can be seen that the insulin levels are above normal following ingestion of food and, further, that they are correlated with the depressed blood glucose levels after 3 hours. That effect is exaggerated in those who consumed a large meal. Thus the data indicate that following a heavy meal a large amount of insulin is secreted. Because the insulin levels are elevated and remain elevated after 3 hours, one would expect the blood levels of glucose to decrease rapidly to below-normal levels. Low blood glucose levels are known to stimulate the "hunger" center in the brain.

35. D   Prostaglandins are "local" hormones produced by most tissues. They have a variety of effects, including the promotion of inflammation, fever, pain, and uterine contractions. The endorphins and enkephalins reduce pain within the central nervous system. Thymosin is produced by the thymus gland and is involved with immunity.

1. Without sufficient iodide, thyroid hormones are not synthesized, resulting in low blood levels of T_3 and T_4. Without the negative-feedback effects of T_3 and T_4, TSH and TRH levels are elevated.

2. During pregnancy the fetus withdraws calcium from the mother's blood and during lactation calcium is lost with the milk. In both cases, lower blood calcium levels stimulate parathyroid hormone release. Consequently, calcium is released from the bones or is absorbed more efficiently from the small intestine and urine, helping to maintain blood calcium levels. The removal of calcium from the bones results in weakened bones.

3. Pheochromocytoma results in overproduction of epinephrine and norepinephrine by the adrenal medulla. These chemicals are released into the blood and produce the same effects as sympathetic nervous system stimulation (see Chapter 16). Consequently, one expects dilated pupils.

4. Addison's disease results from hyposecretion of aldosterone and cortisol. With low levels of cortisol, the adenohypophysis is not inhibited, and ACTH secretion increases.

5. Rapid sexual development in a prepubertal boy is indicative of hypersecretion of sex steroids. The two most likely tissues are the testes and the adrenal glands. Because the testes are of normal size and there is no indication of abnormal testes function, the most logical tissue to suspect is the adrenal glands. It is possible that removal of an adrenal tumor would cure the boy.

6. Anxiety could result in sympathetic system activity. The sympathetic system inhibits insulin secretion and can therefore bias the test results. In addition, sympathetic activity could result in epinephrine release. Epinephrine causes an increase in blood sugar levels that can also bias the test results.

7. One might expect a person blind from birth to enter puberty at a later age than a normally sighted person. A decrease in melatonin production may be involved with the onset of puberty. In a blind person melatonin may be produced at higher than normal levels, because melatonin production increases in the dark.

19 Cardiovascular System: Blood

FOCUS: Blood consists of plasma and formed elements. The plasma is 92% water with dissolved or suspended molecules, including albumin, globulins, and fibrinogen. The formed elements include erythrocytes, leukocytes, and platelets. Erythrocytes contain hemoglobin, which can transport oxygen and carbon dioxide, and carbonic anhydrase, which is involved in carbon dioxide transport. Erythrocyte production is stimulated by renal erythropoietic factor released from the kidneys when blood oxygen levels decrease. Erythrocytes can be typed according to their surface antigens into ABO, Rh, and other blood groups. Leukocytes protect the body against microorganisms and remove dead cells and debris from the body. The different leukocytes include neutrophils, eosinophils, basophils, lymphocytes, and monocytes. Platelets prevent bleeding by forming platelet plugs and by producing factors involved in clotting. Blood clotting can be divided into an extrinsic (initiated by chemicals outside the blood) and intrinsic pathway (initiated by platelets and plasma coagulation factors). Once initiated both pathways cause the activation of thrombin, which converts fibrinogen into fibrin (the clot). Overproduction of clots is prevented by antithrombin and heparin, and clots are dissolved by plasmin.

CONTENT LEARNING ACTIVITY

Functions

"*Blood travels to and from the tissues of the body and plays an important role in homeostasis.*"

Using the terms provided, complete these statements:

Carbon dioxide and waste products
Clotting
Hormones and enzymes
Maintenance
Oxygen and nutrients
Protection

Blood has many functions in the body. It transports _(1)_ to cells for cellular respiration, and _(2)_ are transported away from cells for elimination. _(3)_ of homeostasis is a crucial function of blood. For example, _(4)_ that regulate body processes are found in blood. Blood also plays a role in maintaining body temperature and the normal fluid, electrolyte, and pH balance of tissues. Another important function of blood is _(5)_; for example, cells and chemicals of the blood are part of the immune system, and blood _(6)_ provides protection against excessive fluid and cell loss.

1. _____

2. _____

3. _____

4. _____

5. _____

6. _____

Plasma

"The liquid matrix of the blood is called plasma."

Match these terms with the correct statement or definition:

Colloidal solution   Serum
Plasma   Solutes
Plasma proteins   Water

_____ 1. Pale yellow fluid; makes up slightly more than half of blood volume.

_____ 2. The major (approximately 91%) component of plasma.

_____ 3. Fine particles suspended in a liquid and resistant to sedimentation or filtration.

_____ 4. Albumin, globulins, and fibrinogen.

Formed Elements

"The formed elements of the blood include specialized cells and cell fragments."

Match these terms with the correct statement or definition:

Erythrocytes   Thrombocytes
Leukocytes

_____ 1. Cells that constitute most of the formed elements; red blood cells.

_____ 2. Formed elements with nuclei; white blood cells.

_____ 3. Cell fragments that are part of the formed elements; platelets.

Production of Formed Elements

"All the formed elements of the blood are derived from a single population of stem cells."

Match these terms with the correct statement or definition:

Hematopoiesis   Monoblasts
Hemocytoblasts   Myeloblasts
Lymphoblasts   Proerythroblasts
Megakaryoblasts

_____ 1. Blood cell production process that takes place in red bone marrow and lymphoid tissue after birth; hemopoiesis.

_____ 2. Stem cells that give rise to all the formed elements.

_____ 3. Cells from which granulocytes develop.

_____ 4. Cells from which lymphocytes develop.

_____ 5. Cells from which platelets develop.

In adults, the hematopoietic red marrow is confined to the skull, ribs, sternum, vertebrae, pelvis, proximal femur, and proximal humerus.

Erythrocytes

66_Erythrocytes are the most numerous of the formed elements in the blood._**99**

A. Using the terms provided, complete these statements:

Biconcave   Nucleus
Hemoglobin

1. _____

2. _____

3. _____

An erythrocyte is a _(1)_ disk that loses its _(2)_ and nearly all of its organelles during maturation. The main component of the erythrocyte is the pigmented protein _(3)_, which occupies about one third of the total cell volume.

B. Match these terms with the correct statement or definition:

Bicarbonate ion   Carbonic anhydrase
Carbonic acid   Hemolysis

_____ 1. Process in which erythrocytes rupture and hemoglobin is released.

_____ 2. Enzyme that catalyzes the reaction between carbon dioxide and water.

_____ 3. Product of the reaction between carbon dioxide and water.

_____ 4. Major form of carbon dioxide transported in the blood.

 The primary functions of erythrocytes are to transport oxygen from the lungs to the various tissues of the body and to transport carbon dioxide from the tissues to the lungs.

C. Match these terms with the correct statement or definition:

Carbaminohemoglobin   Heme
Deoxyhemoglobin   Iron
Globin   Oxyhemoglobin

_____ 1. One of four protein chains in the hemoglobin molecule.

_____ 2. Red pigment molecule.

_____ 3. Element in the center of each heme molecule; binds with oxygen.

_____ 4. Hemoglobin that has oxygen associated with each heme group.

_____ 5. Hemoglobin with a darker red color; reduced hemoglobin.

_____ 6. Hemoglobin with carbon dioxide attached to amino groups of the globin molecule.

D. Using the terms provided, complete these statements:

Anemia   Less
Iron   More

Embryonic and fetal globins are __(1)__ effective at binding oxygen than is adult globin. Abnormal globins are __(2)__ effective at attracting oxygen than is normal globin, and may result in __(3)__. __(4)__ is necessary for the normal function of hemoglobin. Dietary iron is absorbed into the circulation in the upper part of the digestive tract. If the iron content of the body is high, __(5)__ iron is absorbed, and as the content drops, __(6)__ iron is absorbed. Iron deficiency can also result in __(7)__.

1. _____
2. _____
3. _____
4. _____
5. _____
6. _____
7. _____

E. Using the terms provided, complete these statements:

Erythrocytes   Proerythroblasts
Erythropoiesis   Reticulocytes
Intermediate erythroblasts

The process by which new erythrocytes are produced is called __(1)__. __(2)__, the cells from which erythrocytes develop, are derived from hemocytoblasts. After several mitotic divisions, proerythroblasts become early erythroblasts, which continue to undergo mitosis and begin to produce hemoglobin. These cells then develop into __(3)__, which have almost a complete complement of hemoglobin. The cells lose their nuclei by a process of extrusion, after which the cells are called __(4)__. The cells are released from red bone marrow into the blood, and within 1 or 2 days they lose their endoplasmic reticulum and become __(5)__.

1. _____
2. _____
3. _____
4. _____
5. _____

F. Match these terms with the correct statement or definition:

Erythropoietin   Oxygen
Iron   Vitamin B_{12} and folic acid.

_____ 1. Necessary for cell division in erythropoiesis.

_____ 2. Humoral factor that stimulates erythropoiesis.

_____ 3. Production of erythropoietin increases when the levels of this substance decrease.

G. Match these terms with the correct statement or definition:

Bilirubin   Macrophages
Biliverdin

_____ 1. Removes old or damaged erythrocytes from the blood.

_____ 2. First breakdown product of heme groups.

_____ 3. Breakdown product of heme groups; excreted by the liver in bile; excess in blood causes jaundice.

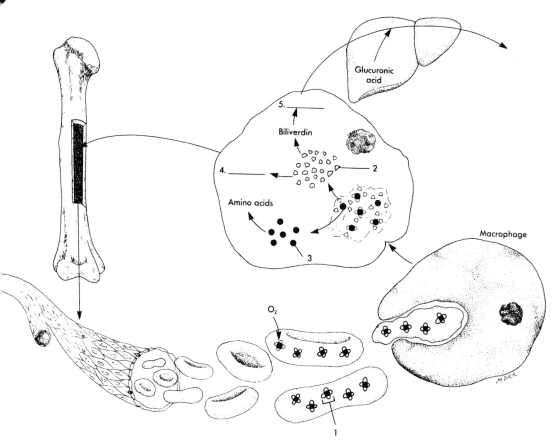

Figure 19-1

H. Match these terms with
the correct parts of the diagram
labeled in Figure 19-1:

Bilirubin   Hemoglobin
Globin   Iron
Heme

1. _____   3. _____   5. _____

2. _____   4. _____

Leukocytes

66*Leukocytes, or white blood cells, are nucleated blood cells that lack hemoglobin.*99

A. Match these terms with the correct statement or definition:

Agranulocytes	Granulocytes
Chemotaxis	Phagocytosis
Diapedesis	Pus

_____ 1. Process by which leukocytes leave the circulation.

_____ 2. Process by which leukocytes find foreign material or dead cells.

_____ 3. Accumulation of dead leukocytes, fluid, and debris.

_____ 4. Leukocytes containing large cytoplasmic granules; includes neutrophils, eosinophils, and basophils.

_____ 5. Leukocytes with very small cytoplasmic granules; includes lymphocytes and monocytes.

B. Match these terms with the correct statement or definition:

Lysozyme	Polymorphonuclear neutrophils
Phagocytosis	

_____ 1. Another name for neutrophils, referring to their trilobed nucleus.

_____ 2. Process used by neutrophils to destroy bacteria, antigen-antibody complexes, and other foreign matter.

_____ 3. Class of enzyme, secreted by neutrophils, capable of destroying certain bacteria.

☞ Neutrophils are by far the most common type of leukocyte in the blood.

C. Match these terms with the correct statement or definition:

Basophils	Heparin
Eosinophils	Histamine

_____ 1. Granulocytes that accept an acid stain.

_____ 2. Granulocytes that produce enzymes that destroy histamine.

_____ 3. Granulocytes that contain histamine and heparin.

_____ 4. Substance that increases inflammation.

_____ 5. Substance that inhibts blood clotting.

D. Match these terms with the
correct statement or definition:

Lymphocyte   Monocyte
Macrophage

_____ 1. Agranular leukocyte that is the smallest of all leukocytes.

_____ 2. Leukocyte that plays a major role in immunity, including antibody production.

_____ 3. This cell enters tissues and is transformed into a macrophage.

_____ 4. Large "eating" cell that phagocytizes bacteria, dead cells, cell fragments, and debris; often associated with chronic infections.

Platelets

"*Platelets play an important role in preventing blood loss.***"**

Match these terms with the
correct statement or definition:

Megakaryocytes
Platelets

_____ 1. Minute fragments of cells consisting of a small amount of cytoplasm surrounded by a cell membrane; thrombocytes.

_____ 2. Cells from which platelets are formed.

Vascular Spasm and Platelet Plug Formation

"*Vascular spasm and platelet plug formation prevent blood loss.***"**

Match these terms with the
correct statement or definition:

ADP   Platelet plug
Endothelin   Thromboxane
Hemostasis   Vascular spasm

_____ 1. Can be divided into three stages: vascular spasm, platelet plug formation, and coagulation.

_____ 2. Immediate but temporary closure of a blood vessel resulting from contraction of smooth muscles within the blood vessel's wall.

_____ 3. Chemical that makes the surface of platelets sticky.

_____ 4. Chemical that induces platelets to release ADP; also stimulates vascular spasms.

_____ 5. Cluster of activated and sticky platelets; seals small holes in blood vessels.

Coagulation

Coagulation prevents blood loss through large tears or cuts in blood vessels.

A. Match these terms with the
   correct statement or definition:

Coagulation   Prothrombin
Fibrin   Thrombin
Fibrinogen   Vitamin K

_____   1. Formation of a blood clot (network of fibrin containing blood cells, platelets, and fluid).

_____   2. Compound required for the production of many blood clotting factors.

_____   3. Soluble plasma protein that is converted to fibrin.

_____   4. Enzyme that catalyzes the reaction that produces fibrin.

B. Match these stages of clotting
   with the processes below:

Stage 1 (extrinsic)   Stage 2
Stage 1 (intrinsic)   Stage 3

_____   1. Tissue factor is produced by damaged cells.

_____   2. Platelet factor XII is activated in the plasma.

_____   3. Prothrombinase is formed.

_____

_____   4. Prothrombin is converted to thrombin.

_____   5. Fibrin is produced from fibrinogen.

C. Match these terms with the
   correct statement or definition:

Anticoagulant   Plasmin
Clot retraction   Serum
Embolus   Thrombus
Fibrinolysis

_____   1. General term for a substance that prevents the initiation of clot formation.

_____   2. Clot within a blood vessel.

_____   3. Clot that has broken loose and is floating through the circulation.

_____   4. Condensation of a clot into a denser, more compact structure.

_____   5. Plasma with the clot-producing proteins removed.

_____   6. Dissolution of a clot.

_____   7. Enzyme that hydrolyzes fibrin.

D. Match these terms with the correct statement or definition:

Antithrombin
EDTA
Heparin
Prostacyclin

Sodium citrate
Streptokinase
Tissue plasminogen activator

_____ 1. Plasma protein produced by the liver; slowly inactivates thrombin.

_____ 2. Increases the effectiveness of antithrombin.

_____ 3. Prostaglandin derivative that counteracts thrombin.

_____ 4. Three chemicals commonly used to prevent blood from clotting outside the body.

_____

_____

_____ 5. Two chemicals used to dissolve clots.

_____

Blood Grouping

66 _Transfusion reactions occur when the donor and recipient have different blood groups._ 99

A. Match these terms with the correct statement or definition:

Agglutination
Antibodies
Antigens

Hemolysis
Transfusion

_____ 1. Transfer of blood or other solutions into the blood.

_____ 2. Substances recognized by the immune system; agglutinogens.

_____ 3. Proteins that react with antigens on blood cells; agglutinins.

_____ 4. Clumping of cells.

B. Match these blood types with the correct statement:

Type A blood
Type AB blood

Type B blood
Type O blood

_____ 1. A person (universal donor) with this type of blood can usually give blood to a person with any blood type.

_____ 2. A person (universal recipient) with this type of blood can usually receive blood from anyone.

_____ 3. These blood types would have type A antibodies.

_____

_____ 4. This blood type is the most common throughout the world.

_____ 5. In this blood type, erythrocytes contain both A and B antigens.

C. Match these blood types with the correct statement:

Rh-negative
Rh-positive

_____ 1. Blood type with no Rh antigens.

_____ 2. Fetus with this type of blood could develop hemolytic disease of the newborn.

_____ 3. A woman with this type of blood will never have a baby who develops hemolytic disease of the newborn.

☞ Hemolytic disease of the newborn is a disorder that occurs when maternal Rh antibodies cross the placenta in large numbers and lyse fetal erythrocytes.

Diagnostic Blood Tests

"Blood tests can prevent transfusion reactions and provide information about a patient's health."

Match these terms with the correct statement or definition:

Blood chemistry
Complete blood count
Hematocrit
Hemoglobin
Platelet count

Prothrombin time
Red blood cell count
Type and cross match
White blood cell count
White blood cell differential

_____ 1. The test used to prevent transfusion reactions.

_____ 2. Includes a red blood cell count, hemoglobin and hematocrit measurements, and white blood cell count.

_____ 3. This test would detect polycythemia.

_____ 4. Measures the percentage of total blood volume composed of erythrocytes.

_____ 5. This test would detect leukemia.

_____ 6. Test that determines the percentages of each of the five kinds of leukocytes.

_____ 7. This test would detect thrombocytopenia.

_____ 8. Measures how long it takes for blood to start clotting.

_____ 9. Determines the composition of materials dissolved or suspended in plasma, such as glucose and bilirubin.

Blood Disorders

"Blood disorders may seriously affect the function of the circulatory system.**"**

A. Match these types of anemia with the correct description:

Aplastic anemia   Pernicious anemia
Hemolytic anemia   Sickle cell anemia
Hemorrhagic anemia   Thalassemia
Iron deficiency anemia

_____ 1. Insufficient erythrocyte production caused by abnormal red bone marrow or destruction of the red bone marrow.

_____ 2. Caused by insufficient vitamin B_{12}.

_____ 3. Deficiency of erythrocytes caused by the loss of large quantities of blood.

_____ 4. Anemia in which erythrocytes rupture or are destroyed at an excessive rate.

_____ 5. Hereditary disorder that causes insufficient production of the globin portion of hemoglobin.

_____ 5. Hereditary disorder that causes the formation of an abnormal hemoglobin.

B. Match these blood disorders with the correct description:

AIDS   Infectious mononucleosis
Hemophilia   Malaria
Hepatitis   Polycythemia
Leukemia   Septicemia

_____ 1. Condition characterized by an overabundance of erythrocytes.

_____ 2. Genetic disorder in which coagulation factors are abnormal or absent.

_____ 3. Type of cancer in which abnormal production of one or more leukocytes occurs.

_____ 4. General term for multiplication of microorganisms in the blood.

_____ 5. Protozoan infection of erythrocytes; symptoms include fever and chills.

_____ 6. Infects the salivary glands and lymphocytes.

_____ 7. Two viral diseases that can be acquired through transfusions.

_____

1. List the three major functions of blood. Give an example of each function.

2. Name the parts of a hemoglobin molecule, give the function of each part, and state the fate of each part when hemoglobin is broken down.

3. List the events that lead to increased red blood cell production when blood oxygen levels decrease.

4. List the three types of granulocytes and give a function of each.

5. List the two types of agranulocytes and give a function of each.

6. Give two ways that platelets prevent blood loss.

7. For stage 1 of the extrinsic and intrinsic clotting mechanism, list the starting and ending chemical.

8. For stages 2 and 3 of the clotting mechanism, list the chemical reactions that occur.

9. List the chemicals that prevent clot formation or dissolve clots.

MASTERY LEARNING ACTIVITY

Place the letter corresponding to the correct answer in the space provided.

_____ 1. Which of the following is a function of blood?
- a. prevents fluid loss
- b. transport of hormones
- c. carries oxygen to cells
- d. involved in regulation of body temperature
- e. all of the above

_____ 2. Which of the following is NOT a component of plasma?
- a. albumin
- b. fibrinogen
- c. platelets
- d. water
- e. glucose

_____ 3. The stem cells that give rise to all the formed elements are
- a. hemocytoblasts.
- b. lymphoblasts.
- c. megakaryoblasts.
- d. monoblasts.
- e. myeloblasts.

_____ 4. Erythrocytes
- a. are the least numerous formed element in the blood.
- b. are cylindrical shaped cells.
- c. are produced in yellow marrow.
- d. do not have a nucleus.
- e. all of the above

_____ 5. Which of the following components of an erythrocyte is correctly matched with its function?
- a. heme group of hemoglobin - oxygen transport
- b. globin portion of hemoglobin - carbon dioxide transport
- c. carbonic anhydrase - carbon dioxide transport
- d. all of the above

_____ 6. Which of the following constituents of blood is used as a sensitive indicator of increased hematopoiesis?
- a. erythrocytes
- b. reticulocytes
- c. neutrophils
- d. platelets
- e. plasma

_____ 7. Which of the components of hemoglobin is correctly matched with its fate following the destruction of an erythrocyte?
- a. heme: reused to form new hemoglobin molecule
- b. globin: broken down into amino acids
- c. iron: mostly secreted in bile
- d. all of the above

_____ 8. Erythropoietin
- a. is produced mainly by the heart.
- b. inhibits the production of erythrocytes.
- c. production increases when blood oxygen decreases.
- d. production is inhibited by testosterone.

_____ 9. Which of the following changes would occur in the blood in response to the initiation of a vigorous exercise program?
- a. increased erythropoietin production
- b. increased concentration of reticulocytes
- c. increased bilirubin formation
- d. a and b
- e. all of the above

_____10. If you lived near the coast and you were training for a track meet in Denver, you would want to spend a few weeks before the meet training at
a. sea level.
b. an altitude similar to Denver's.
c. a facility with a hyperbaric chamber.
d. it doesn't matter

_____11. The blood cells that function to inhibit inflammation are
a. eosinophils.
b. basophils.
c. neutrophils.
d. monocytes.

_____12. The most numerous type of leukocytes, whose primary function is phagocytosis, is
a. macrophages.
b. lymphocytes.
c. neutrophils.
d. thrombocytes.

_____13. Monocytes
a. are the smallest leukocytes.
b. increase in number in chronic infections.
c. give rise to neutrophils.
d. all of the above

_____14. A constituent of blood plasma that forms the network of fibers in a clot is
a. fibrinogen.
b. tissue factor.
c. platelets.
d. thrombin.

_____15. Given the following chemicals:
1. activated factor XII
2. fibrinogen
3. prothrombinase
4. thrombin

Choose the arrangement that lists the chemicals in the order they would be used during clot formation.
a. 1, 3, 4, 2
b. 2, 3, 4, 1
c. 3, 4, 1, 2
d. 3, 4, 2, 1

_____16. The chemical that is involved in the breakdown of a clot (fibrinolysis) is
a. fibrinogen.
b. antithrombin.
c. heparin.
d. plasmin.

_____17. A person with type A blood
a. has type A antibodies.
b. has type B antigens.
c. will have a transfusion reaction if given type B blood.
d. all of the above

_____18. Rh negative mothers who receive a RhoGAM injection are given that injection
a. to initiate the synthesis of Rh antibodies in the mother.
b. to initiate Rh antibody production in the baby.
c. to prevent the mother from producing Rh antibodies.
d. to prevent the baby from producing Rh antibodies.

_____19. The blood test that would distinguish between leukocytosis and leukopenia is
a. type and cross match.
b. hematocrit.
c. platelet count.
d. complete blood count.

_____20. An elevated neutrophil count is usually indicative of
a. an allergic reaction.
b. a bacterial infection.
c. a viral infection.
d. a parasitic infection.
e. increased antibody production.

FINAL CHALLENGES

Use a separate sheet of paper to complete this section.

1. Explain the effect of a decreased intake of iron on hematocrit.

2. Patients with advanced kidney diseases that impair kidney function often become anemic. On the other hand, patients with kidney tumors sometimes develop polycythemia. Can you explain these symptoms? (Hint: Tumors often cause overactivity of the tissue affected.)

3. If the plasma from a person with hemolytic anemia were injected into a normal person, what would be the effect on erythrocyte production in the normal person? Explain.

4. Cigarette smoke produces carbon monoxide. If a nonsmoker smoked a pack of cigarettes a day for a few weeks, what would happen to his reticulocyte count? Explain.

5. A young child has periodic episodes of difficulty in breathing. Because of the patient's history, the physician suspects that the attacks are brought on by a stress-anxiety reaction. However, asthma (an allergic reaction) is also a legitimate possibility. The physician orders a complete blood count. What information in the complete blood count could be used to decide between the two diagnoses.?

6. How could you distinguish between excessive bleeding caused by a vitamin K deficiency and thrombocytopenia?

7. During pregnancy the developing fetus must manufacture many new red blood cells. What precautions should the mother take with her diet to prevent the development of anemia in herself and the fetus?

ANSWERS TO CHAPTER 19

Functions
  1. Oxygen and nutrients; 2. Carbon dioxide and waste products; 3. Maintenance; 4. Hormones and enzymes; 5. Protection; 6. Clotting

Plasma
  1. Plasma; 2. Water; 3. Colloidal solution; 4. Plasma proteins;

Formed Elements
  1. Erythrocytes; 2. Leukocytes; 3. Thrombocytes

Production of Formed Elements
  1. Hematopoiesis; 2. Hemocytoblasts; 3. Myeloblasts; 4. Lymphoblasts; 5. Megakaryoblasts

Erythrocytes
  A. 1. Biconcave; 2. Nucleus; 3. Hemoglobin
  B. 1. Hemolysis; 2. Carbonic anhydrase; 3. Carbonic acid; 4. Bicarbonate ion
  C. 1. Globin; 2. Heme; 3. Iron; 4. Oxyhemoglobin; 5. Deoxyhemoglobin; 6. Carbaminohemoglobin
  D. 1. More; 2. Less; 3. Anemia; 4. Iron; 5. Less; 6. More; 7. Anemia
  E. 1. Erythropoiesis; 2. Proerythroblasts; 3. Intermediate erythroblasts; 4. Reticulocytes; 5. Erythrocytes
  F. 1. Vitamin B_{12} and folic acid; 2. Erythropoietin; 3. Oxygen
  G. 1. Macrophages; 2. Biliverdin; 3. Bilirubin
  H. 1. Hemoglobin; 2. Heme; 3. Globin; 4. Iron; 5. Bilirubin

Leukocytes
  A. 1. Diapedesis; 2. Chemotaxis; 3. Pus; 4. Granulocytes; 5. Agranulocytes
  B. 1. Polymorphonuclear neutrophils; 2. Phagocytosis; 3. Lysozyme
  C. 1. Eosinophils; 2. Eosinophils; 3. Basophils; 4. Histamine; 5. Heparin
  D. 1. Lymphocyte; 2. Lymphocyte; 3 Monocyte; 4. Macrophage

Platelets
  1. Platelets; 2. Megakaryocytes

Vascular Spasm and Platelet Plug Formation
  1. Hemostasis; 2. Vascular spasm; 2. ADP; 4. Thromboxane; 5. Platelet plug

Coagulation
  A. 1. Coagulation; 2. Vitamin K; 3. Fibrinogen; 4. Thrombin
  B. 1. Stage 1 (extrinsic); 2. Stage 1 (intrinsic); 3. Stage 1 (extrinsic) and Stage 1 (intrinsic); 4. Stage 2; 5. Stage 3
  C. 1. Anticoagulant; 2. Thrombus; 3. Embolus; 4. Clot retraction; 5. Serum; 6. Fibrinolysis; 7. Plasmin
  D. 1. Antithrombin; 2. Heparin; 3. Prostacyclin; 4. Heparin, EDTA, sodium citrate; 5. Streptokinase, tissue plasminogen activator

Blood Groups
  A. 1. Transfusion; 2. Antigens; 3. Antibodies; 4. Agglutination
  B. 1. Type O blood; 2. Type AB blood; 3. Type B blood, type O blood; 4. Type O blood; 5. Type AB blood
  C. 1. Rh-negative; 2. Rh-positive; 3. Rh-positive

Diagnostic Blood Tests
  1. Type and cross match; 2. Complete blood count; 3. Red blood cell count; 4. Hematocrit; 5. White blood cell count; 6. White blood cell differential; 7. Platelet count; 8. Prothrombin time; 9. Blood chemistry

Blood Disorders
  A. 1. Aplastic anemia; 2. Pernicious anemia; 3. Hemorrhagic anemia; 4. Hemolytic anemia; 5. Thalassemia; 6. Sickle cell anemia
  B. 1. Polycythemia; 2. Hemophilia; 3. Leukemia; 4. Septicemia; 5. Malaria; 6. Infectious mononucleosis; 7. AIDS, hepatitis

1. Transportation: gases, nutrients, waste products, vitamin D, lactic acid; maintenance: hormones, enzymes, pH, fluid and electrolyte balance, temperature regulation; protection: part of immune system, blood clotting

2. Globin: transports carbon dioxide, broken down to amino acids that are recycled; heme: transports oxygen, broken down to biliverdin, then to bilirubin, which is carried in the plasma to the liver, where it is incorporated into bile; iron is recycled

3. Decreased oxygen, increased erythropoietin production by kidney, erythropoiesis stimulated

4. Neutrophils: phagocytize foreign matter, secrete lysozyme; eosinophils: reduce inflammatory response; basophils: release histamine for inflammatory or allergic response.

5. Lymphocytes: immunity, including antibody production; monocytes: become macrophages

6. Formation of platelet plugs and formation of clots
7. Stage 1 extrinsic: tissue factor to prothrombinase
   Stage 1 intrinsic: Factor XII (Hageman factor) to prothrombinase
8. Stage 2: Prothrombin to thrombin; Stage 3: Fibrinogen to fibrin
9. Heparin, antithrombin and prostacyclin prevent clots; plasmin dissolves clots

MASTERY LEARNING ACTIVITY

1. E. Blood performs many important functions such as transport (oxygen, carbon dioxide, nutrients, waste products, hormones, enzymes), protection against bacteria and other foreign substances, coagulation (prevents fluid loss), temperature regulation, and pH regulation.

2. C. Albumin, fibrinogen, water, and glucose are all important components of plasma. Platelets are formed elements and are not part of the plasma.

3. A. Hemocytoblasts give rise to all the formed elements. Hemocytoblasts give rise to lymphoblasts (from which lymphocytes develop), megakaryoblasts (from which platelets develop), monoblasts (from which monocytes develop), myeloblasts (from which granulocytes develop), and proerythroblasts (from which erythrocytes develop).

4. D. Erythrocytes do not have a nucleus. They are the most numerous formed elements, they are biconcave disks, and they are produced in red bone marrow.

5. D. Heme groups are involved with oxygen transport, and the globins with carbon dioxide transport. Carbonic anhydrase catalyzes the reaction between carbon dioxide and water to form carbonic acid, which dissociates to form bicarbonate and hydrogen ions.

6. B. Reticulocytes are immature red blood cells that have just been released into the blood. An increase in the number of reticulocytes indicates an increased rate of hematopoiesis.

7. B. Globin is broken down into its component amino acids, most of which are used in the production of other proteins. Iron is recycled and used in the production of new hemoglobin. Heme is converted to biliverdin and then to bilirubin, which is secreted in bile.

8. C. Erythropoietin is produced mainly by the kidneys and stimulates hemopoietic bone marrow to increase erythrocyte production. When blood oxygen levels decrease, the kidneys produce erythropoietin, which stimulates erythrocyte production. Testosterone increases erythropoietin production and therefore stimulates erythropoiesis.

9. E. Vigorous exercise reduces blood flow to the kidney (see Chapter 16); therefore there is reduced oxygen delivery. The kidneys release more erythropoietin, which results in increased erythropoiesis, and the reticulocyte count increases. With increased exercise one might expect increased damage to erythrocytes and increased bilirubin formation as hemoglobin was broken down.

10. B. At higher altitudes the blood is less able to pick up oxygen from the air. The resulting decrease in blood oxygen stimulates erythropoiesis. Because the oxygen-carrying capacity of blood increases with an increased erythrocyte count, it is an advantage for a person who is going to compete at a high altitude to be exposed to that altitude long enough to acclimate.

11. A. Eosinophils produce enzymes that destroy inflammatory chemicals such as histamine. Basophils promote inflammation through the release of histamines.

12. C. Neutrophils are the most numerous leukocytes (60% to 70%); their primary function is phagocytosis. Macrophages are also phagocytic but are derived from monocytes that make up 2% to 8% of leukocytes. Thrombocytes or platelets are much more numerous than leukocytes, but they are not considered leukocytes and they are involved with clot formation, not phagocytosis.

13. B. Monocyte numbers typically increase during chronic infection. Monocytes are the largest of the leukocytes (lymphocytes are the smallest), and they give rise to macrophages.

14. A. During the process of clotting fibrinogen gives rise to fibrin, which undergoes polymerization to form the network of fibers that make up a clot.

15. A   Through the intrinsic pathway, factor XII is activated. A series of coagulation factors are then activated, eventually resulting in the formation of prothrombinase. Prothrombinase converts prothrombin to thrombin, and the thrombin converts fibrinogen to fibrin.

16. D.   Plasmin breaks down fibrin. Fibrinogen is converted into fibrin during clot formation. Antithrombin and heparin are anticoagulants that prevent clot formation.

17. C.   A person with type A blood has type A antigens and type B antibodies; one with type B blood has type B antigens and type A antibodies. If the type A person receives type B blood, the A antibodies in the donated blood bind to the A antigen of the person, and a transfusion reaction occurs. Also, the person's type B antibodies can bind to the type B antigens of the donated blood.

18. C.   The RhoGAM injection contains Rh antibodies that remove fetal Rh antigens from the maternal circulation before the mother's immune system recognizes their presence and begins to produce Rh antibodies.

19. D.   A complete blood count includes a red blood cell count, hemoglobin and hematocrit measurements, and a white blood cell count. The white blood cell count would differentiate between leukopenia (lower-than-normal leukocyte numbers) and leukocytosis (higher-than-normal leukocyte numbers).

20. B.   Neutrophils are capable of phagocytizing bacteria, but not viruses. Neutrophil numbers often increase dramatically in response to bacterial infections. Eosinophil numbers increase in response to allergic reactions and certain parasitic infections. Lymphocytes are responsible for antibody production.

 ## FINAL CHALLENGES

1. Inadequate iron in the diet results in decreased hemoglobin production. Because hemoglobin normally makes up about one third of the volume of an erythrocyte, the erythrocytes are smaller than normal. Hematocrit is the percent of the total blood volume that is erythrocytes. With decreased erythrocyte size, the hematocrit is decreased.

2. Impairment of kidney function could result in decreased erythropoietin production. Thus, erythropoiesis decrease and anemia results. If a tumor caused overproduction of erythropoietin, polycythemia could result.

3. In hemolytic anemia erythrocytes are rapidly destroyed. Erythropoietin is released from the kidneys into the plasma in an attempt to stimulate the production of erythrocytes to replace those lost. If the plasma were injected into a normal person, one would expect erythrocyte production to increase.

4. Carbon monoxide binds to the iron of hemoglobin to form carboxyhemoglobin, which does not transport oxygen. The decreased oxygen stimulates erythropoiesis, and the reticulocyte count should be elevated over normal.

5. An elevated eosinophil count could indicate an allergic reaction (asthma) that produces inflammation. The eosinophils function to reduce the severity of the inflammation.

6. A possible solution would be to take a blood sample and count the number of platelets. Thrombocytopenia is characterized by a reduced platelet count. Another possibility might be to give an injection of vitamin K and see if the problem resolves.

7. The mother should include adequate amounts of vitamin B_{12} and folic acid (to ensure erythrocyte production), iron (to ensure hemoglobin production), and vitamin K (to ensure proper blood clotting).

20 Cardiovascular System: The Heart

FOCUS: The heart is surrounded by the pericardium, which anchors the heart in the mediastinum, prevents overdistention of the heart, and protects the heart against friction. The heart has two atria, which receive blood from the body and the lungs, and two ventricles, which pump blood to the body and lungs. Atrioventricular valves and semilunar valves ensure one-way flow of blood through the heart, and heart sounds are produced as these valves close. The conducting system of the heart produces action potentials in the SA node that initiate contraction of the atria. Action potentials are delayed in the AV node, allowing time for the atria to contract and move blood into the ventricles. Then the ventricles contract, starting at the apex of the heart. Cardiac muscle cells are autorhythmic and have a prolonged depolarization (plateau phase), which extends the refractory period and prevents tetanus. The electrocardiogram measures the electrical activity of the heart, i.e., the electrical changes produced by cardiac muscle cell action potentials. The action potentials result in the cardiac cycle, a repetitive sequence of systole (contraction) and diastole (relaxation). During systole, pressure builds within heart chambers and blood is ejected, whereas during diastole, pressure in heart chambers decreases and blood flows into the chambers. The rate and stroke volume (amount of blood ejected per beat) of the heart are regulated. Starling's law of the heart states that stroke volume is equal to venous return. The parasympathetic system inhibits heart rate, whereas the sympathetic system, epinephrine, and norepinephrine increase heart rate and stroke volume. The baroreceptor reflex detects changes in blood pressure and causes alteration of heart rate and stroke volume that results in the return of blood pressure to normal levels.

CONTENT LEARNING ACTIVITY

Size, Form, and Location of the Heart

❝*The adult heart has the shape of a blunt cone and is about the size of a closed fist.*❞

Match these terms with the correct statement or definition:

Apex
Base

Pericardial cavity

_____ 1. Space within the mediastinum that contains the heart.

_____ 2. End of the heart where veins enter and arteries exit; superior part of heart.

 Knowing the exact location of the heart in the thoracic cavity is important for positioning a stethoscope, positioning electrodes to record an electrocardiogram, and performing effective CPR.

Anatomy of the Heart

"_The heart is a muscular pump consisting of four chambers._**"**

A. Match these terms with the correct statement or definition:

Fibrous pericardium	Pericardium
Parietal pericardium	Serous pericardium
Pericardial fluid	Visceral pericardium (epicardium)

_____ 1. Double-layered closed sac that surrounds the heart.

_____ 2. Tough outer layer of the pericardium.

_____ 3. Portion of the serous pericardium that lines the fibrous pericardium.

_____ 4. Fluid that fills the pericardial cavity.

☞ The serous pericardium and pericardial fluid reduce friction as the heart moves within the pericardium.

B. Match these terms with the correct statement or definition:

Crista terminalis	Musculi pectinati
Endocardium	Myocardium
Epicardium	Trabeculae carneae

_____ 1. Thin serous membrane comprising the outer surface of the heart.

_____ 2. Thick middle layer of the heart; composed of cardiac muscle cells.

_____ 3. Smooth inner surface of the heart.

_____ 4. Muscular ridges found in both auricles.

_____ 5. Ridge that separates musculi pectinati from smooth portion of the right atrium.

_____ 6. Ridges and columns in ventricles.

C. Match these vessels or structures with the correct description:

Aorta and pulmonary trunk　　Coronary sulcus
Auricles　　　　　　　　　　Interventricular sulcus
Cardiac veins　　　　　　　　Pulmonary veins
Coronary arteries　　　　　　Venae cavae
Coronary sinus

_____ 1. Flaplike extensions of the atria.

_____ 2. Veins that carry blood from the body to the right atrium.

_____ 3. Veins that carry blood from the lungs to the left atrium.

_____ 4. Large arteries that exit the heart.

_____ 5. Large groove that separates the atria and ventricles and runs obliquely around the heart.

_____ 6. Arise from the aorta; carry blood to the wall of the heart.

_____ 7. Vessels that carry blood from heart walls to coronary sinus.

_____ 8. Large venous cavity that empties into the right atrium; carries blood from the walls of the heart.

D. Match these terms with the correct statement or definition:

Atrioventricular canal　　Interatrial septum
Foramen ovale　　　　　　Interventricular septum
Fossa ovalis

_____ 1. Wall separating the right and left atria.

_____ 2. Oval depression on the right side of the interatrial septum.

_____ 3. Opening between the right and left atria in embryonic and fetal stages of development.

_____ 4. Opening between an atrium and a ventricle.

_____ 5. Wall separating the two ventricles.

E. Match these terms with the correct statement or definition:

Atrioventricular valve　　Papillary muscles
Bicuspid (mitral) valve　　Semilunar valve
Chordae tendineae　　　　Tricuspid valve

_____ 1. General term for a one-way valve between the atrium and ventricle.

_____ 2. Valve between the right atrium and right ventricle.

_____ 3. Cone-shaped muscular pillars in the ventricles.

_____ 4. Connective tissue strings between papillary muscles and atrioventricular valves.

_____ 5. One-way valve in the aorta or pulmonary trunk with three pocketlike cusps.

The papillary muscles and the chordae tendineae function to prevent the atrioventricular valves from opening back into the atria.

F. Match these terms with the correct parts of the diagram labeled in Figure 20-1:

Aorta
Aortic semilunar valve
Bicuspid (mitral) valve
Chordae tendineae
Interventricular septum
Left atrium
Left ventricle
Papillary muscles

Pulmonary semilunar valve
Pulmonary trunk
Pulmonary veins
Right atrium
Right ventricle
Superior vena cava
Tricuspid valve

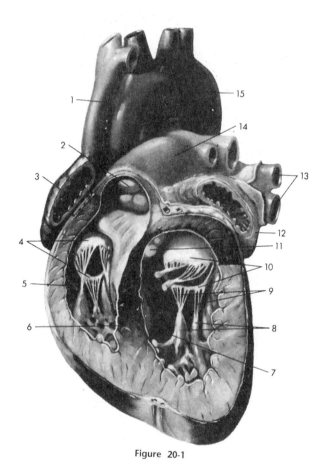

Figure 20-1

1. _____

2. _____

3. _____

4. _____

5. _____

6. _____

7. _____

8. _____

9. _____

10. _____

11. _____

12. _____

13. _____

14. _____

15. _____

Route of Blood Flow Through the Heart

❝_Blood flow through the heart occurs simultaneously in both right and left sides._**❞**

Arrange these terms in sequential order as blood returns to the heart from the body and is pumped through the heart:

Aorta
Aortic semilunar valve
Bicuspid (mitral) valve
Left atrium
Left ventricle

Lungs
Pulmonary arteries
Pulmonary semilunar valve
Pulmonary trunk
Pulmonary veins

Right atrium
Right ventricle
Tricuspid valve
Venae cavae

1. _____

2. _____

3. _____

4. _____

5. _____

6. _____

7. _____

8. _____

9. _____

10. _____

11. _____

12. _____

13. _____

14. _____

Histology

❝_The heart contains cardiac muscle, epithelium, and connective tissue._**❞**

A. Using the terms provided, complete these statements:

Contraction
Desmosomes
Gap junctions
Intercalated disks
Mitochondria
Oxygen debt

Sarcomeres
Sarcoplasmic
   reticulum
Transverse tubules
   (T tubules)

1. _____

2. _____

3. _____

4. _____

5. _____

6. _____

7. _____

8. _____

9. _____

Like skeletal muscle, cardiac muscle contains actin and myosin organized to form _(1)_ and myofibrils. Cardiac muscle also has a smooth _(2)_ that is loosely associated with membranes of the _(3)_. This loose association is partly responsible for the slow onset of _(4)_ in cardiac muscle. The production of energy in cardiac muscles depends on oxygen, and cardiac muscle cannot develop a significant _(5)_. An extensive capillary network and large numbers of _(6)_ help to sustain the normal myocardial energy requirement. Cardiac muscle cells are bound end-to-end and laterally by specialized cell-to-cell contacts called _(7)_, which greatly increase contact between adjacent cells. Specialized cell membrane structures called _(8)_ hold the cells together and _(9)_ function as areas of low electrical resistance between the cells, allowing cardiac muscle cells to function as a single unit.

The skeleton of the heart is a connective tissue plate that electrically isolates the atria from the ventricles and forms the fibrous rings around the heart valves.

B. Match these terms with the correct statement or definition:

Apex   Bundle branches
Atrioventricular (AV) bundle Purkinje fibers
Atrioventricular (AV) node Sinoatrial (SA) node

_____ 1. Cardiac muscle cells that generate spontaneous action potentials with the greatest frequency; the pacemaker.

_____ 2. Modified cardiac muscle cells that delay action potentials between the atria and the AV bundle.

_____ 3. Conducting cells that arise from the AV node.

_____ 4. Right and left subdivisions of the AV bundle.

_____ 5. Inferior, terminal branches of the bundle branches, composed of large-diameter cardiac muscle fibers.

_____ 6. Part of the heart where ventricular contraction begins.

C. Match these terms with the correct parts of the diagram labeled in Figure 20-2:

AV bundle
AV node
Bundle branches
Purkinje fibers
SA node

1. _____

2. _____

3. _____

4. _____

5. _____

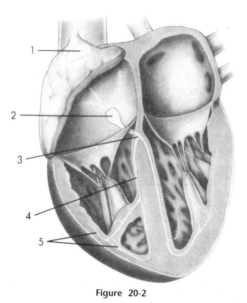

Figure 20-2

Electrical Properties

"Cardiac muscle cells have characteristics in common with other electrically excitable cells."

A. Match these terms with the correct statement or definition:

Calcium slow channels
Plateau phase
Potassium membrane
   channels

Rapid depolarization phase
Rapid repolarization phase
Resting membrane potential
Sodium fast channels

_____ 1. Condition necessary for electrically excitable cells to produce an action potential.

_____ 2. Phase brought about by opening of sodium fast channels in the cardiac muscle membrane.

_____ 3. Prolonged period of depolarization in cardiac muscle cells.

_____ 4. During the plateau phase, these membrane channels for calcium open and keep the cell partly depolarized.

_____ 5. During the plateau phase, these channels close and cause the membrane to remain partly depolarized.

_____ 6. During this phase in the cardiac cell membrane, calcium slow channels close, and the potassium membrane channels open.

B. Match these terms with the correct statement or definition:

Cardiac muscle
Skeletal muscle

_____ 1. Depolarization phase of the action potential occurs because of sodium and calcium ions.

_____ 2. Action potentials are faster.

_____ 3. Action potentials are conducted from cell to cell.

C. Match these terms with the correct statement or definition:

Absolute refractory period
Autorhythmic
AV node

Ectopic pacemakers (foci)
Relative refractory period
SA node

_____ 1. Producing action potentials automatically and at regular intervals.

_____ 2. Area of the heart with the greatest number of calcium slow channels; acts as the pacemaker.

_____ 3. Normally, the area of the heart with the second greatest frequency of action potentials.

_____ 4. Areas that initiate beats, if they are located outside the heart's conductile system.

_____ 5. Time during which a cardiac muscle cell is completely insensitive to further stimulation.

379

D. Match these terms with the
   correct statement or definition:

Electrocardiogram (ECG)   QRS complex
P-Q (P-R) interval   Q-T interval
P wave   T wave

_____ 1. Summated record of all action potentials transmitted through
                           the heart during a given time period.

_____ 2. Record of action potentials that cause depolarization of the
                           atrial myocardium.

_____ 3. Record of action potentials from ventricular depolarization.

_____ 4. Record of repolarization of the ventricles.

_____ 5. Approximate time of ventricular contraction.

E. Match these terms with the correct location
   on the diagram in Figure 20-3:

P-Q (P-R) interval
P wave
QRS complex
Q-T interval
T wave

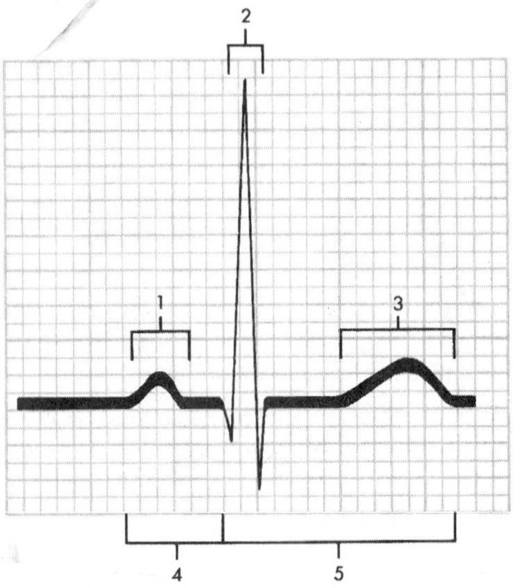

Figure 20-3

1. _____

2. _____

3. _____

4. _____

5. _____

F. Match these terms with the
   correct symptoms:

Atrial fibrillation   Tachycardia
Bradycardia   Ventricular fibrillation
Sinus arrhythmia

_____ 1. Condition in which heart rate is in excess of 100 beats/minute.

_____ 2. Condition in which no P waves are seen on the ECG.

_____ 3. Condition in which no QT complexes are seen on the ECG.

_____ 4. Condition in which heart rate is less than 60 beats/min.

Cardiac Cycle

"The cardiac cycle begins with cardiac muscle contraction and ends with relaxation.**"**

A. Match these terms with the correct statement or definition:

Final one third of ventricular diastole
First one third of ventricular diastole

Second one third of ventricular diastole
Ventricular systole

_____ 1. Contraction of the ventricular myocardium.

_____ 2. 70% of ventricular filling; results from low ventricular pressure.

_____ 3. Little ventricular filling.

_____ 4. 30% of ventricular filling (results from atrial contraction).

B. Match these terms with the correct statement or definition:

a wave
c wave

v wave

_____ 1. Result of atrial contraction.

_____ 2. Result of slight increase in atrial pressure caused by ventricular systole.

_____ 3. Result of continuous blood flow into the atria.

C. Match these time periods with the correct statement or definition:

Ejection
Isometric contraction

Isometric relaxation

_____ 1. Period between atrioventricular valve closure and semilunar valve opening; no movement of blood out of the ventricles.

_____ 2. Period of time that blood flows from the ventricles.

_____ 3. Period between semilunar valve closure and atrioventricular valve opening; no blood flows from atria into the ventricles.

D. Match these terms with the correct statement or definition:

Blood pressure
Cardiac output
Cardiac reserve
End-diastolic volume

End-systolic volume
Dicrotic notch (incisura)
Stroke volume

_____ 1. Volume of blood in the ventricles when they are filled.

_____ 2. Volume of blood pumped during each cardiac cycle.

_____ 3. Total amount of blood pumped per minute.

_____ 4. Force responsible for blood movement in vessels (proportional to cardiac output times the peripheral resistance).

_____ 5. Increase in aortic pressure when the semilunar valve closes and blood flows back toward the ventricle from the aorta.

E. Match these terms with the correct statement or definition:

First heart sound
Second heart sound

Third heart sound

_____ 1. The vibrations associated with the atrioventricular valves closing.

_____ 2. The vibrations associated with the semilunar valves closing.

_____ 3. The sound of blood flowing in a turbulent fashion into the ventricles.

_____ 4. The low-pitched "lubb" sound.

F. Match these terms with the correct location on Graph A in in Figure 20-4:

Atrioventricular valves close
Atrioventricular valves open

Semilunar valves close
Semilunar valves open

Match these terms with the correct location on Graph B in Figure 20-4:

Ejection
End diastolic volume
End systolic volume

Isometric contraction
Isometric relaxation
Stroke volume

1. _____

2. _____

3. _____

4. _____

5. _____

6. _____

The difference between 5 and 6 above:

7. _____

8. _____

9. _____

10. _____

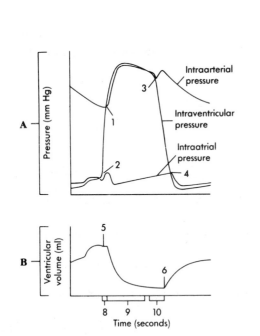

Figure 20-4

G. Match these terms with the correct statement or definition:

Incompetent valve   Stenosed valve
Murmur

_____ 1. Abnormal heart sound.

_____ 2. Valve that leaks significantly.

_____ 3. Valve with an abnormally narrow opening.

☞ In cases of severe dysfunction of the heart, repair may involve angioplasty, coronary bypass, surgical implantation of a pacemaker or artificial valves, or heart transplant.

Regulation of the Heart

❝_The amount of blood pumped by the heart may vary dramatically depending on conditions._**❞**

A. Match these terms with the correct statement or definition:

Extrinsic regulation
Intrinsic regulation

_____ 1. Regulation by the heart's normal functional characteristics.

_____ 2. Regulation of the heart by neural or hormonal control.

_____ 3. Regulation that involves venous return and Starling's law of the heart.

☞ The extent to which the ventricular walls are stretched at end-diastolic volume is called preload; afterload is the pressure the contracting ventricles must produce to overcome the pressure in the aorta.

B. Match these terms with the correct statement or definition:

Decrease   No effect
Increase

_____ 1. Effect of decreased venous return on cardiac output.

_____ 2. Effect of decreased preload on cardiac output.

_____ 3. Effect of stretching the right atrial wall on heart rate.

_____ 4. Effect of changes in afterload on heart pumping efficiency.

_____ 5. Effect of stimulating the vagus nerve on heart rate.

_____ 6. Effect of acetylcholine on permeability of cardiac muscle cell membrane to K^+ ions.

_____ 7. Effect of sympathetic stimulation on heart rate and the force of heart muscle contraction.

_____ 8. Effect on stroke volume if heart rate becomes too great.

_____ 9. Effect of norepinephrine and epinephrine on heart rate and force of contraction.

Heart and Homeostasis

66 *The heart's pumping efficiency plays an important role in the maintenance of homeostasis.* **99**

A. Match these terms with the correct statement or definition:

Baroreceptors
Cardioregulatory center

_____ 1. Sensory receptors that measure blood pressure, located in the walls of the aorta and internal carotid arteries.

_____ 2. Area of the medulla oblongata that integrates sensory information and sends efferent impulses to the heart.

☞ Baroreceptor reflexes detect changes in blood pressure and result in changes in heart rate and in the force of contraction of the heart.

B. Match these terms with the correct statement or definition:

Decreases
Increases

_____ 1. Effect on heart rate when arterial blood pressure increases.

_____ 2. Effect of stimulation of stretch receptors in right atrium on sympathetic stimulation (Bainbridge reflex).

_____ 3. Effect on heart rate of an increase in blood pH and a decrease of blood carbon dioxide level.

_____ 4. Effect on heart rate of a sudden large decrease in blood oxygen level; mediated through carotid body chemoreceptor reflex.

_____ 5. Effect on heart rate of a gradual decrease in blood oxygen level; results from increased respiration rate.

_____ 6. Effect on heart rate and stroke volume of excess potassium ions.

_____ 7. Effect on force of contraction of increased extracellular calcium ions.

_____ 8. Effect on heart rate and force of contraction of increased temperature.

QUICK RECALL

---

1. List the parts of the serous pericardium and describe their function.

2. Name the major veins that enter the right and left atria.

3. Name the four valves that regulate the direction of blood flow in the heart, and give their location.

4. Give the two nodes of the conducting system of the heart and their functions.

5. List four differences between skeletal and cardiac muscle in regard to action potentials.

6. State the cause of the P wave, the QRS complex, and the T wave of the ECG. Name the contraction events associated with each wave.

7. List the two major normal heart sounds, and give the reason for each.

8. Give two intrinsic regulatory mechanisms that cause cardiac output to equal venous return.

9. List the effects of sympathetic and parasympathetic stimulation of the heart.

10. List the locations where the nervous system detects changes in blood pressure, carbon dioxide, pH, and oxygen that affect the heart.

MASTERY LEARNING ACTIVITY

Place the letter corresponding to the correct answer in the space provided.

_____ 1. The fibrous pericardium
   a. is in contact with the heart.
   b. is a serous membrane.
   c. is also known as the epicardium.
   d. forms the outer layer of the pericardium (pericardial sac).

_____ 2. Which of these structures carry blood to the right atrium?
   a. coronary sinus
   b. superior vena cava
   c. inferior vena cava
   d. all of the above

_____ 3. The valve located between the right atrium and the right ventricle is the
   a. aortic semilunar valve.
   b. pulmonary semilunar valve.
   c. tricuspid valve.
   d. bicuspid (mitral) valve.

_____ 4. The papillary muscles
   a. are attached to the chordae tendineae.
   b. are found in the atria.
   c. contract to close the foramen ovale.
   d. are attached to the semilunar valves.

_____ 5. Given these blood vessels:
   1. aorta
   2. inferior vena cava
   3. pulmonary trunk
   4. pulmonary vein

   Choose the arrangement that lists the vessels in the order a red blood cell would encounter them in going from the systemic veins back to the systemic arteries.
   a. 1,3,4,2
   b. 2,3,4,1
   c. 2,4,3,1
   d. 3,2,1,4
   e. 3,4,2,1

_____ 6. The skeleton of the heart
   a. electrically insulates the atria from the ventricles.
   b. provides a rigid site of attachment for the cardiac muscle.
   c. functions to reinforce or support the valve openings.
   d. all of the above

_____ 7. The bulk of the heart wall is
   a. epicardium.
   b. pericardium.
   c. myocardium.
   d. endocardium.

_____ 8. Muscular ridges on the interior surface of the auricles are called
   a. trabeculae carneae.
   b. crista terminalis.
   c. musculi pectinati.
   d. endocardium.

_____ 9. Cardiac muscle has
   a. sarcomeres.
   b. a sarcoplasmic reticulum.
   c. transverse tubules.
   d. all of the above

_____ 10. Electrical impulses are conducted from one cardiac muscle cell to another
   a. because of gap junctions.
   b. because of a special cardiac nervous system.
   c. because of the large voltage of the action potentials.
   d. because of the plateau phase of the action potentials.

_____ 11. During the transmission of action potentials through the conducting system of the heart, there is a temporary delay of transmission at the
   a. bundle branches.
   b. Purkinje fibers.
   c. AV node.
   d. SA node.

12. Given these structures of the conduction system of the heart:
    1. atrioventricular bundle
    2. AV node
    3. bundle branches
    4. Purkinje fibers
    5. SA node

    Choose the arrangement that lists the structures in the order an action potential would pass through them.
    a. 2,5,1,3,4
    b. 2,5,3,1,4
    c. 2,5,4,1,3
    d. 5,2,1,3,4
    e. 5,2,4,3,1

13. Purkinje fibers
    a. are specialized cardiac muscle cells.
    b. conduct impulses much more slowly than ordinary cardiac muscle.
    c. conduct action potentials through the atria.
    d. all of the above

14. The apex of the heart is
    a. located superior to the base of the heart.
    b. where the major blood vessels exit the heart.
    c. where ventricular contraction begins.
    d. all of the above

15. If the SA node is damaged and becomes nonfunctional, which of these is most likely?
    a. The heart will stop.
    b. The ventricles will contract with a greater frequency than the atria.
    c. The AV node will probably become the pacemaker.
    d. The heart will enter fibrillation.
    e. Tachycardia will develop.

16. Cardiac muscle cells will not tetanize because
    a. the nervous stimulation is not sufficiently strong.
    b. the SA node is the pacemaker.
    c. of the intercalated disks.
    d. of the long refractory period.

17. T waves represent
    a. depolarization of the ventricles.
    b. repolarization of the ventricles.
    c. depolarization of the atria.
    d. repolarization of the atria.

18. Which of these conditions observed in an electrocardiogram would suggest that the AV node is not conducting action potentials?
    a. complete lack of the P wave
    b. complete lack of the QRS complex
    c. more QRS complexes than P waves
    d. a prolonged P-R interval
    e. complete asynchrony between the P wave and QRS complexes

19. The greatest amount of ventricular filling occurs during
    a. the first third of diastole.
    b. the second third of diastole.
    c. the last third of diastole.
    d. ventricular systole.

20. While the semilunar valves are open during a normal cardiac cycle,
    a. the pressure in the left ventricle is greater than the pressure in the aorta.
    b. the pressure in the left ventricle is less than the pressure in the aorta.
    c. the pressure in the left ventricle is exactly the same as the pressure in the left atrium.
    d. the pressure in the left ventricle is the same as the pressure in the left atrium.

21. The pressure within the left ventricle fluctuates between
    a. 120 and 80 mm Hg
    b. 120 and 0 mm Hg
    c. 80 and 0 mm Hg
    d. 20 and 0 mm Hg

22. Blood flows neither into nor out of the ventricles during
    a. the period of isometric contraction.
    b. the period of isometric relaxation.
    c. diastole.
    d. systole.
    e. a and b

_____ 23. Stroke volume
   a. is the amount of blood pumped by the heart per minute.
   b. is the difference between end-diastolic and end-systolic volume.
   c. is the difference between the amount of blood pumped at rest and that pumped at maximum output.
   d. is the amount of blood pumped from the atria into the ventricles.

_____ 24. Cardiac output is defined as
   a. blood pressure times peripheral resistance.
   b. peripheral resistance times heart rate.
   c. heart rate times stroke volume.
   d. stroke volume times blood pressure.

_____ 25. Pressure in the aorta is at its lowest
   a. at the time of the first heart sound.
   b. at the time of the second heart sound.
   c. just before the AV valves open.
   d. just before the semilunar valves open.

_____ 26. Just after the dicrotic notch on the aortic pressure curve,
   a. the pressure in the aorta is greater than the pressure in the ventricle.
   b. the pressure in the ventricle is greater than the pressure in the aorta.
   c. the pressure in the left atrium is greater than the pressure in the left ventricle.
   d. the pressure in the left atrium is greater than the pressure in the aorta.
   e. blood flow in the aorta has stopped.

_____ 27. The "lubb" sound (first heart sound) of the heart is caused by the
   a. closing of the AV valves.
   b. closing of the semilunar valves.
   c. blood rushing out of the ventricles.
   d. filling of the ventricles.
   e. ventricular contraction.

_____ 28. Increased venous return results in increased
   a. stroke volume.
   b. heart rate.
   c. cardiac output.
   d. all of the above

_____ 29. Parasympathetic nerve fibers are carried in the _____ nerves and release _____ at the heart.
   a. cardiac, acetylcholine
   b. cardiac, norepinephrine
   c. vagus, acetylcholine
   d. vagus, norepinephrine

_____ 30. Increased parasympathetic stimulation of the heart will
   a. increase the force of ventricular contraction.
   b. increase the rate of depolarization in the SA node.
   c. decrease the heart rate.
   d. increase cardiac output.

_____ 31. Sympathetic stimulation of the heart increases which of the parameters listed below?
   a. heart rate
   b. stroke volume
   c. force of ventricular contraction
   d. all of the above

_____ 32. Epinephrine released from the adrenal medulla
   a. increases the rate of heart contractions.
   b. decreases the force of heart contractions.
   c. produces the same effect as parasympathetic stimulation of the heart.
   d. a and c
   e. b and c

_____ 33. Because of the baroreceptor reflex, when normal arterial blood pressure decreases, you would expect
   a. heart rate to decrease.
   b. stroke volume to decrease.
   c. blood pressure to return to normal.
   d. all of the above

_____34. A decrease in blood pH and an increase in blood carbon dioxide levels result in
a. increased heart rate.
b. increased stroke volume.
c. increased sympathetic stimulation of the heart.
d. all of the above

_____35. An increase in extracellular potassium levels could cause
a. an increase in stroke volume.
b. an increase in the force of contraction.
c. a decrease in heart rate.
d. a and b

FINAL CHALLENGES

Use a separate sheet of paper to complete this section.

1. Skeletal muscle exhibits graded responses because of multiple motor unit summation and multiple wave summation. Explain why the heart does not do this.

2. A friend tells you that her son had an ECG and it revealed that he had a slight heart murmur. Should you be convinced that he has a heart murmur? Explain.

3. Predict the effect on heart rate if the vagus nerves to the heart were cut.

4. Predict the effect on Starling's law of the heart if the vagus nerves to the heart were cut.

5. Predict the effect on heart rate if the glossopharyngeal nerves to the heart were cut.

6. An experiment on a dog was performed in which the mean arterial blood pressure was monitored before and after the common carotid arteries were clamped (at time A). The results are graphed below:

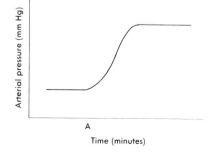

Explain the change in mean arterial blood pressure (hint: baroreceptors are located in the internal carotid arteries, which are superior to the site of clamping of the common carotid arteries).

7. What would happen to blood pressure and heart rate following the ingestion of a large amount of isosmotic fluid? Explain.

8. During hemorrhagic shock (caused by loss of blood) the blood pressure may fall dramatically, although the heart rate is elevated. Explain why the blood pressure falls despite the increase in heart rate.

9. During defecation it is not uncommon to "strain" by holding the breath and compressing the thoracic and abdominal muscles. Assume that the increased pressure in the thoracic cavity compresses the venae cavae. What would happen to blood pressure? Describe the compensatory mechanisms that would be activated to correct the change in blood pressure. Next, assume that the increased thoracic pressure increases the pressure in the aortic arch. What would happen to blood pressure (hint: sensory receptors)?

10. A patient exhibited these symptoms: chest pain, rapid pulse, a greater frequency of P waves than QRS complexes, and disappearance of these symptoms following the administration of a beta adrenergic blocking agent (inhibited the action of norepinephrine on the heart). Does the patient have a myocardial infarct, heart block, or atrial tachycardia? Explain.

ANSWERS TO CHAPTER 20

Size, Form, and Location of the Heart
1. Pericardial cavity; 2. Base

Anatomy of the Heart
A. 1. Pericardium; 2. Fibrous pericardium; 3. Parietal pericardium; 4. Pericardial fluid
B. 1. Epicardium; 2. Myocardium; 3. Endocardium; 4. Musculi pectinati; 5. Crista terminalis; 6. Trabeculae carneae
C. 1. Auricles; 2. Venae cavae; 3. Pulmonary veins; 4. Aorta and pulmonary trunk; 5. Coronary sulcus; 6. Coronary arteries; 7. Cardiac veins; 8. Coronary sinus
D. 1. Interatrial septum; 2. Fossa ovalis; 3. Foramen ovale; 4. Atrioventricular canal; 5. Interventricular septum
E. 1. Atrioventricular valve; 2. Tricuspid valve; 3. Papillary muscles; 4. Chordae tendineae; 5. Semilunar valve
F. 1. Superior vena cava; 2. Pulmonary semilunar valve; 3. Right atrium; 4. Tricuspid valve; 5. Right ventricle; 6. Interventricular septum; 7. Left ventricle; 8. Papillary muscles; 9. Chordae tendineae; 10; Bicuspid (mitral) valve; 11. Aortic semilunar valve; 12. Left atrium; 13. Pulmonary veins; 14. Pulmonary trunk; 15. Aorta

Route of Blood Flow Through the Heart
1. Venae cavae; 2. Right atrium; 3. Tricuspid valve; 4. Right ventricle; 5. Pulmonary semilunar valve; 6. Pulmonary trunk; 7. Pulmonary arteries; 8. Lungs; 9. Pulmonary veins; 10 Left atrium; 11. Bicuspid (mitral) valve; 12. Left ventricle; 13. Aortic semilunar valve; 14. Aorta

Histology
A. 1. Sarcomeres; 2. Sarcoplasmic reticulum; 3. Transverse tubules (T tubules); 4. Contraction; 5. Oxygen debt; 6. Mitochondria; 7. Intercalated disks; 8. Desmosomes; 9. Gap junctions
B. 1. Sinoatrial (SA) node; 2. Atrioventricular (AV) node; 3. Atrioventricular (AV) bundle; 4. Bundle branches; 5. Purkinje fibers; 6. Apex
C. 1. SA node; 2. AV node; 3. AV bundle; 4. Bundle branches; 5. Purkinje fibers

Electrical Properties
A. 1. Resting membrane potential; 2. Rapid depolarization phase; 3. Plateau phase;

4. Calcium slow channels; 5. Potassium membrane channels; 6. Rapid repolarization phase
B. 1. Cardiac muscle; 2. Skeletal muscle; 3. Cardiac muscle
C. 1. Autorhythmic; 2. SA node; 3. AV node; 4. Ectopic pacemakers (foci); 5. Absolute refractory period
D. 1. Electrocardiogram (ECG); 2. P wave; 3. QRS complex; 4. T wave; 5. QT interval
E. 1. P wave; 2. QRS complex; 3. T wave; 4. P-Q (P-R) interval; 5. QT interval
F. 1. Tachycardia; 2. Atrial fibrillation; 3. Ventricular fibrillation; 4. Bradycardia

Cardiac Cycle
A. 1. Ventricular systole; 2. First one third of ventricular diastole; 3. Second one third of ventricular diastole; 4. Final one third of ventricular diastole
B. 1. a wave; 2. c wave; 3. v wave
C. 1. Isometric contraction; 2. Ejection; 3. Isometric relaxation
D. 1. End-diastolic volume; 2. Stroke volume; 3. Cardiac output; 4. Blood pressure; 5. Dicrotic notch (Incisura)
E. 1. First heart sound; 2. Second heart sound; 3. Third heart sound; 4. First heart sound
F. 1. Semilunar valves open; 2. Atrioventricular valves close; 3. Semilunar valves close; 4. Atrioventricular valves open; 5. End diastolic volume; 6. End systolic volume; 7. Stroke volume; 8. Isometric contraction; 9. Ejection; 10. Isometric relaxation
G. 1. Murmur; 2. Incompetent valve; 3. Stenosed valve

Regulation of the Heart
A. 1. Intrinsic regulation; 2. Extrinsic regulation; 3. Intrinsic regulation
B. 1. Decrease; 2. Decrease; 3. Increase; 4. No effect; 5. Decrease; 6. Increase; 7. Increase; 8. Decrease; 9. Increase

Heart and Homeostasis
A. 1. Baroreceptors; 2. Cardioregulatory center
B. 1. Decreases; 2. Increases; 3. Decreases; 4. Decreases; 5. Increases; 6. Decreases; 7. Increases; 8. Increases

1. Parietal pericardium and visceral pericardium - reduce friction

2. Right atrium - inferior and superior venae cavae and coronary sinus; left atrium - four pulmonary veins

3. Tricuspid valve - between right atrium and right ventricle; bicuspid (mitral) valve - between left atrium and left ventricle; aortic semilunar valve - in the aorta; pulmonary semilunar valve - in the pulmonary trunk

4. SA node - pacemaker of the heart; AV node - slows action potentials, allowing atria to contract and move blood into ventricles

5. Depolarization caused by sodium and calcium in cardiac muscle (sodium in skeletal muscle); rate of action potential propagation is slower in cardiac muscle; action potentials are propagated from cell to cell in cardiac muscle (intercalated disks) but not in skeletal muscle; cardiac muscle has spontaneous generation of action potentials, whereas skeletal muscle action potentials result from nervous system stimulation; and cardiac muscle has a prolonged depolarization (plateau) phase

6. P wave: atrial depolarization, atrial systole; QRS complex: ventricular depolarization, ventricular systole; T wave: ventricular repolarization, ventricular diastole

7. First heart sound: closing of atrioventricular valves; second heart sound: closing of semilunar valves

8. Venous return stretches cardiac muscle fibers and causes them to contract with greater force (Starling's law of the heart); it also stretches the SA node and increases heart rate (Bainbridge reflex)

9. Parasympathetic stimulation: decreased heart rate and small decrease in stroke volume; sympathetic stimulation: increased heart rate, force of contraction, and stroke volume

10. Baroreceptors in aorta and carotid arteries monitor blood pressure; chemoreceptors in medulla monitor carbon dioxide and pH changes; chemoreceptors near carotid arteries monitor oxygen changes.

MASTERY LEARNING ACTIVITY

1. D. The pericardium consists of an outer fibrous pericardium and an inner serous pericardium (a serous membrane). The serous pericardium consists of a parietal pericardium, which lines the fibrous pericardium, and a visceral pericardium (also called the epicardium), which covers the heart.

2. D. The coronary sinus, superior vena cava, and inferior vena cava all carry blood to the right atrium. In addition, a number of cardiac veins from the anterior surface of the heart empty into the right atrium.

3. C. The tricuspid valve is between the right atrium and the right ventricle. The bicuspid valve is between the left atrium and the left ventricle. The aortic semilunar valve is between the left ventricle and the aorta. The pulmonary semilunar valve is between the right ventricle and the pulmonary trunk.

4. A. The papillary muscles (located in the ventricles) are attached to the chordae tendineae, which are attached to the atrioventricular valves. Contraction of the papillary muscles helps to prevent the valves from opening back into the atria. The foramen ovale is an opening between the right and left atria during development. It becomes the fossa ovalis.

5. B. The red blood cell would enter the right atrium through the inferior vena cava and pass into the right ventricle. The red blood cell would exit the right ventricle through the pulmonary trunk, pass through the lungs, and return to the left atrium through the pulmonary vein. From the left atrium, the red blood cell would enter the left ventricle and leave the heart through the aorta.

6. D. The skeleton of the heart is composed of fibrous connective tissue. It serves as electrical insulation between the atria and ventricles, provides a site of attachment for cardiac muscles, and forms a support for the valves of the heart (fibrous rings).

7. C. The myocardium is the middle layer of the heart wall, is composed of cardiac muscle, and makes up the bulk of the heart.

8. C. The musculi pectinati are muscular ridges in the auricles and a portion of the right atrial wall. The crista terminalis separates the musculi pectinati of the right atrial wall from the smooth portion of the right atrial wall. The trabeculae carneae are modified ridges in the interior ventricular walls. The endocardium is not muscle.

9. D. Cardiac muscle has many similarities with skeletal muscle. Cardiac muscle has actin and myosin myofilaments arranged to form sarcomeres. The sarcoplasmic reticulum and transverse tubules in cardiac muscle are more loosely arranged than in skeletal muscle.

10. A. The intimate connection between cardiac muscle cells at the gap junctions constitutes an area of low electrical resistance. Action potentials are propagated from one muscle cell to the adjacent muscle cell, resulting in the function of cardiac muscle cells as a single unit.

11. C. The delay in the AV node allows the atria to contract before the ventricles do.

12. D. The SA node generates an action potential that passes to the AV node. From the AV node the action potential passes through the atrioventricular bundle, the bundle branches, and the Purkinje fibers.

13. A. Purkinje fibers are large-diameter cardiac muscle cells specialized to conduct action potentials rapidly. They are found in the ventricles.

14. C. Ventricular contraction begins at the apex of the heart. The result is that blood is pushed superiorly toward the base of the heart where the major blood vessels exit.

15. C. If the SA node becomes nonfunctional, it is unlikely that the heart will either stop or fibrillate, although there is a chance that either condition may occur. Because heart muscle cells are autorhythmic, another region of the heart will probably become the pacemaker, probably the AV node. Other regions of the heart will have an intrinsic rate slower than the SA node, however.

16. D. Cardiac muscles cells do not tetanize because the action potential has a refractory period that is sufficiently long to allow contraction and relaxation before another action potential can be initiated. The plateau phase of the action potential in which depolarization is prolonged is responsible for the prolonged refractory period.

17. B. T waves represent repolarization of the ventricles. The QRS complex represents depolarization of the ventricles; and the P wave, depolarization of the atria. Repolarization of the atria is not seen on the ECG because it is masked by the QRS complex.

18. E. No conduction of action potentials through the AV node is called a complete heart block. This prevents the conduction of action potentials from the atria to the ventricles, and a pacemaker develops in the ventricle. The result is complete asynchrony or a complete lack of a relationship between atrial (P wave) and ventricular depolarization (QRS complex). In addition, the frequency of the P waves would be greater than the frequency of the QRS complexes.

19. A. During the first one third of diastole, the atrioventricular valves open and blood flows into the ventricles under the influence of the small pressure differential between the atria and the ventricles. About 70% of ventricular filling occurs during that period. The remaining 30% of ventricular filling occurs during the last one third of diastole because of atrial contraction.

20. A. The semilunar valves are open during ventricular systole, and blood is flowing from the left ventricle to the aorta. Therefore the pressure in the left ventricle must be higher than the pressure in the aorta.

21. B. Left ventricular pressure reaches 120 mm Hg (normal systolic pressure) during ventricular systole (contraction). During diastole (relaxation) the pressure in the ventricle falls to nearly zero, whereas the pressure within the aorta drops to 80 mm Hg.

22. E. Blood does not leave the ventricles during the period of isometric contraction and does not enter the ventricles during the period of isometric relaxation. During ventricular diastole blood enters the ventricles, and during ventricular systole blood leaves the ventricles.

23. B. Stroke volume is the difference between end-diastolic volume (the amount of blood in a filled ventricle) and end-systolic volume (the amount of blood in a ventricle after contraction has occurred). Therefore stroke volume is the amount of blood ejected per beat of the heart. Cardiac output is the amount of blood pumped by the heart per minute. Cardiac reserve is the difference between the amount of blood pumped at rest and at maximum output.

24. C. Cardiac output is equal to heart rate times stroke volume.

25. D. Following ejection of blood into the aorta, pressure drops until the next ejection of blood. Because the next ejection of blood occurs when the semilunar valves open, the lowest pressure in the aorta occurs just before the semilunar valves open.

26. A. The dicrotic notch represents closure of the aortic semilunar valve. That valve closes when the pressure within the aorta exceeds the pressure within the ventricle.

27. A. The first heart sound is caused by closure of the AV valves, and the second heart sound ("dupp") is caused by closure of the semilunar valves. A third heart sound can sometimes be heard. It is caused by turbulent flow of blood into the ventricles.

28. D.   Increased venous return results in increased stroke volume (Starling's law of the heart) and increased heart rate (because of stretching of the SA node). The increased stroke volume and heart rate result in increased cardiac output.

29. C.   The vagus nerves carry parasympathetic fibers to the heart where the postganglionic parasympathetic neurons release acetylcholine. The cardiac nerves carry sympathetic fibers to the heart where the postganglionic sympathetic neurons release norepinephrine.

30. C.   Increased parasympathetic stimulation decreases heart rate by decreasing the rate of depolarization of the SA node. Parasympathetic stimulation has little inhibitory effect on the force of ventricular contraction. Because heart rate decreases, cardiac output decreases.

31. D.   Sympathetic stimulation of the heart results in an increased heart rate and force of ventricular contraction. Stroke volume increases because of the increased force of ventricular contraction.

32. A.   Epinephrine produces the same effects as sympathetic stimulation of the heart, i.e., increased rate and force of contraction of the heart.

33. C.   When blood pressure decreases, the baroreceptor reflex causes an increase in heart rate and stroke volume (because of increased sympathetic stimulation of the heart). Consequently, blood pressure increases (returns to normal).

34. D.   When pH decreases and carbon dioxide levels increase, chemoreceptors detect the change and reflexes are activated that increase sympathetic stimulation of the heart. Consequently, heart rate and stroke volume increase.

35. C.   Increased extracellular potassium results in a decreased heart rate and stroke volume.

 FINAL CHALLENGES

1. Because cardiac muscle cells are interconnected by intercalated disks, cardiac muscle acts as a single unit and does not exhibit multiple motor unit summation (i.e., recruitment of motor units). Because cardiac muscle action potentials have a prolonged plateau phase, cardiac muscle relaxes before the next contraction and does not exhibit multiple wave summation.

2. An ECG measures the electrical activity of the heart and would not indicate a slight heart murmur. Heart murmurs are detected by listening to the heart sounds. The boy may have a heart murmur, but the mother does not understand the basis for making such a diagnosis.

3. The vagus nerves carry parasympathetic fibers to the heart, which inhibit heart rate. This is called vagal tone. Cutting the vagus nerves removes this inhibition, and heart rate increases. This is called vagal escape.

4. Because Starling's law of the heart is an intrinsic regulatory mechanism not dependent on the nervous system, cutting the vagus nerves should have no effect on Starling's law. Therefore venous return will still equal stroke volume.

5. The glossopharyngeal nerves and the vagus nerves contain afferent neurons from baroreceptors. Cutting the glossopharyngeal nerves reduces the number of afferent impulses sent from the baroreceptors, the same effect produced by a decrease in blood pressure. Consequently, the heart rate increases.

6. When both common carotid arteries are clamped, the blood pressure within the internal carotid arteries drops dramatically. The decreased blood pressure is detected by the baroreceptors, and the baroreceptor reflex causes an increase in heart rate and stroke volume. The resulting increase in cardiac output causes the increase in blood pressure.

7. The ingested fluid increases blood volume and therefore venous return. As venous return increases, cardiac output increases (Starling's law of the heart), producing an increase in blood pressure. This increased blood pressure is detected and by means of the baroreceptor reflex heart rate decreases, returning blood pressure to normal.

8. Venous return declines markedly in hemorrhagic shock because of the loss of blood volume. With decreased venous return, stroke volume decreases (Starling's law of the heart). The decreased stroke volume results in a decreased cardiac output, which produces a decreased blood pressure. In response to the decreased blood pressure, the baroreceptor reflex causes an increase in heart rate in an attempt to restore normal blood pressure. However, with inadequate venous return the increased heart rate is not able to restore normal blood pressure.

9. This act of holding the breath is called the Valsalva maneuver. Increased intrathoracic pressure produced by straining causes compression of the venae cavae, leading to decreased venous return to the heart, a decreased stroke volume (Starling's law of the heart), and a decreased cardiac output. The decreased cardiac output results in a decreased mean arterial blood pressure, which, by means of the baroreceptor reflex, causes an increase in heart rate and stroke volume. The increased heart rate and stroke volume cause an increase in blood pressure. If one strains for too long it is possible to faint because of inadequate delivery of blood to the brain (i.e., the increase in blood pressure is not adequate). Increased intrathoracic pressure could also cause an increase in pressure in the aorta and by means of the baroreceptor reflex a decreased heart rate, stroke volume, and blood pressure. Although either an increase or a decrease in blood pressure is possible with the Valsalva maneuver, usually there is an increase.

10. Atrial tachycardia results in a greater number of P waves than QRS complexes because the ventricles require a greater time to repolarize than the atria. The tachycardia results in inadequate blood pressure within the coronary blood vessels and increases energy requirements of the cardiac muscle. The results are anoxia of the cardiac tissue and chest pain. Decreasing sympathetic stimulation of the heart by administering beta-adrenergic blocking agents is often used as a treatment because these drugs slow the heart rate. One would not expect a rapid pulse with heart block or a greater frequency of P waves than QRS complexes with a myocardial infarct.

21 Cardiovascular System: Peripheral Circulation and Regulation

FOCUS: Blood from the heart flows through elastic arteries, muscular arteries, and arterioles to capillaries in the tissues. From the capillaries blood returns to the heart through venules, small veins, and large veins. Generally, blood vessels consist of endothelium, smooth muscle and elastic tissue, and connective tissue. The pulmonary circulation transports blood from the heart to the lungs and back to the heart, whereas the systemic circulation carries blood from the heart to the body and back to the heart. Blood flow through the blood vessels is mostly laminar. Turbulent blood flow produces Korotkoff sounds, which can be used to estimate aortic blood pressure. Blood pressure is responsible for the movement of blood (Poiseuille's law), for keeping blood vessels from collapsing (law of Laplace), for the volume of blood contained within the vessel (vascular compliance), and for the

movement of fluid out of capillaries. Pressure can be increased or decreased within vessels through the effect of gravity. Blood flow through tissues is controlled by relaxation and contraction of precapillary sphincters in response to vasodilator substances and nutrients, and blood pressure is regulated by nervous control (e.g., baroreceptor reflexes, chemoreceptor reflexes, central nervous system ischemic response) and humoral control (epinephrine, renin-angiotensin-aldosterone, ADH, and atrial natriuretic hormone). The baroreceptor reflexes are the most important short-term regulators of blood pressure, achieving changes in blood pressure by altering heart rate, stroke volume, and peripheral resistance. The kidneys are the most important long-term regulators of blood pressure, because they can control blood volume.

CONTENT LEARNING ACTIVITY

General Features of Blood Vessel Structure

66 *Blood flows from the heart through elastic arteries, muscular arteries, and arterioles to* 99 *capillaries, then to venules, small veins, and large veins to return to the heart.*

A. Match these terms with the correct statement or definition:

Adventitia
Basement membrane

Endothelium
Pericapillary cells

_____ 1. Layer of simple squamous epithelium lining all blood vessels.

_____ 2. Delicate layer of loose connective tissue surrounding the basement membrane of capillaries.

_____ 3. Scattered cells that lie between the basement membrane and endothelial cells.

395

B. Match these terms with the correct statement or definition:

Continuous capillaries   Sinusoids
Fenestrated capillaries   Venous sinuses
Sinusoidal capillaries

_____ 1. Capillaries with no gaps between endothelial cells; present in nervous and muscle tissue.

_____ 2. Capillaries with endothelial cells possessing numerous fenestrae; present in intestinal villi and glomeruli of the kidney.

_____ 3. Capillaries with large diameters, large fenestrae, and a less prominent basement membrane; present in endocrine glands.

_____ 4. Large-diameter sinusoidal capillaries; present in the liver and bone marrow.

_____ 5. Even larger than sinusoids; found in the spleen.

C. Match these terms with the correct statement or definition:

Arterial capillaries   Thoroughfare channels
Metarterioles   Venous capillaries
Precapillary sphincters

_____ 1. End of capillaries closest to the arterioles.

_____ 2. Arterioles with isolated smooth muscle cells along their walls.

_____ 3. Channels through which blood flow is relatively continuous and that extend from a metarteriole to a venule.

_____ 4. Smooth muscle cells that regulate blood flow from the thoroughfare channel into capillaries.

D. Match these terms with the correct statement or definition:

Tunica adventitia   Tunica media
Tunica intima

_____ 1. Tunic closest to the lumen of blood vessels, consisting of endothelium, a basement membrane, lamina propria, and a layer of elastic fibers.

_____ 2. Middle layer of blood vessel walls, consisting of smooth muscle cells and elastic and collagen fibers.

_____ 3. Outer layer of blood vessel walls, composed of connective tissue that varies from dense to loose, depending on the blood vessel.

_____ 4. Regulates blood flow by vasoconstriction and vasodilation.

E. Match these types of arteries with the correct statement or definition:

Arterioles   Medium arteries
Elastic arteries   Small arteries

_____

1. Largest-diameter arteries; often called conducting arteries.

_____

2. Have relatively thick walls because of smooth muscle layers in the tunica media; often called muscular or distributing arteries.

_____

3. Transport blood from the small arteries to the capillaries.

_____

4. Capable of vasoconstriction and vasodilation.

_____

_____

F. Match these terms with the correct statement or definition:

Medium and large veins   Venules
Small veins

_____

1. Structure of these veins is similar to capillaries; the smallest are capable of nutrient exchange.

_____

2. Smallest veins to have a continuous layer of smooth muscle.

_____

3. Most of the veins in this category have valves (folds in the tunica intima that prevent backflow of blood).

G. Match these terms with the correct statement or definition:

Arteriovenous anastomoses   Vasa vasorum
Glomus

_____

1. Small vessels that supply blood to the walls of veins and arteries.

_____

2. Arterioles that allow blood to flow directly into small veins without an intermediate capillary; function in temperature regulation.

_____

3. Arteriovenous anastomosis that consists of arterioles arranged in a convoluted fashion.

H. Match these terms with the correct statement or definition:

Arteriosclerosis   Phlebitis
Atherosclerosis   Varicose veins

_____

1. Dilated veins with incompetent valves.

_____

2. Inflammation of the veins.

_____

3. Degenerative changes in arteries that make them less elastic.

_____

4. Deposition of a fatlike substance containing cholesterol in the walls of arteries to form plaques.

Pulmonary Circulation

"The pulmonary vessels transport blood from the right ventricle through the lungs and back to the left atrium."

Match these terms with the correct statement or definition:

Pulmonary arteries   Pulmonary veins
Pulmonary trunk

_____ 1. Blood passes from the right ventricle directly into this vessel.

_____ 2. These vessels transport blood to each lung.

_____ 3. There are four of these vessels from the lungs, which enter the left atrium.

Systemic Circulation: Arteries

"Oxygenated blood passes from the left ventricle to the aorta and is distributed to all parts of the body."

A. Match these parts of the aorta with the correct statement:

Abdominal aorta   Descending aorta
Aortic arch   Thoracic aorta
Ascending aorta

_____ 1. Portion of the aorta that gives rise to the right and left coronary arteries.

_____ 2. Three major branches of this part of the aorta are the brachiocephalic, the left common carotid, and the left subclavian arteries.

_____ 3. Longest part of the aorta, running from the aortic arch to the common iliac arteries.

_____ 4. Portion of the aorta between the aortic arch and the diaphragm.

_____ 5. Portion of the aorta between the diaphragm and the common iliac arteries.

B. Match these arteries with the
correct parts of the diagram
labeled in Figure 21-1:

Brachiocephalic artery
Left common carotid artery
Left subclavian artery
Left vertebral artery

Right common carotid artery
Right subclavian artery
Right vertebral artery

1. _____

2. _____

3. _____

4. _____

5. _____

6. _____

7. _____

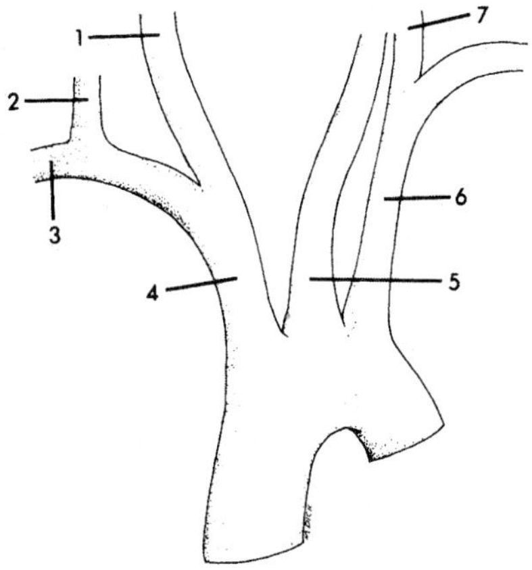

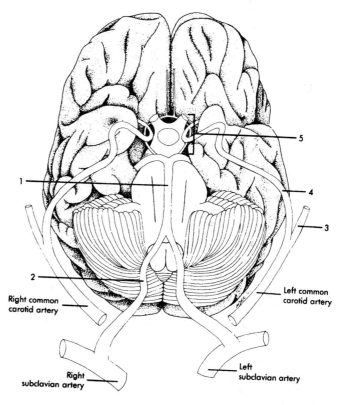

Figure 21-1

C. Match these arteries with the
correct parts of the diagram
labeled in Figure 21-2:

Basilar artery
Cerebral arterial circle
  (circle of Willis)

External carotid artery
Internal carotid artery
Vertebral artery

1. _____

2. _____

3. _____

4. _____

5. _____

Figure 21-2

D. Match these arteries with the
   correct parts of the diagram
   labeled in Figure 21-3:

Axillary artery
Brachial artery
Digital artery
Palmar arches
Radial artery
Ulnar artery

1. _____

2. _____

3. _____

4. _____

5. _____

6. _____

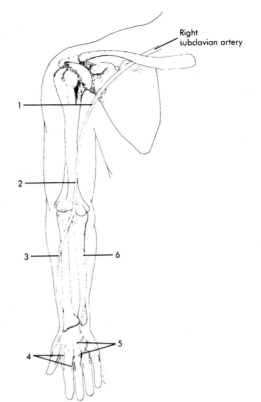

Figure 21-3

E. Match these arteries with the
   correct parts of the diagram
   labeled in Figure 21-4:

Anterior intercostal artery
Internal thoracic artery
Posterior intercostal artery
Visceral arteries

1. _____

2. _____

3. _____

4. _____

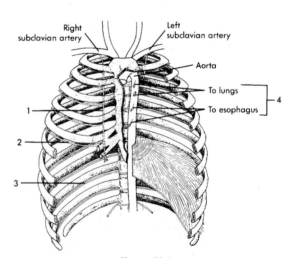

Figure 21-4

F. Match these arteries with the
   correct parts of the diagram
   labeled in Figure 21-5:

Celiac trunk
Common hepatic artery
Common iliac artery
Inferior mesenteric artery
Left gastric artery
Gonadal artery
Renal artery
Splenic artery
Superior mesenteric artery
Suprarenal artery

1. _____

2. _____

3. _____

4. _____

5. _____

6. _____

7. _____

8. _____

9. _____

10. _____

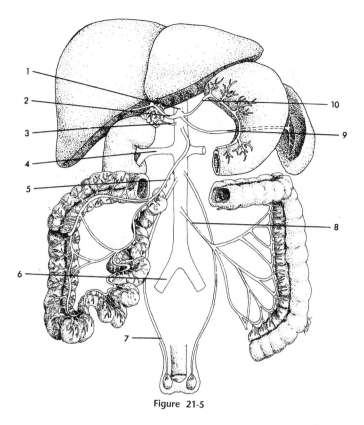

Figure 21-5

G. Match these arteries with the
correct parts of the diagram
labeled in Figure 21-6:

Anterior tibial artery
Digital arteries
Dorsalis pedis artery
Femoral artery
Lateral plantar artery
Medial plantar artery
Fibular (peroneal) artery
Popliteal artery
Posterior tibial artery

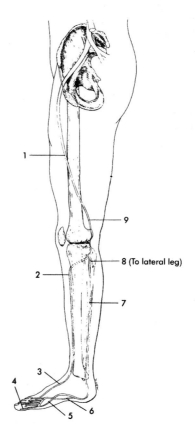

Figure 21-6

1. _____

2. _____

3. _____

4. _____

5. _____

6. _____

7. _____

8. _____

9. _____

Systemic Circulation: Veins

❝Veins transport blood from the capillary beds toward the heart.**❞**

A. Match these major veins with
the correct description:

Coronary sinus
Inferior vena cava
Internal jugular vein

Superior vena cava
Venous sinuses

_____   1. Vein that returns blood from the walls of the heart.

_____   2. Vein that returns blood from the head, neck, thorax, and upper
limbs to the heart.

_____   3. Vein that returns blood from the abdomen, pelvis, and lower
limbs to the heart.

_____   4. Spaces within the dura mater surrounding the brain; the
superior sagittal sinus is an example.

_____   5. Vein that drains blood from the venous sinuses of the brain.

B. Match these veins with the correct parts of the diagram labeled in Figure 21-7:

Inferior vena cava
Pulmonary veins
Right brachiocephalic vein
Right external jugular vein
Right internal jugular vein
Right subclavian vein
Superior vena cava

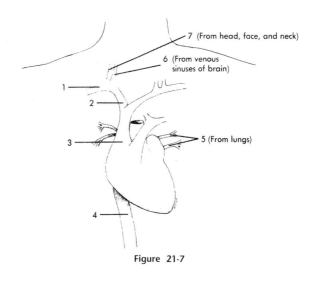

Figure 21-7

1. _____

2. _____

3. _____

4. _____

5. _____

6. _____

7. _____

C. Match these veins with the correct parts of the diagram labeled in Figure 21-8:

Axillary vein
Basilic vein
Brachial veins
Cephalic vein
Median cubital vein
Venous arches

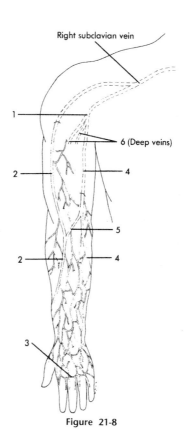

Figure 21-8

1. _____

2. _____

3. _____

4. _____

5. _____

6. _____

D. Match these veins with the
   correct parts of the diagram
   labeled in Figure 21-9:

   Accessory hemiazygos vein
   Azygos vein
   Hemiazygos vein

   1. _____

   2. _____

   3. _____

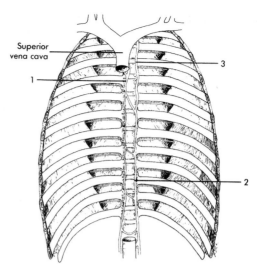

Figure 21-9

E. Match these veins with the
   correct parts of the diagram
   labeled in Figure 21-10:

   Common iliac vein
   External iliac vein
   Gonadal vein
   Internal iliac vein
   Renal vein
   Suprarenal vein

   1. _____

   2. _____

   3. _____

   4. _____

   5. _____

   6. _____

   7. _____

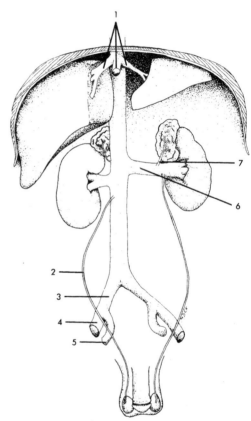

Figure 21-10

F. Match these veins with the correct parts of the diagram labeled in Figure 21-11:

Gastric vein
Hepatic portal vein
Hepatic veins
Inferior mesenteric vein

Inferior vena cava
Splenic vein
Superior mesenteric vein

1. _____

2. _____

3. _____

4. _____

5. _____

6. _____

7. _____

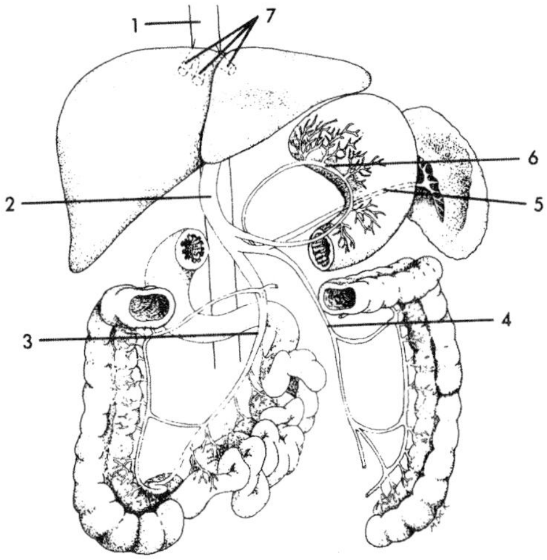

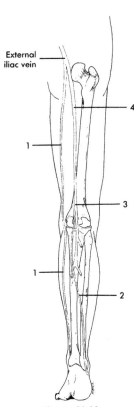

Figure 21-11

G. Match these veins with the correct parts of the diagram labeled in Figure 21-12:

Femoral vein
Great saphenous vein
Popliteal vein
Small saphenous vein

1. _____

2. _____

3. _____

4. _____

External iliac vein

Figure 21-12

Lymphatic Vessels

66 *The lymphatic system, unlike the circulatory system, only carries fluid away from the tissues.* **99**

Match these terms with the correct statement or definition:

Cisterna chyli
Lymph
Lymph capillaries
Lymph nodes

Lymphatic vessels
Right lymphatic duct
Thoracic duct

_____ 1. Interstitial fluid that is returned to the circulatory system through lymphatic vessels.

_____ 2. Vessels that are responsible for collecting interstitial fluid.

_____ 3. Vessels that resemble small veins and contain one-way valves.

_____ 4. Largest lymph duct, which enters the left subclavian vein.

_____ 5. Enlargement of the thoracic duct in the superior abdominal cavity.

_____ 6. Lymph duct that drains the right thorax, right upper limb, and right side of the head and neck.

Physics of Circulation

66 *The physical characteristics of blood and the physical principles affecting the flow of liquids* **99** *through vessels dramatically influence the circulation of blood.*

A. Match these terms with the correct statement or definition:

Laminar flow
Turbulent flow

Viscosity

_____ 1. Measure of the resistance of a liquid to flow.

_____ 2. Tendency for a fluid to flow through tubes as if the fluid is composed of concentric layers.

_____ 3. Numerous small currents flowing crosswise or obliquely to the long axis of a vessel.

☞ As the hematocrit increases, the viscosity of blood increases logarithmically.

B. Match these terms with the correct statement or definition:

Auscultatory
Blood pressure

Korotkoff sounds
Sphygmomanometer

_____ 1. Measure of the force blood exerts against the blood vessel walls.

_____ 2. Device that uses an inflatable cuff to measure blood pressure.

_____ 3. Method of determining blood pressure by listening to the sound of blood flowing through the arteries.

C. Using the terms provided, complete these statements:

1. _____

Decreased   Increased

2. _____

According to Poiseuille's law, blood flow is dramatically
 (1) when the radius of a blood vessel is increased.
Increased blood viscosity causes (2) flow, and decreased
vessel length results in (3) flow. Also, if the pressure
gradient in a vessel is decreased, flow is (4) .

3. _____

4. _____

D. Match these terms with the
   correct statement or definition:

Aneurysm   Increases
Critical closing pressure   The law of Laplace
Decreases   Vascular compliance

_____ 1. Blood pressure below which a blood vessel will collapse.

_____ 2. Force that stretches the vascular wall is proportional to the
                                diameter of the vessel times blood pressure.

_____ 3. What happens to the force acting on the wall of a blood vessel
                                when the diameter of the vessel decreases (e.g., with
                                sympathetic stimulation).

_____ 4. Bulge in a weakened blood vessel wall.

_____ 5. Tendency for blood vessel volume to change as blood pressure
                                increases.

_____ 6. Change in blood vessel volume when blood pressure increases.

☞ Veins have a higher vascular compliance than arteries. A small increase in pressure
   causes a large increase in venous volume, and the veins act as a storage area for blood.

Physiology of Systemic Circulation

❝The anatomy of the circulatory system and the physics of blood flow participate in determining❞
the physiological characteristics of the circulatory system.

A. Match these terms with the
   correct statement or definition:

Aorta   Capillaries
Arteries   Veins
Arterioles

_____ 1. Largest percentage of blood volume is in these vessels.

_____ 2. Vessels with the slowest velocity of blood flow, but the
                                greatest cross-sectional area.

_____ 3. Blood flowing through this vessel has the greatest velocity
                                and the greatest pressure.

_____ 4. Vessels with the highest resistance to flow.

_____ 5. Vessels with the lowest resistance to flow.

B. Using the terms provided, complete the following statements:

Decreased   Pulse pressure
Increased

The difference between the systolic and diastolic pressure is
called the __(1)__. When stroke volume is increased, pulse pressure
is __(2)__. For a given stroke volume, pulse pressure is __(3)__ when
vascular compliance is decreased (e.g., with age). Pulse pressure
produces a pressure wave that can be monitored as the pulse.
Weak pulses usually indicate a(n) __(4)__ stroke volume or a(n)
__(5)__ constriction of the arteries as a result of intense sympathetic
stimulation.

1. _____

2. _____

3. _____

4. _____

5. _____

C. Match these terms with the   Blood pressure   Lymphatic system
   correct statement or definition: Diffusion   Osmotic pressure
                                       Edema

_____   1. Major means by which nutrients and waste products are
                                  exchanged across capillary surfaces.

_____   2. Physical force that moves fluid out of the capillary.

_____   3. At the venous end of the capillary, this force has decreased.

_____   4. At the venous end of the capillary, this force has increased.

_____   5. System that picks up excess tissue fluid and returns it to the
                                  general circulation.

_____   6. Swelling caused by excess tissue fluid accumulation.

D. Match these terms with the   Decreases
   correct statement or definition: Increases

_____   1. Effect on cardiac output when blood volume increases.

_____   2. Effect on venous return to the heart when sympathetic
                                  stimulation increases venous tone.

_____   3. Effect on venous return to the heart when exercising.

_____   4. Effect on venous pressure in the legs when standing still; can
                                  result in edema.

Local Control of Blood Flow by the Tissues

❝*In most tissues, blood flow to the tissues is proportional to the metabolic needs of the tissue.***❞**

Using the terms provided, complete these statements:

Autoregulation
Decrease
substances
Increase
Metabolism
Metarterioles and precapillary sphincters

Nutrients
Vasodilator

Vasomotion

1. _____
2. _____
3. _____
4. _____
5. _____
6. _____
7. _____
8. _____
9. _____

Control of local blood flow occurs through __(1)__. Because there is little innervation of these structures, local factors regulate blood flow. As the rate of __(2)__ of a tissue increases, rate of blood flow through its capillaries increases. __(3)__, including carbon dioxide and lactic acid, increase in extracellular fluid as the rate of metabolism increases. Lack of __(4)__ may also be important in regulating local blood flow; for instance, increased metabolism reduces the amount of oxygen and other nutrients in the tissues. In response to a(n) __(5)__ in vasodilator substances or a(n) __(6)__ in nutrients, metarterioles and precapillary sphincters dilate, resulting in an increased blood flow. Periodic contraction and relaxation of the precapillary sphincters, which is called __(7)__, produces cyclic blood flow through tissues. __(8)__ refers to the local control mechanisms that determine blood flow to tissues, despite relatively large changes in systemic blood pressure. In long-term regulation of blood flow, if the metabolic activity of a tissue remains elevated, the number of capillaries in the tissue will __(9)__.

Nervous Regulation of Local Circulation

❝*Nervous control of arterial blood pressure is important in minute-to-minute***❞** *regulation, and during exercise or shock.*

Match these terms with the correct statement or definition:

Acetylcholine
Epinephrine
Hypothalamus and cerebral cortex
Norepinephrine

Sympathetic
Vasomotor center
Vasomotor tone

_____ 1. Most important ANS division for nervous control of blood flow.

_____ 2. Brain area controlling sympathetic nerve impulses to blood vessels.

_____ 3. Brain areas that can inhibit or stimulate the vasomotor center.

_____ 4. Condition of partial constriction of peripheral blood vessels.

_____ 5. Neurotransmitter causing vasoconstriction in most blood vessels.

_____ 6. Substance causing vasodilation in skeletal muscle blood vessels.

Part of the vasomotor center is excitatory and is tonically active, and part of the vasomotor center is inhibitory and induces vasodilation when it is active.

Regulation of Mean Arterial Pressure

66*Regulatory mechanisms exist that maintain the average systemic blood pressure*99 *at approximately 100 mm Hg.*

Match these terms with the correct statement or definition:

Cardiac output
Decreases
Heart rate
Increases

Mean arterial pressure
Peripheral resistance
Stroke volume

_____ 1. Equal to heart rate times stroke volume.

_____ 2. Volume of blood pumped by the heart during each contraction.

_____ 3. Equal to heart rate x stroke volume x peripheral resistance.

_____ 4. Resistance to the flow of blood in the blood vessels.

_____ 5. Effect on mean arterial pressure when heart rate decreases.

_____ 6. Effect on mean arterial pressure when peripheral resistance increases.

_____ 7. Effect on mean arterial pressure when stroke volume increases.

Nervous Regulation of Blood Pressure

66*Several mechanisms are involved in nervous regulation of blood pressure.*99

A. Match these types of short-term blood pressure regulation with the correct statement or definition:

Baroreceptor reflex
Central nervous system ischemic response

Chemoreceptor reflex

_____ 1. Mechanism that involves sensory receptors sensitive to stretch; receptors are in the carotid sinus and aortic arch.

_____ 2. Mechanism that involves sensory receptors sensitive to decreased oxygen or increased carbon dioxide and hydrogen ion levels (i.e., decreased pH); receptors are in the carotid bodies and aortic bodies.

_____ 3. Mechanism that responds to increased carbon dioxide and hydrogen ions (i.e., decreased pH) in the medulla.

_____ 4. Mechanisms usually active only during emergency situations.

_____

_____ 5. Mechanism that is important in regulating blood pressure on a moment-to-moment basis.

410

B. Match these terms with the correct statements:

Decreased   Vasoconstriction
Increased   Vasodilation

_____ 1. Baroreceptor reflex effect on heart rate and stroke volume following a sudden decrease in arterial pressure.

_____ 2. Effect on blood vessels produced by the baroreceptor reflexes when blood pressure increases.

_____ 3. Effect on blood vessels produced by the chemoreceptor reflexes when arterial oxygen levels decrease markedly.

_____ 4. Effect on blood vessels produced by the CNS ischemic response when medullary carbon dioxide levels increase markedly.

Hormonal Regulation of Blood Pressure

66 *Four important hormonal mechanisms control arterial pressure.* **99**

A. Match these types of hormonal control with the correct statement or definition:

Adrenal medullary mechanism
Atrial natriuretic mechanism
Renin-angiotensin-aldosterone mechanism
Vasopressin mechanism

_____ 1. Involves secretion of epinephrine and norepinephrine.

_____ 2. Involves the release of an enzyme from the juxtaglomerular apparatuses in the kidneys.

_____ 3. Involves a hormone from the posterior pituitary.

_____ 4. Involves the release of a polypeptide from the heart.

B. Using the terms provided, complete the following statements:

Aldosterone   Decreased
Angiotensin II   Increased
Angiotensinogen   Renin

The kidneys release an enzyme called _(1)_ into the circulatory system from the juxtaglomerular apparatuses. Renin acts on a plasma protein called _(2)_ to split a fragment off one end. The fragment, called angiotensin I, has two more amino acids cleaved from it by angiotensin converting enzyme to become _(3)_. Angiotensin II causes _(4)_ vasoconstriction in arterioles and veins, which functions to raise blood pressure. Angiotensin II also stimulates _(5)_ secretion from the adrenal cortex. Aldosterone acts on the kidneys, resulting in _(6)_ urine production. Angiotensin II also stimulates thirst, increases salt appetite, and stimulates ADH secretion. Stimuli that increase renin secretion include _(7)_ blood pressure, _(8)_ potassium ion levels, and _(9)_ sodium ion levels.

1. _____

2. _____

3. _____

4. _____

5. _____

6. _____

7. _____

8. _____

9. _____

411

The renin-angiotensin-aldosterone mechanism is important in maintaining blood pressure under the conditions of circulatory shock.

C. Match these terms with the correct parts of the diagram labeled in Figure 21-13:

Aldosterone
Angiotensin II

Angiotensinogen
Renin

1. _____

2. _____

3. _____

4. _____

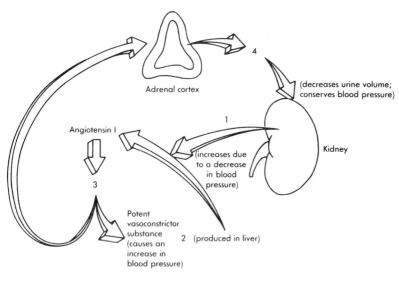

Figure 21-13

D. Match these terms with the correct statements as they apply to hormonal mechanisms that control arterial pressure:

Decrease
Increase

_____ 1. Effect of epinephrine and norepinephrine on heart rate and stroke volume.

_____ 2. Effect of epinephrine on vasoconstriction (of skin and visceral blood vessels).

_____ 3. Effect of ADH on vasoconstriction.

_____ 4. Effect of ADH on urine production.

_____ 5. Effect of a decrease in blood pressure on ADH secretion.

_____ 6. Effect of atrial natriuretic hormone on urine production.

_____ 7. Effect of increased atrial blood pressure on atrial natriuretic hormone secretion.

The Fluid Shift Mechanism and the Stress-Relaxation Response

66*Two mechanisms, in addition to the nervous and hormonal mechanisms,*99
help to regulate blood pressure.

Match these terms with the correct statement or definition: response

Decrease
Fluid shift mechanism

Increase
Stress-relaxation

_____ 1. When blood pressure increases, fluid moves from the blood vessels into the interstitial spaces; when blood pressure decreases, the opposite effect occurs.

_____ 2. When blood pressure declines, smooth muscle cells in blood vessel walls contract; when blood pressure increases, the smooth muscle cells relax.

_____ 3. Change in the amount of interstitial fluid moved into capillaries when blood pressure decreases.

_____ 4. Effect on blood pressure when fluid moves out of tissues into capillaries.

_____ 5. Change in contraction of smooth muscle in blood vessel walls when blood volume decreases.

_____ 6. Effect on blood pressure when smooth muscle in blood vessel walls contracts.

Long-Term Regulation of Blood Pressure

66*Long-term regulation of blood pressure is achieved by the kidneys, which affect blood pressure*99
by changing blood volume.

Match these terms with the correct statement:

Decrease
Increase

_____ 1. Effect on urine production of an increase in blood pressure.

_____ 2. Effect on blood volume of an increase in urine production.

_____ 3. Effect on blood pressure of a decrease in blood volume.

☞ Urine production is regulated by several hormones. Increased blood volume results in increased atrial natriuretic hormone and decreased renin, angiotensin, aldosterone, and ADH. These changes cause increased urine production and a reduction in blood volume.

1. Name the types of blood vessels, starting and ending at the heart.

2. List the three types of capillaries.

3. Name the subdivisions of the aorta.

4. Name the three major arteries that branch from the aorta to supply the head and upper limbs.

5. List the three unpaired arteries and the four major paired arteries that branch from the abdominal aorta.

6. List three veins that return blood to the superior vena cava.

7. Name the veins that drain blood from the head and upper limbs.

8. Name the veins that join the inferior vena cava to return blood from the kidneys and the lower limbs.

9. List the two major superficial veins of the upper limbs and the two major superficial veins of the lower limbs.

10. Name four factors in Poiseuille's law that influence blood flow.

11. List two factors that influence the force acting on the wall of a blood vessel.

12. List three factors that compress lymphatic vessels and move lymph toward the circulatory system.

13. List the major force responsible for the movement of fluid out of capillaries and the major force responsible for the movement of fluid into capillaries.

14. List three nervous mechanisms for short-term regulation of blood pressure.

15. List four hormonal mechanisms for control of arterial pressure.

16. List two mechanisms in addition to nervous and hormonal mechanisms that help to regulate systemic blood pressure.

Place the letter corresponding to the correct answer in the space provided.

_____ 1. Given the following blood vessels:
1. arteriole
2. capillary
3. elastic artery
4. muscular artery
5. vein
6. venule

Choose the arrangement that lists the blood vessels in the order an erythrocyte passes through them as it leaves the heart, travels to a tissue, and returns to the heart.
a. 3, 4, 2, 1, 5, 6
b. 3, 4, 1, 2, 6, 5
c. 4, 3, 1, 2, 5, 6
d. 4, 3, 2, 1, 6, 5

_____ 2. Given the following structures:
1. metarteriole
2. precapillary sphincter
3. thoroughfare channel

Choose the arrangement that lists the structures in the order an erythrocyte encounters them as it passes through a tissue.
a. 1, 3, 2
b. 2, 1, 3
c. 2, 3, 1
d. 3, 2, 1

_____ 3. Which tunic contains most of the smooth muscle in a blood vessel?
a. tunica intima
b. tunica media
c. tunica adventitia
d. tunica muscularis

_____ 4. In which of the following blood vessels would elastic fibers be present in the greatest amounts?
a. large arteries
b. medium arteries
c. arterioles
d. venules
e. large veins

_____ 5. Comparing and contrasting arteries and veins,
a. veins have thicker walls.
b. veins have a greater amount of smooth muscle than arteries.
c. veins have a tunica media and arteries do not.
d. veins have valves and arteries do not.
e. all of the above

_____ 6. Which structure supplies the walls of blood vessels with blood?
a. venous shunt
b. tunic channels
c. arteriovenous anastomoses
d. vasa vasorum

_____ 7. Given the following blood vessels:
1. aorta
2. inferior vena cava
3. pulmonary arteries
4. pulmonary veins

Which vessels carry oxygen-rich blood?
a. 1, 3
b. 1, 4
c. 2, 3
d. 2, 4

_____ 8. Given the following list of arteries:
1. basilar
2. common carotids
3. internal carotids
4. vertebral

Which of the arteries have DIRECT connections with the cerebral arterial circle (circle of Willis)?
a. 2, 1
b. 2, 4
c. 3, 1
d. 3, 4

9. Given the following vessels:
   1. axillary artery
   2. brachial artery
   3. brachiocephalic artery
   4. radial artery
   5. subclavian artery

   Choose the arrangement that lists
   the vessels in order going from the
   aorta to the right hand.
   a. 2, 5, 4, 1
   b. 5, 2, 1, 4
   c. 5, 3, 1, 4, 2
   d. 3, 5, 1, 2, 4

10. Which artery is most commonly used
    to take the pulse near the wrist?
    a. basilar
    b. brachial
    c. cephalic
    d. radial
    e. ulnar

11. A branch of the aorta that supplies
    the liver, stomach, and spleen is the
    a. celiac trunk.
    b. common iliac.
    c. inferior mesenteric.
    d. superior mesenteric.

12. Given the following arteries:
    1. common iliac
    2. external iliac
    3. femoral
    4. popliteal

    Choose the arrangement that lists
    the arteries in order going from the
    aorta to the knee.
    a. 1, 2, 3, 4
    b. 1, 2, 4, 3
    c. 2, 1, 3, 4
    d. 2, 1, 4, 3

13. Given the following veins:
    1. brachiocephalic
    2. internal jugular
    3. superior vena cava
    4. venous sinus

    Choose the arrangement that lists
    the veins in order going from the
    brain to the heart.
    a. 2, 4, 1, 3
    b. 2, 4, 3, 1
    c. 4, 2, 1, 3
    d. 4, 2, 3, 1

14. Blood returning from the arm to the
    subclavian vein would pass through
    which of the following veins?
    a. cephalic
    b. basilic to axillary
    c. brachial to axillary
    d. a and b
    e. all of the above

15. Given the following veins:
    1. accessory hemiazygos
    2. azygos
    3. hemiazygos
    4. inferior vena cava
    5. superior vena cava

    Trace the route an erythrocyte
    would take when returning from the
    LEFT INFERIOR posterior surface of
    the thorax to the heart.
    a. 2, 4
    b. 2, 5
    c. 1, 2, 5
    d. 3, 2, 4
    e. 3, 2, 5

16. Given the following vessels:
    1. inferior mesenteric vein
    2. superior mesenteric vein
    3. hepatic portal vein
    4. hepatic vein

    Choose the arrangement that lists
    the vessels in order going from the
    small intestine to the inferior vena
    cava.
    a. 1, 3, 4
    b. 1, 4, 3
    c. 2, 3, 4
    d. 2, 4, 3

17. Given the following list of veins:
    1. small saphenous
    2. great saphenous
    3. fibular (peroneal)
    4. posterior tibial

    Which are superficial veins?
    a. 1, 2
    b. 1, 3
    c. 2, 3
    d. 2, 4
    e. 3, 4

_____18. Lymph capillaries
a. have a basement membrane.
b. are less permeable than blood capillaries.
c. prevent backflow of lymph into the tissues.
d. all of the above

_____19. Lymph is moved through lymphatic vessels because of
a. contraction of surrounding skeletal muscle.
b. contraction of smooth muscle within the lymphatic vessel wall.
c. pressure changes in the thorax during respiration.
d. all of the above

_____20. Lymphatic vessels
a. do not have valves.
b. empty into lymph nodes.
c. from the right upper limb join the thoracic duct.
d. all of the above

_____21. Korotkoff sounds
a. are caused by closing of the AV valves.
b. are caused by closing of the semilunar valves.
c. are caused by turbulent blood flow in an artery when it is partly occluded.
d. equal the pulse rate multiplied by two.

_____22. If one could increase any of the following factors that affect blood flow by twofold, which one would cause the greatest increase in blood flow?
a. blood viscosity
b. the pressure gradient
c. vessel radius

_____23. According to the law of Laplace,
a. the force that stretches the wall of a blood vessel is proportional to the diameter of the vessel times the blood pressure.
b. as blood pressure decreases, the force acting on the wall of a blood vessel decreases.
c. as the diameter of a blood vessel increases, the force acting on the wall of the blood vessel increases.
d. all of the above

_____24. Vascular compliance is
a. greater in arteries than in veins.
b. the increase in vessel volume divided by the increase in vessel pressure.
c. the pressure at which blood vessels collapse.
d. all of the above.

_____25. The resistance to blood flow is greatest in the
a. aorta.
b. arterioles.
c. capillaries.
d. venules.
e. veins.

_____26. Pulse pressure
a. is the difference between systolic and diastolic pressure.
b. increases when stroke volume increases.
c. increases as vascular compliance decreases.
d. all of the above

_____27. Concerning fluid movement at the capillary level,
a. the amount of fluid leaving the arterial end of a capillary is equal to the amount of fluid entering the venous end.
b. fluid enters the venous end of a capillary primarily because of a decrease in capillary hydrostatic pressure.
c. fluid tends to move into tissues at the arteriole end of a capillary because there is a greater concentration of proteins in the tissue than in the blood.
d. all of the above

28. Veins
    a. increase their volume because of their large compliance.
    b. increase venous return to the heart when they vasodilate.
    c. vasodilate because of increased sympathetic stimulation.
    d. all of the above

29. Which of the following average pressure relationships is correctly matched in a standing person?
    a. 0 mm Hg: pressure in the right atrium
    b. 100 mm Hg: pressure in the aorta by the heart
    c. 110 mm Hg: pressure in an arteriole in the foot
    d. all of the above

30. Local direct control of blood flow through a tissue
    a. maintains a relatively constant rate of flow despite large changes in arterial blood pressure.
    b. results from relaxation and contraction of the precapillary sphincters.
    c. occurs in response to a buildup of carbon dioxide in the tissue.
    d. occurs in response to a decrease of oxygen in the tissue.
    e. all of the above

31. Blood vessels
    a. are continually stimulated by the sympathetic nervous system, accounting for vasomotor tone.
    b. in most tissues constrict in response to norepinephrine.
    c. in skeletal muscle dilate in response to epinephrine.
    d. all of the above

32. An increase in mean arterial pressure may result from
    a. an increase in peripheral resistance.
    b. an increase in heart rate.
    c. an increase in stroke volume.
    d. all of the above

33. In response to an increase in mean arterial pressure, the baroreceptor reflex would cause
    a. an increase in sympathetic nervous system activity.
    b. a decrease in peripheral resistance.
    c. stimulation of the vasomotor center.
    d. vasoconstriction.

34. Through the chemoreceptor reflex, when oxygen levels markedly decrease in the blood, you would expect
    a. peripheral resistance to decrease.
    b. mean arterial blood pressure to increase.
    c. vasomotor tone to decrease.
    d. all of the above

35. Through the central nervous system ischemic response, as
    a. oxygen levels increase, vasoconstriction results.
    b. oxygen levels decrease, vasodilation results.
    c. carbon dioxide levels increase, vasoconstriction results.
    d. carbon dioxide levels decrease, vasoconstriction results.

36. When blood pressure is suddenly decreased a small amount (10 mm Hg), which of the following mechanisms would be activated to restore blood pressure to normal levels?
    a. chemoreceptor reflexes
    b. baroreceptor reflexes
    c. central nervous system ischemic response
    d. all of the above

37. A sudden release of epinephrine from the adrenal medulla would
    a. increase heart rate.
    b. increase stroke volume.
    c. cause vasoconstriction in visceral blood vessels.
    d. all of the above

38. When blood pressure decreases,
    a. renin secretion increases.
    b. angiotensin II formation decreases.
    c. aldosterone secretion decreases.
    d. all of the above

39. In response to an increase in blood pressure,
    a. ADH secretion decreases.
    b. the kidneys increase urine production.
    c. blood volume decreases.
    d. all of the above.

40. In response to a decrease in blood pressure,
    a. more fluid than normal enters tissues (fluid shift mechanism).
    b. smooth muscles in blood vessels relax (stress-relaxation response).
    c. the kidneys retain more salts and water than normal.
    d. all of the above

41. During spinal anesthesia a decrease in the frequency of action potentials carried along sympathetic nerves can occur, especially in the lower region of the spine. Choose the most appropriate consequence from the list provided.
    a. decreased systemic blood pressure
    b. markedly decreased heart rate
    c. increased systemic blood pressure
    d. marked cutaneous vasoconstriction
    e. a and b

42. A patient who has a myocardial infarction that clearly compromises cardiac function will exhibit which of the following?
    a increased sympathetic stimulation of the heart
    b. above normal blood pressure
    c. peripheral vasodilation
    d. all of the above

43. Wilber Merocrine sat in a hot bath for 30 minutes. He then quickly got out of the hot bath and jumped into an ice bath. As a consequence, which of the following occurred?
    a. Wilber's blood pressure increased dramatically.
    b. Wilber's peripheral resistance increased.
    c. Wilber's heart rate increased dramatically.
    d. a and b
    e. all of the above

44. A patient was found to have severe arteriosclerosis of his renal arteries, which reduced renal blood pressure. Which of the following is consistent with that condition?
    a. hypotension
    b. hypertension
    c. decreased vasomotor tone
    d. exaggerated sympathetic stimulation of the heart
    e. a and c

45. During exercise the blood flow through skeletal muscle may increase up to twentyfold. However, the cardiac output does not increase that much. This occurs
    a. because of vasoconstriction in the viscera.
    b. because of vasoconstriction in the skin (at least temporarily).
    c. because of vasodilation of skeletal muscle blood vessels.
    d. a and b
    e. all of the above

Use a separate sheet of paper to complete this section.

1. When cancer of the colon is discovered, a liver scan is often made to see if the patient has cancer of the liver. Explain why this is a reasonable test to make.

2. Why does vasoconstriction of blood vessels increase resistance to blood flow? How can blood flow be maintained when resistance to blood flow increases?

3. In varicose veins the blood vessels become distended and enlarge. Once this process starts, it becomes worse and worse. Explain why. Why does raising the feet help this condition?

4. Just before blast-off an astronaut had her blood pressure measured while lying horizontal. Once in space, under weightless conditions (i.e., no gravity), the measurement was repeated. Would you expect her blood pressure in space to be higher, lower, or unchanged compared with her blood pressure on earth?

5. In cirrhosis of the liver there is often abdominal swelling. Explain (hint: cirrhosis impairs blood flow through the liver and impairs albumin production by the liver).

6. While donating blood a student nurse felt dizzy and faint. Explain why on the basis of the following: blood donation results in loss of blood; emotional reactions associated with donating blood could cause hyperventilation; and emotional reactions can decrease sympathetic and increase parasympathetic nervous system activity.

7. The student nurse in the preceding question was told to lie down and breathe into a paper bag. Explain why these therapies would help her.

8. Private U.P. Wright had to stand at attention in the hot sun for several hours. Eventually he fainted. Explain what happened.

9. Just as Private U.P. Wright faints, his buddies realize what is happening. To save him from the embarrassment of fainting they hold him in a standing position. Explain what is wrong with this treatment. What would you suggest they do?

10. A patient has the following symptoms: below normal blood pressure and edema in the ankles, legs, and hands. Assume that the patient has had a myocardial infarct that has damaged one of the ventricles. On the basis of the symptoms, explain why you believe it is the left or the right ventricle that was damaged.

Did you hear about the anatomy and physiology student who failed the course because she did not attend the lectures on the heart, blood vessels, and other organs? She said it was because she did not like to attend organ recitals.

ANSWERS TO CHAPTER 21

CONTENT LEARNING ACTIVITY

General Features of Blood Vessel Structure
A. 1. Endothelium; 2. Adventitia; 3. Precapillary cells
B. 1. Continuous capillaries; 2. Fenestrated capillaries; 3. Sinusoidal capillaries; 4. Sinusoids; 5. Venous sinuses
C. 1. Arterial capillaries; 2. Metarterioles; 3. Thoroughfare channels; 4. Precapillary sphincters
D. 1. Tunica intima; 2. Tunica media; 3. Tunica adventitia; 4. Tunica media
E. 1. Elastic arteries; 2. Medium arteries; 3. Arterioles; 4. Medium arteries, small arteries, arterioles
F. 1. Venules; 2. Small veins; 3. Medium and large veins
G. 1. Vasa vasorum; 2. Arteriovenous anastomoses; 3. Glomus
H. 1. Varicose veins; 2. Phlebitis; 3. Arteriosclerosis; 4. Atherosclerosis

Pulmonary Circulation
1. Pulmonary trunk; 2. Pulmonary arteries; 3. Pulmonary veins

Systemic Circulation: Arteries
A. 1. Ascending aorta; 2. Aortic arch; 3. Descending aorta; 4. Thoracic aorta; 5. Abdominal aorta
B. 1. Right common carotid artery; 2. Right vertebral artery; 3. Right subclavian artery; 4. Brachiocephalic artery; 5. Left common carotid artery; 6. Left subclavian artery; 7. Left vertebral artery
C. 1. Basilar artery; 2. Vertebral artery; 3. External carotid artery; 4. Internal carotid artery; 5. Cerebral arterial circle (circle of Willis)
D. 1. Axillary artery; 2. Brachial artery; 3. Radial artery; 4. Digital artery; 5. Palmar arches; 6. Ulnar artery
E. 1. Internal thoracic artery; 2. Anterior intercostal artery; 3. Posterior intercostal artery; 4. Visceral arteries
F. 1. Celiac trunk; 2. Common hepatic artery; 3. Suprarenal artery; 4. Renal artery; 5. Superior mesenteric artery; 6. Common iliac artery; 7. Gonadal artery; 8. Inferior mesenteric artery; 9. Splenic artery; 10. Left gastric artery
G. 1. Femoral artery; 2. Anterior tibial artery; 3. Dorsalis pedis artery; 4. Digital arteries; 5. Medial plantar artery; 6. Lateral plantar artery; 7. Posterior tibial artery; 8. Fibular (peroneal) artery; 9. Popliteal artery

Systemic Circulation: Veins
A. 1. Coronary sinus; 2. Superior vena cava; 3. Inferior vena cava; 4. Venous sinuses; 5. Internal jugular vein
B. 1. Right subclavian vein; 2. Right brachiocephalic vein; 3. Superior vena cava; 4. Inferior vena cava; 5. Pulmonary veins; 6. Right internal jugular vein; 7. Right external jugular vein
C. 1. Axillary vein; 2. Cephalic vein; 3. Venous arches; 4. Basilic vein; 5. Median cubital vein; 6. Brachial veins
D. 1. Azygos vein; 2. Hemiazygos vein; 3. Accessory hemiazygos vein
E. 1. Hepatic veins; 2. Gonadal vein; 3. Common iliac vein; 4. External iliac vein; 5. Internal iliac vein; 6. Renal vein; 7. Suprarenal vein
F. 1. Inferior vena cava; 2. Hepatic portal vein; 3. Superior mesenteric vein; 4. Inferior mesenteric vein; 5. Splenic vein; 6. Gastric vein; 7. Hepatic veins
G. 1. Great saphenous vein; 2. Small saphenous vein; 3. Popliteal vein; 4. Femoral vein

Lymphatic Vessels
1. Lymph; 2. Lymph capillaries; 3. Lymphatic vessels; 4. Thoracic duct; 5. Cisterna chyli; 6. Right lymphatic duct

Physics of Circulation
A. 1. Viscosity; 2. Laminar flow; 3. Turbulent flow
B. 1. Blood pressure; 2. Sphygmomanometer; 3. Auscultatory
C. 1. Increased; 2. Decreased; 3. Increased; 4. Decreased
D. 1. Critical closing pressure; 2. The law of Laplace; 3. Decreases; 4. Aneurysm; 5. Vascular compliance; 6. Increases

Physiology of Systemic Circulation
A. 1. Veins; 2. Capillaries; 3. Aorta; 4. Arterioles; 5. Veins
B. 1. Pulse pressure; 2. Increased; 3. Increased; 4. Decreased; 5. Increased
C. 1. Diffusion; 2. Blood pressure; 3. Blood pressure; 4. Osmotic pressure; 5. Lymphatic system; 6. Edema
D. 1. Increases; 2. Increases; 3. Increases; 4. Increases

Local Control of Blood Flow by the Tissues
1. Metarterioles and precapillary sphincters; 2. Metabolism; 3. Vasodilator substances; 4. Nutrients; 5. Increase; 6. Decrease; 7. Vasomotion; 8. Autoregulation; 9. Increase

Nervous Regulation of Local Circulation
1. Sympathetic; 2. Vasomotor center;
3. Hypothalamus and cerebral cortex;
4. Vasomotor tone; 5. Norepinephrine;
6. Epinephrine

Regulation of Mean Arterial Pressure
1. Cardiac output; 2. Stroke volume; 3. Mean arterial pressure; 4. Peripheral resistance;
5. Decreases; 6. Increases; 7. Increases

Nervous Regulation of Blood Pressure
A. 1. Baroreceptor reflex; 2. Chemoreceptor reflex; 3. Central nervous system ischemic response; 4. Chemoreceptor reflex and central nervous system ischemic response;
5. Baroreceptor reflex
B. 1. Increased; 2. Vasodilation;
3. Vasoconstriction; 4. Vasoconstriction

Hormonal Regulation of Blood Pressure
A. 1. Adrenal medullary mechanism; 2. Renin-angiotensin-aldosterone mechanism;
3. Vasopressin mechanism; 4. Atrial natriuretic mechanism
B. 1. Renin; 2. Angiotensinogen; 3. Angiotensin II; 4. Increased; 5. Aldosterone; 6. Decreased;
7. Decreased; 8. Increased; 9. Decreased
C. 1. Renin; 2. Angiotensinogen; 3 Angiotensin II; 4. Aldosterone
D. 1. Increase; 2. Increase; 3. Increase;
4. Decrease; 5. Increase; 6. Increase;
7. Increase

The Fluid Shift Mechanism and the Stress-Relaxation Response
1. Fluid shift mechanism; 2. Stress-relaxation response; 3. Increase; 4. Increase; 5. Increase;
6. Increase

Long-Term Regulation of Blood Pressure
1. Increase; 2. Decrease; 3. Decrease

QUICK RECALL

1. Elastic arteries, muscular arteries, arterioles, capillaries, venules, small veins, medium and large veins
2. Continuous capillaries, fenestrated capillaries, and sinusoidal capillaries
3. Ascending aorta, aortic arch, descending aorta (thoracic aorta and abdominal aorta)
4. Brachiocephalic artery, left common carotid artery, and left subclavian artery
5. Unpaired: celiac trunk, superior mesenteric artery, inferior mesenteric artery; paired: renal arteries, suprarenal arteries, gonadal arteries, common iliac arteries
6. Left and right brachiocephalic veins and azygos vein
7. From head and neck: internal jugular vein, external jugular vein; from upper limbs: subclavian vein
8. From kidneys: renal veins; from lower limbs: common iliac veins

9. Basilic and cephalic veins in the upper limbs, small and great saphenous veins in the lower limbs
10. Viscosity, diameter of blood vessel, length of blood vessel, and pressure gradient
11. Force = diameter X blood pressure (law of Laplace)
12. Skeletal muscle contraction, contraction of lymphatic vessel smooth muscle, and thoracic pressure changes
13. Blood pressure: out of blood vessels; osmosis: into blood vessels
14. Baroreceptor reflex, chemoreceptor reflex, and the central nervous system ischemic response
15. Adrenal medullary mechanism, renin-angiotensin-aldosterone mechanism, vasopressin mechanism, and atrial natriuretic hormone mechanism
16. Fluid-shift mechanism and stress-relaxation response

MASTERY LEARNING ACTIVITY

1. B. The erythrocyte would pass through an elastic artery, muscular artery, arteriole, capillary, venule, and vein.

2. A. Blood flows from arterioles into metarterioles, from which thoroughfare channels extend to venules. Blood flow through the capillaries that branch off the thoroughfare channels is regulated by precapillary sphincters.

3. B. The tunica media consists of circularly arranged smooth muscles and variable amounts of elastic and collagen fibers. The tunica intima consists of endothelium, basement membrane, lamina propria (with a small amount of smooth muscle), and an internal elastic membrane. The tunica adventitia is connective tissue. There is no tunica muscularis.

4. A. The large arteries such as the aorta contain large amounts of elastic connective tissue, which plays an important role in the maintenance of blood pressure during diastole.

5. D. Only veins have valves. Veins are thinner walled with less smooth muscle than arteries. Both veins and arteries have all three tunics.

6. D. Vasa vasorum are small blood vessels that penetrate the walls of blood vessels. Arteriovenous anastomoses allow blood to pass from arteries to veins without passing through capillaries. "Venous shunt" and "tunic channels" are not valid terms.

7. B. Oxygen rich blood returns from the lungs to the heart through the pulmonary veins, passes through the left side of the heart, and exits through the aorta.

8. C. The internal carotids connect to the cerebral arterial circle (circle of Willis) after branching from the common carotids. The basilar artery is formed by the joining of the left and right vertebral arteries. The basilar artery then connects to the cerebral arterial circle.

9. D. Blood going to the right hand passes from the aorta into the brachiocephalic artery and into the right subclavian artery. Blood going to the left hand passes from the aorta into the left subclavian artery. From the subclavian artery (either side) blood passes through the axillary artery, brachial artery, and radial arteries to reach the hand.

10. D. The radial artery passes over the radial bone near the wrist, and the pulse can be easily taken by slightly compressing the artery against the bone.

11. A. The celiac trunk branches to form the hepatic (to the liver), left gastric (to the stomach), and the splenic (to the spleen) arteries.

12. A. The order is common iliac, external iliac, femoral, and popliteal arteries.

13. C. Blood from the brain enters a number of venous sinuses that empty into the internal jugulars. The internal jugulars join the brachiocephalic veins, which combine to form the superior vena cava.

14. E. The basilic and cephalic veins are superficial veins of the forearm and arm. The basilic vein goes deep to become the axillary vein, which becomes the subclavian vein. The cephalic vein goes deep and joins the axillary vein. The brachial veins are deep veins that also join the axillary vein.

15. E. Blood from the posterior surface of the thorax can return through three routes. Blood from the left inferior surface flows through the hemiazygos, azygos, and superior vena cava. Blood from the left superior surface flows through the accessory hemiazygos, azygos, and superior vena cava. Blood from the right surface flows through the azygos to the superior vena cava.

16. C. The superior mesenteric vein drains the small intestine and joins the splenic vein to form the hepatic portal vein, which enters the liver. Blood from the liver flows to the inferior vena cava through the hepatic veins.

17. A. The small and great saphenous veins are the major superficial veins of the lower limb. The anterior tibial, posterior tibial, and fibular (peroneal) veins are the major deep veins of the leg.

18. C. Lymph capillaries are simple squamous epithelium, but unlike most other epithelial tissue, they do not have a basement membrane. In addition, the epithelial cells are loosely attached to each other. Consequently, lymph capillaries are far more permeable than blood capillaries. The lymph capillaries prevent backflow of lymph into the tissue because the epithelial cells overlap each other and function as one-way valves.

19. D. Compression of the lymphatic vessel by surrounding skeletal muscle, smooth muscles of the lymphatic vessel wall, and thoracic pressure changes all cause movement of lymph.

20. B. Most lymphatic vessels empty into a lymph node. Lymphatic vessels have valves that ensure one-way flow of lymph. Lymphatic vessels from the right upper limb, right thorax, and right side of the head and neck join the right lymphatic duct, which empties into the right subclavian vein. The lymphatic vessels from the rest of the body join the thoracic duct, which empties into the left subclavian vein.

21. C. Blood flow in the brachial artery, where blood pressure is determined by the auscultatory method, is normally laminar (streamlined). When a blood pressure measurement is taken, the pressure cuff constricts the brachial artery, resulting in turbulent flow that produces Korotkoff sounds.

22. C.   According to Poiseuille's law, blood flow is directly proportional to the blood vessel radius raised to the fourth power, directly proportional to the pressure gradient, and inversely proportional to viscosity. Changing the blood vessel radius twofold would affect the blood flow more than an identical change in any of the other parameters.

23. D.   The force acting on a vessel wall is proportional to the diameter of the vessel times the blood pressure. Consequently, as pressure decreases the force decreases, and as diameter increases the force increases.

24. B.   Vascular compliance is the tendency for blood vessel volume to increase as the blood pressure increases. It is greater in veins than in arteries. Critical closing pressure is the pressure at which blood vessels collapse.

25. B.   The greatest resistance to blood flow occurs in the arterioles, followed by the capillaries. The resistance to blood flow is relatively low in the aorta, venules, and veins.

26. D.   Pulse pressure is the difference between systolic and diastolic pressure. It increases with increased stroke volume and decreased vascular compliance.

27. B.   Fluid movement out of the capillary results from capillary hydrostatic pressure and a slight negative pressure in the tissue. Fluid movement into the capillary results from osmosis because the blood has a higher concentration of proteins than the tissue. At the arteriole end of the capillary, the forces moving fluid out exceed the forces moving fluid in. At the venous end the situation is reversed, primarily because of a decrease in hydrostatic pressure. However, overall, more fluid moves out of the capillary than returns. The excess fluid in the tissue is removed by means of the lymphatic system.

28. A.   Because of the large compliance of veins, a small change in venous blood pressure results in a large increase in venous volume. Sympathetic stimulation decreases venous compliance and volume, thus increasing venous return to the heart and cardiac output (Starling's law of the heart).

29. D.   Blood pressure in the right atrium is about 0 mm Hg and in the aorta about 100 mm Hg. Because of the effect of gravity, the pressure in a standing person increases below the level of the heart (110 mm Hg in an arteriole of the foot) and decreases above the level of the heart (90 mm Hg in an artery of the head).

30. E.   Local direct control results from the relaxation and contraction of the precapillary sphincters and metarterioles (vasomotion). The effect is to produce a cyclic flow of blood and to maintain a normal flow despite changes in arterial blood pressure (autoregulation). The precapillary sphincters relax in response to a buildup of carbon dioxide and other compounds, called vasodilator substances, or a decrease in nutrients such as oxygen in the tissue.

31. D.   The vasomotor center tonically stimulates most blood vessels through sympathetic fibers. The result is a state of partial contraction of smooth muscle in blood vessels, which is called vasomotor tone. The sympathetic neurotransmitter responsible for vasoconstriction is norepinephrine. Epinephrine from the adrenal gland causes vasodilation in skeletal muscle.

32. D.   Mean arterial pressure in the aorta is proportional to heart rate times stroke volume times peripheral resistance.

33. B.   In response to an increase in mean arterial blood pressure, the baroreceptor reflex inhibits the vasomotor center, resulting in decreased sympathetic nervous system activity. Consequently, blood vessels vasodilate and peripheral resistance decreases.

34. B.   The chemoreceptor reflex causes increased sympathetic stimulation of blood vessels, which vasoconstrict. Consequently, vasomotor tone and peripheral resistance increase. The increase in peripheral resistance causes mean arterial pressure to increase.

35. C.   Increased carbon dioxide or decreased pH levels directly stimulate the vasomotor center, resulting in vasoconstriction.

36. B.   The baroreceptor reflexes are the most important for regulating small, sudden changes in blood pressure. The chemoreceptor reflexes and the central nervous system ischemic response are activated when blood pressure decreases significantly, i.e., emergency conditions.

37. D.   Epinephrine increases heart rate and stroke volume. It also causes vasoconstriction of skin and visceral blood vessels and vasodilation of skeletal muscle blood vessels.

38. A.   A decreased blood pressure is detected by the juxtaglomerular apparatuses of the kidneys, which increase their secretion of renin. The increase in renin results in increased formation of angiotensin I, which increases the formation of angiotensin II. Angiotensin II causes vasoconstriction, which helps to raise blood pressure, and stimulates increased aldosterone secretion. The aldosterone helps to increase blood volume, and therefore blood pressure, by decreasing urine production.

39. D.   An increase in blood pressure is detected by baroreceptors, resulting in decreased ADH secretion. Consequently, the kidneys increase urine production, and blood volume decreases. The decreased blood volume helps to reduce blood pressure.

40. C.   Following a decrease in blood pressure, the kidneys retain more salts and water than normal, resulting in an increase in blood volume and therefore blood pressure. A decrease in blood pressure results in less fluid movement into tissues (fluid shift mechanism) and contraction of blood vessel smooth muscle (stress-relaxation response).

41. A.   Decreased sympathetic stimulation of blood vessels in the lower region of the body results in decreased vasomotor tone. Therefore peripheral resistance, and subsequently blood pressure, decrease. Remember that the sympathetic supply to the heart originates in the thoracic region of the spinal cord. Therefore the baroreceptor reflex is not affected by the spinal anesthesia in this case. The decrease in blood pressure activates the baroreceptor reflex, which causes increased sympathetic stimulation of the heart and an increase in heart rate. Marked cutaneous vasoconstriction does not occur because of the effects of the anesthesia on vasomotor tone.

42. A.   The decreased cardiac function causes a drop in blood pressure. Through the baroreceptor reflex, there is increased sympathetic stimulation of the heart and systemic blood vessels, which results in an increased heart rate and peripheral vasoconstriction.

43. D.   Wilber's peripheral resistance increased dramatically following his immersion in the cold water. Stimulation of cold receptors in the skin results in cutaneous vasoconstriction. As the peripheral resistance increased, Wilber's blood pressure increased markedly. His heart rate decreased because of the sudden increase in blood pressure (baroreceptor reflex).

44. B.   Renal arteriosclerosis causes decreased blood flow through the kidneys. Consequently, the kidneys increase their secretion of renin, which causes the formation of angiotensin I. Angiotensin I is converted into angiotensin II, which causes vasoconstriction and significantly increases aldosterone secretion from the adrenal cortex. The aldosterone increases renal sodium chloride and water retention. The expanded blood volume results in hypertension.

45. E.   During exercise the following events occur with respect to the peripheral vasculature: (1) visceral and cutaneous vasoconstriction and (2) skeletal muscle vasodilation. The result is that blood is shunted from the skin and viscera to the skeletal muscle. With the normal increase in blood pressure and the changes that occur in the peripheral vasculature the blood flow to skeletal muscle may increase twentyfold.

 ★ | FINAL CHALLENGES | ★

1. The abdominal and pelvic viscera drain through the hepatic portal system into the liver. A cancer in these viscera (e.g., the colon) can enter the blood and travel to the liver.

2. According to Poiseuille's law, resistance to blood flow is inversely proportional to the radius of the blood vessel raised to the fourth power. Therefore decreasing the radius greatly increases the resistance. As resistance increases, in order to maintain blood flow, pressure must increase.

3. According to the law of Laplace, the greater the diameter of a vessel, the greater the force acting on the vessel wall. Thus as the vein enlarges, the force causing it to enlarge also increases. The law of Laplace also states that the force acting on the vessel wall increases or decreases with the blood pressure within the vessel. Raising the feet helps because it reduces the hydrostatic pressure produced by gravity.

4. Because the astronaut's blood pressure on earth was taken while she was lying horizontal, there would be no hydrostatic effect from gravity. Therefore there would be no difference between the measurement in space and on earth.

5. In the capillaries, blood pressure forces fluid out of the capillaries and, because of the higher concentration of proteins in the blood than in the tissues, osmosis causes fluid to move into the capillaries. Reduced blood flow through the liver increases capillary blood pressure, so more fluid leaves the capillaries. At the same time, reduced blood proteins (albumin) result in less movement of fluid into the capillaries. The result is abdominal swelling.

6. Loss of blood in the donation causes a reduced blood volume, reduced venous return, and thus reduced cardiac output (Starling's law of the heart). Reduced cardiac output results in reduced blood pressure and inadequate delivery of blood to the brain. Emotional reactions could have caused hyperventilation and a "blow off" of carbon dioxide. The reduced blood carbon dioxide directly inhibits the vasomotor center, resulting in decreased peripheral resistance and thus decreased blood pressure. Emotional reactions through the limbic system could have caused decreased sympathetic activity (resulting in decreased peripheral resistance, decreased venous return, and decreased stroke volume) and increased parasympathetic activity (resulting in decreased heart rate). Consequently, cardiac output and blood pressure are reduced.

7. Because the veins are distensible (have a large compliance), when standing the hydrostatic pressure produced by gravity causes them to expand and the volume of the blood in the veins increases. Upon lying down, the hydrostatic pressure produced by gravity is removed, the venous volume becomes smaller, and venous return to the heart increases. Consequently, cardiac output increases (Starling's law of the heart), resulting in increased blood pressure and better delivery of blood to the brain. Breathing into the paper bag raises carbon dioxide levels in the blood, stimulating the chemoreceptor reflex. This increases heart rate, stroke volume, and peripheral resistance, restoring cardiac output and blood pressure. Raising blood carbon dioxide levels also removes the inhibitory effect of low carbon dioxide on the vasomotor center.

8. Standing at attention in the hot sun reduces venous return to the heart for two reasons. First, loss of sweat results in a reduced blood volume. Second, hydrostatic pressure from gravity causes a "pooling" of blood in the distensible veins. The reduced venous return results in reduced cardiac output, lowered blood pressure, and inadequate delivery of blood to the brain. In addition, in the hot sun the blood vessels in the skin vasodilate as the body attempts to lose heat. The vasodilation lowers peripheral resistance and therefore blood pressure.

9. Holding Private Wright in a standing position maintains the hydrostatic pressure from gravity. His friends should lay him horizontal to eliminate the effect of gravity on the blood. Raising the feet slightly above the head is also helpful. When this is done, venous volume decreases and venous return increases. Thus cardiac output and blood pressure increase, providing better delivery of blood to the brain.

10. Damage to the right ventricle could result in inadequate pumping of blood by the right ventricle. Consequently, there is a decreased delivery of blood to the left side of the heart, resulting in a decreased stroke volume. The decreased stroke volume produces a decreased cardiac output and therefore a below-normal blood pressure. Because the right side of the heart does not pump blood effectively, there is a buildup of blood in the venous system. This increases venous capillary pressure; less fluid returns from the tissues into the capillaries, and edema results.

Lymphatic Organs And Immunity

FOCUS: The lymphatic system performs three major functions: removal of excess fluid from tissues, absorption of fats from the digestive tract, and defense against microorganisms and other foreign substances. Lymph organs include the tonsils, which form a protective ring of lymph nodules around the openings of the nasal and oral cavities; the lymph nodes, which filter lymph and produce lymphocytes; the spleen, which removes foreign substances and worn-out red blood cells from the blood; and the thymus, which processes lymphocytes. Innate immunity includes mechanical mechanisms, chemicals (e.g., complement, interferon, lysozyme, histamine), phagocytic cells (neutrophils and macrophages), cells involved with inflammation (basophils, mast cells, eosinophils), and natural killer (NK) cells that kill tumor and virus-infected cells. Adaptive immunity can recognize and remember specific antigens. Major histocompatibility antigens (MHC) and antigen-presenting cells are important in activating lymphocytes; in many cases, costimulation by cytokines or surface proteins is also required for lymphocyte activation. Antibody-mediated immunity involves B cells and the production of antibodies, which activate mechanisms that result in the destruction of the antigen. Antibody-mediated immunity is most effective against extracellular antigens such as bacteria. Cell-mediated immunity involves T cells, which can lyse tumor or virus-infected cells. T cells also produce cytokines, which activate mechanisms that eliminate the antigen.

CONTENT LEARNING ACTIVITY

Lymphatic System

❝ *The lymphatic system includes lymph, lymphocytes, lymph vessels, lymph nodes, tonsils, spleen, and thymus.* **❞**

Match these terms with the correct statement or definition:

Chyle	Lymph
Lacteal	Lymphocyte

_____ 1. Interstitial fluid that is returned to the circulatory system through lymph vessels.

_____ 2. Special lymph vessel in the small intestine that transports fats.

_____ 3. Lymph with a milky appearance because of its high fat content.

_____ 4. Leukocyte found in lymphatic tissue that is capable of destroying microorganisms and foreign substances.

Lymphatic Organs

"Lymphatic organs contain lymphatic tissue, which consists of many lymphocytes and other cells."

A. Match these terms with the correct statement or definition:

Diffuse Lymphatic tissue	Palatine tonsils
Lingual tonsil	Pharyngeal tonsil
Lymph follicles	Peyer's patches
Lymph nodules	Reticular fibers

_____ 1. Fine collagen fibers that form a network to hold lymphocytes and other cells in place.

_____ 2. Lymphatic tissue that has no clear boundary and blends with surrounding tissues.

_____ 3. Lymphatic tissue that is organized into compact, spherical structures found in loose connective tissue of the digestive, respiratory, and urinary systems.

_____ 4. Lymph nodules in the lymph nodes and spleen.

_____ 5. Aggregations of lymph nodules found in the lower half of the small intestine and the appendix.

_____ 6. Large oval, lymphoid masses on each side of the junction between the oral cavity and the pharynx.

_____ 7. Aggregation of lymphatic tissue near the internal opening of the nasal cavity; called the adenoid when enlarged.

B. Match these terms with the correct statement or definition:

Afferent lymph vessel	Lymphatic tissue
Capsule	Medulla
Cortex	Medullary cords
Efferent lymph vessel	Reticular fibers
Germinal center	Trabeculae
Lymph sinus	

_____ 1. Extensions of capsule; form internal skeleton in lymph nodes.

_____ 2. Structures that extend from the capsule and trabeculae to form a fiber network throughout the lymph node.

_____ 3. Areas of the lymph node in which lymphocytes and macrophages are packed around the reticular fibers.

_____ 4. Lymph node space through which reticular fibers extend.

_____ 5. Lymph node outer layer; contains lymph nodules and sinuses.

_____ 6. Structures consisting of branching, irregular strands of diffuse lymphatic tissue; separated by sinuses.

_____ 7. Vessel that enters the lymph node.

_____ 8. Area of rapid lymphocyte division in the lymph node.

C. Match these terms with the correct parts of the diagram labeled in Figure 22-1:

Afferent lymphatic vessel
Capsule
Cortex
Diffuse lymphatic tissue
Efferent lymphatic vessel

Germinal center
Lymph sinus
Medulla
Medullary cord
Trabecula

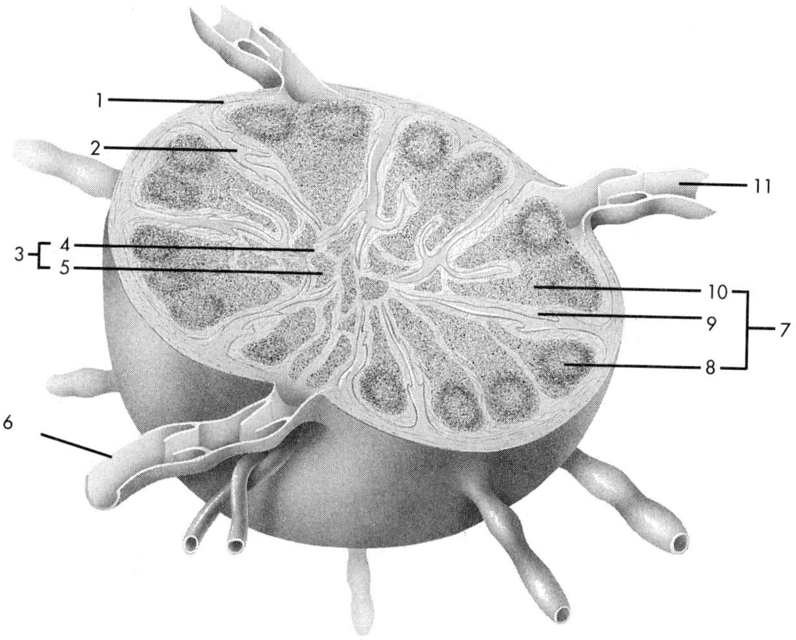

1. _____

2. _____

3. _____

4. _____

5. _____

6. _____

7. _____

8. _____

9. _____

10. _____

11. _____

Figure 22-1

D. Match these terms with the correct statement as it applies to the spleen:

Capsule
Hilum
Periarterial sheath
Red pulp

Splenic cords
Trabeculae
Venous sinuses
White pulp

_____ 1. Fibrous outer covering of the spleen.

_____ 2. Location where splenic arteries enter the spleen.

_____ 3. Diffuse lymphatic tissue around the arteries leaving the trabeculae.

_____ 4. Periarterial sheath and lymph nodules associated with the arteries.

_____ 5. Consists of splenic cords and venous sinuses.

_____ 6. Network of reticular fibers filled with blood cells that have come from the capillaries.

_____ 7. Enlarged blood vessels that separate the splenic cords.

E. Match these terms with the
correct parts of the diagram
labeled in Figure 22-2:

Capsule
Lymph nodule
Periarterial sheath

Red pulp
Trabecula
White pulp

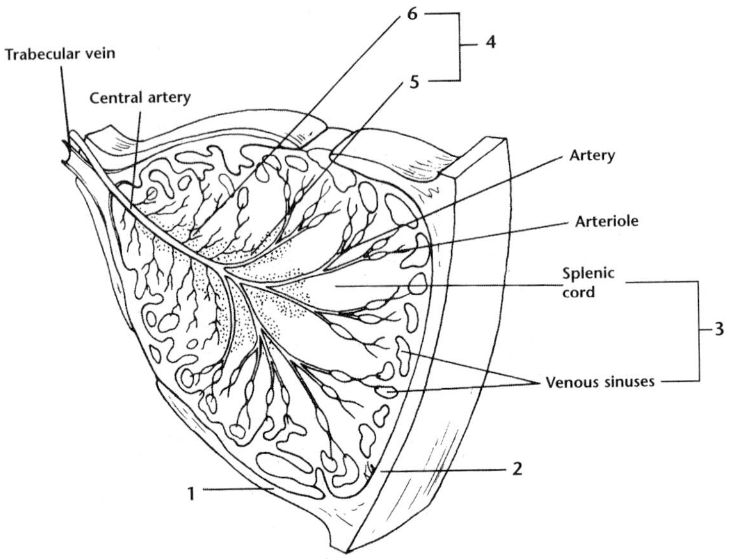

1. _____

2. _____

3. _____

4. _____

5. _____

6. _____

Trabecular vein
Central artery
Artery
Arteriole
Splenic cord
Venous sinuses

Figure 22-2

F. Match these terms with the
correct statement as it
applies to the thymus:

Blood-thymic barrier
Cortex
Lobules

Medulla
Thymic corpuscles

_____

_____

_____

_____

1. Subdivisions of the thymus formed by the inward extension of
trabeculae.

2. Outer portion of the thymus, which contains lymphocytes.

3. Relatively lymphocyte-free core of each lobule of the thymus.

4. Layer of reticular cells that prevents large molecules from
leaving the capillaries in the thymus.

G. Match these lymph organs with
the correct description or function:

Lymph node
Spleen

Thymus

_____

_____

_____

_____

1. Filters lymph and provides a source of lymphocytes; the only
structure that has both afferent and efferent lymph vessels.

2. Detects and responds to foreign substances in the blood,
destroys worn-out red blood cells, and acts as a reservoir for red
blood cells.

3. Produces lymphocytes that move to other lymph tissue where
they can respond to foreign substances.

4. Found primarily in the superior mediastinum.

Immunity

66*Immunity is the ability to resist damage from foreign substances.*99

Match these terms with
the correct description or function:

Adaptive immunity
Innate immunity

_____   1. For this type of immunity, the ability to destroy foreign
                                organisms does not improve each time the body is exposed to
                                them.

_____   2. This type of immunity involves specificity and memory.

Innate Immunity

66*In innate immunity, each time the body is exposed to a microorganism or substance the response*99
is the same.

Using the terms provided, complete these statements:

Cells   Mechanical
Chemicals   mechanisms
                        Inflammation

Innate immunity consists of several important components.
The skin and mucous membrane are _(1)_ that prevent the
entry of microorganisms or other foreign substances, whereas _(2)_ are substances that directly kill microorganisms or
activate other mechanisms that result in the destruction of
microorganisms. Phagocytosis and the production of chemicals is achieved by _(3)_ . _(4)_ mobilizes the immune system and isolates microorganisms until they are destroyed.

1. _____

2. _____

3. _____

4. _____

Mechanical Mechanisms

66*Mechanical mechanisms prevent the entry of microbes into the body or physically remove*99
them from body surfaces.

Using the terms provided, complete these statements:

Chemicals   Skin
Ciliated   Tears
Respiratory   Urine
Saliva

The _(1)_ and mucous membranes form barriers to the entry
of microorganisms and _(2)_ into the body. In addition,
microorganisms and other substances are washed from the
eyes by _(3)_ , from the mouth by _(4)_ , and from the urinary
tract by _(5)_ . In the respiratory tract, mucous membranes
are _(6)_ , and microbes trapped in mucus are swept to the
back of the throat and swallowed. Coughing and sneezing
also remove microorganisms from the _(7)_ tract.

1. _____

2. _____

3. _____

4. _____

5. _____

6. _____

7. _____

Chemical Mediators

❝Chemical mediators are molecules involved with the development of immunity.**❞**

A. Using the terms provided, complete these statements:

Alternate pathway	Complement
Classical pathway	Complement
Inflammation	cascade

_(1)_ is a group of about 20 proteins that make up approximately 10% of the globulin portion of serum. These
proteins become activated in the _(2)_ , a series of reactions
in which each component of the series activates the next
component. These reactions begin through one of two
pathways. The _(3)_ is part of innate immunity, whereas the
_(4)_ is part of adaptive immunity. The alternate pathway is
initiated when complement protein C3 becomes spontaneously activated and combines with foreign substances
Once activated, the complement may form a hole in
bacterial cell membrane, promote phagocytosis, or _(5)_ .

1. _____
2. _____
3. _____
4. _____
5. _____

B. Using the terms provided, complete these statements:

Antiviral	Nucleic acids
Does	Specific
Does not	Viral

Interferons are proteins that protect the body against
_(1)_ infection and perhaps some forms of cancer. When a
virus infects a cell, viral _(2)_ take control of directing the
activities of the cell. Viruses and other substances can also
stimulate cells to produce interferon. Interferon _(3)_ protect the cell that produces it and _(4)_ act directly against
viruses. Instead, interferon binds to the surface of other
cells and causes the production of _(5)_ proteins, which
prevent the production of new viral nucleic acids and
proteins. Interferon viral resistance is not _(6)_ ; the same
interferon acts against many different viruses.

1. _____
2. _____
3. _____
4. _____
5. _____
6. _____

Cells

❝Leukocytes and cells derived from them are the most important cellular components of the immune system.**❞**

A. Match these terms with the
correct statement or definition:

Chemotactic factors	Leukocytes
Chemotaxis	Phagocytosis

_____ 1. Parts of microbes or chemicals that are released by tissue cells
and act as chemical signals to attract leukocytes.

_____ 2. Ability to detect and move toward chemotactic factors.

_____ 3. Endocytosis and destruction of particles by cells.

B. Match these terms with the correct statement or definition:

    Macrophages   Mononuclear phagocytic system
    Neutrophils

_____ 1. Usually the first cells to enter infected tissues; the primary constituent of pus.

_____ 2. Cells derived from monocytes that leave the blood and enlarge and increase their numbers of lysosomes and mitochondria.

_____ 3. Responsible for most phagocytosis in the late stages of infection.

_____ 4. Leukocytes that reside beneath free surfaces or in sinuses where they trap and destroy microbes.

_____ 5. The collective name for monocytes and macrophages.

C. Match these types of leukocytes with the correct function or description:

    Basophils   Mast cells
    Eosinophils   Natural killer cells

_____ 1. Motile leukocytes that become activated and secrete chemicals that promote inflammation or cause smooth muscle contraction.

_____ 2. Nonmotile leukocytes located in connective tissue; when activated produce chemicals that promote inflammation or cause smooth muscle contraction.

_____ 3. Leukocytes that produce enzymes that inhibit inflammation, and secrete enzymes that kill some parasites.

_____ 4. Lymphocytes that kill certain tumor and virus-infected cells; they exhibit no memory response.

Inflammatory Response

❝_The inflammatory response is a complex sequence of events._**❞**

A. Using the terms provided, complete these statements:

Chemical mediators   Fibrin
Chemotactic   Vascular
Complement   permeability
                     Vasodilation

1. _____
2. _____
3. _____
4. _____
5. _____
6. _____

Inflammation results when a microbe or damage to tissues causes the release or activation of __(1)__ such as histamine, prostaglandins, leukotrienes, and complement. Mediators cause __(2)__, and this increased blood flow brings phagocytes and other leukocytes to the area. Some of the mediators are __(3)__ factors that stimulate phagocytes to leave the blood. Mediators also increase __(4)__, allowing fibrin and complement to enter the tissue. __(5)__ prevents the spread of infection by walling off the infected area. __(6)__ enhances the inflammatory response and attracts additional phagocytes.

B. Using the terms provided, complete these statements:

Blood flow
Local inflammation
Loss of function
Neutrophils
Pain

Pyrogens
Systemic
  inflammation
Vascular
  permeability

1. _____

2. _____

3. _____

4. _____

5. _____

6. _____

7. _____

8. _____

An inflammatory response confined to a specific area of the body is a __(1)__. Symptoms of local inflammation include redness, heat, and swelling, which result from increased __(2)__ and increased vascular permeability. Other symptoms are __(3)__, resulting from swelling and chemicals acting on nerve endings, and __(4)__ caused by tissue destruction and swelling. An inflammatory response that occurs in many parts of the body is a __(5)__. In addition to local symptoms, additional features may be present in systemic inflammation. These include production and release of large numbers of __(6)__ that promote phagocytosis, release of __(7)__ by microorganisms or leukocytes to produce fever, and in severe cases a great increase in __(8)__, which causes a large amount of fluid loss from the blood, and may lead to shock and death.

Adaptive Immunity

66 *Adaptive immunity involves the ability to recognize, respond to, and remember a particular substance.* 99

A. Match these terms with the correct statement or definition:

Antigen
Foreign antigen

Hapten
Self-antigen

_____

_____

_____

1. General term for substance that stimulates adaptive immunity.

2. Small molecule capable of combining with a larger molecule to stimulate adaptive immunity.

3. Molecule produced by the body that stimulates adaptive immunity; stimulates autoimmune disease.

B. Match these terms with the correct statement or definition:

Antibody-mediated immunity
B cells
Cell-mediated immunity

Effector T cells
Regulatory T cells

_____

_____

_____

_____

1. Lymphocytes that produce antibodies.

2. Immunity produced by plasma antibodies; humoral immunity.

3. Lymphocytes that produce the cell-mediated immunity; include cytotoxic T cells and delayed hypersensitivity T cells.

4. Lymphocytes that control the activities of cell-mediated or antibody-mediated immunity; include helper T cells and suppressor T cells.

Origin and Development of Lymphocytes

"All blood cells, including lymphocytes, are derived from stem cells in red bone marrow.**"**

Match these terms with the correct statement or definition:

Clones
Positive selection
Negative selection

Red bone marrow
Thymus

_____ 1. Location where hormones are produced for T cell maturation.

_____ 2. Location where pre-B cells are processed into B cells.

_____ 3. Small groups of identical lymphocytes produced during embryonic development.

_____ 4. Elimination or suppression of clones acting against self-antigens.

☞ T cell receptors have a single variable region, whereas B cell receptors have two identical variable regions.

Activation of Lymphocytes

"To produce adaptive immunity, lymphocytes must be activated by an antigen.**"**

A. Match these terms with the correct statement or definition:

Antigen receptors
Antigenic determinants

_____ 1. Specific regions of a given antigen that activate a lymphocyte; an epitope.

_____ 2. Proteins on the surface of lymphocytes that combine with antigenic determinants.

B. Match these terms with the correct statement or definition:

Antigen-presenting cells   MHC class II antigen
Major histocompatibility   MHC restricted
    complex (MHC) antigens
MHC class I antigen

_____ 1. General term for cell surface glycoproteins that are involved in lymphocyte activation.

_____ 2. Antigen that combines with antigens produced inside the cell; the combined antigen complex is then displayed on the cell surface.

_____ 3. Process in which the MHC I/antigen complex is required to activate a T cell.

_____ 4. Cells specialized to take in foreign antigens, process the antigens, and display the antigens to other immune system cells.

_____ 5. B cells, macrophages, monocytes, and dendritic cells.

_____ 6. Antigen that combines with foreign antigen taken in by antigen-presenting cells; found in antigen complex that is displayed on the cell surface.

_____ 7. An antigen displayed with this type of MHC antigen results in the destruction of the displaying cell.

_____ 8. An antigen displayed with this type of MHC antigen results in the activation of other immune system cells.

☞ Dendritic cells are large, motile cells with long cytoplasmic extensions scattered throughout most tissues; dendritic cells in the skin are often called Langerhans cells.

C. Match these terms with the correct statement or definition:

CD4 or CD8   Cytokines
Costimulation

_____ 1. Proteins or peptides secreted by one cell as a regulator of neighboring cells; also called lymphokines.

_____ 2. A response from a B cell or T cell that requires both cytokines and a MHC II/antigen complex.

_____ 3. Surface proteins involved in costimulation; helps connect T cells to other cells.

Inhibition of Lymphocytes

❝_Both the quantity and quality of the immune system's response are regulated._**❞**

Using the terms provided, complete these statements:

Activation   Self-reactive
Anergy   Suppressor T cells
Costimulation   Tolerance
Self-antigens

A general term for a state of unresponsiveness of lymphocytes to a specific antigen is (1) . The most important
function of tolerance is to prevent the immune system from
responding to (2) . One way tolerance can be induced is
deletion of (3) lymphocytes during prenatal development
and after birth. Also, blocking, altering, or deleting an
antigen receptor prevents (4) of lymphocytes. A condition
of inactivity in which a B cell or T cell does not respond to
an antigen is called (5) ; this condition occurs when a
MHC/antigen complex binds to an antigen receptor and
there is no (6) . In addition, (7) can prevent the activity
of helper T cells, B cells and effector T cells.

1. _____
2. _____
3. _____
4. _____
5. _____
6. _____
7. _____

Antibody-Mediated Immunity

❝_Exposure of the body to an antigen can lead to activation of B cells and to the production of antibodies._**❞**
Antibody-mediated immunity is most effective against extracellular microorganisms.

A. Match these terms with the
   correct statement or definition:

Constant region   Opsonins
Gamma globulins   Variable region
Immunoglobulins

_____ 1. Other terms used for antibodies.

_____
_____ 2. Part of the antibody that combines with the antigenic
                            determinant of the antigen; determines the specificity of the
                            antibody.

_____ 3. Part of the antibody that is responsible for the ability to activate
                            complement, attach the antibody to cells such as macrophages,
                            basophils, mast cells, and eosinophils, or to initiate
                            inflammation.

_____ 4. Substances that make an antigen more susceptible to
                            phagocytosis.

B. Match these terms with the correct parts of the diagram labeled in Figure 22-3:

Antigen binding site
Complement binding site
Constant region
Heavy chain

Light chain
Macrophage binding site
Variable region

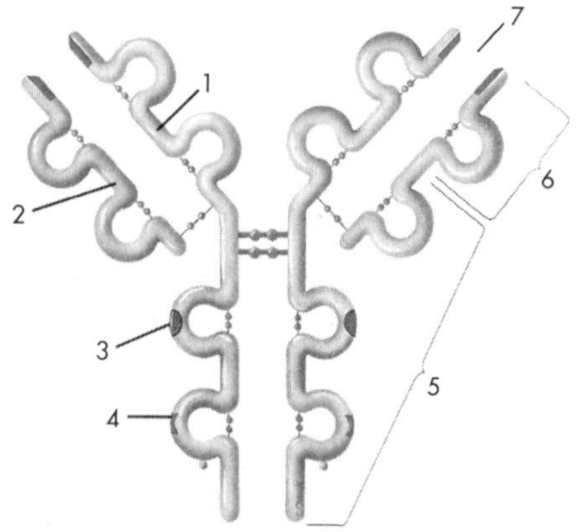

1. _____
2. _____
3. _____
4. _____
5. _____
6. _____

Figure 22-3

C. Match these terms with the correct statement or definition:

Memory B cells
Plasma cells

Primary response
Secondary (memory) response

_____ 1. Cell division, cell differentiation, and antibody production from the first exposure of a B cell to an antigen.

_____ 2. Derived from activated B cells, these enlarged lymphocytes produce antibodies.

_____ 3. Cells that divide and produce plasma and memory cells when exposed to a previously encountered antigen.

_____ 4. Antibody-mediated response that is the fastest and produces the most antibodies.

Cell-Mediated Immunity

❝Cell-mediated immunity is a function of T cells and is most effective against intracellular microorganisms such as viruses, fungi, intracellular bacteria, and parasites.❞

Using the terms provided, complete these statements:

Allergic
Cytokines
Effector T cells
Inflammation

Lyse
Memory T cells
Secondary response

1. _____
2. _____
3. _____
4. _____
5. _____
6. _____

Once activated, T cells undergo a series of divisions and produce effector T cells and (1) . Memory T cells can provide a (2) and long-lasting immunity. Cytotoxic T cells have two main effects: they (3) cells and they release (4) that activate additional components of the immune system. Delayed hypersensitivity T cells respond to antigens by releasing cytokines; consequently, they promote phagocytosis and (5) , especially in (6) reactions.

Immunotherapy treats disease by stimulating or inhibiting the immune system.

Acquired Immunity

❝_There are four ways to acquire specific immunity._**❞**

Match these terms with the correct statement or definition:

Active artificial immunity   Passive artificial immunity
Active natural immunity   Passive natural immunity
Immunization

_____

1. The general term that includes deliberate introduction of an antigen or antibody into the body.

_____

2. Natural exposure to an antigen that causes the body to mount an immune system response against the antigen.

_____

3. Deliberate introduction of an antigen into the body; the type of immunity produced by vaccination.

_____

4. Type of immunity produced by transfer of antibodies from mother to child.

_____

5. Type of immunity produced by antivenins, antisera, and antitoxins.

QUICK RECALL

1. List three basic functions of the lymphatic system.

2. List the functions of the lymph nodes, spleen, and thymus.

3. Name the two ways that complement is activated.

4. List the steps that occur when a cell is protected against a viral infection by interferon.

441

5. List the two cell types responsible for most of the phagocytosis in the body.

6. Name the function that basophils and mast cells have in common. Name the cell type that counteracts this function.

7. Name the cell type that produces antibodies and the cell type that produces most cytokines.

8. Explain the function of MHC class I and class II antigens.

9. List four types involved of antigen-presenting cells.

10. Contrast costimulation and anergy.

11. Name three ways to achieve immunologic tolerance.

12. Give the two basic ways that antibody-mediated immunity (antibodies) act against an antigen.

13. Name the two cell types responsible for the secondary response.

14. Give the two basic effects of cytotoxic cells.

MASTERY LEARNING ACTIVITY

Place the letter corresponding to the correct answer in the space provided.

_____ 1. The lymphatic system
   a. removes excess fluid from tissues.
   b. absorbs fats from the digestive tract.
   c. defends the body against microorganisms and other foreign substances.
   d. all of the above

_____ 2. The tonsils
   a. consist primarily of diffuse lymphatic tissue.
   b. consist of three groups.
   c. are located in the nasal cavity.
   d. increase in size in adults.

_____ 3. Lymph nodes
   a. filter lymph.
   b. produce lymphocytes.
   c. contain a network of reticular fibers.
   d. all of the above

_____ 4. The spleen
   a. contains a dense accumulation of lymphocytes called red pulp.
   b. has red pulp tissue surrounding arteries.
   c. destroys worn-out red blood cells.
   d. is surrounded by trabeculae.

_____ 5. The thymus
   a. increases in size in adults.
   b. produces lymphocytes that move to other lymphatic tissue.
   c. responds to foreign substances in the blood.
   d. all of the above

_____ 6. Which of these is an example of innate immunity?
   a. Tears and saliva wash away microorganisms.
   b. Basophils and mast cells release histamine and leukotrienes.
   c. Neutrophils phagocytize a microorganism.
   d. The complement cascade is activated.
   e. all of the above

_____ 7. Neutrophils
   a. enlarge to become macrophages.
   b. account for most of the dead cells in pus.
   c. are usually the last cell type to enter infected tissues.
   d. are usually located in lymph and blood sinuses.

_____ 8. Macrophages
   a. are large phagocytic cells that outlive neutrophils.
   b. develop from mast cells.
   c. often die after a single phagocytic event.
   d. have the same function as eosinophils.
   e. all of the above

_____ 9. Which of these cells is most important in the release of histamine, which promotes inflammation?
   a. monocyte
   b. macrophage
   c. eosinophil
   d. mast cell

_____ 10. Which of these is a symptom of systemic inflammation?
   a. large numbers of neutrophils are produced and released
   b. pyrogens stimulate fever production
   c. greatly increased vascular permeability
   d. all of the above

_____ 11. During the inflammatory response
   a. histamine and other chemical mediators are released.
   b. chemotaxis of phagocytes occurs.
   c. fibrin enters tissue from the blood.
   d. blood vessels vasodilate.
   e. all of the above

_____ 12. Antigens
   a. are foreign substances introduced into the body.
   b. are molecules produced by the body.
   c. stimulate a specific immune system response.
   d. all of the above

_____ 13. B cells
   a. are processed in the thymus.
   b. originate in red bone marrow.
   c. once released into the blood stay in the blood.
   d. all of the above

_____ 14. MHC antigens
   a. are glycoproteins.
   b. attach to the plasma membrane.
   c. have a variable region that can bind to foreign and self-antigens.
   d. may for a complex that activates T cells.
   e. all of the above

_____ 15. The antigen that forms a complex with antigens produced inside a cell is a
   a. MHC class I antigen.
   b. MHC class II antigen.
   c. foreign antigen.
   d. self-antigen.
   e. a and b

_____ 16. Antigen-presenting cells
   a. take in foreign antigens.
   b. process antigens.
   c. use MHC class II antigens to display the antigen.
   d. stimulate other immune system cells.
   e. all of the above

_____ 17. The most important function of tolerance is to
   a. increase lymphocyte activity.
   b. increase complement activation.
   c. prevent the immune system from responding to self-antigens.
   d. prevent excessive immune system response to foreign system.
   e. a and b

_____ 18. Variable amino acid sequences on the arms of the antibody molecule
   a. make the antibody specific for a given antigen.
   b. enable the antibody to activate complement.
   c. enable the antibody to attach to basophils and mast cells.
   d. all of the above

_____ 19. Antibodies
   a. prevent antigens from binding together.
   b. promote phagocytosis.
   c. inhibit inflammation.
   d. block complement activation.

_____ 20. The secondary antibody response
   a. is slower than the primary response.
   b. produces less antibodies than the primary response.
   c. prevents the appearance of disease symptoms.
   d. a and b

_____ 21. The type of lymphocyte that is responsible for the secondary antibody response?
   a. memory B cell
   b. B cell
   c. T cell
   d. helper T cell

_____ 22. Antibody-mediated immunity
   a. works best against intracellular antigens.
   b. is more important than cell-mediated immunity in graft rejection.
   c. cannot be transferred from one person to another person.
   d. is responsible for immediate hypersensitivity reactions.

_____ 23. The activation of T cells can result in the
   a. lysis of virus infected cells.
   b. production of cytokines.
   c. production of memory T cells.
   d. all of the above

_____ 24. Cytokines
   a. inhibit T helper cells.
   b. activate macrophages.
   c. block interferon production.
   d. isolate the antigen from neutrophils.

_____ 25. Delayed hypersensitivity is
   a. caused by activation of B cells.
   b. a result of antibodies reacting with an allergen.
   c. mediated by T cells.
   d. all of the above

FINAL CHALLENGES

Use a separate sheet of paper to complete this section.

1. Ivy Mann developed a poison ivy rash after a camping trip. His doctor prescribed a cortisone ointment to relieve the inflammation. A few weeks later Ivy scraped his elbow, which became inflamed. Because he had some of the cortisone ointment left over, he applied it to the scrape. Explain why the ointment was or was not a good idea for the poison ivy and for the scrape.

2. Explain why booster shots are given.

3. A young woman has just had her ears pierced. To her dismay she finds that, when she wears inexpensive (but tasteful) jewelry, by the end of the day there is an inflammatory (allergic) reaction to the metal in the jewelry. Does this result from antibodies or cytokines?

4. A child appears to be healthy until about 9 months of age. Then he develops severe bacterial infections, one after another. Fortunately the infections are successfully treated with antibiotics. When infected with the measles and other viral diseases the child recovers without unusual difficulty. Explain the different immune system response to these infections. Why did it take so long for this disorder to become apparent (hint: IgG)?

5. A baby was born with severe combined immunodeficiency disease (SCID). In an attempt to save the patient's life a bone marrow transplant was performed. Explain why the bone marrow transplant might help the patient. Unfortunately there was a graft rejection, and the patient died. Explain what happened.

6. A patient is taking a drug to prevent graft rejection. Would you expect the patient to be more or less likely to develop cancer? Explain.

7. Sometimes antilymphocyte serum is given to prevent rejection of an organ transplant. The serum consists of antibodies that can bind to the patient's lymphocytes. Explain why the serum prevents graft rejection and why it could produce harmful side effects.

445

Why don't anteaters ever
catch a cold?

Because they're full of little
antie-bodies.

ANSWERS TO CHAPTER 22

Lymphatic System
1. Lymph; 2. Lacteal; 3. Chyle; 4. Lymphocyte

Lymphatic Organs
- A. 1. Reticular fibers; 2. Diffuse lymphatic tissue;
  3. Lymph nodules; 4. Lymph follicles;
  5. Peyer's patches; 6. Palatine tonsils;
  7. Pharyngeal tonsils
- B. 1. Trabeculae; 2. Reticular fibers;
  3. Lymphatic tissue; 4. Lymph sinus; 5. Cortex;
  6. Medullary cords; 7. Afferent lymph vessel;
  8. Germinal center
- C. 1. Capsule; 2. Trabecula; 3. Medulla; 4. Lymph
  sinus; 5. Medullary cord; 6. Efferent lymphatic
  vessel; 7. Cortex; 8. Germinal center; 9. Lymph
  sinus; 10. Diffuse lymphatic tissue;
  11. Afferent lymphatic vessel
- D. 1. Capsule; 2. Hilum; 3. Periarterial sheath;
  4. White pulp; 5. Red pulp; 6. Splenic cords;
  7. Venous sinuses
- E. 1. Trabecula; 2. Capsule; 3. Red pulp; 4. White
  pulp; 5. Periarterial sheath; 6. Lymph nodule
- F. 1. Lobules; 2. Cortex; 3. Medulla; 4. Blood-
  thymic barrier
- G. 1. Lymph node; 2. Spleen; 3. Thymus;
  4. Thymus

Immunity
1. Innate immunity; 2. Adaptive immunity

Innate Immunity
1. Mechanical mechanisms; 2. Chemicals; 3. Cells;
4. Inflammation

Mechanical Mechanisms
1. Skin; 2. Chemicals; 3. Tears; 4. Saliva; 5. Urine;
6. Ciliated; 7. Respiratory

Chemical Mediators
- A. 1. Complement; 2. Complement cascade;
  3. Alternate pathway; 4. Classical pathway;
  5. Inflammation
- B. 1. Viral; 2. Nucleic acids; 3. Does not; 4. Does
  not; 5. Antiviral; 6. Specific

Cells
- A. 1. Chemotactic factors; 2. Chemotaxis;
  3. Phagocytosis
- B. 1. Neutrophils; 2. Macrophages;
  3. Macrophages; 4. Macrophages;
  Mononuclear phagocytic system
- C. 1. Basophils; 2. Mast cells; 3. Eosinophils;
  4. Natural killer cells

Inflammatory Response
- A. 1. Chemical mediators; 2. Vasodilation;
  3. Chemotactic; 4. Vascular permeability;
  5. Fibrin; 6. Complement
- B. 1. Local inflammation; 2. Blood flow; 3. Pain;
  4. Loss of function; 5. Systemic inflammation;
  6. Neutrophils; 7. Pyrogens; 8. Vascular
  permeability

Adaptive Immunity
- A. 1. Antigen; 2. Hapten; 3. Self-antigen
- B. 1. B cells; 2. Antibody-mediated immunity;
  3. Effector T cells; 4. Regulatory T cells

Origin and Development of Lymphocytes
1. Thymus; 2. Red bone marrow; 3. Clones;
4. Negative selection

Activation of Lymphocytes
- A. 1. Antigenic determinants; 2. Antigen
  receptors
- B. 1. Major histocompatibility complex (MHC)
  antigens; 2. MHC class I antigen; 3. MHC
  restricted; 4. Antigen-presenting cells;
  5. Antigen-presenting cells; 6. MHC class II
  antigens; 7. MHC class I antigen; 8. MHC class
  II antigen
- C. 1. Cytokines; 2. Costimulation; 3. CD4 or CD8

Inhibition of Lymphocytes
1. Tolerance; 2. Self-antigens; 3. Self-reactive;
4. Activation; 5. Anergy; 6. Costimulation;
7. Suppressor T cells

Antibody-Mediated Immunity
- A. 1. Gamma globulins or immunoglobulins;
  2. Variable region; 3. Constant region;
  4. Opsonins
- B. 1. Heavy chain; 2. Light chain;
  3. Complement binding site; 4. Macrophage
  binding site; 5. Constant region; 6. Variable
  region; 7. Antigen-binding site
- C. 1. Primary response; 2. Plasma cells;
  3. Memory B cells; 4. Secondary (memory)
  response

Cell-Mediated Immunity
1. Memory T cells; 2. Secondary response; 3. Lyse;
4. Cytokines; 5. Inflammation; 6. Allergic

Acquired Immunity
1. Immunization; 2. Active natural immunity;
3. Active artificial immunity; 4. Passive natural
immunity; 5. Passive artificial immunity

1. Maintains fluid balance, absorbs fat and other substances from the intestines, and defends against microorganisms and other foreign substances
2. Lymph nodes: filter lymph and remove substances by phagocytosis, stimulate and release lymphocytes.
   Spleen: foreign substances stimulate lymphocytes in white pulp, foreign substances are phagocytized in the red pulp
   Thymus: processes T cells, and produces and releases T cells
3. Alternate pathway: when activated, C3 protein combines with a foreign substance; classical pathway: by antibody activation
4. Interferon is produced by a virally infected cell, moves to other cells, and protects them from viral infection.
5. Neutrophils and macrophages
6. Basophils and mast cells release inflammatory chemicals and cause smooth muscle contraction. Eosinophils release enzymes that reduce inflammation.
7. B cells (plasma cells) produce antibodies. T cells produce most cytokines.
8. MHC class I antigen: combines with antigens produced inside a cell and the combined antigen complex is displayed on the cell surface; results in destruction of the displaying cell. MHC class II antigen: combines with foreign antigen taken in by antigen-presenting cells and forms an antigen complex that is displayed on the cell surface; results in activation of other immune system cells.
9. B cells, macrophages, monocytes, and dendritic cells
10. Costimulation is activation of a B cell or T cell that requires both cytokines and a MHC II/antigen complex; anergy is a condition of inactivity in which a B cell or T cell does not respond to an antigen. Anergy develops when an MHC/antigen complex binds to an antigen receptor but there is no costimulation.
11. Deletion of self-reactive lymphocytes, preventing activation of lymphocytes, and suppression of lymphocytes by suppressor T cells
12. Direct effects: inactivate antigen or bind antigens together. Indirect effects (activates other mechanisms): act as opsonins, activate complement, and increase inflammation
13. Memory B cells and memory T cells
14. Effector T cells lyse cells and produce cytokines.

1. D. The lymphatic system is involved in tissue fluid balance, fat absorption, and defense against microorganism.

2. B. There are three groups of tonsils: the palatine, pharyngeal, and lingual tonsils. The tonsils consist of large groups of lymph nodules, and they decrease in size in adults and may disappear. The tonsils are found in the back of throat (i.e., oropharynx and nasopharynx).

3. D. Lymph nodes contain a network of reticular fibers. Within lymph sinuses the reticular fiber network and phagocytic cells filter the lymph to remove microorganisms and other foreign substances. Within the lymphatic tissue of the lymph node, lymphocytes are produced in germinal centers.

4. C. The spleen detects and responds to foreign substances in the blood, destroys worn-out red blood cells, and acts as a reservoir for red blood cells. The spleen is surrounded by a capsule and subdivided by trabeculae. White pulp, dense accumulations of lymphocytes, is found surrounding arteries. Red pulp is associated with venous sinuses and consists of a network of reticular fibers filled with blood cells and macrophages.

5. B. The thymus produces lymphocytes that move to other lymph tissues. The blood-thymic barrier prevents the thymus from responding to foreign substances in the blood. With increasing age in the adult, the thymus decreases in size.

6. E. All of the answers are examples of innate immunity.

7. B. Neutrophils are phagocytic cells that usually die after a single phagocytic event. They are usually the first cell type to enter infected tissues, and they account for most of the dead cells in pus. Monocytes enter tissues to become macrophages that can be found in lymph and blood sinuses and beneath the free surfaces of the body (e.g., skin, mucous membranes, serous membranes). Macrophages appear in infected tissue after neutrophils and are responsible for the cleanup of neutrophils and other cellular debris.

8. A. Macrophages are large phagocytic cells that outlive neutrophils. They develop from monocytes and can ingest more and larger particles than neutrophils, which often die after a single phagocytic event. Macrophages help clean up cellular debris, and also produce chemicals such as interferon, prostaglandins, and complement that enhance the immune system response. Eosinophils produce enzymes that reduce the inflammatory response. Mast cells are nonmotile cells in connective tissue that release mediators of inflammation.

9. D. Mast cells and basophils release histamine, which promotes inflammation. Eosinophils release enzymes that break down histamine and inhibit inflammation.

10. D. All are symptoms of systemic inflammation.

11. E. Damage to tissue causes the release of histamine and other chemical mediators that cause vasodilation, attract phagocytes (chemotaxis), and increase vascular permeability (allowing the entry of fibrin, which walls off infected areas).

12. D. Antigens are large molecular weight molecules that stimulate a specific immune system response. They can be foreign substances (foreign antigens) or molecules produced by the body (self-antigens).

13. B. B cells originate and are processed in red bone marrow. T cells originate in red bone marrow and are processed in the thymus. Once B and T cells are released into the blood, they circulate between the blood and lymphatic tissues.

14. E. MHC antigens are glycoproteins attached to the cell membrane; they have a variable region that can bind to foreign or self-antigens. The MHC/antigen complex can activate B cells or T cells.

15. A. The antigen that combines with antigens produced inside a cell is a MHC class I antigen. MHC class I antigens may combine with either foreign antigens or self-antigens produced inside a cell. MHC class II antigens are found on antigen-presenting cells; they combine with foreign antigens.

16. E. Antigen-presenting cells are specialized to take in foreign antigens, process the antigens, and to use MHC class II antigens to display the antigen to other immune system cells. This stimulates other immune system cells to respond to the antigen.

17. C. Tolerance is a state of unresponsiveness to a specific antigen. Its most important function is to prevent the immune system from responding to self-antigens. Increasing lymphocyte activity and increasing complement activation both increase the immune system response.

18. A. The variable region, a sequence of amino acids on the antibody molecule, makes the antibody specific for a given antigen. The constant region is involved with the activation of complement and enables the antibody to attach to various cells.

19. B. Antibodies promote phagocytosis, activate the complement cascade, stimulate inflammation, and bind antigens together.

20. C. The secondary response is more rapid and produces more antibodies than the primary response. Thus the secondary response effectively destroys the antigen and prevents the appearance of disease symptoms.

21. A. During the primary response memory B cells are formed. They are responsible for the secondary response.

22. D. Antibody-mediated immunity is responsible for immediate hypersensitivity reactions, and it works best against extracellular antigens. It can be transferred to another person through antibodies in plasma. Cell-mediated immunity is more important in graft rejection and control of intracellular antigens.

23. D. When activated, T cells produce cytotoxic T cells that can directly lyse target cells and can produce cytokines. Also formed are memory T cells, which are responsible for a secondary response (rapid, strong) to the antigen.

24. **B.** Cytokines promote phagocytosis and inflammation. They attract macrophages, neutrophils, and other cells and convert ordinary macrophages into more effective phagocytic cells. Interferon is one type of cytokine. Cytokines do not inhibit T helper cells or block interferon production.

25. **C.** Delayed hypersensitivity is a type of cell-mediated immunity that is mediated by delayed hypersensitivity T cells. Immediate hypersensitivity is caused by antibodies (produced by B cells) reacting to an allergen.

 FINAL CHALLENGES

1. The ointment was a good idea for the poison ivy, which caused a delayed hypersensitivity reaction, i.e., too much inflammation. For the scrape it is a bad idea, because a normal amount of inflammation is beneficial and helps to fight infection in the scrape.

2. The first shot produces a primary response. The booster shot produces a secondary response with a higher production of antibodies. This provides greater, longer-lasting immunity.

3. Because antibodies and cytokines both produce inflammation, the fact that the metal in the jewelry resulted in inflammation is not enough information to answer the question. However, the fact that it took most of the day (many hours) to develop the reaction would indicate a delayed hypersensitivity reaction and therefore cytokines.

4. The child's antibody-mediated system is not functioning properly, whereas the child's cell-mediated system is functioning properly. This explains the susceptibility to extracellular bacterial infections and the resistance to intracellular viral infections. It took so long to become apparent because IgG from the mother crossed the placenta and provided the infant with protection. Once these antibodies were degraded, the infant began to get sick.

5. Bone marrow is the source of lymphocytes. If a bone transplant were successful, the baby would have started producing lymphocytes and would have had a functioning immune system. In this case there was a graft versus host rejection in which the transplanted lymphocytes in the red marrow mounted an immune system attack against the baby's tissues, resulting in death.

6. The drug is probably suppressing cell-mediated immunity, which is responsible for rejecting grafts and is involved in tumor control. One might expect a higher incidence of cancer in such a patient, and there is evidence that this is true for some cancers. However, it is not true for many cancers, which would indicate that other components of the immune system are involved in tumor control.

7. When the antibodies bind to the lymphocytes, mechanisms are activated that destroy the lymphocytes, e.g., opsonization or activation of complement. Without the lymphocytes an effective immune system response to the graft does not occur. The side effect could be an increased susceptibility to infections.

23 Respiratory System

FOCUS: Respiration is the exchange of oxygen and carbon dioxide between the atmosphere and the body cells. The respiratory system consists of the nasal cavity, which warms and filters air; the pharynx, which connects the nasal cavity to the larynx; the larynx, which contains the vocal cords; an air-conducting system consisting of the trachea, bronchi, and bronchioles; and the site of gas exchange with the blood, the alveoli. The diaphragm and thoracic wall muscles change the volume of the thoracic cavity, producing pressure gradients responsible for the movement of air into and out of the lungs. The elastic recoil of lung tissue and water surface tension make the lungs collapse, a tendency that is opposed by intrapleural pressure and surfactant. Movement of a gas through the respiratory membrane is influenced by the partial pressure of the gas, the diffusion coefficient of the gas, and the thickness and surface area of the respiratory membrane. Once in the blood, oxygen is mostly transported bound to hemoglobin. The oxygen-hemoglobin dissociation curve shows how hemoglobin is saturated with oxygen in the lungs and how hemoglobin releases oxygen in the tissues. Most carbon dioxide is transported in the blood as bicarbonate ions. Neural control of respiration is achieved by the inspiratory, expiratory, pneumotaxic, and apneustic centers in the brain and by the Hering-Breuer reflex. The most important regulators of resting respiration are blood carbon dioxide and pH levels, although low blood oxygen levels can increase respiration. Respiration during exercise is mostly determined by the cerebral motor cortex and by feedback from proprioceptors.

CONTENT LEARNING ACTIVITY

Nose and Nasal Cavity

Air enters the external nares and passes through the nasal cavity into the pharynx.

A. Match these terms with the correct statement or definition:

Conchae
External nares (nostrils)
Hard palate
Internal nares (choanae)
Meatus
Nasal septum
Vestibule

_____ 1. Posterior openings from the nasal cavity into the pharynx.

_____ 2. Anterior part of the nasal cavity, just inside the external nares.

_____ 3. Floor of the nasal cavity.

_____ 4. Three bony ridges on the lateral wall of each nasal cavity.

_____ 5. Passageway between the conchae.

The openings for the paranasal sinuses are found in the superior and median meatus, and the opening of the nasolacrimal duct is found in the inferior meatus.

B. Match these terms with the correct statement or definition:

Cilia   Mucus

Mucous membrane   Nasal hair

_____ 1. Structures located in the vestibule that trap large dust particles.

_____ 2. Substance secreted by goblet cells that traps debris in the incoming air.

_____ 3. Structures on the surface of the mucous membrane that move mucus posteriorly to the pharynx.

_____ 4. Structure responsible for the addition of moisture and warmth to inspired air.

Pharynx

" *The pharynx is the common opening of both the digestive and respiratory systems.* **"**

A. Match these terms with the correct statement or definition:

Fauces   Nasopharynx

Laryngopharynx   Oropharynx

_____ 1. Superior portion of the pharynx that extends from the internal nares to the uvula.

_____ 2. Portion of the pharynx that extends from the uvula to the epiglottis.

_____ 3. Portion of the pharynx that contains the openings of the auditory tubes.

_____ 4. Portion of the pharynx that contains two sets of tonsils.

_____ 5. Opening of the oral cavity into the oropharynx.

B. Match these terms with the correct parts labeled on the diagram in Figure 23-1:

Conchae
Epiglottis
External naris (nostril)
Hard palate
Internal naris (choana)
Laryngopharynx
Larynx

Meatus
Nasopharynx
Oropharynx
Soft palate
Uvula
Vestibule

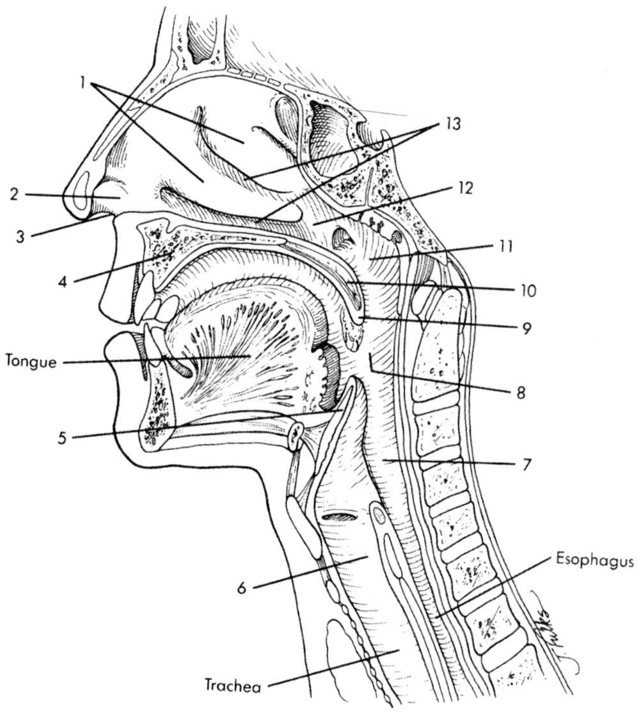

1. _____

2. _____

3. _____

4. _____

5. _____

6. _____

7. _____

8. _____

9. _____

10. _____

11. _____

12. _____

13. _____

Figure 23-1

Larynx

66 *The larynx consists of an outer covering of nine cartilages that are connected* 99
to each other by muscles and ligaments.

A. Match these terms with the correct statement or definition:

Cricoid cartilage
Epiglottis
Laryngitis

Thyroid cartilage
True vocal cords
Vestibular folds

_____ 1. Largest, most superior laryngeal cartilage; the Adam's apple.

_____ 2. Covers the opening into the larynx during swallowing.

_____ 3. Superior pair of ligaments that close to prevent the movement of air, food, or liquids through the larynx.

_____ 4. Inferior pair of ligaments involved in sound production; these ligaments and the space between them are called the glottis.

_____ 5. Inflammation of the mucosal epithelium of the vocal cords.

453

B. Match these terms with the correct parts labeled on the diagram in Figure 23-2:

Arytenoid cartilage
Corniculate cartilage
Cricoid cartilage
Cuneiform cartilage
Epiglottis
Hyoid bone
Thyroid cartilage
Tracheal cartilage

1. _____
2. _____
3. _____
4. _____
5. _____
6. _____
7. _____

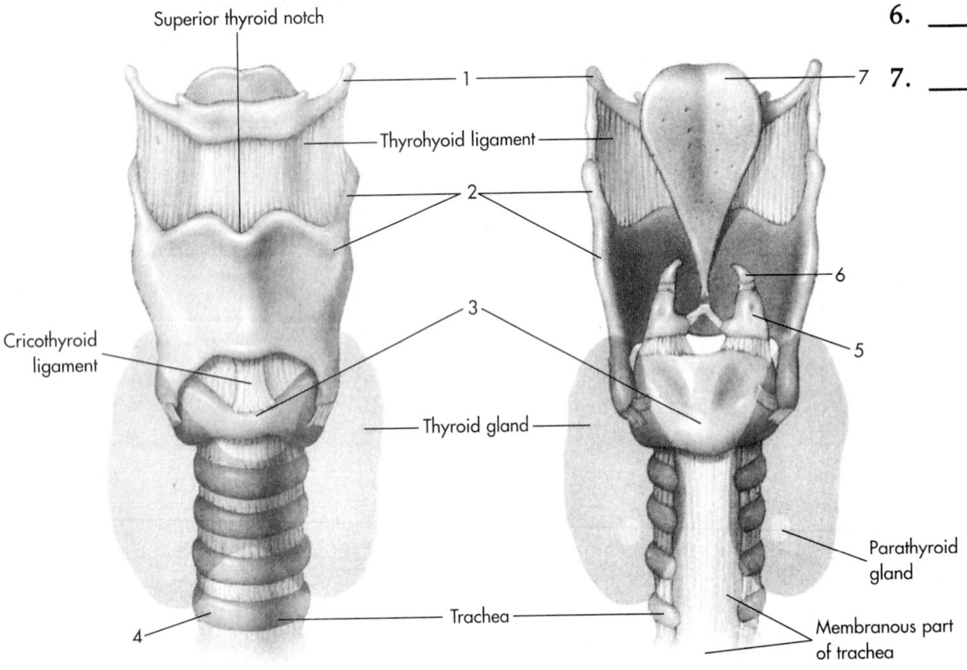

Superior thyroid notch
Thyrohyoid ligament
Cricothyroid ligament
Thyroid gland
Trachea
Parathyroid gland
Membranous part of trachea

Figure 23-2

Trachea, Bronchi, and Lungs

" *The trachea and bronchi serve as passageways for air between the larynx and lungs; the lungs are the principal organs of respiration.* **"**

A. Match these terms with the correct statement or definition:

Alveolar sac
Alveoli
Asthma attack
C-shaped cartilage rings
Cartilage plates
Hilum
Lobes
Lobules
Pseudostratified ciliated columnar epithelium
Simple squamous epithelium
Smooth muscle

_____ 1. Structures that reinforce the walls of the trachea and primary bronchi.

_____ 2. Support structures present in the walls of secondary and tertiary bronchi.

_____ 3. Cell type lining the trachea and bronchi.

_____ 4. Sections of the lung separated by connective tissue but not visible as surface fissures.

454

_____ 5. Supplied by tertiary bronchi.

_____ 6. Single air sacs in the lungs.

_____ 7. Two or more alveoli that share a common opening.

_____ 8. Result of forceful contraction of smooth muscle in the
   bronchiole walls.

☞ The nasal cavity, pharynx, and associated structures are referred to as the upper
respiratory tract, whereas the lower respiratory tract includes the larynx, trachea,
bronchi, and lungs.

B. Using Figure 23-3, arrange the structures in the order that a
molecule of oxygen would pass through them to enter the blood.

Alveolar duct
Alveolar sac
Alveolus
Primary bronchi
Respiratory bronchiole
Terminal bronchiole
Trachea

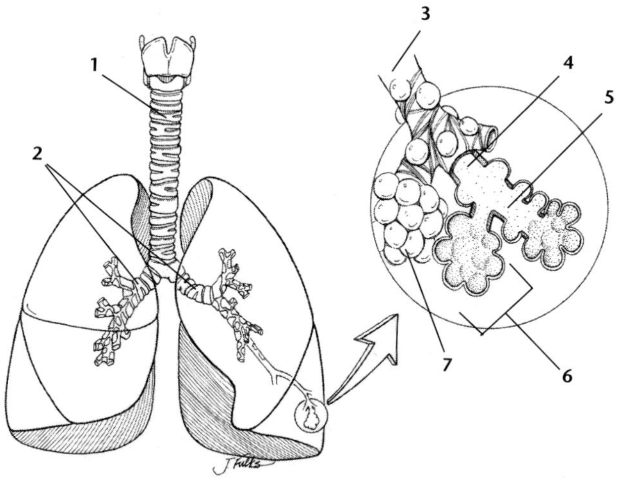

Figure 23-3

1. _____

2. _____

3. _____

4. _____

5. _____

6. _____

7. _____

Pleura

❝_The serous membranes lining the pleural cavity and covering the lungs are called pleura._❞

Match these terms with the
correct statement or definition:

Parietal pleura   Pleural fluid
Pleural cavity   Visceral pleura

_____ 1. Cavity surrounding each lung.

_____ 2. Covers the thoracic wall, superior surface of the diaphragm,
   and the mediastinum.

_____ 3. Covers the surface of the lungs.

_____ 4. Lubricant for the lungs that helps hold the pleural membranes
   together.

Blood Supply

"There are two major blood flow routes to the lung."

Match these terms with the correct statement or definition:

Bronchial arteries   Pulmonary arteries
Bronchial veins   Pulmonary veins

_____ 1. Blood vessels that carry deoxygenated blood to the lungs.

_____ 2. Blood vessels that branch from the thoracic aorta and carry oxygenated blood to the bronchi.

_____ 3. Blood vessels that carry oxygenated blood from lungs to the heart.

_____ 4. Blood vessels that carry deoxygenated blood from the bronchi to the azygos system.

☞ A small amount of deoxygenated blood from the bronchial veins mixes with the oxygenated blood of the pulmonary veins.

Muscles of Respiration and the Thoracic Wall

"The diaphragm and thoracic wall form the boundaries of the thoracic cavity."

Match these terms with the correct statement or definition:

Costal cartilages   Internal intercostal muscles
Diaphragm   External intercostal muscles

_____ 1. Large dome of skeletal muscles that, when contracted, expands the superior-inferior dimension of the thoracic cavity.

_____ 2. Structures that allow lateral rib movement and lateral expansion of the thoracic cavity.

_____ 3. Muscles of inspiration that, when contracted, elevate the ribs.

_____ 4. Muscles of expiration that depresses the ribs during labored breathing.

☞ During quiet breathing, expiration occurs when the muscles of inspiration relax and a passive decrease in thoracic volume occurs; during labored breathing, the muscles of expiration can contract, producing a more rapid and greater decrease in thoracic volume.

Pressure Differences and Air Flow

66 *Air flow into the lungs requires a pressure gradient from the outside of the body to the alveoli.* 99

Match these terms with
the statements below:

Decreases
Increases

_____ 1. Effect of inspiration on thoracic volume.

_____ 2. Effect of increased thoracic volume on lung volume.

_____ 3. Effect of expansion of the lungs on alveolar volume.

_____ 4. Effect on intrapulmonary pressure if alveolar volume increases.

_____ 5. Effect on air movement into the alveoli if intrapulmonary pressure decreases below atmospheric pressure.

Collapse of the Lungs

66 *Two factors tend to make the lungs collapse, and two factors keep the lungs from collapsing.* 99

Match these terms with the
correct statement or definition:

Elastic recoil
Intrapleural pressure
Intrapulmonary pressure

Surface tension of alveolar
fluid
Surfactant

_____ 1. Two factors that cause the lungs to tend to collapse.

_____

_____ 2. Two factors that keep the lungs from collapsing.

_____

_____ 3. Mixture of lipoprotein molecules produced by the secretory cells of the alveolar epithelium.

_____ 4. Pressure within the pleural cavity.

Compliance of the Lungs and Thorax

66 *Compliance is a measure of the expansibility of the lungs and thorax.* 99

Match these terms with the
correct statement :

Decreases
Increases

_____ 1. Effect on the ease of lung expansion when compliance increases.

_____ 2. Effect on compliance when the collapsing force of the lungs increases, as in respiratory distress syndrome.

_____ 3. Effect of emphysema on compliance.

Pulmonary Volumes and Capacities

66*Factors such as sex, age, body size, and physical conditioning cause variations in respiratory*99
volumes and capacities from one individual to another.

Match these terms with the
correct definition and Figure 23-4:

Expiratory reserve volume
Functional residual capacity
Inspiratory capacity
Inspiratory reserve volume
Residual volume
Tidal volume
Total lung capacity
Vital capacity

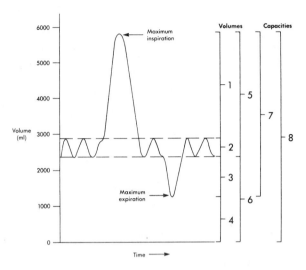

Figure 23-4

1. The volume of air that can be forcefully inspired after inspiration of the normal tidal volume.

2. The volume of air inspired or expired in a normal inspiration or expiration.

3. The volume of air that can be forcefully expired after expiration of the normal tidal volume.

4. The volume of air still in the respiratory passages after the most forceful expiration.

5. The tidal volume plus the inspiratory reserve volume.

6. The expiratory reserve volume plus the residual volume.

7. The sum of expiratory reserve, tidal, and inspiratory reserve volumes.

8. The sum of the inspiratory reserve, expiratory reserve, tidal, and residual volumes.

☞ The rate at which lung volume changes during direct measurement of the vital capacity is called the forced expiratory vital capacity.

Minute Respiratory Volume and Alveolar Ventilation Rate

"*Minute respiratory volume and alveolar ventilation rate are measurements of gas movement in the respiratory system.*"

Match these terms with the
correct statement or definition:

Alveolar ventilation rate   Minute respiratory volume
Anatomical dead air space   Physiological dead air space

_____

1. Total amount of air moved in and out of the respiratory system each minute; tidal volume times respiratory rate.

_____

2. Nasal cavity, pharynx, larynx, trachea, bronchi, bronchioles, and terminal bronchioles; gas exchange does not take place.

_____

3. Anatomical dead air space plus the volume of any nonfunctional alveoli.

_____

4. Volume of air that is available for gas exchange each minute.

Partial Pressure

"*It is traditional to designate the partial pressure of individual gases in a mixture as* PN_2, PO_2, *or* PCO_2, *for example.*"

A. Match these terms with the
correct statement or definition:

Dalton's law   Vapor pressure
Partial pressure

_____

1. In a mixture of gases the portion of the total pressure resulting from each type of gas is determined by the percentage of the total volume represented by each gas type.

_____

2. Pressure exerted by each type of gas in a mixture.

_____

3. Partial pressure of water molecules in the gaseous form.

B. Using the terms provided, complete these statements:

Greater
Smaller

1. _____

2. _____

3. _____

Compared to atmospheric air, alveolar air contains a
_(1)_ partial pressure of oxygen, _(2)_ partial pressure of
carbon dioxide, and _(3)_ vapor pressure.

Diffusion of Gases Through Liquids

66 *When a gas comes into contact with a liquid such as water there is a tendency for the gas* 99
to dissolve in the liquid.

Match these terms with the
correct statement or definition:

Henry's law
Solubility coefficient

_____ 1. Concentration of a dissolved gas is its partial pressure
multiplied by its solubility coefficient.

_____ 2. Measure of how easily a gas dissolves in a liquid.

Diffusion of Gases Through the Respiratory Membrane

66 *The respiratory membranes are in the respiratory bronchioles, alveolar ducts, and alveoli.* 99

A. Using the diagram in Figure 23-5,
arrange each of the structures of
the respiratory membrane in the
correct order from inside the alveoli
to the blood:

Alveolar epithelium
Alveolar epithelium basement
  membrane
Capillary endothelium
Capillary endothelium basement
  membrane
Thin fluid layer
Thin interstitial space

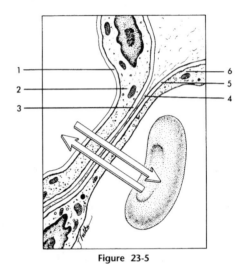

Figure 23-5

1. _____ 3. _____ 5. _____

2. _____ 4. _____ 6. _____

B. Match these terms with the
correct statements:

Decreases
Increases

_____ 1. Effect on diffusion rate if the thickness of the respiratory
membrane is increased.

_____ 2. Effect on diffusion rate if diffusion coefficient decreases.

_____ 3. Effect on diffusion rate if respiratory surface area increases.

_____ 4. Effect on diffusion rate if partial pressure gradient increases.

Relationship Between Ventilation and Capillary Blood Flow

66*There are two ways that the relationship between ventilation and blood flow can be affected.***99**

Match these problems with the correct statement or definition:

Inadequate cardiac output
Shunted blood

_____

1. In this situation, ventilation may exceed the ability of the blood to pick up oxygen.

_____

2. In this situation, ventilation is not great enough to oxygenate the blood in the alveolar capillaries.

 The physiological shunt is the deoxygenated blood from the alveoli plus the deoxygenated blood returning from the bronchi and bronchioles.

Oxygen and Carbon Dioxide Diffusion Gradients

66*To diffuse efficiently, carbon dioxide partial pressure gradients do not have to be as great***99** *as the partial pressure gradients for oxygen.*

Match these partial pressures with the correct location in Figure 23-6:

104 mm Hg   45 mm Hg
95 mm Hg   40 mm Hg

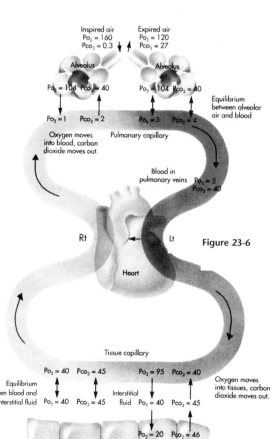

Figure 23-6

1. _____

2. _____

3. _____

4. _____

5. _____

461

Hemoglobin and Oxygen Transport

66*About 97% of the oxygen transported in the blood from the lungs to the tissues is transported in*99 *combination with the hemoglobin in the erythrocytes and the remaining 3% is dissolved in the water portion of plasma.*

Match these terms with the correct statement or definition:

Bohr effect	Saturated hemoglobin
Decreases	Shift to the left
Increases	Shift to the right

_____ 1. Hemoglobin molecule with an oxygen molecule bound to each of its four heme groups.

_____ 2. Change in the oxygen-hemoglobin dissociation curve resulting from changes in blood pH.

_____ 3. Shift of the oxygen-hemoglobin dissociation curve resulting in a lower affinity of hemoglobin for oxygen; normally occurs in the tissues.

_____ 4. Shift of the oxygen-hemoglobin dissociation curve resulting in a higher affinity of hemoglobin for oxygen; normally occurs in the lungs.

_____ 5. Effect on the amount of oxygen bound to hemoglobin when pH decreases (PCO_2 increases).

_____ 6. Effect on the amount of oxygen bound to hemoglobin when temperature increases.

 The substance 2,3-bisphosphoglycerate, which is a by-product of glucose metabolism in erythrocytes, decreases the ability of oxygen to bind to hemoglobin.

Transport of Carbon Dioxide

66*Carbon dioxide is transported in the blood in three major ways.*99

A. Match these terms with the correct statement or definition:

Bicarbonate ions	Dissolved in plasma
Carbamino compounds (including carbamino-hemoglobin)	Haldane effect

_____ 1. 72% of carbon dioxide is transported in this form.

_____ 2. 20% of carbon dioxide is transported in this form.

_____ 3. 8% of carbon dioxide is transported in this form.

_____ 4. Hemoglobin that has released oxygen binds more readily to carbon dioxide than hemoglobin that has oxygen bound to it.

 Carbon dioxide binds to the globin portion of hemoglobin.

B. Using the terms provided, complete these statements:

Bicarbonate ions   Chloride ions
Buffer   Chloride shift
Carbonic acid   Hemoglobin
Carbonic anhydrase

1. _____

2. _____

3. _____

4. _____

5. _____

6. _____

7. _____

Most of the carbon dioxide inside the erythrocyte reacts with water to form _(1)_ ; the reaction is catalyzed by _(2)_ inside the erythrocyte. The carbonic acid then dissociates to form _(3)_ and hydrogen ions. Because more bicarbonate ions are inside the cells than outside, the bicarbonate ions readily diffuse out of the erythrocytes into the plasma. In response to this movement of negatively charged ions out of the erythrocytes, _(4)_ move from the plasma into the erythrocytes. The exchange of chloride ions for the bicarbonate ions across the membranes of the erythrocytes is called the _(5)_ . The hydrogen ions formed by the dissociation of carbonic acid bind to _(6)_ in the erythrocytes. This binding prevents hydrogen ions from leaving the cells; thus hemoglobin acts as a _(7)_ to prevent a decrease in blood pH.

Nervous Control of Rhythmic Ventilation

"Nerve impulses responsible for controlling the respiratory muscles originate within** "** neurons of the medulla oblongata and pons.

Match these terms with the correct statement or definition:

Apneustic center   Inspiratory center
Expiratory center   Pneumotaxic center
Hering-Breuer reflex

_____ 1. Spontaneously active; establishes the basic rhythm of respiration.

_____ 2. Remains inactive during quiet respiration, but becomes active when the rate and depth of ventilation are increased.

_____ 3. Stimulates the inspiratory center.

_____ 4. Has an inhibitory effect on the inspiratory and apneustic center.

_____ 5. Occurs when action potentials from stretch receptors in the bronchi and bronchioles inhibit the inspiratory center.

☞ Through the cerebral cortex it is possible to consciously increase or decrease the rate and depth of the respiratory movements.

Chemical Control of Respiration

"The respiratory system maintains the concentration of oxygen and carbon dioxide and the pH of the body fluids within a normal range of values."

A. Match these terms with the correct statement or definition:

Central
Decreases
Increases
Hypercapnia

Hypocapnia
Hypoxia
Peripheral

_____ 1. Chemoreceptors of the chemosensitive area of the medulla; detects changes in cerebrospinal fluid pH.

_____ 2. Chemoreceptors in the carotid and aortic bodies; detect changes in blood pH.

_____ 3. Chemoreceptors responsible for most of the response to changes in pH or PCO_2.

_____ 4. Lower-than-normal carbon dioxide levels.

_____ 5. Effect of increased blood carbon dioxide on breathing rate and depth.

_____ 6. Effect of decreased blood pH on breathing rate and depth.

_____ 7. Effect of greatly decreased blood oxygen level on breathing rate and depth.

☞ Small changes in blood carbon dioxide and blood pH levels produce much larger changes in respiratory rate than do small changes in blood oxygen levels.

Effect of Exercise on Respiratory Movements

"Neural impulses from the cerebral cortex and from proprioceptors act as the major regulators of respiration during exercise."

Match these terms with the correct statements:

Decreases
Increases

_____ 1. Effect of action potentials, traveling from collateral fibers of motor pathways, on breathing rate during exercise.

_____ 2. Effect of stimulation of proprioceptors on respiratory rate during exercise.

1. Name three things that happen to inspired air on its way through the nasal cavity.

2. Trace the path of inspired air from the trachea to the alveoli by naming the structures through which the air passes.

3. Describe the relationship between the volume and the pressure of a gas.

4. List two factors that tend to cause the lungs to collapse and two factors that prevent the lungs from collapsing.

5. List the four pulmonary volumes and define pulmonary capacity.

6. Give the formula for calculating alveolar ventilation rate.

7. List the six layers of a respiratory membrane through which gases must diffuse.

8. Name four factors that influence the rate of gas diffusion across the respiratory membrane.

9. List two ways oxygen is transported in the blood, and state their relative importance.

10. List three ways that carbon dioxide is transported in the blood, and indicate their relative importance.

11. List the events that result in a decrease in blood pH when blood carbon dioxide levels increase.

12. Name the factors that have the greatest effect on the regulation of respiration at rest and during exercise.

MASTERY LEARNING ACTIVITY

Place the letter corresponding to the correct answer in the space provided.

_____ 1. The nasal cavity
a. is mostly lined with pseudostratified ciliated epithelium.
b. has a vestibule, which contains the olfactory epithelium.
c. is connected to the pharynx by the external nares.
d. has passageways called conchae.

_____ 2. The nasopharynx
a. has openings from the paranasal sinuses.
b. contains the pharyngeal tonsil.
c. opens into the oral cavity through the fauces.
d. extends to the tip of the epiglottis.

_____ 3. The larynx
a. connects the laryngopharynx to the trachea.
b. has three unpaired cartilages and six paired cartilages.
c. contains the vocal cords.
d. all of the above

_____ 4. The trachea possesses
a. skeletal muscle.
b. pleural fluid glands.
c. C-shaped rings of cartilage.
d. walls with stratified epithelium.

5. Gas exchange between the air in the lungs and the blood takes place in the
   a. alveoli.
   b. bronchi.
   c. terminal bronchioles.
   d. trachea.

6. During an asthma attack, the patient has difficulty breathing because of constriction of the
   a. trachea.
   b. bronchi.
   c. terminal bronchioles.
   d. alveoli.
   e. respiratory membrane.

7. The parietal pleura
   a. covers the surface of the lung.
   b. covers the inner surface of the thoracic cavity.
   c. is the connective tissue membrane that divides the thoracic cavity into the right and left pleural cavities.
   d. covers the inner surface of the alveoli.
   e. is the membrane across which gas exchange occurs.

8. Given the following muscles:
   1. diaphragm
   2. external intercostals
   3. internal intercostals
   4. rectus abdominis

   Which muscles are involved with increasing the volume of the thorax?
   a. 1, 2
   b. 1, 3
   c. 2, 3
   d. 2, 4
   e. 3, 4

9. During the process of inspiration, which of the following pressures decrease when compared with the resting condition?
   a. intrapulmonary
   b. intrapleural
   c. atmospheric
   d. a and b
   e. all of the above

10. During the process of expiration the pressure within the alveoli must be
    a. greater than the pressure in the pleural spaces.
    b. greater than atmospheric pressure.
    c. less than atmospheric pressure.
    d. a and b

11. Contraction of the bronchiolar smooth muscle has which of the following effects?
    a. decreased resistance to gas flow through the respiratory tree
    b. a smaller pressure gradient is required to get the same rate of gas flow when compared with the normal bronchioles
    c. increased resistance to gas flow through the respiratory tree
    d. does not affect gas flow through the respiratory tree
    e. a and b

12. Because of _____, the lung does not normally collapse
    a. surfactant
    b. intrapleural pressure
    c. elastic recoil
    d. a and b
    e. all of the above

13. Immediately after the creation of an opening through the thorax into the pleural cavity,
    a. air would flow into the pleural cavity through the hole.
    b. air would flow out of the pleural cavity through the hole.
    c. air would neither flow out nor in through the hole.
    d. the lung would be forced to protrude through the hole.
    e. b and d

14. Compliance of the lungs and thorax is
    a. the volume by which the lungs and thorax change for each unit change of intrapulmonary pressure.
    b. increased in emphysema.
    c. decreased because of lack of surfactant.
    d. all of the above

15. A patient expires normally; then, using forced expiration, he blows as much air as possible into a spirometer. This would measure the
    a. inspiratory reserve.
    b. expiratory reserve.
    c. residual volume.
    d. tidal volume.
    e. vital capacity.

16. Given the following lung volumes:
    1. tidal volume = 500 ml
    2. residual volume = 1000 ml
    3. inspiratory reserve = 2500 ml
    4. expiratory reserve = 1000 ml
    5. dead air space = 1000 ml

    The vital capacity would be
    a. 3000 ml
    b. 3500 ml
    c. 4000 ml
    d. 5000 ml
    e. 6000 ml

17. The alveolar ventilation rate is
    a. tidal volume times respiratory rate.
    b. the minute respiratory volume plus the dead air space.
    c. the amount of air available for gas exchange in the lungs.
    d. all of the above

18. If the total pressure of a gas is 760 mm Hg and its composition is 20% oxygen, 0.04% carbon dioxide, 75% nitrogen, and 5% water vapor, the partial pressure of oxygen is
    a. 15.2 mm Hg.
    b. 20 mm Hg.
    c. 148 mm Hg.
    d. 152 mm Hg.
    e. 740 mm Hg.

19. Which of the following layers must gases cross to pass from the alveolus to the blood within the alveolar capillaries?
    a. endothelium
    b. basement membrane
    c. simple squamous epithelium
    d. a and b
    e. all of the above

20. Which of the following would increase the rate of gas exchange across the respiratory membrane?
    a. increase in thickness of the respiratory membrane
    b. decrease in surface area of the respiratory membrane
    c. increase in partial pressure differences of gases across the respiratory membrane
    d. all of the above

21. Which gas diffuses most rapidly across the respiratory membrane?
    a. carbon dioxide
    b. oxygen

22. In which of the following sequences does PO_2 progressively decrease?
    a. arterial blood, alveolar air, body tissues
    b. body tissues, arterial blood, alveolar air
    c. body tissues, alveolar air, arterial blood
    d. alveolar air, arterial blood, body tissues

23. Oxygen is mostly transported in the blood
    a. dissolved in plasma.
    b. bound to blood proteins.
    c. within bicarbonate ions.
    d. bound to the heme portion of hemoglobin.

24. If the alveolar partial pressure of oxygen decreased to 77 mm Hg,
    a. significantly less oxygen would be bound to hemoglobin.
    b. the subject would rapidly die of asphyxiation.
    c. approximately 50% of the hemoglobin would be saturated with oxygen.
    d. nearly all of the hemoglobin would be saturated with oxygen.

25. The oxygen-hemoglobin dissociation
curve is adaptive because it
    a. shifts to the right in the alveolar
       capillaries and to the left in the
       tissue capillaries.
    b. shifts to the left in the alveolar
       capillaries and to the right in
       the tissue capillaries.
    c. doesn't shift.

26. Carbon dioxide is mostly transported
in the blood
    a. dissolved in plasma.
    b. bound to blood proteins.
    c. within bicarbonate ions.
    d. bound to the globin portion of
       hemoglobin.

27. The partial pressure of carbon
dioxide in the venous blood is
    a. greater than in the tissue spaces.
    b. less than in the tissue spaces.
    c. less than in the alveoli.
    d. less than in arterial blood.

28. After passing through the lungs,
hemoglobin is better able to combine
with carbon dioxide because of the
    a. Bohr effect.
    b. Haldane effect.
    c. a and b

29. The chloride shift
    a. occurs primarily in alveolar
       capillaries.
    b. occurs when chloride ions replace
       bicarbonate ion within
       erythrocytes.
    c. results in an increase in blood pH.
    d. all of the above

30. If one could directly stimulate
neurons within the inspiratory
center, the frequency of action
potentials within which of the
following nerves would immediately
increase?
    a. intercostal nerves
    b. phrenic nerves
    c. vagus nerves
    d. glossopharyngeal nerves
    e. a and b

31. The pneumotaxic center
    a. stimulates the inspiratory center.
    b. inhibits the apneustic center.
    c. is located in the medulla
       oblongata.
    d. all of the above

32. The Hering-Breuer reflex
    a. decreases inspiratory volume.
    b. increases inspiratory volume.
    c. occurs in response to changes in
       carbon dioxide levels in the
       blood.
    d. a and b
    e. b and c

33. The chemosensitive area
    a. stimulates the respiratory center
       when blood carbon dioxide levels
       increase.
    b. stimulates the respiratory center
       when blood pH increases.
    c. is located in the pons.
    d. all of the above

34. Blood oxygen levels
    a. are more important than carbon
       dioxide in the regulation of
       respiration.
    b. need to change only slightly to
       cause a change in respiration.
    c. are detected by sensory receptors
       in the carotid and aortic bodies.
    d. all of the above

35. During exercise respiration rate and
depth increases primarily because of
    a. increased blood carbon dioxide
       levels.
    b. decreased blood oxygen levels.
    c. decreased blood pH.
    d. input to the respiratory center
       from the cerebral motor cortex
       and from proprioceptors.

FINAL CHALLENGES

Use a separate sheet of paper to complete this section.

1. Marty Blowhard used a spirometer with the following results:
   a. After a normal inspiration, a normal expiration was 500 ml.
   b. Following a normal expiration, he was able to expel an additional 1000 ml.
   c. Taking as deep a breath as possible then forcefully exhaling all the air possible yielded an output of 4500 ml.

   On the basis of these measurements, what is Marty's inspiratory reserve?

2. A patient has pneumonia, and fluid builds within the alveoli. Explain why this results in an increased rate of respiration that can be returned to normal with oxygen therapy.

3. A patient has severe emphysema that has extensively damaged the alveoli and reduced the surface area of the respiratory membrane. Although the patient is receiving oxygen therapy, he still has a tremendous urge to take a breath, i.e., he does not feel as if he is getting enough air. Why does this occur?

4. Consider the oxygen-hemoglobin dissociation curve graphed below:

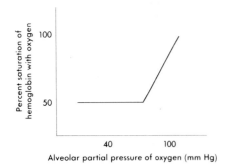

   If an animal has such an oxygen-hemoglobin dissociation curve, would you expect to find it living at sea level or high altitudes? Explain.

5. A patient has a bronchial tumor that greatly restricts air flow into his left lung. If the oxygen-hemoglobin dissociation curve of this patient's left lung were compared with the oxygen-hemoglobin dissociation curve of his right lung (which is normal), would it show a shift to the left or right (Bohr effect)? Why would this be an advantage or disadvantage?

6. President Reagan was shot with a 22-caliber pistol in 1981. The bullet entered the thoracic cavity on his left side and lodged in his mediastinum. Explain why his respiratory depth increased before he arrived at the hospital (hint: the bullet caused a pneumothorax).

7. Patients with diabetes mellitus may occasionally take too little insulin. The result is the rapid metabolism of lipids and the accumulation of acidic by-products of lipid metabolism in the circulatory system. What effect would this have on respiration? Why is this change beneficial?

8. A hysterical patient is hyperventilating. The doctor makes the patient breathe into a paper bag. Because you are an especially astute physiology student, you say to the doctor, "When the patient was hyperventilating, he was blowing off carbon dioxide and blood carbon dioxide levels decreased. When he breathed into the paper bag, carbon dioxide was trapped, and as a result, blood carbon dioxide levels increased. When blood carbon dioxide levels increase, the 'urge' to breathe should also increase, and breathing into a paper bag should make the patient hyperventilate more. So why is breathing into a paper bag recommended for hyperventilation?" How do you think the doctor would respond?

9. The effects of changes of arterial oxygen concentration on ventilation rates were determined under two different circumstances:

1. Arterial carbon dioxide and pH levels were held at a constant value while arterial oxygen concentration was varied.
2. Arterial carbon dioxide and pH levels were allowed to change while arterial oxygen concentration was varied.

The results of these two experiments are graphed below:

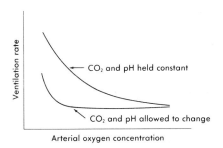

These data show that, when blood carbon dioxide and pH levels are allowed to fluctuate as normally occurs in the body (2), arterial oxygen concentration exerts very little effect on ventilation rate until very low blood oxygen levels are reached. On the other hand, when blood carbon dioxide and pH levels are held at a constant level (1), arterial oxygen concentration can significantly affect ventilation rate. Propose an explanation.

10. Shown below is the blood pH of a runner following the start of a race:

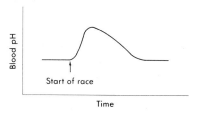

Two physiology students were puzzled over these results. They knew that ventilation rate increases with exercise, and they knew that decreased pH is responsible for stimulating the respiratory centers and causing an increase in respiration rate. Propose an explanation that accounts for the increased pH values following the start of the race (hint: don't say the instructor is wrong!).

ANSWERS TO CHAPTER 23

Nose and Nasal Cavity
- A. 1. Internal nares (choanae); 2 Vestibule; 3. Hard palate; 4. Conchae; 5. Meatus
- B. 1. Nasal hair; 2. Mucus; 3. Cilia; 4. Mucous membrane

Pharynx
- A. 1. Nasopharynx; 2. Oropharynx; 3. Nasopharynx; 4. Oropharynx; 5. Fauces
- B. 1. Conchae; 2. Vestibule; 3. External naris (nostril); 4. Hard palate; 5. Epiglottis; 6. Larynx; 7. Laryngopharynx; 8. Oropharynx; 9. Uvula; 10. Soft palate; 11. Nasopharynx; 12. Internal naris (choana); 13. Meatus

Larynx
- A. 1. Thyroid cartilage; 2. Epiglottis; 3. Vestibular folds; 4. True vocal cords; 5. Laryngitis
- B. 1. Hyoid bone; 2. Thyroid cartilage; 3. Cricoid cartilage; 4. Tracheal cartilage; 5. Arytenoid cartilage; 6. Corniculate cartilage; 7. Epiglottis

Trachea, Bronchi, and Lungs
- A. 1. C-shaped cartilage rings; 2. Cartilage plates; 3. Pseudostratified ciliated columnar epithelium; 4. Lobules; 5. Lobules; 6. Alveoli; 7. Alveolar sac; 8. Asthma attack
- B. 1. Trachea; 2. Primary bronchi; 3. Terminal bronchiole; 4. Respiratory bronchiole; 5. Alveolar duct; 6. Alveolar sac; 7. Alveolus

Pleura
- 1. Pleural cavity; 2. Parietal pleura; 3. Visceral pleura; 4; Pleural fluid

Blood Supply
- 1. Pulmonary arteries; 2. Bronchial arteries; 3. Pulmonary veins; 4. Bronchial veins

Muscles of Respiration and Thoracic Wall
- 1. Diaphragm; 2. Costal cartilages; 3. External intercostal muscles

Pressure Differences and Air Flow
- 1. Increases; 2. Increases; 3. Increases; 4. Decreases; 5. Increases

Collapse of the Lungs
- 1. Elastic recoil and surface tension of alveolar fluid; 2. Surfactant and intrapleural pressure; 3. Surfactant; 4. Intrapleural pressure

Compliance of the Lungs and Thorax
- 1. Increases; 2. Decreases; 3. Increases

Pulmonary Volumes and Capacities
- 1. Inspiratory reserve volume; 2. Tidal volume; 3. Expiratory reserve volume; 4. Residual volume; 5. Inspiratory capacity; 6. Functional residual capacity; 7. Vital capacity; 8. Total lung capacity

Minute Respiratory Volume and Alveolar Ventilation Rate
- 1. Minute respiratory volume; 2. Anatomical dead air space; 3. Physiological dead air space; 4. Alveolar ventilation rate

Partial Pressure
- A. 1. Dalton's law; 2. Partial pressure; 3. Vapor pressure
- B. 1. Smaller; 2. Greater; 3. Greater

Diffusion of Gases Through Liquids
- A. 1. Henry's law; 2. Solubility coefficient

Diffusion of Gases Through the Respiratory Membrane
- A. 1. Thin fluid layer; 2. Alveolar epithelium; 3. Alveolar epithelium basement membrane; 4. Thin interstitial space; 5. Capillary endothelium basement membrane; 6. Capillary endothelium
- B. 1. Decreases; 2. Decreases; 3. Increases; 4. Increases

Relationship Between Ventilation and Capillary Blood Flow
- 1. Inadequate cardiac output; 2. Shunted blood

Oxygen and Carbon Dioxide Diffusion Gradients
- 1. 40 mm Hg.; 2. 45 mm Hg; 3. 104 mm Hg; 4. 40 mm Hg; 5. 95 mm Hg

Hemoglobin and Oxygen Transport
- 1. Saturated hemoglobin; 2. Bohr effect; 3. Shift to the right; 4. Shift to the left; 5. Decreases; 6. Decreases

Transport of Carbon Dioxide
- A. 1. Bicarbonate ions; 2. Carbamino compounds (including carbamino-hemoglobin); 3. Dissolved in plasma; 4. Haldane effect
- B. 1. Carbonic acid; 2. Carbonic anhydrase; 3. Bicarbonate ions; 4. Chloride ions; 5. Chloride shift; 6. Hemoglobin; 7. Buffer

Nervous Control of Rhythmic Ventilation
- 1. Inspiratory center; 2. Expiratory center; 3. Apneustic center; 4. Pneumotaxic center; 5. Hering-Breuer reflex

Chemical Control of Respiration
- 1. Central; 2. Peripheral; 3. Central; 4. Hypocapnia; 5. Increases; 6. Increases; 7. Increases

Effect of Exercise on Respiratory Movements
- 1. Increases; 2. Increases

1. Filtered, warmed, and moistened
2. Trachea, primary bronchus, secondary bronchus, tertiary bronchus, bronchiole, terminal bronchiole, respiratory bronchiole, alveolar duct, alveolar sac, alveolus
3. Increasing volume decreases pressure, and decreasing volume increases pressure
4. Recoil of elastic fibers and water surface tension cause the lungs to collapse; surfactant and intrapleural pressure prevent the collapse of the lungs
5. The pulmonary volumes are tidal volume, inspiratory reserve volume, expiratory reserve volume, and residual volume. Pulmonary capacities are the sum of two or more pulmonary volumes
6. Alveolar ventilation rate = respiratory rate times the difference between the tidal volume and the dead air space. AVR = RR(TV - DAS)
7. Thin layer of fluid, alveolar epithelium, alveolar epithelium basement membrane, interstitial space, basement membrane of capillary endothelium, and capillary endothelium
8. Thickness of membrane, diffusion coefficient, surface area of membrane, and partial pressure gradient
9. Hemoglobin 97%, dissolved in plasma 3%
10. Bicarbonate ions 72%, blood proteins 20%, and dissolved in plasma 8%
11. Carbon dioxide and water combine to form carbonic acid, which dissociates into hydrogen ions and bicarbonate ions
12. At rest: changes in pH (and carbon dioxide) detected by central chemoreceptors; during exercise: input from the motor cortex and proprioceptors

MASTERY LEARNING ACTIVITY

1. A The pseudostratified ciliated epithelium of the nasal cavity catches debris in the air and moves it to the pharynx (through the internal nares). The vestibule is the most anterior portion of the nasal cavity and is lined with stratified squamous epithelium. The olfactory epithelium is located in the most superior part of the nasal cavity. The conchae are bony ridges that form passageways called meatus.

2. B The pharyngeal tonsil (adenoid) is located in the nasopharynx. The auditory tubes open into the nasopharynx (the paranasal sinuses open into the nasal cavity). The oropharynx opens into the oral cavity through the fauces and extends to the tip of the epiglottis.

3. D The larynx contains the vocal cords. There are three unpaired cartilages (thyroid cartilage, cricoid cartilage, and epiglottis) and six paired cartilages (arytenoid, corniculate, and cuneiform cartilages). The larynx is located between the laryngopharynx and the trachea.

4. C The trachea is held open by C-shaped cartilage rings. The ends of the rings are held together by smooth muscle, which can alter the diameter of the trachea by contracting. The walls are lined with pseudostratified ciliated columnar epithelium that contains many mucus-secreting goblet cells.

5. A Gas exchange takes place mainly in the alveoli and to a lesser extent in the respiratory bronchioles.

6. C The terminal bronchioles have smooth muscle and no cartilage in their walls. Contraction of the smooth muscle, which can occur during an asthma attack, can impede air flow.

7. B The parietal pleura covers the inner surface of the thoracic cavity. It is continuous at the hilum with the visceral pleura, which covers the surface of the lung. The space between the parietal and visceral pleura is a potential space and contains a small amount of pleural fluid, which reduces friction and helps to hold the pleural membranes together.

8. A The diaphragm and the external intercostals increase thoracic volume, whereas the internal intercostals and rectus abdominis decrease thoracic volume.

9. D During inspiration the intrapleural and intrapulmonary volumes increase. The result is that the pressure in those areas decreases (P=RT/V). Thus a pressure gradient is established between the atmospheric pressure and the intrapulmonary pressure, and air flows into the lungs (flow = pressure difference/resistance).

10. B. The relationship that describes pressure changes within the alveoli is $P = nRT/V$. During expiration the volume of the thorax and the lungs decreases. The pressure within the alveoli therefore increases during expiration. Gas then flows out of the lungs because the intrapulmonary pressure increases above atmospheric pressure.

11. C. Flow is directly proportional to the pressure gradient and inversely proportional to the resistance. Contraction of the bronchiolar smooth muscle increases the resistance to gas flow and therefore, for a given pressure gradient, decreases gas flow.

12. D. Surfactant, which reduces the surface tension of the film of water that lines the alveoli, and intrapleural pressure prevent the lungs from collapsing. Elastic recoil caused by elastic fibers in the alveoli and water surface tension cause the lungs to collapse.

13. A. Because the pressure within the pleural cavity is less than atmospheric pressure, a hole as described allows air to flow into the pleural cavity. This condition, called pneumothorax, results in collapse of the lung.

14. D. Compliance is a measure of the expansibility of the lungs. The lungs are more expansible in emphysema because of the destruction of elastic tissue. Compliance is decreased when the lungs are less expansible, as when surfactant levels are too low.

15. B. The expiratory reserve is the amount of air that can be forcefully expired after expiration of normal tidal volume.

16. C. Vital capacity is the sum of the inspiratory reserve plus the tidal volume plus the expiratory reserve (2500 ml + 500 ml + 1000 ml = 4000 ml).

17. C. The alveolar ventilation rate is the respiratory rate times the tidal volume minus the dead air space, i.e., the amount of air available for gas exchange in the lungs. The minute respiratory volume is the tidal volume times the respiratory rate, i.e., the amount of air moved into and out of the respiratory system each minute.

18. D. The partial pressure of a gas can be determined by multiplying the percent composition of the gas by the total pressure. In this case, 760 mm Hg x 0.20 = 152 mm Hg.

19. E. Gases within the alveolus must cross the respiratory membrane: the fluid that lines the alveolus, the epithelial wall of the alveolus, a basement membrane, a thin interstitial space, a basement membrane, and the endothelium of the alveolar capillary. The alveolar wall and the endothelium are simple squamous epithelium.

20. C. The greater the difference in partial pressure of a gas across the respiratory membrane, the greater the rate of gas exchange. Increasing the thickness or decreasing the surface area of the respiratory membrane decreases the rate of gas exchange.

21. A. Carbon dioxide has a diffusion coefficient that is 20 times greater than that of oxygen; therefore carbon dioxide crosses the respiratory membrane more rapidly than oxygen does.

22. D. The partial pressure of O_2 decreases as it moves from the alveoli to the arterial blood and from the arterial blood to the body tissues. These partial pressure gradients are responsible for the diffusion of oxygen.

23. D. About 97% of the oxygen transported in blood is bound to the heme portion of hemoglobin inside erythrocytes. About 3% is transported dissolved in plasma.

24. D. Review the oxygen-hemoglobin dissociation curve. The partial pressure of oxygen must decrease below about 70 mm Hg before the hemoglobin becomes significantly less than 100% saturated with oxygen.

25. B. A shift of the oxygen-hemoglobin dissociation curve to the left results in an increased tendency of hemoglobin to become saturated with oxygen at a lower PO_2. A shift to the right results in a decreased tendency of oxygen to become saturated with oxygen at a lower PO_2 or increases the likelihood that hemoglobin will release its oxygen. These tendencies are adaptive because they result in increased oxygen bound to hemoglobin in the lung and an increased tendency for hemoglobin to release oxygen bound to it in the tissue capillaries. Therefore a greater amount of oxygen is made available to cells.

26. C. About 72% of the carbon dioxide in blood is transported in the form of bicarbonate ions, 20% is in combination with blood proteins (mostly the globin portion of hemoglobin), and 8% is dissolved in the plasma.

27. B.   The partial pressure of carbon dioxide is normally greatest in the tissue spaces, less in the venous blood, and still less in the alveoli. Carbon dioxide moves by diffusion from the tissue spaces into the capillaries and finally from the alveolar capillaries into the alveoli. Diffusion occurs from areas of higher to lower partial pressure of carbon dioxide.

28. C.   According to the Haldane effect the ability of hemoglobin to combine with carbon dioxide increases after hemoglobin releases oxygen. The Bohr effect in the lungs increases the release of oxygen from hemoglobin.

29. B.   The chloride shift occurs when bicarbonate ions diffuse out of erythrocytes and chloride ions diffuse into erythrocytes. This normally takes place within tissues (i.e., not alveolar capillaries, where the reverse movements occur). At the same time, the movement of hydrogen ions out of the erythrocyte can decrease blood pH.

30. E.   Nerves that innervate muscles of inspiration are stimulated if the inspiratory center is stimulated. Of those listed, the intercostal and phrenic nerves would carry a greater frequency of action potentials to stimulate the external intercostal muscles and diaphragm, respectively.

31. B.   The pneumotaxic center inhibits the apneustic and inspiratory centers. The pneumotaxic center is located in the pons.

32. A.   The Hering-Breuer reflex decreases inspiratory volume. It occurs in response to stretch of the lungs during inspiration.

33. A.   The chemosensitive area stimulates the respiratory center (resulting in increased rate and depth of respiration) when blood carbon dioxide levels increase or blood pH decreases. The chemosensitive area is located in the medulla oblongata.

34. C.   Blood oxygen levels are detected by peripheral chemoreceptors in the carotid and aortic bodies. However, the arterial partial pressure of oxygen must decrease about 50% before a large stimulatory effect on respiration results. pH (and carbon dioxide) are normally much more important than oxygen in regulating respiration.

35. D.   Normally there is little change in blood carbon dioxide, oxygen, and pH levels during exercise; i.e., homeostasis is maintained. Increased respiration results from impulses from the cerebral motor cortex and proprioceptors.

 ☆ FINAL CHALLENGES ☆

1.   The inspiratory reserve is 3000 ml. It is equal to the vital capacity minus the sum of the tidal volume and the expiratory reserve, that is, 4500 - (500 + 1000) = 3000.

2.   The buildup of fluid increases the thickness of the respiratory membrane. This reduces the exchange of gases between the alveolar air and the blood. However, because carbon dioxide diffuses across the respiratory membrane 20 times more rapidly than oxygen, only the levels of oxygen are significantly affected. The decreased oxygen levels stimulate the increased respiration. With oxygen therapy, blood oxygen levels are restored because of the increased partial pressure gradient for oxygen that results when the concentration of oxygen in alveolar air increases.

3.   The lungs have been sufficiently damaged that diffusion of oxygen and carbon dioxide is inadequate. The oxygen therapy addresses only the oxygen problem. With a buildup of carbon dioxide in the blood, the patient feels that he needs to breathe more, even though he is getting enough oxygen.

4.   The normal oxygen-hemoglobin dissociation curve indicates that the alveolar partial pressure of oxygen can decrease to about 70 mm Hg before hemoglobin becomes significantly less than 100% saturated with oxygen. The curve presented indicates that even a slight decrease in alveolar partial pressure of oxygen results in far less than 100% of the hemoglobin being saturated with oxygen. Because the total atmospheric pressure decreases with altitude, the partial pressure of oxygen within the alveoli also decreases with altitude. Therefore the likelihood of surviving at high altitudes is not good, and one would expect to find the animal at sea level.

5.   In a patient with such a bronchial tumor, carbon dioxide builds up in the left lung because of inadequate ventilation. Consequently, there is a shift of the oxygen-hemoglobin dissociation curve to the right. This is disadvantageous because the blood picks up less oxygen in the lungs than normal.

6. The bullet caused a pneumothorax, and the left lung collapsed. Because the lung was collapsed, the stretch receptors in his left lung were not stimulated during inspiration. This interruption of the Hering-Breuer reflex on the left side caused his respiratory depth to increase. Because the other lung wasn't damaged, he probably was not suffering from lack of oxygen or increased carbon dioxide levels.

7. The acidic by-products of lipid metabolism cause a decrease in blood pH. The decreased pH stimulates the chemosensitive area of the medulla, resulting in an increased rate and depth of respiration. Consequently, carbon dioxide is lost from the blood at a faster rate, and pH increases. The changes in respiration help to correct the acid increase produced by the lipid by-products.

8. Emotional excitement or hysteria can cause hyperventilation because higher brain centers stimulate the respiratory centers. The results are decreased blood carbon dioxide levels, inhibition of the vasomotor center, decreased peripheral resistance, decreased blood pressure, inadequate delivery of blood to the brain, and dizziness or a faint feeling (see Chapter 21). The dizziness could further contribute to the hysteria. Breathing into the paper bag causes blood carbon dioxide levels to increase and helps to correct the feeling of dizziness. Concentrating on breathing might also take the patient's mind off the emotional stimulus that caused the hyperventilation. It is not likely that raising blood carbon dioxide levels from below normal to normal is going to significantly affect respiration rate until the influence of higher brain centers returns to normal.

9. This phenomenon is known as the braking effect of carbon dioxide and pH on the regulation of respiration. As oxygen decreases, chemoreceptor reflexes cause the respiration rate to increase. However, this results in a "blow off" of carbon dioxide and an increase in blood pH, which inhibits the respiratory center and thus opposes an increase in respiration rate. Therefore, when carbon dioxide and pH are allowed to fluctuate, oxygen levels are much less potent in changing the respiration rate.

10. At the start of the race, respiration rate initially increases more than is necessary to meet the requirement of exercise. Thus hyperventilation occurs. This removes carbon dioxide from the blood, and pH increases. The increased pH quickly inhibits the respiratory center and brings respiration rates back down to the appropriate level, and pH levels decrease. The increase in respiration rate when exercise is initiated is controlled by cerebral motor areas and by proprioceptors.

24

Digestive System

FOCUS: The digestive system is specialized to ingest food, propel the food through the digestive tract, and digest and absorb the food. In the oral cavity the teeth mechanically break up the food and salivary amylase from the salivary glands begins the digestion of carbohydrates. The bolus of food formed is swallowed (partially a voluntary and partially a reflex activity) and passes through the esophagus to the stomach. The stomach stores the food, begins protein digestion (pepsin), and converts the food into chyme. The chyme enters the small intestine, where bile from the liver and enzymes from the pancreas and the lining of the small intestine complete the digestive process. Most of the products of digestion are absorbed in the small intestine. In the large intestine, water and vitamins produced by intestinal bacteria are absorbed, and feces is formed. The feces is eliminated by means of the defecation reflex. Regulation of the digestive tract is controlled through the nervous system (local reflexes and central nervous system reflexes) and hormones.

CONTENT LEARNING ACTIVITY

General Overview

“ *The primary functions of the digestive system are to provide the body with water, electrolytes, and other nutrients.* ”

Using the terms provided, complete these statements:

Accessory organs Gastrointestinal
tract
Alimentary Tube

The digestive system consists of the digestive tract, a __(1)__ extending from the mouth to the anus, and its associated __(2)__ (primarily glands), which secrete fluids into the digestive tract. The digestive tract is also called the __(3)__ canal. The term __(4)__ technically refers only to the stomach and intestines but is often used as a synonym for the digestive tract.

1. Tube
2. Accessory Organs tract
3. Alimentary canal
4. Gastrointestinal tract

Anatomy Overview

" *The digestive system consists of the digestive tract (a tube), and its associated accessory organs.* **"**

A. Match these terms with the correct parts of the diagram labeled in Figure 24-1:

Anal canal
Ascending colon
Cecum
Descending colon
Duodenum
Esophagus
Gallbladder
Jejunum and ileum
Liver

Pancreas
Pharynx
Rectum
Salivary glands
Sigmoid colon
Stomach
Transverse colon
Vermiform appendix

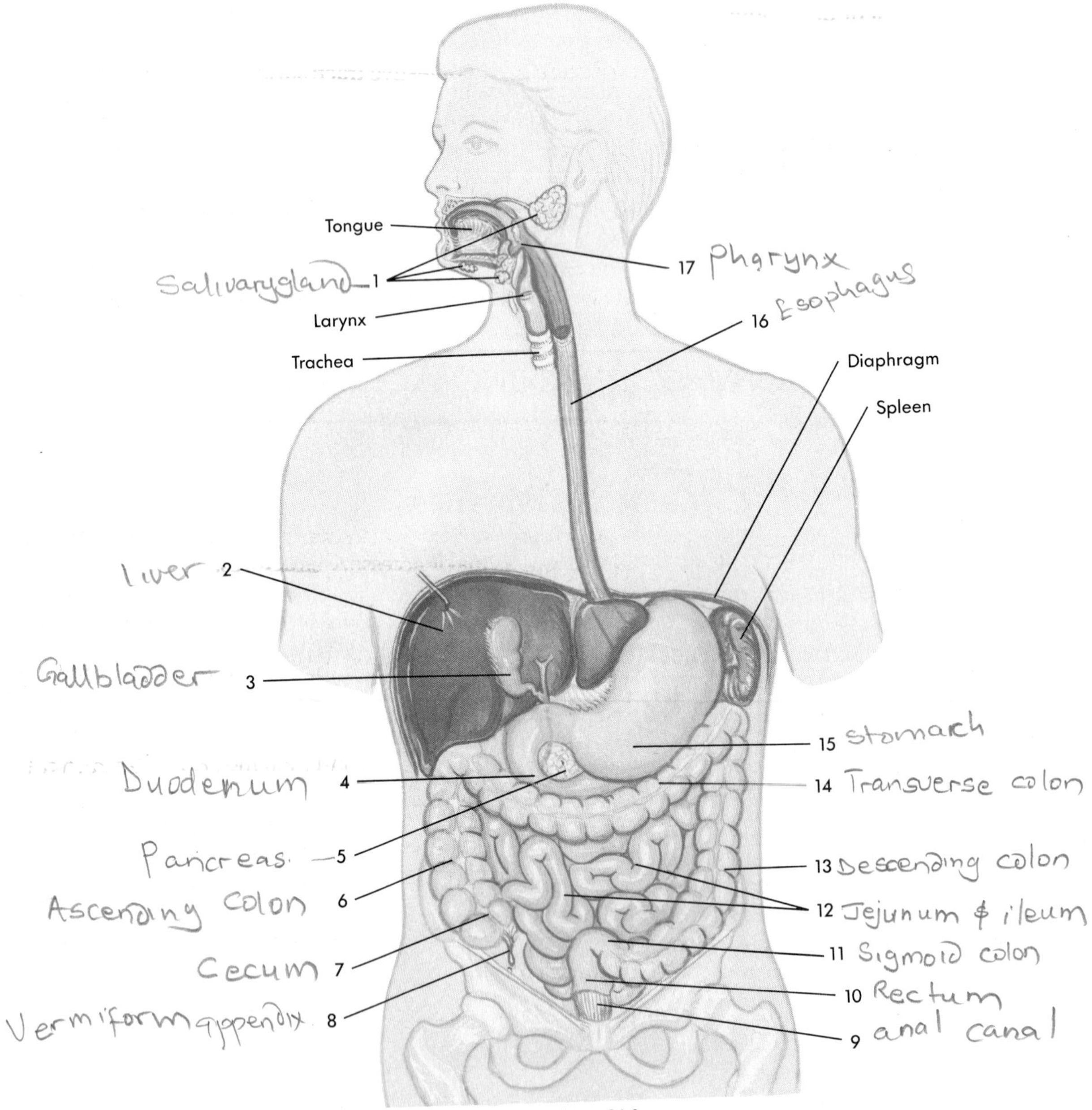

Figure 24-1

1. _____	7. _____	13. _____
2. _____	8. _____	14. _____
3. _____	9. _____	15. _____
4. _____	10. _____	16. _____
5. _____	11. _____	17. _____
6. _____	12. _____	

B. Match these terms with the correct statement or definition:

Esophagus   Salivary glands and tonsils
Oral cavity   Stomach

Oral cavity

1. First section of the digestive tract; surrounded by the lips, cheeks, teeth, and palate.

Salivary glands $ tonsils

2. Accessory glands of the oral cavity.

Esophagus

3. Structure that connects the pharynx and stomach.

Stomach.

4. Part of digestive tract with many tubelike glands that release acid and enzymes; mixes acid, enzymes, and food together.

C. Match these terms with the correct statement or definition:

Anus   Ileum
Cecum   Jejunum
Colon   Rectum
Duodenum

_____

1. First segment of the small intestine, with the liver, gallbladder, and pancreas as major accessory structures.

_____

2. Second segment of the small intestine; major site of absorption.

_____

3. Third segment of the small intestine; more mucus secretion and less absorption.

_____

4. First segment of the large intestine; a blind pouch to which the vermiform appendix attaches.

_____

5. Part of the large intestine with ascending, transverse, descending, and sigmoid segments.

_____

6. Part of large intestine between the sigmoid colon and the anal canal.

_____

7. Inferior termination of the digestive tract.

Histology Overview

The digestive tube consists of four layers or tunics.

A. Match these terms with the correct statement or definition:

Mucosa   Serosa (adventitia)

Muscularis   Submucosa

_____ 1. Innermost tunic, consisting of three layers.

_____ 2. Thick connective tissue tunic between the mucosa and the muscularis tunics.

_____ 3. Tunic that consists of two layers of smooth muscle, located between the submucosa and serosa tunics.

_____ 4. Connective tissue tunic that forms the outermost layer of the digestive tract.

B. Match these parts of the tunics with the correct statement or definition:

Intramural plexus   Myenteric plexus

Lamina propria   Submucosal plexus

Mucous epithelium   Visceral peritoneum

Muscularis mucosae

_____ 1. Layer of mucosa composed of squamous or columnar epithelial cells.

_____ 2. Layer of mucosa composed of loose, irregular connective tissue.

_____ 3. Layer of mucosa composed of a thin, smooth muscle layer.

_____ 4. Nerve plexus consisting of nerve fibers and parasympathetic cell bodies that is located between smooth muscle layers.

_____ 5. Collective name for the submucosal and myenteric plexuses.

_____ 6. Outermost serosa layer; protrudes into the peritoneal cavity.

Physiology Overview

The digestive system has numerous functions for which it is specialized.

Match these terms with the correct statement or definition:

Absorption   Mastication

Defecation   Peristalsis

Deglutition   Regulation

Digestion   Secretion

Ingestion   Segmental contraction

Mass movements

_____ 1. Introduction of solid or liquid food into the stomach.

_____ 2. Process by which food is taken into the mouth and chewed.

_____ 3. The act of swallowing; moves a bolus from the oral cavity into the esophagus.

_____ 4. Contraction of circular and longitudinal muscles in waves; mixes food or moves food through the digestive tract.

_____ 5. The adding of mucus, water, and enzymes to the contents of the digestive tract.

_____ 6. Breakdown, mechanically or chemically, of organic molecules.

_____ 7. Movement of molecules from the digestive tract into the circulatory system.

_____ 8. Elimination of semisolid waste from the digestive tract.

Oral Cavity

❝_The first section of the digestive tract is the oral cavity._❞

A. Match these terms with the correct statement or definition:

Buccal fat pad   Lingual tonsil
Buccinator muscle   Mastication
Extrinsic muscles   Palate
Fauces   Speech
Frenulum   Terminal sulcus
Intrinsic muscles   Vestibule

_____ 1. Posterior boundary of the oral cavity; the opening into the pharynx.

_____ 2. Space between the lips or cheeks and the alveolar processes.

_____ 3. Muscle that flattens the cheek against the teeth.

_____ 4. Structure that rounds out the profile of the side of the face.

_____ 5. Two important functions of the lips, cheeks, and tongue.

_____
_____ 6. Thin fold of tissue that attaches the tongue to the floor of the mouth.

_____ 7. Muscles attached to the tongue; responsible for moving the tongue.

_____ 8. Groove that divides the tongue into two portions.

_____ 9. Aggregation of lymphatic tissue on the posterior one third of the tongue.

B. Match these numbers with the correct description:

1   3
2

_____ 1. Number of incisors in each quadrant of the mouth.

_____ 2. Number of canines in each quadrant of the mouth.

_____ 3. Number of premolars in each quadrant of the mouth.

_____ 4. Number of molars in each quadrant of the mouth.

☞ The third molars are referred to as wisdom teeth.

C. Match these terms with the correct statement or definition:

Alveoli   Periodontal membrane
Gingiva (gum)   Primary teeth
Periodontal ligament   Secondary teeth

_____ 1. Deciduous teeth; also called milk teeth.

_____ 2. Sockets containing the teeth.

_____ 3. Dense fibrous connective tissue and stratified squamous epithelium that covers the alveolar ridges.

_____ 4. Connective tissue that holds the teeth in the alveoli.

_____ 5. Membrane that lines the alveolar walls.

D. Match these terms with the correct statement or definition:

Apical foramen   Neck
Cementum   Pulp
Crown   Pulp cavity
Dentin   Root
Enamel   Root canal

_____ 1. Cutting or chewing surface of tooth, with one or more cusps

_____ 2. Blood vessels, nerves, and connective tissue in center of tooth

_____ 3. Pulp cavity within the root of the tooth.

_____ 4. Living, calcified tissue; surrounds the pulp cavity of the tooth

_____ 5. Extremely hard, nonliving, acellular substance; covers dentin and protects tooth from abrasion and acids.

_____ 6. Substance that covers the root; helps anchor tooth in the jaw

E. Match these terms with the correct parts of the diagram labeled in Figure 24-2:

Apical foramen   Neck
Cementum   Periodontal ligaments
Crown   Pulp cavity
Dentin   Root
Enamel   Root canal

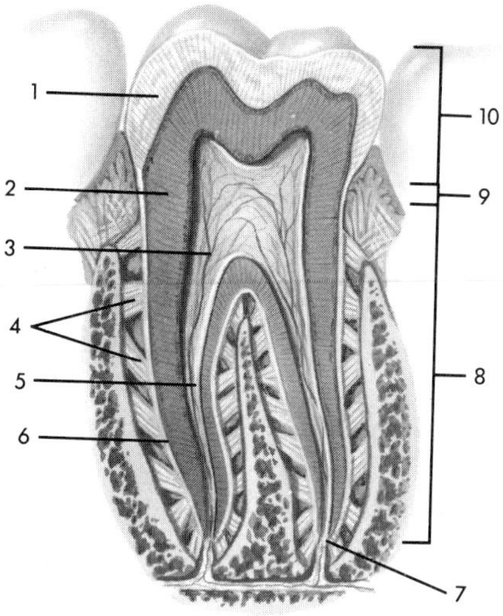

1. _____

2. _____

3. _____

4. _____

5. _____

6. _____

7. _____

8. _____

9. _____

10. _____

Figure 24-2

F. Match these terms with the correct statement or definition:

Hard palate   Soft palate
Muscles of mastication   Uvula
Palatine tonsils

_____ 1. Responsible for movement of the mandible.

_____ 2. Anterior bony structure that separates the nasal and oral cavities.

_____ 3. Posterior, skeletal muscle and connective tissue structure that separates the nasal and oral cavities.

_____ 4. Projection from the posterior edge of the soft palate.

_____ 5. Collection of lymphoid tissue, located in the lateral walls of the fauces.

 The muscles of mastication include the temporalis, masseter, and medial and lateral pterygoid muscles.

G. Match these terms with the correct statement or definition:

Parotid glands   Submandibular glands
Sublingual glands

_____ 1. Largest salivary glands, located just anterior to the ear; produces mostly serous fluid.

_____ 2. Salivary glands that become infected with mumps.

_____ 3. Salivary glands located along the inferior border of the posterior half of the mandible, produce more serous fluid than mucus.

_____ 4. Salivary glands located immediately below the mucous membrane of the floor of the mouth; produce mostly mucus.

Pharynx and Esophagus

" *In the digestive tract, the pharynx communicates anteriorly with the oral cavity and* "
posteriorly with the esophagus.

Match these terms with the correct statement or definition:

Esophagus   Oropharynx
Laryngopharynx   Pharyngeal constrictors
Nasopharynx

_____ 1. Portions of the pharynx that normally transmit food.

_____

_____ 2. Portion of the pharynx superior to the oropharynx.

_____ 3. Portion of the digestive tract that extends between the pharynx and stomach.

_____ 4. Three muscles located in the posterior walls of the oropharynx and laryngopharynx.

 An upper esophageal sphincter and lower esophageal sphincter regulate movement of materials into and out of the esophagus.

Stomach

❝*The stomach is an enlarged segment of the digestive tract in the left superior part of the abdomen.***❞**

A. Match these terms with the correct parts of the diagram labeled in Figure 24-3:

Body
Cardiac region
Fundus
Gastroesophageal (cardiac) opening

Pyloric opening
Pyloric sphincter
Rugae

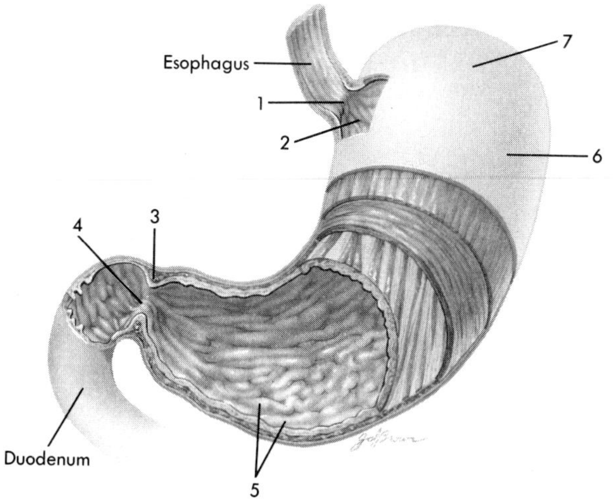

1. _____
2. _____
3. _____
4. _____
5. _____
6. _____
7. _____

Esophagus
7
1
2
6
3
4
Duodenum
5

Figure 24-3

B. Match these terms with the correct statement or definition:

Muscularis layer
Pyloric sphincter

Rugae

_____

_____

_____

1. Relatively thick ring of smooth muscle that surrounds the opening between the stomach and the small intestine.

2. Part of the stomach wall that consists of longitudinal, circular, and oblique layers of smooth muscle.

3. Large mucosal folds formed when the stomach is empty.

C. Match these terms with the correct statement or definition:

Chief cells
Endocrine cells
Gastric glands
Gastric pits

Mucous neck cells
Parietal cells
Surface mucous cells

_____

_____

_____

_____

1. Tubelike openings in the mucosal surface of the stomach.

2. Glands in the stomach that open into the gastric pits.

3. Epithelial cells in the gastric glands that produce hydrochloric acid and intrinsic factor.

4. Epithelial cells in the gastric glands that produce pepsinogen.

Small Intestine

The small intestine consists of three portions: the duodenum, the jejunum, and the ileum.

A. Match these terms with the correct statement or definition:

Greater duodenal papilla
Hepatopancreatic ampulla
Hepatopancreatic ampullar sphincter
Lesser duodenal papilla

_____ 1. Larger mound where the hepatopancreatic ampulla empties into the duodenum.

_____ 2. Smaller mound where the accessory pancreatic duct opens into the duodenum.

_____ 3. Formed from the junction of the common bile duct and the pancreatic duct.

_____ 4. Smooth muscle that keeps the hepatopancreatic ampulla closed.

B. Match these terms with the correct statement or definition:

Absorptive cells
Duodenal glands
Endocrine cells
Goblet cells
Granular cells
Intestinal glands
Lacteal
Microvilli
Plicae circulares (circular folds)
Villi

_____ 1. Folds formed from the mucosa and submucosa that run perpendicular to the long axis of the duodenum.

_____ 2. Tiny, fingerlike projections of the mucosa of the duodenum.

_____ 3. Lymph capillary found in a villus.

_____ 4. Cytoplasmic extensions of villi; collectively they form the brush border.

_____ 5. Simple columnar epithelial cells in the duodenum that are specialized to produce digestive enzymes and absorb food.

_____ 6. Simple columnar epithelial cells in the duodenum that produce a protective mucus.

_____ 7. Cells that may help protect the intestinal epithelium from bacteria.

_____ 8. Tubular invaginations of the mucosa at the base of the villi.

_____ 9. Coiled, tubular mucous glands in the submucosa of the duodenum.

 The plicae circulares, villi, and microvilli function to increase surface area in the small intestine.

C. Match these terms with the correct statement or definition:

Duodenum
Ileocecal sphincter
Ileocecal valve

Ileum
Jejunum
Peyer's patches

_____

_____

1. Two parts of the small intestine that are the major sites of nutrient absorption.

_____

2. Aggregations of lymph nodules in the ileum.

_____

3. Ring of smooth muscle located at the junction of the ileum and the large intestine (ileocecal junction).

_____

4. One-way valve at the ileocecal junction.

Liver and Gallbladder

66 *The liver is the largest internal organ of the body.* 99

A. Match these terms with the correct statement or definition:

Caudate and quadrate
Common bile duct
Common hepatic duct
Cystic duct

Gallbladder
Left and right
Porta

_____

1. Two minor lobes of the liver.

_____

2. Located on the inferior surface of the liver where various vessels, ducts, and nerves enter and exit the liver.

_____

3. Duct formed from the junction of the right and left hepatic ducts.

_____

4. Duct from the gallbladder.

_____

5. Duct formed from the junction of the common hepatic duct and the cystic duct.

_____

6. Small sac on the inferior surface of the liver that stores bile.

B. Match these terms with the correct statement or definition:

Bare area
Endothelial and phagocytic cells
Hepatic cords
Hepatic sinusoids
Hepatocytes
Lobules
Portal triad

_____ 1. Small area on the diaphragmatic surface of the liver not covered by capsule and visceral peritoneum.

_____ 2. Portions of the liver divided by connective tissue septa.

_____ 3. Corner of a liver lobule where three vessels are commonly located.

_____ 4. Structures located between the central vein and the septa of each lobule; consist of hepatocytes.

_____ 5. Functional cells of the liver; produce bile.

_____ 6. Blood channels that separate the hepatic cords.

_____ 7. Cells that line the liver sinusoids.

C. Using the terms provided, complete these statements:

Bile canaliculi
Central vein
Hepatic ducts
Hepatic portal vein
Hepatic veins

In the liver, blood from the _(1)_ and the hepatic artery flows into the sinusoids and becomes mixed. The mixed blood then flows to the _(2)_, where it exits the lobule and then exits the liver through the _(3)_. Bile, which is produced by the hepatocytes, flows through the _(4)_ toward the triad and exits the liver through the _(5)_.

1. _____

2. _____

3. _____

4. _____

5. _____

 The wall of the gallbladder has rugae, which allow gallbladder expansion, and smooth muscle, which enables the gallbladder to contract.

Pancreas

The pancreas is a complex organ composed of both endocrine and exocrine tissues that perform several functions.

A. Using the terms provided, complete these statements:

Acini
Head
Intercalated ducts
Interlobular ducts
Intralobular ducts

Main pancreatic duct
Pancreatic islets (islets of Langerhans)
Tail

1. _____
2. _____
3. _____
4. _____
5. _____
6. _____
7. _____
8. _____

The pancreas consists of a __(1)__ located within the curvature of the duodenum, a body, and a __(2)__, which extends to the spleen. The exocrine portion of the pancreas consists of __(3)__, which produce digestive enzymes. Clusters of acini are connected by small __(4)__ to __(5)__, which leave the lobules to join __(6)__ between the lobules. The interlobular ducts attach to the __(7)__, which joins the common hepatic duct at the hepatopancreatic ampulla. Insulin and glucagon are produced by cells within the __(8)__.

B. Match these terms with the correct parts of the diagram labeled in Figure 24-4:

Common bile duct
Common hepatic duct
Cystic duct

Hepatic ducts
Hepatopancreatic ampulla
Pancreatic duct

1. _____
2. _____
3. _____
4. _____
5. _____
6. _____

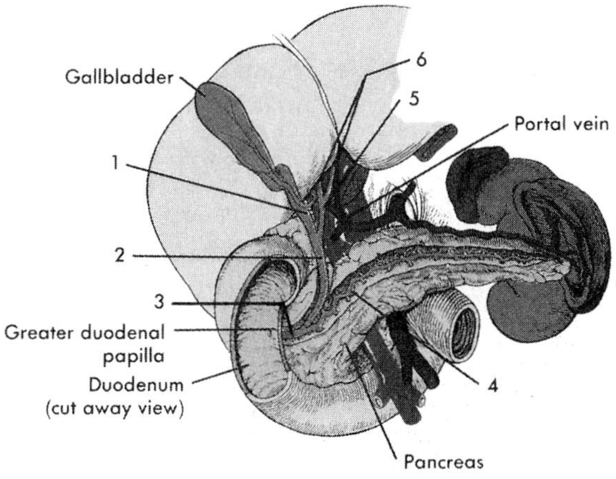

Gallbladder
6
5
Portal vein
1
2
3
Greater duodenal papilla
Duodenum (cut away view)
4
Pancreas

Figure 24-4

Large Intestine

66 *The large intestine extends from the ileocecal junction to the inferior end of the rectum.* 99

Match these terms with the
correct statement or definition:

Anal canal   Haustra
Anus   Rectum
Ascending colon   Sigmoid colon
Crypts   Teniae coli
Descending colon   Transverse colon
Epiploic appendages Vermiform appendix

_____ 1. Portion of the colon that extends from the right colic flexure to the left colic flexure.

_____ 2. Portion of the colon that forms an S-shaped tube that ends at the rectum.

_____ 3. Three bands of longitudinal smooth muscle that run the length of the colon.

_____ 4. Pouches formed in the colon when the teniae coli contract.

_____ 5. Small, fat-filled connective tissue pouches attached to the outer surface of the colon.

_____ 6. Straight tubular glands in the epithelium of the large intestine.

_____ 7. Straight, muscular tube between the sigmoid colon and anal canal.

_____ 8. Last 2 to 3 cm of the digestive tube.

☞ Smooth muscle forms the internal anal sphincter at the superior end of the anal canal, and skeletal muscle forms the external anal sphincter at the inferior end of the anal canal.

Peritoneum

66 *The body walls and organs of the abdominal cavity are lined with serous membranes.* 99

Match these terms with the
correct statement or definition:

Coronary ligament   Omental bursa
Falciform ligament   Parietal peritoneum
Greater omentum   Retroperitoneal
Lesser omentum   Transverse and sigmoid
Mesenteries   mesocolons
Mesentery proper   Visceral peritoneum
Mesoappendix

_____ 1. Serous membrane that covers the abdominal organs.

_____ 2. General term for connective tissue sheets that hold abdominal organs in place.

_____ 3. Abdominal organs that lie against the abdominal wall and have no mesenteries.

_____ 4. Mesentery connecting the lesser curvature of the stomach to the liver and diaphragm.

_____ 5. Cavity or pocket formed in the greater omentum as it extends inferiorly.

_____ 6. Mesentery that attaches the liver to the anterior abdominal wall.

_____ 7. Mesenteries of the small intestine.

_____ 8. Mesenteries of the colon.

Functions of the Digestive System

66 *Each segment of the digestive tract is specialized to assist in moving its contents from the oral end* 99 *to the anal end.*

Using the terms provided, complete these statements:

Effector organ	Nervous
Intramural plexus	Receptors
Local reflex	

The processes of secretion, movement, and absorption are regulated by elaborate _(1)_ and hormonal mechanisms. The digestive tract has a unique regulatory mechanism, called the _(2)_ , which does not involve the spinal cord or brain. Stimuli (e.g., distention of the digestive tract) activate _(3)_ within the wall of the digestive tract, and action potentials are generated in the neurons of the _(4)_ . The action potentials travel up or down the intramural plexus and produce a response in a(n) _(5)_ (e.g., smooth muscle or gland).

1. _____
2. _____
3. _____
4. _____
5. _____

Oral Cavity

66 *Food is masticated in the mouth, as secretions are added to the food.* 99

A. Match these terms with the correct statement or definition:

Chewing (mastication) reflex   Salivary amylase
Mucin

_____ 1. Digestive enzyme in the serous portion of saliva.

_____ 2. Proteoglycan secreted by the submandibular and sublingual glands that adds lubrication to the saliva.

_____ 3. Reflex integrated in the medulla oblongata that controls chewing.

B. Using the terms provided, complete these statements:

Antibacterial
Immunoglobulin A
Incisors and canines

Parasympathetic
Premolars and
   molars
Starch

1. _____

2. _____

3. _____

4. _____

5. _____

6. _____

The amylase in saliva starts the digestive process by breaking the covalent bonds between glucose molecules in _(1)_ and other polysaccharides. In addition, saliva prevents bacterial infection in the mouth by washing the oral cavity, and it contains substances (e.g., lysozyme) with weak _(2)_ action. Saliva also contains _(3)_, which helps prevent bacterial infection. Salivary gland secretion is stimulated mainly by the _(4)_ nervous system, but also by the sympathetic nervous system. Tactile simulation, certain tastes, odors, and higher brain centers also affect the activity of the salivary glands. Food taken into the mouth is chewed, or masticated, by the teeth. The most anterior teeth, the _(5)_, primarily cut and tear food, whereas the _(6)_ primarily crush and grind the food.

Deglutition

66 *Deglutition, or swallowing, can be divided into three different phases.* 99

A. Match these phases of deglutition with the correct statement or definition:

Esophageal phase
Pharyngeal phase
Voluntary phase

_____

_____

_____

1. Phase of swallowing that involves forcing a bolus of food into the oropharynx.

2. Phase of swallowing that involves closing the nasopharynx, forcing food through the pharynx, and covering the opening into the larynx.

3. Phase of swallowing that is responsible for moving food from the pharynx to the stomach.

B. Match these terms with the correct statement or definition:

Epiglottis
Peristaltic waves

Pharyngeal constrictor muscles
Swallowing center

_____

_____

_____

_____

1. Area in the medulla oblongata that controls swallowing.

2. Muscles that contract and force food through the pharynx.

3. Part of the larynx that covers the opening into the larynx.

4. Muscular contractions of the esophagus.

492

The presence of food in the esophagus stimulates the intramural plexus, which initiates the peristaltic waves; the peristaltic contractions cause relaxation of the lower esophageal sphincter.

Stomach

❝The stomach functions primarily as a storage and mixing chamber for ingested food.❞

A. Match these terms with the correct statement or definition:

Gastrin   Mucus
Hydrochloric acid   Pepsin
Intrinsic factor   Pepsinogen

_____ 1. Viscous, alkaline substance that covers the surface of epithelial cells.

_____ 2. Glycoprotein secreted by parietal cells that binds with vitamin B_{12} and makes it more readily absorbed in the ileum.

_____ 3. Inactive form of pepsin, packaged in zymogen granules; secreted by chief cells.

_____ 4. Stomach enzyme that catalyzes the cleavage of some covalent bonds in proteins.

_____ 5. Hormone that increases gastric secretion and increases stomach emptying.

B. Match these phases of stomach secretion with the correct statement:

Cephalic phase
Gastric phase
Intestinal phase

_____ 1. Phase of gastric secretion that responds to taste, smell, and the sensations of chewing and swallowing.

_____ 2. Phase of gastric secretion that is stimulated by food in the stomach.

_____ 3. Phase of gastric secretion that is stimulated by the entrance of acidic chyme into the duodenum.

C. Using the terms provided, complete these statements:

Decrease   Increase

Several mechanisms regulate gastric secretions. Through
the medulla the smell, taste, or thought of food can
_(1)_ parasympathetic stimulation of parietal cells, chief
cells, and endocrine cells. The endocrine cells secrete
gastrin, which travels in the blood back to the stomach
mucosa and causes a(n) _(2)_ in hydrochloric acid secretion.
In the stomach, distention and the presence of amino acids
and peptides activate CNS and local reflexes that
_(3)_ gastric secretions. Partially digested proteins, caffeine,
and alcohol can also_(4)_ secretions of gastrin. Increased
acidity in the duodenum stimulates the secretion of secretin,
which acts to _(5)_ gastric acid secretion; fatty acids in the
duodenum stimulate the secretion of cholecystokinin and
gastric inhibitory peptide, which function to _(6)_ gastric
secretions. Also in the duodenum, distention and increased
acidic chyme can activate the enterogastric reflex and cause
a(n) _(7)_ in gastric secretions.

1. _____

2. _____

3. _____

4. _____

5. _____

6. _____

7. _____

D. Match these terms with the   Chyme   Peristaltic waves
   correct statement or definition: Mixing waves   Pyloric pump

_____   1. Semifluid material formed from ingested food mixed with
                                    stomach gland secretions.

_____   2. Strong waves of contraction that force chyme toward the pyloric
                                    sphincter.

_____   3. Movement of chyme through the partially closed pylorus by the
                                    force of peristaltic contraction.

☞ Motility in the stomach is regulated by many of the hormonal and neural mechanisms that
   stimulate stomach secretions. For instance, distention of the stomach stimulates local
   reflexes, central nervous system reflexes, and the release of gastrin, all of which increase
   stomach motility and relaxation of the pyloric sphincter.

Small Intestine

"The small intestine is the site where the greatest amount of digestion and absorption occurs.**"**

A. Match these terms with the   Disaccharidases   Peptidases
   correct description: Mucus   Secretin and cholecystokinin
                                        Nucleases

_____   1. Secreted by duodenal glands, goblet cells, and within intestinal
                                    glands.

_____   2. Hormones released from intestinal mucosa that stimulate
                                    hepatic and pancreatic secretions.

_____ 3. Enzymes on the intestinal microvilli that break down disaccharides to monosaccharides.

_____ 4. Enzymes on the intestinal microvilli that break peptide bonds between amino acid chains.

_____ 5. Enzymes on the intestinal microvilli that break down nucleic acids.

☞ Duodenal gland secretion is stimulated by the vagus nerve, secretin, and chemical or tactile irritation of the duodenal mucosa. Goblet cells produce mucus in response to tactile and chemical stimulation of the mucosa.

B. Match these terms with the correct statement or definition:

Decreases   Peristaltic contractions
Increases   Segmental contractions

_____ 1. Propagated for only short distances, they mix the intestinal contents.

_____ 2. Effect of small intestine distention on intestinal smooth muscle contraction.

_____ 3. Effect of low pH, amino acids, and peptides on small intestine contraction.

_____ 4. Effect of parasympathetic stimulation on small intestine contraction.

_____ 5. Effect of cecal distention on constriction of the ileocecal sphincter.

Liver and Gallbladder

❝ *The liver produces bile, whereas the gallbladder stores and concentrates bile.* ❞

A. Match these terms with the correct statement:

Inhibits
Stimulates

_____ 1. Effect of secretin on bile secretion.

_____ 2. Effect of parasympathetic stimulation or increased blood flow through the liver on bile secretion.

_____ 3. Effect of bile salts on bile secretion.

_____ 4. Effect of cholecystokinin on contraction of the gallbladder.

_____ 5. Effect of vagal stimulation on contraction of the gallbladder.

_____ 6. Effect of cholecystokinin on the hepatopancreatic ampullar sphincter.

Most bile salts are reabsorbed in the ileum and are carried back to the liver in the blood, where they stimulate further bile secretion.

B. Using the terms provided, complete these statements:

Bile pigments   Phagocytize
Bile salts   Store
Detoxify   Synthesize
Glycogen   Urea
Interconversion

1. _____

2. _____

3. _____

4. _____

5. _____

6. _____

7. _____

8. _____

9. _____

Although bile does not contain digestive enzymes, it does have (1) , which emulsify fats. Bile also contains excretory products such as (2) and cholesterol. Hepatocytes can (3) fat, vitamins A, D, E, and K, copper, and iron. The liver is an important regulator of blood sugar levels because hepatocytes can remove sugar from the blood and store it as (4) , or they can break down glycogen into sugar that is released into the blood. Another function that hepatocytes perform is (5) of nutrients, in which the proportion of nutrients is controlled by changing one type of nutrient into another (e.g., amino acids into glucose). Substances not readily usable by cells are transformed within the liver. For example, vitamin D is converted to its active form by hepatocytes. The hepatocytes also (6) many harmful substances by altering their structure, such as converting ammonia to (7) . Hepatic phagocytic cells (8) worn out and dying red and white blood cells, bacteria, and other debris. The liver can also (9) its own unique new compounds, including albumins and other blood proteins.

Pancreas

"The exocrine secretions of the pancreas are called pancreatic juice."

A. Match these components of pancreatic juice with the correct statement or definition:

Aqueous component
Enzymatic component

_____

_____

_____

_____

_____

1. Portion of pancreatic juice that contains sodium ions, potassium ions, and bicarbonate ions.

2. Portion of pancreatic juice produced by the epithelial cells of the small pancreatic ducts.

3. Portion of pancreatic juice that is important for the digestion of food.

4. Portion of pancreatic juice produced by the acinar cells of the pancreas.

5. Portion of pancreatic juice that has an increased secretion rate because of secretin.

_____ 6. Portion of pancreatic juice that has an increased secretion rate because of cholecystokinin.

_____ 7. Portion of pancreatic juice that has an increased secretion rate because of parasympathetic stimulation.

B. Match these terms with the correct statement or definition:

Amylase	Enterokinase
Carboxypeptidase	Lipase
Chymotrypsin	Ribonuclease
Deoxyribonuclease	Trypsin

_____ 1. Major proteolytic enzymes in pancreatic juice.

_____

_____

_____ 2. Proteolytic enzyme that cleaves trypsinogen to trypsin; produced by the brush border of the small intestine.

_____ 3. Enzyme that digests polysaccharides.

_____ 4. Enzyme that digests lipids.

_____ 5. Enzyme that digests DNA to its component nucleotides.

Large Intestine

❝While in the colon, chyme is converted to feces, where it is stored until eliminated by defecation.**❞**

A. Match these terms with the correct statement or definition:

Bicarbonate ions	Mucus
Flatus	Water and salts
Microorganisms	

_____ 1. Substances absorbed by the colon.

_____ 2. Substance secreted by goblet cells in the colon.

_____ 3. Neutralize the acid produced by bacteria in the colon.

_____ 4. Source of vitamin K synthesis, and 30% of the dry weight of feces.

_____ 5. Gases produced by bacterial action in the colon.

B. Match these terms with the correct statement or definition:

Defecation reflex
Duodenocolic reflex
Gastrocolic reflex
Mass movements

_____ 1. Strong peristaltic contractions of the transverse and descending colon.

_____ 2. Strong peristaltic contractions of the colon initiated by stomach.

_____ 3. Distention of the rectal wall by feces initiates this reflex.

_____ 4. Results in reinforcement of peristaltic contraction in the lower colon and rectum and relaxation of the internal anal sphincter.

☞ Local reflexes cause weak contractions and relaxation of the internal sphincter. Parasympathetic reflexes cause stronger contractions and are normally responsible for most of the defecation reflex.

Digestion, Absorption, and Transport

66 *Digestion, absorption, and transport are components of nutrition.* 99

Match these terms with the correct statement or definition:

Absorption
Digestion
Ions and water-
    soluble substances
Lipids and lipid-soluble
    substances
Transport

_____ 1. Begins in the mouth.

_____ 2. Most occurs in the duodenum and jejunum, although some occurs in the ileum.

_____ 3. Transported through the hepatic portal system to the liver.

_____ 4. Transported into lacteals, through the lymphatic system to the left subclavian vein, and then to the liver or adipose tissue.

Carbohydrates

66 *Ingested carbohydrates consist primarily of starches, glycogen, sucrose, lactose, glucose, and fructose.* 99

Match these terms with the correct statement or definition:

Disaccharidases
Glucose
Insulin
Pancreatic amylase
Salivary amylase

_____ 1. Enzyme that digests starch and is secreted into the oral cavity.

_____ 2. Enzymes bound to the microvilli of the intestinal epithelium.

_____ 3. Sugar transported by the circulatory system to cells that need energy.

_____ 4. Hormone that greatly increases the rate of glucose transport into most types of cells.

Glucose enters the cells by the process of facilitated diffusion.

Lipids

66*Lipids are molecules that are insoluble or only slightly soluble in water.*99

Using the terms provided, complete these statements:

Bile salts   Lipase
Chylomicrons   Liver
Emulsification   Micelles
Lacteal   Triacylglycerols

Lipids include triacylglycerols, phospholipids, steroids, and fat-soluble vitamins. __(1)__ consist of one glycerol molecule and three fatty acids covalently bound together. The first step in lipid digestion is __(2)__, which is the transformation of large lipid droplets into much smaller droplets. This process is accomplished by __(3)__ secreted by the liver. __(4)__ secreted by the pancreas digests lipid molecules. Once lipids are digested in the intestine, bile salts aggregate around the small droplets to form __(5)__. When these structures come into contact with epithelial cells of the small intestine, their contents pass through the cell membrane of the epithelial cell by the process of simple diffusion. Within the smooth endoplasmic reticulum of the intestinal epithelial cells, free fatty acids combine with monoacylglycerol to form triacylglycerol. Proteins in the epithelial cells coat droplets of triacylglycerols, phospholipids, and cholesterol to form __(6)__, which leave the epithelial cell to enter a __(7)__. From there the chylomicrons are transported to the blood and are carried to adipose tissue or the __(8)__.

1. _____
2. _____
3. _____
4. _____
5. _____
6. _____
7. _____
8. _____

Proteins

66*Proteins are taken into the body from a variety of dietary sources.*99

Match these terms with the correct statement or definition:

Pepsin   Trypsin
Peptidase

_____

_____

_____

1. Enzyme in the stomach that catalyzes the cleavage of covalent bonds in proteins.

2. Enzyme produced by the pancreas that continues the digestion of proteins started in the stomach; produces small peptide chains.

3. Enzyme bound to the microvilli and found inside intestinal epithelial cells; completes the breakdown of small peptide chains.

Active transport of amino acids into the cells of the body is stimulated by growth hormone and insulin.

Water and Ions

66 *Water can move in either direction across the wall of the small intestine.* 99

Match these terms with the
correct statement or definition:

Active transport   Osmosis
Into circulation   Simple diffusion
Into lumen of intestine

_____   1. Mechanism responsible for water movement across the wall of the small intestine.

_____   2. Mechanism that moves sodium, potassium, magnesium, calcium, and phosphate into the epithelial cells of the small intestine.

_____   3. Passive movement of negative ions (e.g., chloride ions) as they follow positive ions (e.g., sodium ions) into intestinal epithelial cells.

_____   4. Direction of water movement when the chyme is very concentrated.

_____   5. Direction of water movement as nutrients are absorbed.

QUICK RECALL

1. Name the four layers or tunics of the digestive tract.

2. Group the four types of teeth found in humans, according to their function.

3. List the three large pairs of multicellular salivary glands, and name the digestive enzyme found in saliva.

4. Name six sphincters that control movement of materials through the digestive tract.

5. Name the five types of epithelial cells in the stomach, and list their secretions.

6. List the three structural modifications that increase surface area in the small intestine.

7. List the three major types of cells found in the intestinal mucosa.

8. Name the three phases of swallowing.

9. List the types of contraction (movement) that occur in the stomach, small intestine, and large intestine.

10. List the three phases of gastric secretion.

11. Name a substance found in pancreatic juice that is responsible for each of these activities: neutralizes acid, digests proteins, digests fats, digests carbohydrates.

12. List three functions of bile.

13. List four major functions of the liver in addition to the production of bile.

14. List three major functions of the colon.

15. In these table, indicate if the control mechanism stimulates (S), inhibits (I), or has no effect (O) on the activity:

	GASTRIN	CHOLE-CYSTO-KININ	SECRETIN	PARA-SYMPA-THETIC
Stomach secretion	_____	_____	_____	_____
Bile secretion	_____	_____	_____	_____
Pancreas secretion	_____	_____	_____	_____
Contraction of gallbladder	_____	_____	_____	_____
Gastric motility	_____	_____	_____	_____

16. List the breakdown products of carbohydrates, proteins, and triacylglycerols.

17. List the locations in the digestive tract where carbohydrate digestion, lipid digestion, and protein digestion occur.

18. Name the routes by which water-soluble and lipid-soluble molecules leave the intestinal epithelial cells.

MASTERY LEARNING ACTIVITY

Place the letter corresponding to the correct answer in the space provided.

_____ 1. Which layer of the digestive tract is in direct contact with the food that is consumed?
  a. mucosa
  b. muscularis
  c. serosa
  d. submucosa

_____ 2. The intramural plexus is found in the
  a. submucosa.
  b. muscularis.
  c. serosa.
  d. a and b
  e. all of the above

_____ 3. The tongue
  a. holds food in place during mastication.
  b. is involved in swallowing.
  c. helps to form words during speech.
  d. all of the above

_____ 4. Dentin
  a. forms the surface of the crown of teeth.
  b. holds the teeth to the periodontal membrane.
  c. is found in the pulp cavity.
  d. makes up most of the structure of teeth.
  e. is harder than enamel.

_____ 5. The number of premolar, deciduous teeth is
  a. 0.
  b. 4.
  c. 8.
  d. 12.

_____ 6. Which of these glands secrete saliva into the oral cavity?
  a. submandibular glands
  b. sublingual glands
  c. parotid glands
  d. all of the above

_____ 7. The stomach
  a. has large folds in the submucosa and mucosa called rugae.
  b. has two layers of smooth muscle in the muscularis layer.
  c. opening into the small intestine is the cardiac opening.
  d. all of the above

_____ 8. Which of these stomach cell types is NOT correctly matched with its function?
  a. surface mucous cells: produce mucus
  b. parietal cells: produce hydrochloric acid
  c. chief cells: produce intrinsic factor
  d. endocrine cells: produce regulatory hormones

_____ 9. Which of these structures function to increase the mucosal surface of the small intestine?
  a. plicae circulares
  b. villi
  c. microvilli
  d. the length of the small intestine
  e. all of the above

_____ 10. Given these parts of the small intestine:
  1. duodenum
  2. ileum
  3. jejunum

  Choose the arrangement that lists the parts in the order food would encounter them as the food passes from the stomach through the small intestine.
  a. 1, 2, 3
  b. 1, 3, 2
  c. 2, 1, 3
  d. 2, 3, 1

11. Which structures release digestive enzymes in the small intestine?
    a. duodenal glands
    b. goblet cells
    c. endocrine cells
    d. absorptive cells

12. The hepatic sinusoids
    a. receive blood from the hepatic artery.
    b. receive blood from the hepatic portal vein.
    c. empty into the central veins.
    d. all of the above

13. Given these ducts:
    1. common bile duct
    2. common hepatic duct
    3. cystic duct
    4. hepatic ducts

    Choose the arrangement that lists the ducts in the order bile would pass through them when moving from the bile canaliculi of the liver to the small intestine.
    a. 3, 4, 2
    b. 3, 2, 1
    c. 4, 2, 1
    d. 4, 1, 2

14. Given these structures:
    1. ascending colon
    2. descending colon
    3. sigmoid colon
    4. transverse colon

    Choose the arrangement that lists the structures in the order food would encounter them as food passes from the small intestine to the rectum.
    a. 1, 2, 3, 4
    b. 1, 4, 2, 3
    c. 2, 3, 1, 4
    d. 2, 4, 1, 3

15. Given these structures:
    1. coronary ligament
    2. greater omentum
    3. lesser omentum
    4. transverse mesocolon

    Choose the arrangement that lists the structures in the order they would be encountered as one moves along the mesenteries from the diaphragm to the posterior abdominal wall.
    a. 1, 2, 3, 4
    b. 1, 3, 2, 4
    c. 2, 3, 4, 1
    d. 3, 2, 4, 1

16. A local reflex
    a. involves the spinal cord.
    b. occurs in the intramural plexus.
    c. causes mesenteries to contract.
    d. is the relaxation of smooth muscle when it is suddenly stretched.

17. The portion of the digestive tract in which digestion begins is the
    a. oral cavity.
    b. esophagus.
    c. stomach.
    d. duodenum.
    e. jejunum.

18. During swallowing,
    a. movement of food results primarily from gravity.
    b. the swallowing center in the medulla oblongata is activated.
    c. food is pushed into the oropharynx during the pharyngeal phase.
    d. the soft palate closes off the opening to the larynx.

19. HCl
    a. is an enzyme.
    b. creates the acid condition necessary for pepsin to work.
    c. is secreted by the small intestine.
    d. all of the above

_____ 20. Why doesn't the stomach digest itself?
   a. The stomach wall is not composed of protein, so there are no digestive enzymes to attack it.
   b. The digestive enzymes in the stomach are not efficient enough.
   c. The lining of the stomach is too tough to be attacked by digestive enzymes.
   d. The stomach wall is protected by large amounts of mucus.

_____ 21. Which of these hormones stimulate stomach secretions?
   a. cholecystokinin
   b. gastric inhibitory peptide
   c. gastrin
   d. secretin

_____ 22. Which of these phases of stomach secretion is correctly matched?
   a. cephalic phase: the largest volume of secretion is produced
   b. gastric phase: gastrin secretion is inhibited by distention of the stomach
   c. gastric phase: initiated by chewing, swallowing, or thinking of food
   d. intestinal phase: stomach secretions are first stimulated, then inhibited

_____ 23. Which of these statements accurately reflects the role that the stomach plays?
   a. It receives relatively large amounts of materials during short periods of time and slowly releases them over a longer time.
   b. It is the primary organ of digestion.
   c. It performs no function other than storage.
   d. It is the primary site of fat digestion.

_____ 24. The function of peristaltic waves in the stomach is to
   a. move chyme into the small intestine.
   b. increase the secretion of HCl.
   c. empty the haustra.
   d. all of the above

_____ 25. Which of these would occur if a person suffered from a severe case of hepatitis that impaired liver function?
   a. Fat digestion might be hampered.
   b. By-products of hemoglobin break-down might accumulate in the blood.
   c. Plasma proteins might decrease in concentration.
   d. b and c
   e. all of the above

_____ 26. The gallbladder
   a. produces bile.
   b. stores bile.
   c. contracts and releases bile in response to secretin.
   d. contracts and releases bile in response to sympathetic stimulation.

_____ 27. The aqueous component of pancreatic secretions
   a. is secreted by the islets of Langerhans.
   b. contains bicarbonate ions.
   c. is released primarily in response to cholecystokinin.
   d. all of the above

_____ 28. Which of these statements is consistent with inflammation of the pancreas (pancreatitis)?
   a. Abdominal pain is enhanced because of the escape of activated digestive enzymes from the pancreas into surrounding tissues.
   b. Release of digestive enzymes into the duodenum is reduced, resulting in disturbed digestion, nausea, and vomiting.
   c. Production of hepatic enzymes compensates for the reduction in pancreatic enzyme production.
   d. Pancreatic enzymes accumulate in the gallbladder until the inflammation subsides.
   e. a and b

_____ 29. Which of these is a function of the large intestine?
   a. storage of wastes
   b. absorption of certain vitamins
   c. absorption of water and salts
   d. production of mucus
   e. all of the above

_____ 30. Defecation
   a. can be initiated by stretch of the rectum.
   b. can occur as a result of mass movements.
   c. involves local reflexes.
   d. involves parasympathetic reflexes mediated by the spinal cord.
   e. all of the above

_____ 31. The enzyme responsible for the digestion of carbohydrates is produced by the
   a. salivary glands.
   b. pancreas.
   c. lining of the small intestine.
   d. a and b
   e. all of the above

_____ 32. Bile
   a. is an important enzyme for the digestion of fats.
   b. is made by the gallbladder.
   c. contains breakdown products from hemoglobin.
   d. emulsifies fats.
   e. c and d

_____ 33. Micelles are
   a. lipids surrounded by bile salts.
   b. produced by the pancreas.
   c. released into lacteals.
   d. all of the above

_____ 34. If the thoracic duct were tied off, which of these classes of nutrients would NOT enter the circulatory system at their normal rate?
   a. amino acids
   b. glucose
   c. lipids
   d. fructose

_____ 35. Two enzymes involved in the digestion of proteins are
   a. pepsin and lipase.
   b. trypsin and hydrochloric acid.
   c. pancreatic amylase and bile.
   d. pepsin and trypsin.

☆ ———————— FINAL CHALLENGES ———————— ☆

Use a separate sheet of paper to complete this section.

1. When you chew bread for a few minutes, it tastes sweet. Explain how this happens.

2. Explain how drugs that bind to bile salts in the small intestine can cause a decrease in blood cholesterol levels.

3. You and your anatomy and physiology instructor are lost in the desert without water. Your instructor suggests that you place a pebble in your mouth. What would you do and why?

4. Suppose you were given a histological slide from the digestive tract with these characteristics: (1) mucosal epithelial cells were columnar, (2) many goblet cells, (3) the surface was highly folded to form leaflike projections, and (4) each fold contained a capillary and a lymphatic vessel. What portion of the digestive tract did the slide come from?

5. Why might cutting the vagal nerve supply to the stomach help someone with a peptic ulcer? What side effects might be expected from such a procedure?

6. If a friend had a peptic ulcer, would you recommend a diet high in fats or high in proteins?

7. If a friend had a duodenal peptic ulcer, would you recommend two large meals or six small meals per day? Explain.

8. An experimenter performed two experiments to determine the factors that affect stomach secretion. In the first experiment he measured the amount of gastric juice secreted as a function of the concentration of fats entering the duodenum. The results are graphed below:

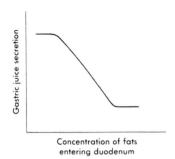

In experiment two, animal A was given food that contained a large amount of fat and animal B was given food that contained a low amount of fat. Then, plasma was removed from animal A and injected into animal C, and plasma from animal B was injected into animal D. Finally, the gastric juice secretion in animals C and D was determined with these results:

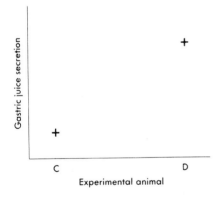

On the basis of these data, what conclusions can be reached regarding the regulation of gastric juice secretion?

9. Upon autopsy, a 50-year-old male was found to have an abnormal liver. It was yellowish and enlarged and appeared to be infiltrated with connective tissue. Before his death this person probably suffered from which of these symptoms: jaundice, edema, easy fatigue, or weight loss?

10. Explain why an enema results in defecation.

11. Many people have a bowel movement following a meal, especially breakfast. Why does this occur?

12. A patient has a spinal cord injury that has damaged the sacral region of the spinal cord. How will this affect her ability to defecate?

ANSWERS TO CHAPTER 24

General Overview
1. Tube; 2. Accessory organs; 3. Alimentary;
4. Gastrointestinal tract

Anatomy Overview
A. 1. Salivary glands; 2. Liver; 3. Gallbladder;
4. Duodenum; 5. Pancreas; 6. Ascending colon; 7. Cecum; 8. Vermiform appendix;
9. Anal canal; 10. Rectum; 11. Sigmoid colon;
12. Jejunum and ileum; 13. Descending colon;
14. Transverse colon; 15. Stomach;
16. Esophagus; 17. Pharynx
B. 1. Oral cavity; 2. Salivary glands and tonsils;
3. Esophagus; 4. Stomach
C. 1. Duodenum; 2. Jejunum; 3. Ileum; 4. Cecum;
5. Colon; 6. Rectum; 7. Anus

Histology Overview
A. 1. Mucosa; 2. Submucosa; 3. Muscularis;
4. Serosa (adventitia)
B. 1. Mucous epithelium; 2. Lamina propria;
3. Muscularis mucosae; 4. Myenteric plexus;
5. Intramural plexus; 6. Visceral peritoneum

Physiology Overview
1. Ingestion; 2. Mastication; 3. Deglutition;
4. Peristalsis; 5. Secretion; 6. Digestion;
7. Absorption; 8. Defecation

Oral Cavity
A. 1. Fauces; 2. Vestibule; 3. Buccinator muscle;
4. Buccal fat pad;   5. Mastication and speech;
6. Frenulum; 7. Extrinsic muscles; 8. Terminal sulcus; 9. Lingual tonsil
B. 1. 2; 2. 1; 3. 2; 4. 3
C. 1. Primary teeth; 2. Alveoli; 3. Gingiva (gum);
4. Periodontal ligament; 5. Periodontal membrane
D. 1. Crown; 2. Pulp; 3. Root canal; 4. Dentin;
5. Enamel; 6. Cementum
E. 1. Enamel; 2. Dentin; 3. Pulp cavity;
4. Periodontal ligaments; 5. Root canal ;
6. Cementum; 7. Apical foramen; 8. Root;
9. Neck; 10. Crown
F. 1. Muscles of mastication; 2. Hard palate;
3. Soft palate; 4. Uvula; 5. Palatine tonsils
G. 1. Parotid glands; 2. Parotid glands;
3. Submandibular glands; 4. Sublingual glands

Pharynx and Esophagus
1. Oropharynx and laryngopharynx;
2. Nasopharynx; 3. Esophagus; 4. Pharyngeal constrictors

Stomach
A. 1. Gastroesophageal opening; 2. Cardiac region; 3. Pyloric sphincter; 4. Pyloric opening;
5. Rugae; 6. Body; 7. Fundus
B. 1. Pyloric sphincter; 2. Muscularis layer;
3. Rugae

C. 1. Gastric pits; 2. Gastric glands; 3. Parietal cells; 4. Chief cells

Small Intestine
A. 1. Greater duodenal papilla; 2. Lesser duodenal papilla; 3. Hepatopancreatic ampulla; 4. Hepatopancreatic ampullar sphincter
B. 1. Plicae circulares (circular folds); 2. Villi;
3. Lacteal; 4. Microvilli; 5. Absorptive cells;
6. Goblet cells; 7. Granular cells; 8. Intestinal glands; 9. Duodenal glands
C. 1. Duodenum and jejunum; 2. Peyer's patches;
3. Ileocecal sphincter; 4. Ileocecal valve

Liver and Gallbladder
A. 1. Caudate and quadrate; 2. Porta; 3 Common hepatic duct; 4. Cystic duct; 5. Common bile duct; 6. Gallbladder
B. 1. Bare area; 2. Lobules; 3. Portal triad;
4. Hepatic cords; 5. Hepatocytes; 6. Hepatic sinusoids; 7. Endothelial and phagocytic cells
C. 1. Hepatic portal vein; 2. Central vein;
3. Hepatic veins; 4. Bile canaliculi; 5. Hepatic ducts

Pancreas
A. 1. Head; 2. Tail; 3. Acini; 4. Intercalated ducts;
5. Intralobular ducts; 6. Interlobular ducts;
7. Main pancreatic duct; 8. Pancreatic islets (islets of Langerhans)
B. 1. Cystic duct; 2. Common bile duct;
3. Hepatopancreatic ampulla; 4. Pancreatic duct; 5. Common hepatic duct; 6. Hepatic ducts

Large Intestine
1. Transverse colon; 2. Sigmoid colon; 3. Teniae coli; 4. Haustra; 5. Epiploic appendages; 6. Crypts;
7. Rectum; 8. Anal canal

Peritoneum
1. Visceral peritoneum; 2. Mesenteries;
3. Retroperitoneal; 4. Lesser omentum; 5. Omental bursa; 6. Falciform ligament; 7. Mesentery proper;
8. Transverse and sigmoid mesocolons

Functions of the Digestive System
1. Nervous; 2. Local reflex; 3. Receptors;
4. Intramural plexus; 5. Effector organ

Oral Cavity
A. 1. Salivary amylase; 2. Mucin; 3. Chewing (mastication) reflex
B. 1. Starch; 2. Antibacterial; 3. Immunoglobulin A; 4. Parasympathetic; 5. Incisors and canines;
6. Premolars and molars

Deglutition
A. 1. Voluntary phase; 2. Pharyngeal phase;
3. Esophageal phase

B. 1. Swallowing center; 2. Pharyngeal constrictor muscles; 3. Epiglottis; 4. Peristaltic waves

Stomach
A. 1. Mucus; 2. Intrinsic factor; 3. Pepsinogen; 4. Pepsin; 5. Gastrin
B. 1. Cephalic phase; 2. Gastric phase; 3. Intestinal phase
C. 1. Increase; 2. Increase; 3. Increase; 4. Increase; 5. Decrease; 6. Decrease; 7. Decrease
D. 1. Chyme; 2. Peristaltic waves; 3. Pyloric pump

Small Intestine
A. 1. Mucus; 2. Secretin and cholecystokinin; 3. Disaccharidases; 4. Peptidases; 5. Nucleases
B. 1. Segmental contractions; 2. Increases; 3. Increases; 4. Increases; 5. Increases

Liver and Gallbladder
A. 1. Stimulates; 2. Stimulates; 3. Stimulates; 4. Stimulates; 5. Stimulates; 6. Inhibits
B. 1. Bile salts; 2. Bile pigments; 3. Store; 4. Glycogen; 5. Interconversion; 6. Detoxify; 7. Urea; 8. Phagocytize; 9. Synthesize

Pancreas
A. 1. Aqueous component; 2. Aqueous component; 3. Enzymatic component; 4. Enzymatic component; 5. Aqueous component; 6. Enzymatic component; 7. Enzymatic component

B. 1. Chymotrypsin, trypsin, and carboxypeptidase; 2. Enterokinase; 3. Amylase; 4. Lipase; 5. Deoxyribonuclease

Large Intestine
A. 1. Water and salts; 2. Mucus; 3. Bicarbonate ions; 4. Microorganisms; 5. Flatus
B. 1. Mass movements; 2. Gastrocolic reflex; 3. Defecation reflex; 4. Defecation reflex

Digestion, Absorption, and Transport
1. Digestion; 2. Absorption; 3. Ions and water-soluble substances; 4. Lipids and lipid-soluble substances

Carbohydrates
1. Salivary amylase; 2. Disaccharidase; 3. Glucose; 4. Insulin

Lipids
1. Triacylglycerols; 2. Emulsification; 3. Bile salts; 4. Lipase; 5. Micelles; 6. Chylomicrons; 7. Lacteal; 8. liver

Proteins
1. Pepsin; 2. Trypsin; 3. Peptidase

Water and Ions
1. Osmosis; 2. Active transport; 3. Simple diffusion; 4. Into lumen of intestine; 5. Into circulation

---

QUICK RECALL

1. Mucosa, submucosa, muscularis, and serosa or adventitia
2. Incisors and canines: cutting and tearing food; molars and premolars: crushing and grinding food
3. Parotid, submandibular, and sublingual glands. Amylase is the enzyme in saliva.
4. Upper esophageal sphincter, lower esophageal sphincter, pyloric sphincter, ileocecal sphincter, internal anal sphincter, external anal sphincter
5. Surface mucous cells: mucus; mucous neck cells: mucus; parietal cells: hydrochloric acid and intrinsic factor; chief cells: pepsinogen; endocrine cells: gastrin
6. Plicae circulares, villi, and microvilli
7. Absorptive cells, goblet cells, and endocrine cells
8. Voluntary, pharyngeal, and esophageal
9. Stomach: mixing waves and peristaltic waves; small intestine: segmental contractions and peristaltic contractions; large intestine: segmental movements and mass movements
10. Cephalic, gastric, and intestinal

11. Bicarbonate ions neutralize acid; trypsin, chymotrypsin, and carboxypeptidase digest protein; lipase digests fats; and amylase digests starch.
12. Neutralizes stomach acids, emulsifies fats, carries out excretory products
13. Storage, nutrient interconversion, detoxification, phagocytosis, and synthesis
14. Reabsorb water and salts, secretion of mucus, absorption of vitamins produced by microorganisms, storage of feces
15. Stomach secretion: S, I, I, S
    Bile secretion: O, O, S, S
    Pancreas secretion: O, S, S, S
    Contraction of gallbladder: O, S, O, S
    Gastric motility: S, I, I, S
16. Carbohydrates: monosaccharides; proteins: amino acids; triacylglycerols: fatty acids and monoacylglycerol
17. Carbohydrate digestion: mouth, small intestine; lipid digestion: small intestine; protein digestion: stomach, small intestine
18. Water-soluble molecules enter the hepatic portal system; lipid-soluble molecules enter the lacteals.

1. A. From the inner lining of the digestive tract to the outer layer is the mucosa, submucosa, muscularis, and serosa.

2. D. The intramural plexus consists of the submucosa plexus (located in the submucosal layer) and the myenteric plexus (found in the muscularis layer).

3. D. The tongue moves food about and helps (with the lips and cheeks) to hold the food between the teeth during mastication. Movement of the tongue is also involved with swallowing and speech. The tongue contains taste buds, which are involved with the sense of taste.

4. D. Most of the tooth is made up of dentin. Dentin is a compound of hydroxyapatite crystals that is generally less dense than bone. Enamel is very hard and covers the outer surface of the tooth (crown). The periodontal membrane secretes cementum, which helps to hold the tooth in its socket. The pulp cavity is found inside the tooth. It contains blood vessels, nerves, and pulp.

5. A. The typical number of teeth is as follows:

Tooth	Deciduous	Permanent
Incisors	8	8
Canines	4	4
Premolars	0	8
Molars	8	12
Total	20	32

6. D. There are three sets of salivary glands: the parotid, submandibular, and sublingual salivary glands. They secrete the saliva that enters the oral cavity.

7. A. The folds in the stomach, which allow the stomach to stretch, are called rugae. The muscularis of the stomach has three layers of smooth muscle because of the presence of an additional oblique layer of smooth muscle. The pyloric opening is the opening between the small intestine and the stomach (the cardiac or gastroesophageal opening is the opening between the stomach and esophagus).

8. C. Chief cells produce pepsinogen. Intrinsic factor is produced by parietal cells.

9. E. All of the structures listed increase the mucosal surface area of the small intestine.

10. B. Food passes from the stomach into the duodenum. From the duodenum food passes through the jejunum to the ileum, which empties into the large intestine.

11. D. The absorptive cells produce digestive enzymes that break down food. The digested food is then absorbed. Duodenal glands and goblet cells produce mucus, and endocrine cells release regulatory hormones.

12. D. Blood from the hepatic artery and hepatic portal vein mixes in the hepatic sinusoids and then flows to the central veins. The central veins join the hepatic veins, which exit the liver.

13. C. Bile would pass through the hepatic ducts, the common hepatic duct, and the common bile duct. Bile could pass through the cystic duct into the gallbladder and then back out into the common bile duct, but that route was not given.

14. B. The order would be ascending, transverse, descending, and sigmoid colon.

15. B. The coronary ligament connects the diaphragm and liver, the lesser omentum connects the liver and stomach, the greater omentum connects the stomach and the large intestine, and the transverse mesocolon connects the large intestine and abdominal wall.

16. B. A local reflex occurs in the intramural plexus and does not involve the spinal cord or brain. Stimulation of receptors within the digestive tract wall results in action potentials that are propagated by means of intramural plexus neurons to effector organs in the digestive tract wall, such as smooth muscle and glands. Neither c nor d are true.

17. A. Both mechanical and chemical digestion begin in the oral cavity. Mastication begins the process of mechanical digestion, and salivary amylase begins the process of chemical digestion.

18. B. Swallowing involves a voluntary phase during which food is pushed into the oropharynx. In the pharyngeal and esophageal phases the swallowing center is activated, and muscular reflexes push the food toward the stomach (gravity assists but is not as important as the muscular contractions). The soft palate closes off the nasopharynx (the epiglottis closes off the larynx).

19. B. HCl is a strong acid secreted by the stomach. The pH in the stomach is reduced by HCl, which stops carbohydrate digestion and creates the acid conditions necessary for pepsin to work.

20. D. The stomach is protected from digestive enzymes by the large amount of mucus that is secreted by the epithelial cells of the stomach lining.

21. C. Gastrin stimulates hydrochloric acid secretion by parietal cells. The other hormones inhibit gastric secretions.

22. D. The intestinal phase first stimulates, then inhibits, gastric secretions. The cephalic phase occurs as a result of chewing, swallowing, or thinking of food. The gastric phase produces the largest volume of secretions, and is stimulated by the presence of food in the stomach.

23. A. Food enters the stomach during a meal. That material is mixed with gastric secretions and is then slowly released, in a regulated fashion, into the duodenum. Although protein digestion begins in the stomach, most digestion takes place in the duodenum.

24. A. Peristaltic waves move chyme from the stomach into the small intestine. The mixing waves of the stomach function to mix the secretions of the stomach and the food within it. Haustra are found in the large intestine, not the stomach. Peristaltic waves do not stimulate HCl production.

25. E. The liver functions to produce many of the plasma proteins. It produces bile salts, which are important in fat emulsification in the small intestine, and eliminates bile pigments, which are by-products of hemoglobin breakdown. Therefore all of the conditions listed would occur if the liver were inflamed and its functioning impaired.

26. B The gallbladder stores and concentrates bile. The bile is produced by the liver. The gallbladder releases bile in response to cholecystokinin and parasympathetic stimulation.

27. B. The aqueous component contains bicarbonate ions that neutralize stomach acid. It is produced by the cells lining the smaller ducts of the pancreas and is released in response to secretin. The enzymatic component contains digestive enzymes, is produced by acinar cells, and is released in response to cholecystokinin. The islets of Langerhans are the endocrine portion of the pancreas, and they produce hormones (e.g., insulin and glucagon).

28. E. Pancreatitis, inflammation of the pancreas, is accompanied by the escape of digestive enzymes from pancreatic tissue into the surrounding tissues. The enzymes digest the tissues with which they come into contact and aggravate the inflammation, which results in severe abdominal pain. Also, the reduction in pancreatic enzymes delivered to the duodenum results in hampered digestion. The disrupted digestion and the abdominal pain produce nausea and vomiting. The liver does not produce digestive enzymes, nor does the gallbladder function to store digestive enzymes.

29. E. The large intestine stores wastes until they are eliminated. Water, salts, and vitamins (e.g., vitamin K produced by intestinal bacteria) are absorbed. Mucus protects the large intestine and provides adhesion of fecal matter.

30. E. Stretch of the rectum initiates local and spinal cord parasympathetic reflexes that result in defecation. Often the stretch results from the movement of feces into the rectum as a result of a mass movement.

31. E. The salivary glands produce salivary amylase, the pancreas produces pancreatic amylase, and the lining of the small intestine produces a number of disaccharidases.

32. E. Bile is secreted by the liver and stored in the gallbladder; bile enters the duodenum through the common bile duct. It contains bile pigments (breakdown products of hemoglobin) and bile salts that emulsify fats. Bile has no enzymatic activity.

33. A. Bile salts surround lipids within the small intestine to form micelles. The lipids diffuse from the micelles into intestinal cells, are coated with proteins, and are released into lacteals as chylomicrons.

34. C. Large amounts of lipids enter the lacteals, are carried to the thoracic duct, and finally enter the venous circulation at the base of the left subclavian vein. If the thoracic duct were tied off, the movements of lipids into the general circulation would be hampered.

35. D. Pepsin (from the stomach) and trypsin (from the pancreas) are enzymes that digest proteins. Amylase digests starches, and lipase digests lipids. Hydrochloric acid and bile have no enzymatic activity.

1. Salivary amylase digests the starch in the bread, breaking the starch down to maltose and isomaltose. These sugars are responsible for the sweet taste.

2. The bile salts normally emulsify fats, which greatly increases the efficiency of the enzymes that digest fats. Bile salts (micelles) also transport fats to the intestinal wall, where the fats are absorbed. A drug that binds to bile salts and prevents these functions would prevent the absorption of fats such as cholesterol.

3. Tactile stimulation in the mouth increases saliva production, making the mouth feel less dry.

4. The slide is from the small intestine. The stomach, small intestine, and large intestine are lined by columnar epithelial cells (the esophagus and rectum are lined with moist, stratified squamous epithelium). Villi are the folds that contain a capillary and a lymphatic vessel, and they are lined with columnar epithelial cells, many of which are goblet cells. The stomach is not lined with goblet cells, and the large intestine does not have villi.

5. Cutting the vagus nerves would eliminate the parasympathetic stimulation of the stomach, reducing stomach acid secretion. The reduction in hydrochloric acid production could help in the treatment of the peptic ulcer. However, after a few months acid secretion often increases, and an ulcer may develop. Because elimination of parasympathetic simulation also decreases stomach movements, food might be retained in the stomach. In some cases the stomach never really empties.

6. A diet high in fats would be most logical, because fatty acids in the duodenum stimulate the secretion of gastric inhibitory polypeptide and cholecystokinin, both of which inhibit gastric secretions. On the other hand, proteins (amino acids and polypeptides) in the stomach stimulate the release of gastrin, which increases gastric secretions.

7. Large meals would not be recommended because distension of the stomach promotes acid secretion and rapid stomach emptying. This would result in large amounts of acidic chyme entering the duodenum and aggravating the duodenal peptic ulcer. With smaller meals there would be less acid production, slower movement of chyme into the duodenum, and an increased ability to neutralize the chyme once it enters the duodenum. Note, however, that although these arguments seem logical, there is at present no conclusive proof that spreading meals throughout the day is truly effective.

8. The first experiment shows that, as the fat content within the duodenum is increased, gastric acid secretion decreases. The second experiment suggests that fats entering the duodenum cause the release of a hormone into the blood that inhibits gastric secretion. Possible hormones could include gastric inhibitory polypeptide and cholecystokinin. Although the data do not indicate it, there is also evidence for an enterogastric reflex that slows gastric secretion in response to large amounts of fat entering the small intestine.

9. Gross observation of the liver suggests that the man suffered from cirrhosis of the liver. He probably experienced symptoms associated with compromised liver function. Normally the liver releases bilirubin (from the breakdown of hemoglobin) into the bile. Jaundice would be expected because normal bile flow would be impaired and the bilirubin levels would increase in the blood. Edema would occur because the liver would not produce adequate plasma proteins. The man would tire easily because nutrients normally processed and stored in the liver would not be available. Weight loss might occur because reduced emulsification of fat resulting from lack of bile salts would hamper digestion.

10. The enema causes stretch of the rectum, which initiates local and central nervous system reflexes that cause defecation.

11. Following a meal the gastrocolic and duodenocolic reflexes initiate mass movement in the colon. Feces moves into the rectum, and the resulting stretch activates the defecation reflex.

12. Damage to the sacral region would eliminate the central nervous system reflex involved in defecation. Also lost would be voluntary control of defecation. However, local reflexes would still be functional, and defecation would occur, although not as well as before the injury.

Nutrition, Metabolism, and Temperature Regulation

FOCUS: Metabolism is the total of all the chemical changes occurring in the body. Nutrients are the chemicals taken into the body that are used to produce energy, provide building blocks for new molecules, or function in other chemical reactions. The six major classes of nutrients are carbohydrates, proteins, lipids, vitamins, minerals, and water. Glucose is converted to pyruvic acid in glycolysis. In anaerobic respiration (without oxygen), the pyruvic acid becomes lactic acid with a net gain of 2 ATP molecules. In aerobic respiration (with oxygen), 38 ATP molecules are produced when the pyruvic acid is converted to acetyl-CoA, which enters the citric acid cycle. NADH and $FADH_2$ molecules produced in aerobic respiration are used to produce ATP molecules in the electron transport chain. Lipids can be broken down by beta-oxidation to yield acetyl-CoA molecules that can be used to produce ATP molecules or ketones (ketogenesis). Amino acids can be used as a source of energy in oxidative deamination reactions. When there is too much blood glucose, it can be stored (glycogenesis); when there is too little blood glucose, it can be produced (glycogenolysis and gluconeogenesis). The energy produced from foods is used for basal metabolism, physical activity, and the assimilation of food. Heat, produced by metabolism, and heat exchanged with the environment by radiation, conduction, and convection, are responsible for body temperature.

CONTENT LEARNING ACTIVITY

Nutrition

"The six major nutrients are carbohydrates, proteins, lipids, vitamins, minerals, and water."

Match these terms with the correct statement or definition:

calorie	Kilocalorie (Calorie)
Carbohydrate	Nutrition
Essential nutrients	Protein
Fat	

_____ 1. Process by which food is obtained and used by the body.

_____ 2. Certain amino acids and fatty acids, most vitamins, minerals, water, and a minimum amount of carbohydrates are required.

_____ 3. Amount of energy required to raise the temperature of 1 g of water from 14° C to 15° C.

_____ 4. Used to express the amount of energy in food; 1000 calories.

_____ 5. Contains approximately 9 kcal per gram.

Carbohydrates

" *Carbohydrates include monosaccharides, disaccharides, and polysaccharides.* **"**

Match these terms with the
correct statement or definition:

Cellulose   Glucose
Complex carbohydrates   Glycogen
Disaccharides   Starch
Fructose   Sucrose

_____ 1. An isomer of glucose; converted into glucose by the liver.

_____ 2. The primary energy source for most cells.

_____ 3. Energy storage molecule produced from glucose in animals.

_____ 4. Table sugar; a disaccharide of glucose and fructose.

_____ 5. Not digestible by humans; provides "roughage."

_____ 6. Category to which sucrose, lactose, and maltose belong.

_____ 7. Category to which starch, glycogen, and cellulose belong;
                        polysaccharides.

Lipids

" *Approximately 95% of the lipids in the human diet are triacylglycerides.* **"**

Match these terms with the
correct statement or definition:

Adipose   Saturated fat
Cholesterol   Unsaturated fat
Phospholipid

_____ 1. Have only single covalent bonds between their carbon atoms.

_____ 2. Have one or more double covalent bonds between carbon atoms.

_____ 3. Excess triacylglycerides are stored in this tissue.

_____ 4. Part of plasma membrane, modified to form bile salts and
                        steroid hormones.

☞ Triacylglycerides are often referred to as fats.

Proteins

"Proteins are chains of amino acids."

Match these terms with the
correct statement or definition:

Antibody   Essential
Collagen   Hemoglobin
Complete   Incomplete
Enzymes   Nonessential

_____ 1. Type of amino acids that can be manufactured by the body.

_____ 2. Provides structural strength.

_____ 3. Regulate the rate of chemical reactions.

_____ 4. Transports oxygen and carbon dioxide.

_____ 5. Food that contains all eight essential amino acids.

Sources and Recommended Requirements

"A variety of foods provide the recommended requirement of carbohydrates, lipids, and proteins."

A. Match these terms with the
correct statement or definition:

Carbohydrate   Fat (saturated)
Cholesterol   Fat (unsaturated)
Fat (all)   Protein

_____ 1. Approximately 125 to 175 g are needed every day; otherwise acidosis or breakdown of muscle tissue occurs.

_____ 2. Should account for 30% or less of the total kilocaloric intake.

_____ 3. Should contribute no more than 10% of total fat intake.

_____ 4. Should be limited to 300 mg or less per day.

_____ 5. A person is in nitrogen balance if the amount of nitrogen in this ingested food is equal to the amount of nitrogen excreted in urine or feces.

B. Match these terms with the
correct statement or definition:

Carbohydrate
Cholesterol
Monounsaturated fat
Polyunsaturated fat

Protein (complete)
Protein (incomplete)
Saturated fat

_____ 1. Fruit, cereal, lactose (in milk).

_____ 2. Lipid in fats of meat, whole milk, cheese, butter, coconut oil, and palm oil.

_____ 3. Lipid in olive and peanut oil.

_____ 4. Lipid in fish, safflower, sunflower, and corn oils.

_____ 5. Lipid in high concentration in brain, liver, and egg yolks.

_____ 6. Proteins in meat, fish, poultry, milk, cheese, and eggs.

_____ 7. Proteins in leafy green vegetables, grains, peas, and beans.

Vitamins

"*Vitamins exist in minute quantities in food and are essential to normal metabolism.***"**

A. Match these terms with the
correct statement or definition:

Fat-soluble vitamins
Provitamins

Water-soluble vitamins

_____ 1. Portions of vitamins that can be assembled or modified by the body into functional vitamins; carotene is an example.

_____ 2. B-complex vitamins and vitamin C.

_____ 3. Vitamins A, D, E, and K.

_____ 4. Vitamins that can be stored in the body.

B. Match these vitamins with the
correct deficiency symptom:

A (retinol)
B_{12} (cobalamin)
C (ascorbic acid)

D (cholecalciferol)
K (phylloquinone)

_____ 1. Scurvy; defective collagen formation and poor wound healing.

_____ 2. Night blindness, retarded growth, and skin disorders.

_____ 3. Excessive bleeding resulting from retarded blood clotting.

_____ 4. Pernicious anemia and nervous system disorders.

Minerals

" *Minerals are inorganic nutrients necessary for normal metabolic functions.* **"**

Match these minerals with
the correct function:

Calcium   Phosphorus
Chlorine   Potassium
Iodine   Sodium
Iron

_____ 1. Bone and teeth formation, blood clotting, muscle activity, and nerve function.

_____ 2. Osmotic pressure regulation; nerve and muscle function.

_____ 3. Blood acid-base balance; hydrochloric acid production in the stomach.

_____ 4. Bone and teeth formation; important in ATP formation; component of nucleic acids.

_____ 5. Component of hemoglobin; ATP production in electron transport system.

_____ 6. Thyroid hormone production; maintenance of normal metabolic rate.

Metabolism

" *The total of all chemical reactions in the cell are often referred to as cellular metabolism.* **"**

Match these terms with the
correct statement or definition:

Anabolism   Oxidation-reduction
ATP   Oxidized
Catabolism   Reduce
Metabolism

_____ 1. Total of all the chemical changes that occur in the body.

_____ 2. Energy-requiring process by which small molecules are joined to form larger molecules.

_____ 3. Energy-releasing process by which large molecules are broken down into smaller molecules.

_____ 4. Energy currency of the cell; used to drive cell activities.

_____ 5. Type of chemical reaction responsible for the transfer of energy from the chemical bonds of nutrient molecules to ATP molecules.

_____ 6. Molecule that has gained electrons, hydrogen ions, and energy.

517

Glycolysis

❝*Glycolysis is a series of chemical reactions that results in the breakdown of glucose to two pyruvic acid molecules.*❞

Match these terms with the
correct statement or definition:

ATP	One
Four	Phosphorylation
NAD$^+$	Two
NADH	

_____ 1. Process of attaching a phosphate group to a molecule.

_____ 2. Reduced form of nicotinamide adenine dinucleotide.

_____ 3. Number of ATP molecules required to start glycolysis for one glucose molecule.

_____ 4. Net number of ATP molecules produced from one glucose molecule by glycolysis.

_____ 5. Number of NADH molecules produced from one glucose molecule by glycolysis.

_____ 6. Number of pyruvic acid molecules produced from one glucose molecule by glycolysis.

Anaerobic Respiration

❝*Anaerobic respiration is the breakdown of glucose in the absence of oxygen.*❞

Match these terms with the
correct statement or definition:

ATP	NADH
Cori cycle	Oxygen debt
Lactic acid	

_____ 1. Net energy gain from the anaerobic respiration of one molecule of glucose is two of these molecules.

_____ 2. Formed by the reduction of pyruvic acid.

_____ 3. Two of these molecules are produced in glycolysis and used (oxidized) when pyruvic acid is reduced.

_____ 4. Lactic acid released from cells is transported to the liver; the lactic acid is converted to glucose that is transported back to cells.

_____ 5. Oxygen necessary for the synthesis of the ATP used to convert lactic acid to glucose.

Aerobic Respiration

❝Aerobic respiration is the breakdown of glucose in the presence of oxygen to produce❞
carbon dioxide, water, and ATP molecules.

A. Match these terms with the correct statement or definition:

Acetyl-CoA formation
Aerobic respiration
Chemiosmotic model

Citric acid (Krebs) cycle
Electron-transport chain
Glycolysis

_____ 1. First phase of aerobic respiration; produces two ATP and two NADH molecules per glucose molecule.

_____ 2. Second phase of aerobic respiration in which pyruvic acid is modified to form acetyl-CoA; produces two NADH and two carbon dioxide molecules per glucose molecule.

_____ 3. Third phase of aerobic respiration, in which acetyl-CoA is combined with oxaloacetic acid to form citric acid; citric acid is then converted by a series of reactions into oxaloacetic acid; produces six NADH, two $FADH_2$, four carbon dioxide, and two ATP molecules per glucose molecule.

_____ 4. Produces 38 (or 36) ATP molecules for each glucose molecule broken down.

_____ 5. Process produces three ATP molecules for every NADH oxidized and two ATP molecules for every $FADH_2$ oxidized; occurs within the mitochondria.

_____ 6. Process uses oxygen as a final electron acceptor, producing water.

_____ 7. Hydrogen ions from NADH and $FADH_2$ are actively pumped out of the inner mitochondrial compartment; diffusion of the hydrogen ions back into the inner mitochondrial compartment provides energy for ATP production.

B. Match these terms with the correct parts of the diagram labeled in Figure 25-1:

Acetyl-CoA
ADP
Aerobic respiration
Anaerobic respiration
ATP

Citric acid cycle
Glycolysis
H_2O
Lactic acid
NADH

1. _____
2. _____
3. _____
4. _____
5. _____
6. _____
7. _____
8. _____
9. _____
10. _____

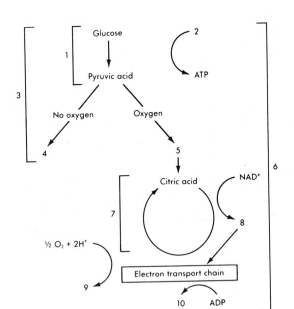

Figure 25-1

Lipid Metabolism

"Lipids are the body's main energy storage molecule."

Match these terms with the correct statement or definition:

Beta-oxidation
Free fatty acids
Ketogenesis

Ketone bodies
Triacylglycerides

_____

_____

_____

_____

_____

1. Primary storage form of lipids in adipose tissue.

2. Fatty acids released into the blood from the breakdown of triacylglycerides; used as an energy source by muscle and liver cells.

3. Series of reactions in which two carbons are removed from the end of a fatty acid chain to form acetyl-CoA.

4. Formation of ketone bodies from acetyl-CoA.

5. Acetoacetic acid, beta-hydroxybutyric acid, and acetone; used as an energy source, especially by skeletal muscle.

Protein Metabolism

"Once absorbed by the body, amino acids are quickly taken up by cells, especially in the liver.**"**

Using the terms provided, complete the following statements:

Citric acid (Krebs) cycle   Proteins
Keto acid   Stored
NADH   Transamination
Oxidative deamination   Urea

Amino acids can be used to synthesize _(1)_ or as a source of energy, but unlike carbohydrates and lipids, amino acids are not _(2)_ in the body. The synthesis of a nonessential amino acid usually begins with a _(3)_, which is usually converted to an amino acid. This process, called _(4)_, involves the transfer of an amine group from an amino acid to the keto acid. Amino acids can be used as a source of energy in a(n) _(5)_ reaction. In this reaction an amine group is removed from an amino acid, leaving ammonia, a keto acid, and _(6)_ that can be used to produce ATP. The ammonia is converted to _(7)_, which is eliminated by the kidney. Keto acids can also enter the _(8)_ or be converted into pyruvic acid or acetyl-CoA.

1. _____
2. _____
3. _____
4. _____
5. _____
6. _____
7. _____
8. _____

Intraconversion of Nutrient Molecules

"Many nutrient molecules may be converted to other nutrients as needed by the body.**"**

A. Using the terms provided, complete the following statements:

Gluconeogenesis   Glycogenolysis
Glycogenesis   Lipogenesis

If there is excess glucose, it can be used to form glycogen through a process called _(1)_. Once glycogen stores are filled, glucose and amino acids are used to synthesize lipids, a process called _(2)_. When glucose is needed, glycogen can be broken down into glucose 6-phosphate through a set of reactions called _(3)_. When liver glycogen levels are inadequate to supply glucose, amino acids from proteins and glycerol from triacylglycerides are used to produce glucose in a process called _(4)_.

1. _____
2. _____
3. _____
4. _____

B. Match these terms with the correct parts of the diagram labeled in Figure 25-2:

Gluconeogenesis
Glycogenesis

Glycogenolysis
Glycolysis

1. _____

2. _____

3. _____

4. _____

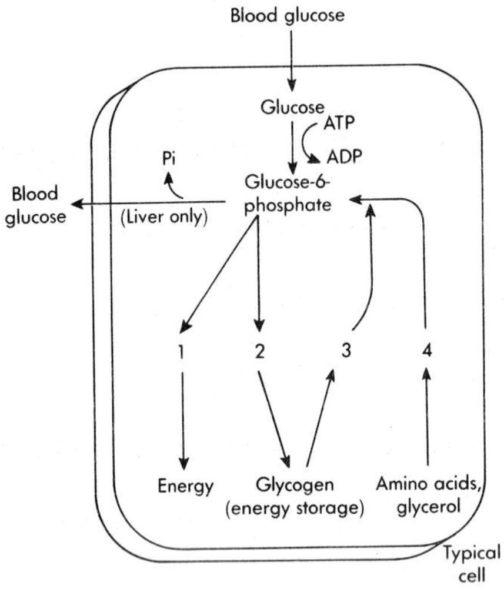

Figure 25-2

Metabolic States

66 *There are two major metabolic states in the body; the absorptive state is the period* 99 *immediately after a meal when nutrients are being absorbed, and the postabsorptive state occurs after the absorptive state has concluded.*

Match these terms with the correct statement or definition:

Absorptive state
Postabsorptive state

_____

_____

_____

1. During this state most of the glucose entering the circulation is used by cell; the remainder is converted into glycogen or fat.

2. During this state blood glucose levels are maintained by converting other molecules to glucose.

3. During this state glycogen is used preferentially, then fats and ketones, and then proteins.

Metabolic Rate

❝*The metabolic rate is the total amount of energy produced and used by the body per unit of time.*❞

Match these terms with the correct statement or definition:

Basal metabolic rate (BMR)　Increases
calorie　　　　　　　　　　 Kilocalorie
Decreases

_____ 1. Energy produced and used by the body at rest; calculated in kilocalories per square meter of body surface per hour.

_____ 2. Effect of increased muscle tissue on BMR.

_____ 3. Effect of increasing age on BMR.

_____ 4. Effect of dieting or fasting on BMR.

_____ 5. Effect of physical activity on expenditure of energy.

_____ 6. Effect of the thermic effect of feeding on expenditure of energy.

☞ For every 3500 kcal above the necessary energy requirement, a pound of body fat can be gained; for every 3500 kcal below the necessary energy requirement, a pound of body fat can be lost.

Body Temperature Regulation

❝*We can regulate our body temperature rather than have our body temperature adjusted by the external environment.*❞

Match these terms with the correct statement or definition:

Conduction　　　 Free energy
Convection　　　 Homeotherms
Decrease　　　　 Increase
Evaporation　　　 Radiation

_____ 1. Animals that can regulate their body temperatures.

_____ 2. Total amount of energy that can be liberated by the complete catabolism of food.

_____ 3. Exchange of heat between objects in direct contact with each other.

_____ 4. Exchange of heat between the body and the air.

_____ 5. Loss of water, which carries heat with it, from the body.

_____ 6. Effect of vasodilation on skin temperature.

_____ 7. Effect of vasoconstriction on skin temperature.

☞ A negative-feedback system accomplishes body temperature regulation. Specific body temperature is maintained by a set point in the hypothalamus.

1. List the six major classes of nutrients.

2. List the main functions of carbohydrates in the body.

3. List the main function of lipids in the body.

4. Name the main functions of proteins in the body.

5. Give the total energy gain to the cell from the breakdown of one molecule of glucose during anaerobic respiration and aerobic respiration.

6. Name three energy-storing compounds produced in aerobic respiration.

7. List the end products formed by anaerobic and aerobic respiration.

8. Name the three phases of aerobic respiration. Assuming the electron transport chain is operating, give the number of ATP molecules produced from a glucose molecule by each phase.

9. Name the chemical reactions by which fatty acids and amino acids are used as a source of energy.

10. Name four processes that involve intraconversion of nutrient molecules.

11. List three ways that metabolic energy can be used.

MASTERY LEARNING ACTIVITY

Place the letter corresponding to the correct answer in the space provided.

_____ 1. Which of the following statements concerning kilocalories is true?
  a. A kilocalorie is the amount of energy required to raise the temperature of 1 g of water from 14° C to 15° C.
  b. There are 4 kcal in a gram of ethanol.
  c. There are 9 kcal in a gram of protein.
  d. A pound of body fat contains 3500 kcal.

_____ 2. Complex carbohydrates
  a. include sucrose.
  b. can be found in large amounts in milk.
  c. form an energy storage molecule in plants and animals.
  d. all of the above

_____ 3. The primary energy source in the typical American diet is
  a. carbohydrates.
  b. fats.
  c. proteins.
  d. cellulose.

_____ 4. A good source of monounsaturated fats is
  a. the fats of meats.
  b. egg yolks.
  c. whole milk.
  d. fish oil.
  e. olive oil.

_____ 5. A complete protein food
  a. provides the daily amount (grams) of protein recommended in a healthy diet.
  b. can be used to synthesize the nonessential amino acids.
  c. contains all 20 amino acids.
  d. includes beans, peas, and leafy green vegetables.

_____ 6. Concerning vitamins,
  a. most can be synthesized by the body.
  b. they are normally broken down before they can be used by the body.
  c. A, D, E, and K are water-soluble vitamins.
  d. they function as coenzymes.

7. Minerals
   a. are inorganic nutrients.
   b. compose about 4% to 5% of total body weight.
   c. act buffers, and osmotic regulators.
   d. are components of enzymes.
   e. all of the above

8. Glycolysis
   a. is the breakdown of glucose to two pyruvic acid molecules.
   b. requires the input of two ATP molecules.
   c. produces two NADH molecules.
   d. does not require oxygen.
   e. all of the above

9. Anaerobic respiration occurs in the _____ of oxygen and produces _____ energy (ATP) for the cell than does aerobic respiration.
   a. absence, less
   b. absence, more
   c. presence, less
   d. presence, more

10. Which of the following reactions take place in both anaerobic and aerobic respiration?
    a. glycolysis
    b. citric acid cycle
    c. electron-transport chain
    d. a and b
    e. all of the above

11. A molecule that moves electrons from the citric acid cycle to the electron-transport chain is
    a. tRNA
    b. mRNA
    c. ADP
    d. NADH

12. The production of ATP molecules by the electron-transport chain is accompanied by the production of
    a. alcohol.
    b. water.
    c. oxygen.
    d. lactic acid.
    e. glucose.

13. The carbon dioxide you breathe out comes from
    a. glycolysis.
    b. electron-transport chain.
    c. anaerobic respiration.
    d. the food you eat.

14. Lipids are
    a. stored primarily as triacylglycerides.
    b. synthesized by beta-oxidation.
    c. broken down by oxidative deamination.
    d. all of the above

15. Amino acids
    a. are classified as essential or nonessential.
    b. are synthesized in a transamination reaction.
    c. can be used as a source of energy.
    d. all of the above

16. Ammonia is
    a. a by-product of lipid metabolism.
    b. formed during ketogenesis.
    c. converted into urea in the liver.
    d. all of the above

17. The conversion of amino acids and glycerol into glucose is called
    a. gluconeogenesis.
    b. glycogenolysis.
    c. glycogenesis.
    d. ketogenesis.

18. Which of the following events takes place during the absorptive state?
    a. Glycogen is converted into glucose.
    b. Glucose is converted into fats.
    c. Ketones are produced.
    d. Proteins are converted into glucose.

19. The major use of energy by the body is in
    a. basal metabolism.
    b. physical activity.
    c. the thermic effect of feeding.

20. The loss of heat because of loss of water from the body's surface is
    a. radiation.
    b. evaporation.
    c. conduction.
    d. convection.

Use a separate sheet of paper to complete this section.

1. Two ATP molecules are required to start beta-oxidation of a fatty acid chain. For each acetyl-CoA formed (except for the last one), one NADH and one $FADH_2$ are also produced during beta-oxidation. Given this information, does the metabolism of a six-carbon fatty acid yield more or less energy than the aerobic metabolism of a six-carbon molecule of glucose?

2. When a person is trying to lose weight, a reduction in caloric input and exercise is recommended. Give three reasons why exercise helps to reduce weight.

3. It is recommended that a person on a diet drink six to eight glasses of cool water per day. How could this practice help a person to lose weight?

4. Suppose a typical male and female of the same weight went on a diet to lose weight. If they both ate the same amount and kind of food and did the same amount and kind of exercise, would they both lose weight at the same rate?

5. In some diseases an infection results in a high fever. The patient is on the way to recovery when the crisis is over and body temperature begins to return to normal. If you were looking for symptoms in a patient who had just passed through the crisis state, would you look for a dry, pale skin or for a wet, flushed skin? Explain.

ANSWERS TO CHAPTER 25

Nutrition
1. Nutrition; 2. Essential nutrients; 3. calorie;
4. Kilocalorie (Calorie); 5. Fat

Carbohydrates
1. Fructose; 2. Glucose; 3. Glycogen; 4. Sucrose;
5. Cellulose; 6. Disaccharides; 7. Complex
carbohydrates

Lipids
1. Saturated fat; 2. Unsaturated fat; 3. Adipose;
4. Cholesterol

Proteins
1. Nonessential; 2. Collagen; 3. Enzymes;
4. Hemoglobin; 5. Complete

Sources and Recommended Requirements
A. 1. Carbohydrate; 2. Fat (all); 3. Fat (saturated);
4. Cholesterol; 5. Protein
B. 1. Carbohydrate; 2. Saturated fat;
3. Monounsaturated fat; 4. Polyunsaturated
fat; 5. Cholesterol; 6. Protein (complete);
7. Protein (incomplete)

Vitamins
A. 1. Provitamins; 2. Water-soluble vitamins;
3. Fat-soluble vitamins; 4. Fat-soluble
vitamins
B. 1. C (ascorbic acid); 2. A (retinol);
3. K (phylloquinone); 4. B_{12} (cobalamin)

Minerals
1. Calcium; 2. Sodium; 3. Chlorine; 4. Phosphorus;
5. Iron; 6. Iodine

Metabolism
1. Metabolism; 2. Anabolism; 3. Catabolism;
4. ATP; 5. Oxidation-reduction; 6. Reduced

Glycolysis
1. Phosphorylation; 2. NADH; 3. Two; 4. Two;
5. Two; 6. Two

Anaerobic Respiration
1. ATP; 2. Lactic acid; 3. NADH; 4. Cori cycle;
5. Oxygen debt

Aerobic Respiration
A. 1. Glycolysis; 2. Acetyl-CoA formation;
3. Citric acid (Krebs) cycle; 4. Aerobic
respiration;
5. Electron-transport chain; 6. Electron-
transport chain; 7. Chemiosmotic model
B. 1. Glycolysis; 2. ADP; 3. Anaerobic
respiration; 4. Lactic acid; 5. Acetyl-CoA;
6. Aerobic respiration; 7. Citric acid cycle;
8. NADH; 9. H_2O; 10. ATP

Lipid Metabolism
1. Triacylglycerides; 2. Free fatty acids; 3. Beta-
oxidation; 4. Ketogenesis; 5. Ketone bodies

Protein Metabolism
1. Proteins; 2. Stored; 3. Keto acid;
4. Transamination; 5. Oxidative deamination;
6. NADH; 7. Urea; 8. Citric acid (Krebs) cycle

Intraconversion of Nutrient Molecules
A. 1. Glycogenesis; 2. Lipogenesis;
3. Glycogenolysis; 4. Gluconeogenesis
B. 1. Glycolysis; 2. Glycogenesis;
3. Glycogenolysis; 4. Gluconeogenesis

Metabolic States
1. Absorptive state; 2. Postabsorptive state;
3. Postabsorptive state

Metabolic Rate
1. Basal metabolic rate (BMR); 2 Increases;
3. Decreases; 4. Decreases; 5. Increases;
6. Increases

Body Temperature Regulation
1. Homeotherms; 2. Free energy; 3. Conduction;
4. Convection; 5. Evaporation; 6. Increase;
7. Decrease

1. Carbohydrates, lipids, proteins, vitamins,
minerals, and water
2. Main energy source of body, minor energy
storage (e.g., glycogen), and structural (e.g.,
DNA, RNA, ATP)
3. Energy source and major energy storage (e.g.,
adipose tissue)
4. Structural (e.g., collagen), regulatory (e.g.,
enzymes, hormones, buffers), transport (e.g.,
hemoglobin, carrier molecules), protection (e.g.,
antibodies), and energy source
5. Anaerobic respiration: two ATP molecules;
aerobic respiration: 38 ATP molecules

6. NADH, $FADH_2$, and ATP
7. Anaerobic respiration: lactic acid; aerobic
respiration: carbon dioxide and water
8. Glycolysis: eight ATP ; acetyl-CoA formation: six
ATP; citric acid cycle: 24 ATP
9. Fatty acids: beta-oxidation; amino acids:
oxidative deamination
10. Glycogenesis, glycogenolysis, lipogenesis, and
gluconeogenesis
11. Basal metabolic rate, muscular energy, and
thermic effect of feeding

1. D.   There are 3500 kcal in a pound of body fat. A calorie is the amount of energy required to raise the temperature of 1 g of water from 14 C to 15 C. There are 1000 calories in a kilocalorie. There are 7 kcal in a gram of ethanol, and 9 kcal in a gram of fat.

2. C.   Complex carbohydrates are large polysaccharides and include starch (energy storage in plants), glycogen (energy storage in animals), and cellulose (roughage). Sucrose (table sugar) and lactose (milk sugar) are disaccharides.

3. A.   The typical American diet (by percent kilocalories) is 50% to 60% carbohydrates, 35% to 45% fats, and 10% to 15% protein. Cellulose is not digestible by humans.

4. E.   Olive and peanut oils are good sources of monounsaturated fats. Polyunsaturated fats are in fish, safflower, sunflower, and corn oils. Saturated fats are in the fats of meats, whole milk, cheese, butter, eggs, nuts, coconut oil, and palm oil. Egg yolks are high in cholesterol.

5. B.   A complete protein food contains the eight essential amino acids, from which the nonessential amino acids can be synthesized. Complete protein foods include meat, fish, poultry, milk, cheese, and eggs; incomplete protein foods include beans, peas, leafy green vegetables, and grains.

6. D.   Vitamins function as coenzymes, parts of coenzymes, or parts of enzymes. Most vitamins cannot be synthesized by the body, and they are not broken down before use. Vitamins A, D, E, and K are fat-soluble vitamins.

7. E.   Minerals are inorganic nutrients that are necessary for normal metabolic functions. They are components of enzymes and function as buffers or osmotic regulators. Minerals compose about 4% to 5% of body weight.

8. E.   Glycolysis is the breakdown of glucose to two pyruvic acid molecules. It requires the input of two ATP molecules and produces two NADH and four ATP molecules. Because two ATP molecules are input and four are produced, there is a net production of two ATP molecules. Glycolysis does not require oxygen.

9. A.   Anaerobic respiration, by definition, occurs in the absence of oxygen. It produces a net gain of two ATP molecules for every molecule of glucose that is degraded. Aerobic respiration, which takes place in the presence of oxygen, produces 38 ATP molecules.

10. A.   Glycolysis is the breakdown of glucose to pyruvic acid. In anaerobic respiration, pyruvic acid is converted to lactic acid. In aerobic respiration, pyruvic acid is converted to acetyl-CoA, which enters the citric acid cycle.

11. D.   NADH is the transport molecule. RNA is involved in protein synthesis, and ADP combines with a phosphate group to form ATP.

12. B.   At the end of the electron-transport chain, hydrogen combines with oxygen to produce water (often called metabolic water).

13. D.   Remember that glucose (and other organic food molecules) is made up of carbon atoms bonded together. At the end of respiration, all the glucose is gone. The carbon atoms combine with oxygen, and you breathe it out as a waste product, carbon dioxide. These reactions take place after glycolysis and in the citric acid cycle. Therefore part of the food you eat is breathed out in carbon dioxide.

14. A.   Lipids are stored primarily as triacylglycerides. The fatty acids of lipids are broken down by beta-oxidation into acetyl-CoA molecules.

15. D.   Amino acids are essential (must be ingested) or nonessential (synthesized by transamination reaction). Amino acids can be used as an energy source in an oxidative deamination reaction.

16. C.   Ammonia is a by-product of oxidative deamination (the breakdown of amino acids for energy). Ammonia is toxic to cells and is converted into the less toxic urea in the liver. The urea is then eliminated by the kidneys.

17. A.   Proteins (amino acids) and lipids (glycerol) can be used as a source of glucose as a result of gluconeogenesis. Glycogenesis is the conversion of glucose into glycogen, and glycogenolysis is the conversion of glycogen into glucose. Ketogenesis is the formation of ketones during lipid metabolism.

18. B.   In the absorptive state glucose is used as an energy source or is stored (glycogen or fats). The other events listed take place in the postabsorptive state.

19. A.   Basal metabolism accounts for 60% of energy expenditure, physical activity 30%, and assimilation of food (the thermic effect of feeding) 10%.

20. B.   Evaporation is the loss of water (i.e., heat). Radiation is the loss of heat by infrared radiation, conduction is the exchange of heat between objects in contact, and convection is the transfer of heat between the body and the air.

☆ FINAL CHALLENGES ☆

1. Beta-oxidation of the six-carbon fatty acid chain produces three acetyl-CoA molecules. For all but the last acetyl-CoA molecule formed, an NADH and an $FADH_2$ are produced. The electron-transport chain converts each NADH to three ATP molecules and each $FADH_2$ to two ATP molecules. Thus for each acetyl-CoA (except the last) there are five ATP molecules produced. For the six-carbon fatty acid therefore there are 10 ATP molecules produced from the NADH and the $FADH_2$ molecules. In addition, each acetyl-CoA can enter the citric acid cycle to yield 12 ATP molecules. Because there are three acetyl-CoA molecules, this yields 36 ATP molecules. Remembering that it takes two ATP molecules to start beta-oxidation, the total production of ATP molecules from the six-carbon fatty acid chain is 44 ATP molecules (10 + 36 - 2). Beta-oxidation of the fatty acid chain produces more ATP molecules than the metabolism of glucose, which yields 38 ATP molecules. In other words, fats can store more energy than carbohydrates can.

2. First, exercise increases energy (kilocalorie) usage. Second, following exercise the basal metabolic rate is elevated because of elevated body temperature and repayment of the oxygen debt. Third, exercise increases the proportion of muscle tissue to adipose tissue in the body. Because muscle tissue is metabolically more active than adipose tissue, basal metabolic rate increases. Fourth, during exercise epinephrine levels increase, resulting in increased blood sugar levels (see Chapter 18), which can depress the hunger center in the brain and reduce food consumption following exercise.

3. Drinking cool water could help in two ways. Because the water is cool, raising the water to body temperature requires the expenditure of calories. Also, stretch of the stomach decreases appetite (see Chapter 24).

4. All else being equal the typical male loses weight more rapidly because males have a higher basal metabolic rate than females.

5. During fever production the body produces heat by shivering. The body also conserves heat by constriction of blood vessels in the skin (producing pale skin) and by reduction in sweat loss (producing dry skin). When the fever breaks, i.e., "the crisis is over," heat is lost from the body to lower body temperature to normal. This is accomplished by dilation of blood vessels in the skin (producing flushed skin) and increased sweat loss (producing wet skin).

26 Urinary System

FOCUS: The urinary system consists of the kidneys, ureters, urinary bladder, and urethra. The kidneys function to remove waste products from the blood, produce vitamin D, stimulate red blood cell production, and regulate blood volume, electrolyte levels, and pH. The ureters carry urine from the kidneys to the urinary bladder, where the urine is stored. Stretch of the urinary bladder initiates reflexes that cause the smooth muscles of the bladder to contract, and the urine passes to the outside of the body through the urethra. The functional unit of the kidney is the nephron, which consists of the glomerulus (a tuft of capillaries) and a tubule.

Materials enter the tubule from the glomerulus by passing through a filtration membrane, which prevents the entry of blood cells and large molecules. As the filtrate passes through the tubule, most of the useful materials are reabsorbed, leaving waste products in the urine. In some parts of the tubule hydrogen and potassium ions are secreted into the urine. The concentration of urine is regulated by aldosterone, which increases sodium reabsorption, and ADH, which increases water reabsorption. In addition, renin secreted by the kidneys results in increased angiotensin II production, which causes a decrease in urine volume.

CONTENT LEARNING ACTIVITY

Introduction

66 *The urinary system participates with other organs to maintain homeostasis in the body by regulating* 99
the interstitial fluid composition within a narrow range of values.

Using the terms provided, complete these statements:

Blood	Red blood cell
Ions	Toxic
pH	Vitamin D

The kidneys remove wastes, many of which are __(1)__ , from the blood and play a major role in controlling __(2)__ volume, the concentration of __(3)__ in the blood, and the __(4)__ of the blood. The kidneys are also involved in the control of __(5)__ production and __(6)__ metabolism.

1. _____
2. _____
3. _____
4. _____
5. _____
6. _____

Kidneys

"The kidneys are bean-shaped organs, each about the size of a tightly clenched fist."

A. Match these terms with the correct statement or definition:

Cortex   Renal fascia
Hilum   Renal fat pad
Medulla   Renal pyramids
Medullary rays Renal sinus
Renal columns

_____ 1. Fibrous connective tissue that surrounds each kidney.

_____ 2. Dense deposit of adipose tissue surrounding the renal capsule.

_____ 3. Thin layer of connective tissue that attaches the kidneys and surrounding adipose tissue to the abdominal wall.

_____ 4. Cavity filled with fat and connective tissue into which the hilum opens.

_____ 5. Outer portion of the kidney.

_____ 6. Triangular structures found in the medulla of the kidney.

_____ 7. Extensions from the base of the renal pyramid that are located in the cortex.

_____ 8. Extensions of the cortex that project between the pyramids.

B. Match these terms with the correct statement or definition:

Major calyces   Renal pelvis
Minor calyces   Ureter
Renal papilla

_____ 1. Apex of the renal pyramid, found in the medulla.

_____ 2. Funnel-shape structures that surround the renal papillae.

_____ 3. Larger funnels that converge to form the renal pelvis.

_____ 4. Enlarged urinary channel that is found in the center of the renal sinus.

_____ 5. Tube that extends from the renal pelvis to the urinary bladder.

C. Match these terms with the
   correct parts of the diagram
   labeled in Figure 26-1:

Major calyx   Renal papilla
Minor calyx   Renal pelvis
Renal capsule   Renal pyramid
Renal column   Ureter

1. _____

2. _____

3. _____

4. _____

5. _____

6. _____

7. _____

8. _____

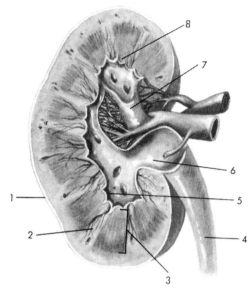

Figure 26-1

D. Match these terms with the
   correct statement or definition:

Bowman's capsule   Nephron
Filtration membrane   Podocytes
Glomerulus   Renal corpuscle
Juxtamedullary nephrons

_____ 1. Basic histological and functional unit of the kidney.

_____ 2. Nephrons that lie near the medulla.

_____ 3. Capillary portion of the renal corpuscle.

_____ 4. Tubule portion of the renal corpuscle, composed of a parietal
                              and visceral layer; proximal end of the nephron.

_____ 5. Specialized cells found in the visceral layer of Bowman's
                              capsule; gaps between their processes are the filtration slits.

_____ 6. Collective name for the capillary epithelium, basement
                              membrane, and podocytes.

E. Match these terms with the
   correct statement or definition:

Ascending limb   Juxtaglomerular apparatus
Collecting duct   Juxtaglomerular cells
Descending limb   Macula densa
Distal convoluted tubule Proximal convoluted tubule

_____ 1. Smooth muscle cells; form a cuff around the afferent arteriole.

_____ 2. Specialized tubule cells found in the distal convoluted tubule, where it is adjacent to the afferent and efferent arterioles.

_____ 3. Part of nephron between Bowman's capsule and loop of Henle.

_____ 4. Portion of loop of Henle connected to distal convoluted tubule.

_____ 5. Duct to which distal convoluted tubules of many nephrons join.

F. Match these terms with the
   correct parts of the diagram
   labeled in Figure 26-2:

Afferent arteriole   Loop of Henle
Collecting duct   Proximal convoluted tubule
Distal convoluted tubule Renal corpuscle
Efferent arteriole

1. _____

2. _____

3. _____

4. _____

5. _____

6. _____

7. _____

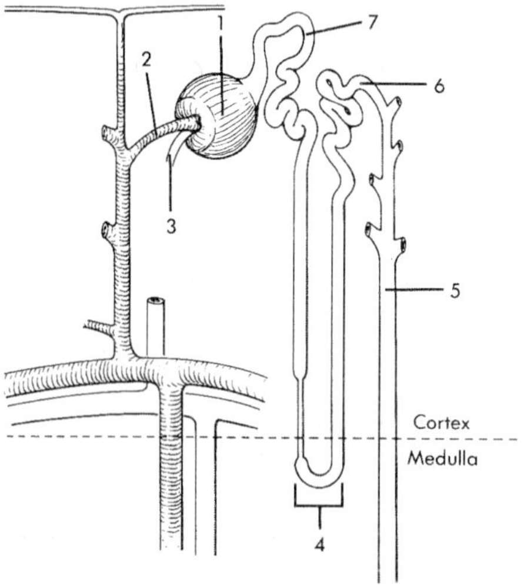

Figure 26-2

G. Place these vessels in
   the correct sequence that blood
   would pass through them,
   from the abdominal aorta
   to the interlobular veins.

Afferent arteriole   Interlobular artery
Arcuate artery   Peritubular capillaries
Efferent arteriole   Renal artery
Glomerulus   Segmental artery
Interlobar artery

1. Abdominal aorta_____ 5. _____ 9. _____

2. _____ 6. _____ 10. _____

3. _____ 7. _____ 11. Interlobular veins_____

4. _____ 8. _____

The vasa recta are specialized portions of the peritubular capillaries that dip into the medulla, along with the loops of Henle.

Ureters and Urinary Bladder

66 *The ureters extend from the renal pelvis to reach the urinary bladder, which functions to store urine.* **99**

Match these terms with the correct statement or definition:

External urinary sphincter
Internal urinary sphincter
Smooth muscle

Transitional epithelium
Trigone

_____

1. Triangular area of the bladder wall between the two ureters posteriorly and the urethra anteriorly.

_____

2. Thickness of the walls of the urinary bladder results mainly from layers of these cells.

_____

3. This internal lining permits changes in size of the urinary bladder and ureter.

_____

4. Skeletal muscle that surrounds the urethra as it extends through the pelvic floor.

Urine Production

66 *Because nephrons are the smallest structural components capable of producing urine,* **99** *they are called the functional units of the kidney.*

A. Match these terms with the correct statement or definition:

Filtration
Reabsorption

Secretion

_____

1. Movement of plasma across the filtration membrane because of a pressure gradient.

_____

2. Movement of substances from the filtrate back into the blood.

_____

3. Active transport of substances into the nephron.

 Urine consists of constituents filtered and secreted into the nephron minus those substances that are reabsorbed.

B. Match these terms with the correct statement or definition:

Filtration fraction
Glomerular filtration rate

Renal blood flow rate
Renal fraction

_____

1. Portion of the total cardiac output that passes through the kidneys; used to calculate renal blood flow rate.

_____

2. Portion of the plasma volume that is filtered through the filtration membrane.

_____

3. Amount of filtrate produced per minute.

The filtration barrier of the renal corpuscle is 100 to 1000 times more permeable than a typical capillary.

C. Match these terms with the correct statement or definition:

Capsule pressure   Filtration pressure
Colloid osmotic pressure   Glomerular capillary pressure

_____ 1. Blood pressure within the glomerulus.

_____ 2. Pressure of filtrate already in Bowman's capsule; opposes the glomerular capillary pressure.

_____ 3. Pressure caused by unfiltered plasma proteins remaining within the glomerular capillary.

_____ 4. Net pressure gradient that forces fluid from the glomerular capillary through the filtration membrane into Bowman's capsule.

D. Using the terms provided, complete these statements:

Bowman's capsule   Increased
Decreased   Peritubular
capillaries

As the diameter of a vessel decreases, the resistance to flow through the vessel increases. Because the efferent arteriole has a small diameter, there is a(n) _(1)_ resistance to blood flow and blood pressure within the glomerulus is high. Consequently, filtration pressure is high, and filtrate moves from the glomerulus into _(2)_. After blood passes through the efferent arteriole, blood pressure is _(3)_; therefore fluid moves out of the interstitial spaces and into the _(4)_. Changing the diameter of the efferent arteriole can alter filtration pressure. For example, constriction of the efferent arteriole would result in _(5)_ filtration pressure, and _(6)_ production of urine.

1. _____

2. _____

3. _____

4. _____

5. _____

6. _____

E. Using the terms provided, complete these statements:

Active transport	Interstitial
Antidiuretic hormone (ADH)	Into
Carrier molecules	Osmosis
Chloride ions	Out of
Cotransport	Permeable
Diffusion	Sodium ions

1. _____
2. _____
3. _____
4. _____
5. _____
6. _____
7. _____
8. _____
9. _____
10. _____
11. _____
12. _____

Amino acids, glucose, and fructose, as well as sodium, potassium, calcium, bicarbonate, and chloride ions, all move from the lumen of the nephron to the _(1)_ spaces. These substances are transported across the membrane of the epithelial cells of the nephron through a process called _(2)_. Within the membrane of the nephron cells, there are _(3)_ that bind to a substance to be transported and to sodium ions. The concentration gradient for sodium ions provides the energy for movement of both sodium ions and other molecules or ions _(4)_ the cell. Because the proximal convoluted tubule is permeable to water, as solute particles are transported to the interstitial spaces, water follows by _(5)_. As filtrate moves through the thin segment of the descending limb of the loop of Henle, water moves _(6)_ the nephron, and solutes diffuse _(7)_ the nephron. The cells of the ascending limb of the loop of Henle actively transport _(8)_ from the lumen of the tubule into the cells. Chloride ions and potassium ions move by _(9)_ into the cells of the ascending limb. Then sodium ions, potassium ions, and chloride ions diffuse from the cells to the interstitial fluid. Water does not follow them because the ascending limb of the loop of Henle is not _(10)_ to water. Sodium ions and chloride ions are actively transported across the wall of the distal convoluted tubule and collecting duct. The permeability of the distal convoluted tubule and the collecting duct is controlled by _(11)_. When this substance increases, water moves by osmosis _(12)_ the distal convoluted tubule and collecting duct.

☞ Approximately 99% of the water that enters the filtrate is reabsorbed. Urea, urate ions, creatinine, sulfates, phosphates, and nitrates are reabsorbed, but at a much lower rate than water; therefore a greater percentage of them are eliminated in the urine.

F. Match these terms with the correct statements as they pertain to tubular secretion:

Active transport
Diffusion

_____ 1. Method of movement of hydrogen ions, potassium ions, and penicillin into the distal convoluted tubules and the collecting ducts.

_____ 2. Method of movement of ammonia into the lumen of the nephron.

Urine Concentration Mechanism

“The urine concentrating mechanism is regulated so that either dilute or concentrated urine is**”** formed by the kidney.

A. Match these terms with the correct statement or definition:

Loop of Henle   Vasa recta
Urea

_____ 1. Countercurrent system that supplies blood to the kidney medulla; carries away extra water and solutes from the medulla without disturbing the interstitial fluid concentration.

_____ 2. Countercurrent multiplier system that adds solutes to the interstitial fluid of the medulla, increasing the concentration of the interstitial fluid.

_____ 3. Molecules responsible for a substantial portion of the high osmolality in the kidney medulla.

☞ The formation of a concentrated urine depends on the production and maintenance of a high interstitial fluid concentration in the medulla. This is accomplished by the loop of Henle, the vasa recta, and urea.

B. Match these terms with the correct statement or definition:

Ascending limb of the   Distal convoluted tubule
   loop of Henle   Proximal convoluted tubule
Descending limb of the
   loop of Henle

_____ 1. 65% of the filtrate is reabsorbed at this location.

_____ 2. As the filtrate moves through these, water moves out and solutes diffuse into the nephron; osmolality increases up to about 1200 mOsm/L.

_____ 3. This part of the nephron is not permeable to water, but sodium ions, chloride ions, and potassium ions move into the interstitial fluid.

_____ 4. In this part of the nephrons and in collecting ducts, if ADH is present, water diffuses out of the nephron and collecting ducts, resulting in a concentrated urine.

Regulation of Urine Concentration and Volume

“Regulation of urine production involves hormonal mechanisms, autoregulation,**”** and sympathetic nervous system stimulation.

A. Match these terms with the correct statement:

Decreases
Increases

_____ 1. Effect of increased blood osmolality on ADH secretion.

_____ 2. Effect of increased ADH secretion on blood osmolality.

_____   3. Effect of increased blood pressure on ADH secretion.

_____   4. Effect of decreased ADH secretion on blood pressure.

_____   5. Effect of increased blood pressure in right atrium on atrial natriuretic hormone secretion.

_____   6. Effect of atrial natriuretic hormone on ADH secretion.

_____   7. Effect of large increases in arterial blood pressure on urine production.

_____   8. Effect of constriction of the afferent arteriole on renal blood flow and filtration pressure.

_____   9. Effect of sympathetic stimulation on renal blood flow and filtrate formation.

B. Match these terms with   Decreases
   the correct statement: Increases

_____   1. Effect of aldosterone on sodium and chloride ion transport into the blood.

_____   2. Effect of the hyposecretion of aldosterone on urine volume and on concentration of solutes in urine.

_____   3. Effect of hypersecretion of aldosterone on potassium and hydrogen ion secretion.

_____   4. Effect of increased concentrations of potassium ions in the plasma or interstitial fluid on aldosterone secretion.

_____   5. Effect of angiotensin II on the secretion of aldosterone.

C. Match these terms with the correct location in these   1. _____
   statements as they pertain to the renin-angiotensin-
   aldosterone system of the kidney: 2. _____

   Decrease(s)   3. _____
   Increase(s)
                                                                   4. _____
If blood pressure in the afferent arteriole decreases, the rate
of renin secretion by the juxtaglomerular apparatus (1) .   5. _____
Renin converts angiotensinogen to angiotensin I, and angiotensin converting enzyme converts angiotensin I to angio-   6. _____
tensin II. Increased angiotensin II (2) blood pressure in
two ways. First, angiotensin II is a potent vasoconstrictor,   7. _____
causing a(n) (3) in peripheral resistance, which causes an
increases in blood pressure. Second, it (4) the rate of
aldosterone secretion, which leads to a(n) (5) in water
retention by the kidney, and a(n) (6) in urine volume. An
increase in sodium and chloride ions in the filtrate passing
through the juxtaglomerular apparatus (7) renin secretion.

Autoregulation, which involves changes in the degree of constriction of the afferent and efferent arterioles, maintains a relatively constant filtration rate despite comparatively large changes in systemic blood pressure.

Clearance and Tubular Maximum

"_Clearance and tubular maximum are concerned with removal of substances from the plasma_**"** _and filtrate._

Match these terms with the correct statement or definition:

Plasma clearance   Tubular maximum
Tubular load

_____ 1. The volume of plasma that is cleared of a specific substance each minute.

_____ 2. Total amount of a substance that filters through the filtration membrane into the nephrons each minute.

_____ 3. Maximum rate at which a substance can be actively reabsorbed.

 If the tubular load for a substance exceeds its tubular maximum, the excess amount of that substance will remain in the urine.

Urine Movement

"_Urine is produced in the nephrons of the kidney, moves into the renal pelvis, then into the ureters,_**"** _next into the urinary bladder, and out through the urethra._

A. Match these terms with the correct statement or definition:

Hydrostatic pressure
Peristaltic contractions

_____ 1. Mechanism responsible for urine flow from the nephron into the renal pelvis.

_____ 2. Mechanism responsible for urine flow through the ureters into the urinary bladder.

B. Match these terms with the correct statement or definition:

External urinary sphincter   Micturition reflex
Higher brain centers

_____ 1. Reflex initiated by stretching of the bladder wall, which results in contraction of the bladder and inhibition of the urinary sphincters.

_____ 2. Part of the nervous system responsible for inhibition or stimulation of the micturition reflex.

_____ 3. This structure is kept tonically contracted by the higher brain centers.

1. List five major parts of a nephron.

2. Name the two parts of the juxtaglomerular apparatus.

3. List the three steps in urine formation.

4. Write the formula for filtration pressure.

5. Complete these table by placing a " + " under the location where the condition exists and a "—" under the location where the condition does not exist.

CONDITION	Proximal Convoluted Tubule	Descending Limb	Ascending Limb	Distal Convoluted Tubule
Na$^+$ reabsorbed from tubule (active transport/ cotransport)	_____	_____	_____	_____
Na$^+$ passively diffuses into tubule	_____	_____	_____	_____
Water moves out of tubule by osmosis	_____	_____	_____	_____

6. Name two ions actively secreted into the distal convoluted tubule.

7. Complete these table by placing a "+" in each column where the filtrate concentration exists and a " —" where the condition does not exist.

FILTRATE	Proximal Convoluted Tubule	Tip of Loop of Henle	Distal Convoluted Tubule
100 mOsm	_____	_____	_____
300 mOsm	_____	_____	_____
1200 mOsm	_____	_____	_____

8. Complete these table by placing a + in each column where the correct percentage of filtrate volume reduction occurs.

FILTRATE VOLUME REDUCED BY:	Proximal Convoluted Tubule	Descending Limb	Ascending Limb	Distal Convoluted Tubule	Collecting Duct
65%	_____	_____	_____	_____	_____
15%	_____	_____	_____	_____	_____
10%	_____	_____	_____	_____	_____
9% (if ADH is present	_____	_____	_____	_____	_____

9. List three hormones that affect urine production, and give the major effect of each on urine production.

10. Following stretch of the bladder, list the events that result in micturition.

MASTERY LEARNING ACTIVITY

Place the letter corresponding to the correct answer in the space provided.

_____ 1. The kidney functions in
   a. disposal of nitrogenous wastes (such as urea).
   b. maintaining blood volume.
   c. maintaining blood pH.
   d. maintaining blood osmotic pressure.
   e. all of the above

_____ 2. The cortex of the kidney contains the
   a. hilus.
   b. glomeruli.
   c. renal fat pad.
   d. renal pyramids.

_____ 3. The portion of the kidney that consists primarily of vasa recta, loops of Henle, and collecting ducts is the
   a. renal cortex.
   b. renal pelvis.
   c. renal fat pad.
   d. renal pyramids.

_____ 4. Given these structures:
   1. major calyx
   2. minor calyx
   3. renal papilla
   4. renal pelvis

   Choose the arrangement that lists the structures in order as urine leaves the collecting tubule and travels to the ureter.
   a. 1,4,2,3,
   b. 2,3,1,4
   c. 3,2,1,4
   d. 4,1,3,2

_____ 5. Which of these structures contains blood?
   a. glomerulus
   b. vasa recta
   c. convoluted tubule
   d. Bowman's capsule
   e. a and b

_____ 6. Given these vessels:
   1. arcuate artery
   2. interlobar artery
   3. interlobular artery

   A red blood cell has just passed through the renal artery into a segmental artery. Choose the path blood would take to reach the afferent arteriole.
   a. 2,1,3
   b. 2,3,1
   c. 3,1,2
   d. 3,2,1

_____ 7. The urinary bladder is
   a. made up of skeletal muscle.
   b. lined by simple columnar epithelium.
   c. connected to the outside of the body by the ureter.
   d. located in the pelvic cavity.

_____ 8. Kidney function is accomplished by which of these mechanisms?
   a. filtration
   b. secretion
   c. reabsorption
   d. a and b
   e. all of the above

_____ 9. The amount of plasma that enters Bowman's capsule per minute is the
   a. glomerular filtration rate.
   b. renal plasma flow.
   c. renal blood flow times the hematocrit.
   d. renal plasma flow times the percent of filtrate that is not reabsorbed into the blood.

10. Given these structures:
    1. basement membrane
    2. fenestra
    3. filtration slit

    Choose the arrangement that lists the structures in the order a molecule of glucose would encounter them as the glucose passes through the filtration membrane to enter Bowman's capsule.
    a. 1,2,3
    b. 2,1,3
    c. 3,1,2
    d. 3,2,1

11. If the glomerular capillary pressure is 40 mm Hg, the capsule pressure is 10 mm Hg, and the colloid osmotic pressure within the glomerulus is 30 mm Hg, the filtration pressure would be
    a. -20 mm Hg.
    b. 0 mm Hg.
    c. 20 mm Hg.
    d. 60 mm Hg.
    e. 80 mm Hg.

12. Which of these conditions would tend to reduce filtration pressure in the glomerulus?
    a. elevated blood pressure
    b. constriction of the afferent arterioles
    c. decreased plasma protein concentrations
    d. dilation of the afferent arterioles

13. Glucose usually is completely reabsorbed from the urine by the time the urine has reached the
    a. end of the proximal convoluted tubule.
    b. tip of the loop of Henle.
    c. end of the distal convoluted tubule.
    d. end of the collecting duct.

14. The greatest volume of water is reabsorbed from the nephron by the
    a. proximal convoluted tubule.
    b. loop of Henle.
    c. distal convoluted tubule.
    d. collecting duct.

15. Water leaves the nephron by
    a. active transport.
    b. filtration into the capillary network.
    c. osmosis.
    d. facilitated diffusion.

16. Potassium ions enter the _____ by _____.
    a. proximal convoluted tubule, diffusion
    b. proximal convoluted tubule, active transport
    c. distal convoluted tubule, diffusion
    d. distal convoluted tubule, active transport

17. The countercurrent mechanism that produces a hyperosmotic environment in the medulla of the kidney results from the
    a. effects of ADH on water permeability of the ascending limb of the loop of Henle.
    b. impermeability of the ascending limb of the loop of Henle to water.
    c. cotransport of Na^+, K^+, and Cl^- out of the ascending limb of the loop of Henle.
    d. a and c
    e. b and c

18. At which of these sites is the osmotic pressure lowest (lowest concentration)?
    a. glomerular capillary
    b. proximal convoluted tubule
    c. bottom of the loop of Henle
    d. initial section of the distal convoluted tubule
    e. collecting duct

19. Increased aldosterone causes
    a. decreased reabsorption of sodium.
    b. decreased secretion of potassium.
    c. decreased reabsorption of chloride.
    d. increased permeability of the distal convoluted tubule to water.
    e. decreased volume of urine.

_____ 20. Juxtaglomerular cells are involved in the secretion of
a. ADH.
b. oxytocin.
c. renin.
d. aldosterone.

_____ 21. ADH governs the
a. sodium pump of the proximal convoluted tubules.
b. pressure of the blood in the afferent arterioles.
c. sodium pump of the vasa recta.
d. water permeability of the distal convoluted tubules and collecting ducts.

_____ 22. A decrease in blood osmolality would result in which of these?
a. increased ADH secretion
b. increased permeability of the collecting ducts
c. decreased urine osmolality
d. all of the above

_____ 23. If the blood pressure increased by 50 mm Hg,
a. the afferent arterioles would constrict.
b. glomerular capillary pressure would increase 50 mm Hg.
c. glomerular filtration rate would increase dramatically.
d. all of the above

_____ 24. The amount of a substance that filters through the filtration membrane per minute is the
a. renal plasma flow.
b. tubular load.
c. plasma clearance.
d. tubular maximum.

_____ 25. Given these events:
1. loss of voluntary motor control of urination
2. loss of the sensation of the need to urinate
3. loss of reflex emptying of the urinary bladder

Following transection of the spinal cord at level L5, which of the events would be expected to occur?
a. 1
b. 2
c. 1,2
d. 2,3
e. 1,2,3

FINAL CHALLENGES

Use a separate sheet of paper to complete this section.

1. A man develops arteriosclerosis of the renal arteries, reducing his renal fraction from 21% to 18%. Assuming that the man has a cardiac output of 5600 ml blood per minute, a hematocrit of 45, a filtration fraction of 19%, and 0.8% of the filtrate is not reabsorbed into the blood, how much urine should he produce in one day (24 hours)?

2. What effect would renal arteriosclerosis have on renin, ADH, and aldosterone levels? Explain.

3. Given the information that alcohol inhibits ADH secretion and caffeine causes dilation of the afferent arterioles, explain the effects on urine production if Mr. I.P. Daily drinks several alcoholic beverages and then tries to sober up by drinking several cups of coffee.

4. Which of these symptoms are consistent with a diagnosis of hypersecretion of aldosterone (primary aldosteronism): polydipsia (excessive drinking), polyuria (excessive urine production), blood pressure 15% higher than normal, and diarrhea? Explain.

5. A patient has severe edema. Diuretics are drugs that increase water loss through the kidneys. Describe all the ways that a diuretic could increase water loss.

6. To be effective, why should a diuretic not only increase water loss, but also increase sodium loss into the urine?

7. It is well known that some patients with hypertension are kept on a low-salt (sodium) diet. Propose an explanation for this therapy.

8. In studies of mammalian kidney function, it is known that animals with kidneys having relatively thicker medullas have a greater ability to conserve water. Explain why this is so.

ANSWERS TO CHAPTER 26

Introduction
1. Toxic; 2. Blood; 3. Ions; 4. pH; 5. Red blood cell; 6. Vitamin D

Kidneys
A. 1. Renal capsule; 2. Renal fat pad; 3. Renal fascia; 4. Renal sinus; 5. Cortex; 6. Renal pyramids; 7. Medullary rays; 8. Renal columns
B. 1. Renal papilla; 2. Minor calyces; 3. Major calyces; 4. Renal pelvis; 5. Ureter
C. 1. Renal capsule; 2. Renal column; 3. Renal pyramid; 4. Ureter; 5. Minor calyx; 6. Renal pelvis; 7. Major calyx; 8. Renal papilla
D. 1. Nephron; 2. Juxtamedullary nephrons; 3. Glomerulus; 4. Bowman's capsule; 5. Podocytes; 6. Filtration membrane
E. 1. Juxtaglomerular cells; 2. Macula densa; 3. Proximal convoluted tubule; 4. Ascending limb; 5. Collecting duct
F. 1. Renal corpuscle; 2. Afferent arteriole; 3. Efferent arteriole; 4. Loop of Henle; 5. Collecting duct; 6. Distal convoluted tubule; 7. Proximal convoluted tubule
G. 2. Renal artery; 3. Segmental artery; 4. Interlobar artery; 5. Arcuate artery; 6. Interlobular artery; 7. Afferent arteriole; 8. Glomerulus; 9. Efferent arteriole; 10. Peritubular capillaries

Ureters and Urinary Bladder
1. Trigone; 2. Smooth muscle; 3. Transitional epithelium; 4. External urinary sphincter

Urine Production
A. 1. Filtration; 2. Reabsorption; 3. Secretion
B. 1. Renal fraction; 2. Filtration fraction; 3. Glomerular filtration rate

C. 1. Glomerular capillary pressure; 2. Capsule pressure; 3. Colloid osmotic pressure; 4. Filtration pressure
D. 1. Increased; 2. Bowman's capsule; 3. Decreased; 4. Peritubular capillaries; 5. Increased; 6. Increased
E. 1. Interstitial; 2. Cotransport; 3. Carrier molecules; 4. Into; 5. Osmosis; 6. Out of; 7. Into; 8. Sodium ions; 9. Cotransport; 10. Permeable; 11. Antidiuretic hormone (ADH); 12. Out of
F. 1. Active transport; 2. Diffusion

Urine Concentration Mechanism
A. 1. Vasa recta; 2. Loop of Henle; 3. Urea
B. 1. Proximal convoluted tubule; 2. Descending limb of the loop of Henle; 3. Ascending limb of the loop of Henle; 4. Distal convoluted tubule

Regulation of Urine Concentration and Volume
A. 1. Increases; 2. Decreases; 3. Decreases; 4. Decreases; 5. Increases; 6. Decreases; 7. Increases; 8. Decreases; 9. Decreases
B. 1. Increases; 2. Increases; 3. Increases; 4. Increases; 5. Increases
C. 1. Increases; 2. Increases; 3. Increase; 4. Increases; 5. Increase; 6. Decrease; 7. Decreases

Clearance and Tubular Maximum
1. Plasma clearance; 2. Tubular load; 3. Tubular maximum

Urine Movement
A. 1. Hydrostatic pressure; 2. Peristaltic contractions
B. 1. Micturition reflex; 2. Higher brain centers; 3. External urinary sphincter

1. Glomerulus, Bowman's capsule, proximal convoluted tubule, loop of Henle, and distal convoluted tubule
2. Macula densa and juxtaglomerular cells of the afferent arterioles
3. Filtration, reabsorption, and secretion
4. Filtration pressure equals glomerular capillary pressure minus capsule pressure minus colloid osmotic pressure.

5.

	PCT	DL	AL	DCT
Na^+ reabsorbed	+	-	+	+
Na^+ passively diffuses in	-	+	-	-
Water moves out by osmosis	+	+	-	+

6. Potassium ions and hydrogen ions

7.

	PCT	TLH	DCT
100 mOsm	-	-	+
300 mOsm	+	-	-
1200 mOsm	-	+	-

8.

	PCT	DL	AL	DCT	CD
65%	+	-	-	-	-
15%	-	+	-	-	-
10%	-	-	-	+	-
9% (if ADH present)	-	-	-	-	+

9. Aldosterone: increased sodium ion reabsorption, resulting in decreased urine concentration and volume; ADH: decreased urine volume; atrial natriuretic factor: inhibits ADH production, resulting in increased urine volume.

10. Stretch of bladder, reflex stimulated, bladder contracts, and urinary sphincters inhibited.

MASTERY LEARNING ACTIVITY

1. E. The kidneys perform all of these functions.

2. B. The glomeruli are found primarily in the cortex (outer layer) of the kidney. The renal pyramids are found in the medulla (inner layer) of the kidney. The renal fat pad surrounds the kidney. The hilum is an indentation through which arteries enter the kidney and veins and the ureter leave.

3. D. The vasa recta and the loops of Henle pass from the cortex into the medulla and then back into the cortical area. The collecting ducts pass through the medulla as they project toward the pelvis. Thus the vasa recta, the loops of Henle, and the collecting ducts are found primarily in the renal pyramids and make up the greatest portion of the renal medulla.

4. C. The collecting tubules converge at the renal papilla and the urine enters the minor calyx. The minor calyx empties into a major calyx, which connects to the renal pelvis. The ureter drains the renal pelvis.

5. E. The glomerulus is a knot of coiled capillaries from which materials enter the tubule portion of nephrons by filtration. The vasa recta is a capillary system that supplies the loop of Henle. The convoluted tubule and Bowman's capsule are parts of the nephron.

6. A. The interlobar arteries branch from the segmental arteries and pass through the renal columns. The arcuate arteries branch from the interlobar arteries and arch over the base of the renal pyramids. The interlobular arteries branch from the arcuate arteries and extend into the cortex. The afferent arterioles are derived from the interlobular arteries.

7. D. The urinary bladder is located in the pelvic cavity. It is connected to the kidneys by the ureters and to the outside of the body by the urethra. It is a hollow organ consisting of smooth muscle and lined by transitional epithelium.

8. E. Filtration occurs when materials move from the glomerulus into Bowman's capsule. Reabsorption (e.g., sodium, glucose) and secretion (e.g., potassium and hydrogen ions) occur throughout the tubule system.

9. A. The glomerular filtration rate is the amount of plasma (filtrate) that enters Bowman's capsule per minute.

10. B. The glucose would pass through the fenestra of the capillary endothelium, the basement membrane, and the filtration slit formed by the processes of the podocytes.

11. B. The pressure gradient that develops across the glomerular membrane is responsible for glomerular filtration. Glomerular capillary pressure (40 mm Hg) is opposed by the capsule pressure within Bowman's capsule (10 mm Hg) and the colloid osmotic pressure of the plasma (30 mm Hg). Therefore the net filtration pressure is 0 mm Hg (i.e., 0 = 40-30-10).

12. B.   Recall that filtration pressure equals glomerular capillary pressure minus capsule pressure minus colloid osmotic pressure. Constriction of the afferent arterioles reduces glomerular capillary pressure (blood pressure) in the glomeruli, thus reducing filtration pressure. Elevated blood pressure or dilation of the afferent arteriole raises the glomerular capillary pressure in the glomeruli, causing an increase in filtration pressure. An increase in filtration pressure also results when plasma protein concentrations decrease because this decreases the colloid osmotic pressure.

13. A.   Normally all of the glucose, proteins, amino acids, and vitamins that enter Bowman's capsule are reabsorbed by active processes (i.e., cotransport) in the proximal convoluted tubule.

14. A.   Approximately 65% of the volume of the filtrate is reabsorbed in the proximal convoluted tubule. Reabsorption in the proximal convoluted tubule is regulated by neither aldosterone nor ADH. It is obligatory reabsorption resulting from such factors as the active transport of sodium, amino acids, and glucose from the proximal convoluted tubule.

15. C.   This is an important fact to know if you are to understand renal physiology. Sodium ions move out of the tubules by active transport, and water passively follows by osmosis because of the resulting osmotic gradient.

16. D.   Potassium is secreted into the distal convoluted and collecting tubules by active transport. Potassium is removed from the proximal convoluted tubule by cotransport.

17. E.   To produce a hyperosmotic solution in the interstitial spaces of the medulla, solutes such as Na^+, K^+, and Cl^- ions must be added. This process must necessarily involve active transport, because movement is against the concentration gradient. Na^+ ions are actively transported out of the cells into the interstitial fluid; K^+ and Cl^- ions are cotransported with Na^+ ions into the cell from the lumen of the tubules and then move by diffusion into the interstitial fluid. If the ascending limb were permeable to water, water would follow the Na^+, Cl^-, and K^+ ions, resulting in dilution and destroying the hyperosmotic environment. Hence the necessity of the impermeability of the ascending limb to water.

18. D.   Filtrate in the proximal convoluted tubule is isosmotic with the plasma in the glomerulus. As the filtrate moves down to the bottom of the loop of Henle, it becomes hyperosmotic because of the countercurrent multiplier system in Henle's loop. By the time the distal convoluted tubule is reached, the filtrate is hyposmotic to blood.

19. E.   As aldosterone concentrations increase, potassium secretion increases and sodium and chloride reabsorption increase. As sodium moves from the urine back into the blood, water passively follows. This increases the extracellular fluid volume and decreases urine volume.

20. C.   Renin is secreted by the juxtaglomerular apparatus, ADH and oxytocin by the neurohypophysis, and aldosterone by the adrenal cortex.

21. D.   ADH governs the permeability of the distal convoluted tubules and collecting ducts and does not directly affect any of the other choices. When ADH levels increase, permeability increases.

22. C.   When blood osmolality decreases, this inhibits the osmoreceptors in the hypothalamus and there is less ADH secretion. As a consequence, the permeability of the collecting tubules decreases and less water is reabsorbed, producing a less concentrated urine.

23. A.   Despite large changes in blood pressure, there is usually a small change in glomerular capillary pressure and therefore only a slight increase in glomerular filtration rate. A large change in glomerular capillary pressure is prevented by constriction of the afferent arterioles; this is called autoregulation.

24. B.   The tubular load is the amount of a substance that enters Bowman's capsule. Tubular maximum is the fastest rate at which a substance is reabsorbed from the nephron. Plasma clearance is the volume of plasma from which a substance is completely removed in a minute.

25. C.   The transection would eliminate nervous impulses to and from the brain; thus voluntary control and the sensation of the need to urinate would be lost. The spinal cord center for urination would still be functional, so reflex emptying of the urinary bladder would still occur.

1. Renal blood flow would be 1008 ml of blood per minute (5600 x 0.18), renal plasma flow would be 554 ml of plasma per minute (1008 x 0.55), glomerular filtration rate would be 105 ml of filtrate per minute (554 x 0.19), and urine production would be 0.84 ml of urine per minute (105 x 0.008) or 1.2 L of urine per day (0.84 x 1.44).

2. Arteriosclerosis would reduce blood pressure in the afferent arterioles and increase renin secretion by the juxtaglomerular apparatus. Renin converts angiotensinogen to angiotensin I, which is converted into angiotensin II. Angiotensin II increases aldosterone secretion. The decreased urine production that results from the increased angiotensin II and the aldosterone causes an increased blood volume and blood pressure. Through baroreceptors the increased blood pressure inhibits ADH secretion.

3. Urine production should increase for several reasons. First the increased fluid intake would increase blood volume, blood pressure, and glomerular capillary pressure. Vasodilation of the afferent arterioles by the caffeine would also increase glomerular capillary pressure. As a result of the increased glomerular capillary pressure, the filtration pressure would increase, producing an increased glomerular filtration rate and therefore a greater volume of urine. Second, inhibition of ADH secretion would decrease the permeability of the collecting ducts to water, so a larger amount of filtrate would pass through as urine.

4. Excess aldosterone would lead to increased sodium and water reabsorption. Because of the water increase, blood volume would increase, causing a rise in blood pressure. Eventually, however, the increase in blood volume is opposed by mechanisms that regulate blood volume and pressure. The increased sodium in the extracellular fluid stimulates the thirst center, resulting in excessive drinking followed by excessive urine production. Diarrhea would be a symptom associated with hyposecretion of aldosterone. When aldosterone is absent or present in low quantities, sodium reabsorption from the intestines is very poor and the intestinal contents become hyperosmotic; water is retained, and diarrhea results.

5. There are several ways that a diuretic could increase water loss from the kidneys.
   A. It could block sodium reabsorption or stimulate sodium excretion. Because water follows sodium, producing a higher than normal amount of sodium in the tubules would result in a larger urine volume. The drug could act directly on the sodium pump mechanism, block the effects of aldosterone on the tubule, or even block aldosterone production.
   B. It could inhibit ADH production or interfere with the effects of ADH on the collecting ducts.
   C. It could add a solute (glycerol or mannitol) that is filtered but not reabsorbed. The solute then osmotically obligates large amounts of tubular water. In the descending limb of the loop of Henle, the water moves into the interstitial fluid, overloading the countercurrent mechanism. The osmolality of the interstitial fluid is reduced because of the influx of water; as a result, less water will be reabsorbed from the collecting ducts.

6. If only water were lost, blood osmolality would increase and stimulate ADH secretion. The ADH would increase the permeability of the collecting ducts to water, increasing water reabsorption and reducing the effectiveness of the diuretic. If sodium is also lost, blood osmolality does not increase, and the ADH response does not occur.

7. A low-salt diet would tend to reduce the osmolality of the blood. Consequently, ADH secretion would be inhibited, producing a dilute urine and thus eliminating water. This in turn would reduce blood volume and blood pressure.

8. As the loops of Henle become longer, the countercurrent multiplier system becomes more and more effective, raising the concentration of the interstitial fluid. The maximum concentration for urine is determined by the concentration of the interstitial fluid at the ends of the loops of Henle. Thus the longer the loops of Henle, the greater the concentration of urine it is possible to produce.

27 Water, Electrolytes, and Acid-Base Balance

FOCUS: Intracellular fluid (i.e., the fluid inside cells) is different from the extracellular fluid (i.e., the fluid outside cells). Sodium ions are responsible for most of the osmotic pressure of extracellular fluid. Consequently, mechanisms such as ADH levels that regulate blood volume or blood osmolality are important for the regulation of blood sodium ion concentrations. Potassium ion concentrations affect resting membrane potentials and are regulated by aldosterone. Calcium ion concentrations, which are regulated by parathyroid hormone and vitamin D, affect the electrical activity of cells. Water levels are regulated by a balance between water intake (thirst) and water output (evaporation, feces, and urine). Blood pH is normally 7.4. A decrease in blood pH below normal is called acidosis, and an increase above normal is termed alkalosis. Blood pH is controlled by buffers, chemicals that resist a change in pH; by the respiratory system, which changes pH by changing blood carbon dioxide levels; and by the kidneys, which can produce an acidic or alkaline urine.

CONTENT LEARNING ACTIVITY

Body Fluids

“There are two major fluid compartments in the body.”

Match these terms with the correct statement or definition:

Extracellular fluid   Intracellular fluid
Interstitial fluid   Plasma

_____ 1. Accounts for about 40% of the total body weight and includes the small amount of fluid in trillions of cells.

_____ 2. Accounts for about 20% of the total body weight and includes lymph, cerebrospinal fluid, plasma, synovial fluid, and interstitial fluid.

_____ 3. Extracellular fluid that occupies spaces outside blood vessels.

_____ 4. Extracellular fluid that occupies space within blood vessels.

 The osmotic concentration of most extracellular fluid compartments is approximately equal.

Regulation of Intracellular and Extracellular Fluid Composition

"The composition of intracellular fluid is substantially different from that of extracellular fluid."

Using the terms provided, complete these statements:

Decreases   Osmosis
Electrolytes   Proteins
Increases

Cell membranes are differentially permeable, being relatively impermeable to __(1)__, whereas the movement of __(2)__ is determined by transport processes, as well as electrical charge differences. Water movement across the cell membrane by __(3)__ is influenced by the composition of the extracellular fluid. During conditions of dehydration the concentration of solutes in the extracellular fluid __(4)__, resulting in the movement of water from the intracellular space into the extracellular space. When water intake increases following a period of dehydration, the concentration of solutes in the extracellular fluid __(5)__, resulting in the movement of water into the cells.

1. _____

2. _____

3. _____

4. _____

5. _____

☞ Maintenance of homeostasis requires that the intake of substances must equal their elimination.

Regulation of Ion Concentrations

"The regulation of water and electrolytes involves the coordinated participation of several organ systems."

A. Match these terms with the correct statement or definition:

Antidiuretic hormone (ADH)   Kidneys
Atrial natriuretic hormone   Sodium ions
Aldosterone   Sweat

_____ 1. Dominant extracellular positively charged ion; more than 90% of the osmotic pressure results from these ions and the negatively charged ions associated with them.

_____ 2. Major route by which sodium ions are secreted.

_____ 3. Hormone that increases sodium ion reabsorption from the distal convoluted tubule and collecting duct.

_____ 4. Hormone that increases water reabsorption from the distal convoluted tubule and collecting duct.

_____ 5. Hormone that decreases sodium ion reabsorption from the distal convoluted tubule and collecting duct.

☞ The most abundant negatively charged ion in the extracellular fluid is chloride; it is regulated by the same mechanisms that regulate sodium ions.

B. Match these terms with the correct location in Figure 27-1 by placing a D in the blank for those substances that are decreased and an I for those that are increased.

Decreased
Increased

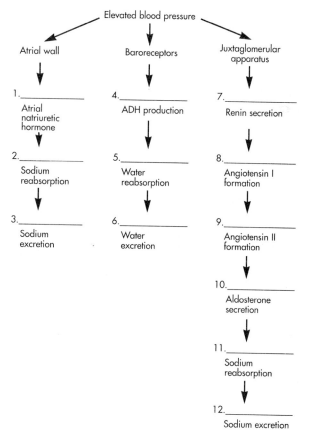

Elevated blood pressure

Atrial wall

1._____
Atrial natriuretic hormone

2._____
Sodium reabsorption

3._____
Sodium excretion

Baroreceptors

4._____
ADH production

5._____
Water reabsorption

6._____
Water excretion

Juxtaglomerular apparatus

7._____
Renin secretion

8._____
Angiotensin I formation

9._____
Angiotensin II formation

10._____
Aldosterone secretion

11._____
Sodium reabsorption

12._____
Sodium excretion

Figure 27-1

C. Using the terms provided, complete these statements:

Decrease
Depolarization
Distal convoluted tubule
Hypokalemia

Hyperkalemia
Hyperpolarization
Increase
Proximal convoluted tubule

1. _____
2. _____
3. _____
4. _____
5. _____
6. _____
7. _____
8. _____

The extracellular concentration of potassium ions must be maintained within a narrow range of concentrations. An increase in the extracellular potassium ion concentration leads to _(1)_ of the resting membrane potential, and a decrease in extracellular potassium ion concentration leads to _(2)_ of the resting membrane potential. Potassium ions are actively reabsorbed in the _(3)_ and are actively secreted in the _(4)_ . This secretion is primarily responsible for controlling the extracellular concentration of potassium ions. The hormone aldosterone plays a major role in regulating extracellular potassium ions by causing a(n) _(5)_ in the rate of potassium ion secretion. Aldosterone secretion is regulated by blood potassium levels; in response to a(n) _(6)_ in blood potassium levels, aldosterone secretion is increased. Abnormally high levels of potassium ions in the extracellular fluid is called _(7)_ , whereas abnormally low levels is _(8)_ .

D. Match these terms with the correct statement or definition:

Decreases   Hypocalcemia
Increases   Hypercalcemia

_____ 1. Effect of reduced extracellular calcium concentration on spontaneous action potential generation.

_____ 2. Effect of parathyroid hormone on extracellular calcium ion levels.

_____ 3. Effect of decreased calcium levels on the secretion of parathyroid hormone.

_____ 4. Effect of parathyroid hormone on calcium reabsorption from the kidneys.

_____ 5. Effect of parathyroid hormone on production of active vitamin D.

_____ 6. Effect of vitamin D on calcium absorption in the gastrointestinal tract.

_____ 7. Effect of calcitonin on extracellular calcium concentration.

_____ 8. Effect of elevated extracellular calcium levels on secretion of calcitonin.

_____ 9. Above normal levels of calcium ions in the extracellular fluid.

Regulation of Water Content

"_Water loss from the body occurs in urine, through evaporation, and in the feces._**"**

A. Match these terms with the correct location in Figure 27-2 by placing a D in the blank for those substances that are decreased and an I for those substances that are increased.

Decreased
Increased

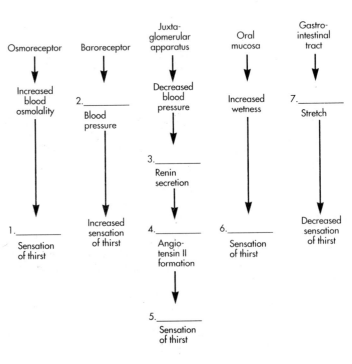

Figure 27-2

B. Using the terms provided, complete these statements:

Body temperature   Kidneys
Decrease   Plasma volume
Feces   Sensible perspiration
Increase   Solutes
Insensible perspiration

The water lost through simple evaporation through the
skin is called __(1)__ . Sweat, or __(2)__ , is secreted by the sweat
glands and, in contrast to insensible perspiration, contains
__(3)__ . Evaporation of water is a major mechanism by which
__(4)__ is regulated. The loss of a large volume of sweat
causes a(n) __(5)__ in plasma volume and a(n) __(6)__ in
hematocrit. Most water is lost in the urine and through
evaporation, but a small amount is lost in the __(7)__ . The
__(8)__ are the primary organs that regulate the composition
and volume of body fluids by controlling the volume of
water and the concentration of solutes excreted in the form
of urine.

1. _____
2. _____
3. _____
4. _____
5. _____
6. _____
7. _____
8. _____

C. Using the terms provided, complete these statements:

Decrease(s)   Increase(s)

Increased extracellular fluid osmolality __(1)__ ADH
secretion from the posterior pituitary and __(2)__ aldosterone
release from the adrenal cortex. Decreased blood pressure
__(3)__ ADH secretion and renin secretion from the kidneys.
These hormonal changes __(4)__ water and sodium
reabsorption and __(5)__ urine volume and the concentration
of solutes in the urine. Conversely, a(n) __(6)__ in
extracellular osmolality or a(n) __(7)__ in blood pressure
causes hormonal changes that result in the formation of a
large volume of dilute urine.

1. _____
2. _____
3. _____
4. _____
5. _____
6. _____
7. _____

Regulation of Acid-Base Balance

❝The maintenance of hydrogen ion concentration within a narrow range of values is essential❞
for normal metabolic reactions.

A. Match these terms with the   Acids   Neutral
   correct statement or definition: Bases   pH scale

_____   1. Measurement of the acidity of a solution; numbers above 7 are
                               basic, whereas those below 7 are acidic.

_____   2. Substance with a pH of 7.

_____   3. Substances that release hydrogen ions into a solution.

👉 Strong acids completely dissociate in a solution so that all the hydrogen ions are released into the solution; weak acids release hydrogen ions but do not dissociate completely.

B. Using the terms provided, complete these statements:

Acidosis   Hyperexcitability
Alkalosis   Increased
Bicarbonate ions   Lactic acid
Decreased   Urine
Depression

When body fluid pH is below 7.35, the condition is referred to as _(1)_ ; when the pH is above 7.45, the condition is called _(2)_ . The major effect of acidosis is _(3)_ of the central nervous system, whereas a major effect of alkalosis is _(4)_ of the nervous system. Respiratory acidosis may occur with _(5)_ elimination of carbon dioxide from the body fluids through the respiratory system. Metabolic acidosis may occur through loss of _(6)_ through diarrhea or vomiting; ingestion of acidic drugs; untreated diabetes mellitus; or _(7)_ buildup from severe exercise, heart failure, or shock. Respiratory alkalosis may occur with _(8)_ elimination of carbon dioxide from the body fluids through the respiratory system. Metabolic alkalosis may occur when loss of large amounts of acidic stomach contents occurs, when alkaline substances are ingested, or when there is a higher-than-normal loss of hydrogen ions in the _(9)_ .

1. _____
2. _____
3. _____
4. _____
5. _____
6. _____
7. _____
8. _____
9. _____

C. Match these terms with the correct statement or definition:

Bicarbonate buffer system   Protein buffer system
Phosphate buffer system

_____

_____

_____

1. Provides three fourths of the buffer system of the body; includes plasma proteins and hemoglobin.

2. Plays an exceptionally important role in controlling the pH of extracellular fluid; involves carbonic acid.

3. An important intracellular buffer system, but concentrations of this system are low compared with those of the other two buffer systems.

👉 Buffers resist changes in pH by chemically binding hydrogen ions when they are added to a solution or by releasing hydrogen ions when their concentration in a solution decreases.

D. Match these terms with the correct statement or definition:

Respiratory system
Urinary system

_____

_____

1. System that responds most rapidly to pH change.

2. System with the greatest regulatory capacity for acid-base balance but with a slower response to pH change.

E. Using the terms provided, complete these statements:

Carbonic acid-bicarbonate   Decreases
Carbonic anhydrase   Increases

The respiratory system regulates acid-base balance through the _(1)_ buffer system. The reaction between carbon dioxide and water is catalyzed by _(2)_, which is found in high concentrations within erythrocytes. As carbon dioxide levels increase, pH of the body fluids _(3)_, neurons in the medullary respiratory center of the brain are stimulated, and the rate and depth of ventilation _(4)_. Carbon dioxide elimination _(5)_, and the concentration of carbon dioxide in the body fluids _(6)_. This causes hydrogen ions to combine with bicarbonate ions to form carbonic acid, which then dissociates to form carbon dioxide and water, and pH _(7)_ to its normal range.

1. _____
2. _____
3. _____
4. _____
5. _____
6. _____
7. _____

F. Using the terms provided, complete these statements:

Bicarbonate ions   Hydrogen ions
Carbonic acid   Increase
Decrease   Sodium ions

Cells in the distal portion of the nephron can secrete _(1)_ into the urine. Within these cells carbon dioxide and water combine to form _(2)_, which dissociates into hydrogen and bicarbonate ions. The hydrogen ions are secreted into the filtrate by an active transport pump that exchanges _(3)_ for hydrogen ions. The _(4)_ pass into the extracellular fluid and combine with hydrogen ions. The loss of hydrogen ions into the filtrate and the combination of bicarbonate ions with hydrogen ions in the extracellular fluid cause the body fluid pH to _(5)_.

1. _____
2. _____
3. _____
4. _____
5. _____

G. Using the terms provided, complete these statements:

Bicarbonate ions   Carbonic acid
Carbon dioxide   Hydrogen ions

Bicarbonate ion loss in the urine is decreased when the body pH is elevated. This occurs because bicarbonate ions in the filtrate combine with _(1)_ to form carbonic acid, which dissociates into carbon dioxide and water. The _(2)_ diffuses into tubule cells and combines with water to form _(3)_, which dissociates into hydrogen ions and bicarbonate ions. The _(4)_ are secreted into the filtrate, and the _(5)_ reenter the extracellular fluid. If body pH is elevated, not enough _(6)_ are secreted into the filtrate to combine with bicarbonate ions, and they are eliminated in the urine. Excretion of excess bicarbonate ions results in a decrease in extracellular fluid pH.

1. _____
2. _____
3. _____
4. _____
5. _____
6. _____

 Buffers (e.g., bicarbonate ions and ammonia) in the urine combine with secreted hydrogen ions to increase the pH of the urine. The increased pH allows additional secretion of hydrogen ions.

1. List three hormones that influence sodium ion concentration, and state if they cause an increase or decrease in blood sodium ion concentration.

2. Name the major hormone that regulates potassium ion concentration, and describe its effect.

3. List three compounds and their influence on extracellular calcium levels.

4. List three ways the sensation of thirst is increased and two ways it is decreased.

5. Name two stimuli that increase ADH, aldosterone, and renin secretion; state the effect of these substances on urine production and concentration.

6. Name three important buffer systems in the body.

7. State the effect on blood pH when respiration rate increases above normal and decreases below normal.

8. State the effect on blood pH when the acidity of the urine increases or decreases.

9. List two mechanisms that influence the number of hydrogen ions secreted into the distal part of the nephron.

MASTERY LEARNING ACTIVITY

Place the letter corresponding to the correct answer in the space provided.

_____ 1. Extracellular fluid
   a. is much like intracellular fluid in composition.
   b. includes interstitial fluid.
   c. osmotic concentration tends to be very different in the different fluid compartments of the body.
   d. all of the above

_____ 2. Which of the following would result in an increased blood sodium concentration?
   a. decrease in ADH secretion
   b. decrease in aldosterone secretion
   c. increase in atrial natriuretic hormone
   d. decrease in renin secretion

_____ 3. A decrease in extracellular potassium
   a. produces hypopolarization of cell membrane.
   b. results when aldosterone levels increase.
   c. occurs when tissues are damaged (e.g., in burn patients).
   d. all of the above

_____ 4. Calcium ion concentration in the blood decreases when
   a. vitamin D levels are lower than normal.
   b. calcitonin secretion decreases.
   c. parathyroid hormone secretion increases.
   d. all of the above

_____ 5. The sensation of thirst increases when
   a. the levels of angiotensin II increase.
   b. the osmolality of the blood decreases.
   c. blood pressure increases.
   d. renin secretion decreases.

_____ 6. Insensible perspiration
   a. is lost through sweat glands.
   b. results in heat loss from the body.
   c. increases when ADH secretion increases.
   d. results in the loss of solutes such as sodium and chloride.

_____ 7. An acid
   a. solution has a pH that is greater than 7.
   b. is a substance that releases hydrogen ions into a solution.
   c. is considered weak if it completely dissociates in water.
   d. all of the above

_____ 8. Buffers
   a. release hydrogen ions when pH increases.
   b. resist changes in the pH of a solution.
   c. include the proteins in blood.
   d. all of the above

_____ 9. Which of these systems regulating blood pH is the fastest acting?
   a. respiratory
   b. kidney

_____ 10. An increase in blood carbon dioxide levels is followed by a (an) _____ in hydrogen ions and a (an) _____ in blood pH.
   a. decrease, decrease
   b. decrease, increase
   c. increase, increase
   d. increase, decrease

_____ 11. High levels of bicarbonate ions in the urine indicate
   a. a low level of hydrogen ion secretion into the urine.
   b. that the kidneys are causing blood pH to decrease.
   c. that urine pH is increasing.
   d. all of the above

_____ 12. High levels of ammonium ions in the urine indicate
   a. a high level of hydrogen ion secretion into the urine.
   b. that the kidneys are causing blood pH to decrease.
   c. that urine pH is too alkaline.
   d. all of the above

_____ 13. Blood plasma pH is normally
   a. slightly acidic.
   b. strongly acidic.
   c. slightly alkaline.
   d. strongly alkaline.
   e. neutral.

_____ 14. Respiratory alkalosis is caused by _____ and can be compensated for by production of a more _____ urine.
   a. hypoventilation, alkaline
   b. hypoventilation, acidic
   c. hyperventilation, alkaline
   d. hyperventilation, acidic

_____ 15. Acidosis
   a. increases neuron excitability.
   b. can produce tetany by affecting the peripheral nervous system.
   c. may produce convulsions through the central nervous system.
   d. may lead to coma.

Use a separate sheet of paper to complete this section.

1. A researcher knows that under normal circumstances the level of sodium in the plasma is very precisely regulated, rarely rising or falling more than 1% from day to day. To determine the effect of the ADH-thirst mechanism versus aldosterone on control of sodium levels, the following two experiments were performed:
    A. Experiment 1: after the aldosterone system was blocked, sodium plasma levels were measured as daily sodium intake was increased.
    B. Experiment 2: after the ADH-thirst mechanism was blocked, sodium plasma levels were measured as daily sodium intake was increased.
    The results of these two experiments are graphed below:

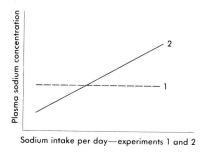

Sodium intake per day—experiments 1 and 2

    On the basis of these two experiments, is aldosterone more important than the ADH-thirst mechanism in the regulation of plasma sodium levels? Explain .

2. A patient has suffered lesions in the hypothalamus and has lost the ability to produce ADH. Despite this loss, normal or near-normal levels of sodium are maintained in the blood plasma. Can you conclude that ADH is unimportant for the regulation of plasma sodium levels? Explain.

3. Would the pH in skeletal muscle be higher, lower, or the same as the pH of lung tissue?

4. John Uptight has a gastric ulcer. One day he consumes 10 packages of antacid tablets (mainly sodium bicarbonate). What effect would their consumption have on blood pH, urine pH, and respiration rate?

5. Many animals use panting as a mechanism to keep cool. Panting, of course, increases the respiratory rate dramatically. Despite this increase, serious respiratory alkalosis does not develop. Another student suggests to you that respiratory alkalosis is prevented by the production of an alkaline urine. Explain why you would agree or disagree with this explanation.

6. Chlorothiazide produces diuresis by inhibiting sodium loss from the ascending limb of the loop of Henle. Thus large amounts of sodium enter the distal convoluted tubule. Much of this sodium cannot be reabsorbed in the distal convoluted tubule, resulting in the retention of water in the tubule and the diuretic effect. However, reabsorption of sodium in the distal convoluted tubule is more rapid than normal because of the higher concentration of sodium present. This rapid reabsorption leaves an excess of buffer ions (e.g., phosphate) in the tubule. These ions are then able to combine with hydrogen ions. What effect does chlorothiazide have on blood pH, urine pH, and respiration rate?

ANSWERS TO CHAPTER 27

Body Fluids
1. Intracellular fluid; 2. Extracellular fluid;
3. Interstitial fluid; 4. Plasma

Regulation of Intracellular and Extracellular Fluid Composition
1. Proteins; 2. Electrolytes; 3. Osmosis;
4. Increases; 5. Decreases;

Regulation of Ion Concentrations
A. 1. Sodium ions; 2. Kidneys; 3. Aldosterone;
4. Antidiuretic hormone (ADH); 5. Atrial natriuretic hormone
B. 1. Increased; 2. Decreased; 3. Increased;
4. Decreased; 5. Decreased; 6. Increased;
7. Decreased; 8. Decreased; 9. Decreased;
10. Decreased; 11. Decreased; 12. Increased
C. 1. Depolarization; 2. Hyperpolarization;
3. Proximal convoluted tubule; 4. Distal convoluted tubule; 5. Increase; 6. Increase;
7. Hyperkalemia; 8. Hypokalemia
D. 1. Increases; 2. Increases; 3. Increases;
4. Increases; 5. Increases; 6. Increases;
7. Decreases; 8. Increases; 9. Hypercalcemia

Regulation of Water Content
A. 1. Increased; 2. Decreased; 3. Increased;
4. Increased; 5. Increased; 6. Decreased;
7. Increased

B. 1. Insensible perspiration; 2. Sensible perspiration; 3. Solutes; 4. Body temperature;
5. Decrease; 6. Increase; 7. Feces; 8. Kidneys
C. 1. Increases; 2. Increases; 3. Increases;
4. Increase; 5. Decrease; 6. Decreases;
7. Increase

Regulation of Acid-Base Balance
A. 1. pH scale; 2. Neutral; 3. Acids
B. 1. Acidosis; 2. Alkalosis; 3. Depression;
4. Hyperexcitability; 5. Decreased;
6. Bicarbonate ions; 7. Lactic acid;
8. Increased; 9. Urine
C. 1. Protein buffer system; 2. Bicarbonate buffer system; 3. Phosphate buffer system
D. 1. Respiratory system; 2. Urinary system
E. 1. Carbonic acid-bicarbonate; 2. Carbonic anhydrase; 3. Decreases; 4. Increases;
5. Increases; 6. Decreases; 7. Increases
F. 1. Hydrogen ions; 2. Carbonic acid; 3. Sodium ions; 4. Bicarbonate ions; 5. Increase
G. 1. Hydrogen ions; 2. Carbon dioxide;
3. Carbonic acid; 4. Hydrogen ions;
5. Bicarbonate ions;
6. Hydrogen ions

1. Aldosterone: increases blood sodium ion concentration (increases sodium ion reabsorption); ADH: decreases blood sodium ion concentration (increases water reabsorption); atrial natriuretic hormone: decreases blood sodium ion concentration (inhibits sodium ion reabsorption)
2. Aldosterone increases potassium secretion in the distal convoluted tubule.
3. Parathyroid hormone: increases extracellular calcium levels; calcitonin: decreases extracellular calcium levels; vitamin D: increases extracellular calcium by increasing calcium uptake in the intestine
4. Increased osmolality of body fluid, reduction in plasma volume, and decrease in blood pressure cause increased thirst. Wetting of the oral mucosa and stretching the gastrointestinal tract decrease thirst.

5. Decreased blood pressure and increased tissue osmolality cause an increase in ADH, aldosterone, and renin production, producing less urine that is more highly concentrated.
6. Plasma protein, carbonic acid-bicarbonate, and phosphate buffer systems
7. Above normal: increase in blood pH; below normal: decrease in blood pH
8. If the acidity of the urine increases, the blood pH increases; if the acidity of the urine decreases, the blood pH decreases.
9. Aldosterone and body fluid pH

1. B. Extracellular fluid is all of the fluid outside of cells and includes interstitial fluid. Most extracellular fluids have approximately the same osmotic concentration. However, extracellular fluid composition is substantially different from that of intracellular fluid.

2. A. Decreased ADH secretion would result in increased water loss in the urine. As blood volume decreased, sodium concentration would increase. The other changes listed would decrease blood sodium concentration.

3. B. An increase in aldosterone results in more secretion of potassium ions into the urine. Increased extracellular potassium produces hyperpolarization and occurs when cells are damaged and release their potassium.

4. A. Vitamin D is required for the normal absorption of calcium from the small intestine. A decrease in calcitonin or an increase in parathyroid hormones causes an increase in blood calcium levels.

5. A. Angiotensin II stimulates thirst by acting on the brain. A decrease in blood pressure stimulates renin secretion, which leads to the production of angiotensin II. An increased blood osmolality also stimulates thirst.

6. B. Water loss by insensible perspiration results in heat loss. Insensible perspiration is water loss from the respiratory passages and skin (but not sweat glands). There is no loss of solutes.

7. B. Acids release hydrogen ions into a solution, have a pH that is less than 7, and are considered strong acids if they completely dissociate (i.e., release all their hydrogen ions).

8. D. Buffers resist changes in pH by releasing or binding hydrogen ions. Blood proteins are responsible for about three fourths of the buffering ability of the blood.

9. A. The respiratory system takes several minutes, and the kidney system requires several hours or even days to respond. The buffer system reacts within fractions of a second.

10. D. When blood carbon dioxide levels increase, the carbon dioxide combines with water to produce carbonic acid. The carbonic acid dissociates to form hydrogen ions and bicarbonate ions. The increase of hydrogen ions lowers the pH, i.e., increases the acidity of the blood.

11. D. High levels of bicarbonate ions occur in the urine when hydrogen ion secretion is low. Consequently, blood pH decreases and urine pH increases.

12. A. Ammonia acts as a buffer in the urine by combining with hydrogen ions to form ammonium ions. High levels of ammonium ions indicate a high level of hydrogen ions in the urine and therefore an acidic urine. Excretion of large amounts of hydrogen ions causes blood pH to increase.

13. C. Blood pH is normally about 7.4 or slightly alkaline.

14. C. Hyperventilation reduces plasma carbon dioxide levels, resulting in alkalosis. The kidneys compensate by producing a more alkaline urine. This conserves hydrogen ions and restores pH to normal levels.

15. D. One of the effects of acidosis is a depression of the central nervous system, which can lead to coma and death. Alkalosis causes excitation of the nervous system. Usually, in alkalosis, the peripheral nervous system is affected first, producing tetany because of repetitive firing of the neurons. Tetany of the respiratory muscles can cause death. Overexcitation of the central nervous system leads to extreme nervousness and convulsions.

1. When the ADH-thirst mechanism is blocked but the aldosterone system is still operating (experiment 2), the levels of plasma sodium increase when daily sodium intake increases. This indicates that aldosterone plays a very minor role in regulation of plasma sodium levels. Indeed, even when the aldosterone system is blocked (experiment 1), plasma sodium levels stay relatively constant, despite increasing sodium intake. These data suggest that the ADH-thirst mechanism is more important than aldosterone in regulating plasma sodium levels. It is consistent with the data to assume that the ADH-thirst mechanism maintains sodium levels as follows: if sodium levels rise, the thirst center is stimulated, fluids are ingested, and sodium concentration decreases. If sodium levels fall, ADH secretion is inhibited, a hyposmotic urine is produced, water is lost, and sodium concentration increases. It seems paradoxical that aldosterone, which controls sodium reabsorption, is not effective in regulating plasma sodium levels, whereas ADH, which regulates water reabsorption, is effective. The ineffectiveness of aldosterone results from the following effect: when aldosterone increases, sodium reabsorption increases, water follows sodium, and extracellular fluid volume increases, producing a rise in arterial blood pressure. The increased blood pressure increases the glomerular filtration rate, and the rapid movement of filtrate through the tubule compensates for the increased reabsorption of sodium caused by the aldosterone.

2. The conclusion might seem reasonable at first. However, careful thought may lead to the opposite conclusion. All that has been lost is the ability to produce ADH. If plasma osmolality increases, the thirst mechanism is activated and the patient drinks more fluids, thus restoring plasma osmolality to normal levels. Conversely, if plasma osmolality decreases, the patient reduces fluid intake and plasma osmolality increases. Thus the thirst mechanism plays an important role in regulating sodium levels. Because there is no ADH present, the patient will produce a dilute urine, and the daily urine production may increase to 5 to 15 L/day. This, of course, requires a corresponding increase of fluid intake. This condition is called diabetes insipidus. The importance of ADH, coupled with the thirst mechanism, can be demonstrated when there is inadequate intake of water (such as during unconsciousness). Then plasma sodium increases tremendously, because without ADH the kidneys are unable to conserve water.

3. The skeletal muscle have a slightly higher carbon dioxide content than lung tissue. Consequently, the pH of skeletal muscle is slightly lower than the pH of lung tissue.

4. The overdose of the alkaline antacid tablets raises blood pH, producing metabolic alkalosis. The secretion of fewer hydrogen ions results in a more alkaline urine, and hydrogen ion concentration in the blood increases, compensating for the alkalosis. Respiration rate also decreases to compensate. Reduced respiration increases plasma CO_2 levels, which reduces the pH through the production of carbonic acid.

5. Although production of an alkaline urine helps to compensate for respiratory alkalosis, it takes the kidneys hours or even days to exert their effect. Instead, panting animals prevent respiratory alkalosis by decreasing tidal volume. As a result, with each breath very little "new" air enters the alveoli. This prevents the loss of excessive amounts of carbon dioxide and the development of respiratory alkalosis.

6. The hydrogen ions combine with the buffer and are excreted as an acidic urine. The loss of hydrogen ions produces alkalosis. The subsequent increase in pH inhibits the respiratory center and lowers respiration rate.

28 Reproductive System

FOCUS: The male reproductive system consists of the testes, where sperm cells and testosterone are produced; the scrotum and cremaster muscles, which help keep the testes at a temperature suitable for sperm cell development; a series of ducts that carry the sperm cells to the outside of the body; glands that produce secretions that provide energy for the sperm cells and neutralize the acidic environment of the male urethra and female reproductive tract; and the penis, which when erect functions in the transfer of sperm cells to the female vagina. The functions of the male reproductive system are under hormonal and neural control.. The hypothalamus releases GnRH, which stimulates the adenohypophysis to secrete FSH and LH. FSH stimulates spermatogenesis, and LH stimulates testosterone production. Testosterone is responsible for the development of male sexual characteristics. The nervous system controls the process of erection, emission, ejaculation, and orgasm. The female reproductive system consists of the ovaries, where the secondary oocyte is formed and hormones such as estrogen and progesterone are produced,; the uterine tubes, where the secondary oocyte is normally fertilized and which carry the developing organism to the uterus; the uterus, where implantation followed by embryonic and fetal development occurs; the vagina, which receives the penis during intercourse and serves as a passageway to the outside of the body for the fetus; the external genitalia, which form a cover for the openings to the vagina and urinary tract; and the breast, where milk is produced for the infant. The ovaries undergo a cyclic production of oocytes and hormones. Hormones act on the uterine lining to prepare the uterus for implantation. During pregnancy hormones produced by the placenta (HCG, estrogen, progesterone) prevent a spontaneous abortion. Parturition involves changes in estrogen, progesterone, and oxytocin levels. Lactation results from increased prolactin, which stimulates milk production, and increased oxytocin, which stimulates milk letdown. Menopause occurs when ovaries stop producing oocytes and hormones.

CONTENT LEARNING ACTIVITY

Scrotum

66*The scrotum is important in regulation of temperature in the testes.*99

Match these terms with the correct statement or definition:

Cremaster muscles   Raphe
Dartos muscle

_____ 1. Irregular ridge on the midline of the scrotum.

_____ 2. Layer of smooth muscle surrounding the scrotum; contracts and causes skin of the scrotum to become firm and wrinkled.

_____ 3. Abdominal muscles; contract and pull testes closer to the body.

Both the dartos muscle and the cremaster muscle operate to keep the testes at the proper temperature for sperm cell production to occur.

Perineum

❝*The area between the thighs, which is bounded by the pubis anteriorly, the coccyx posteriorly, and ischial tuberosities laterally, is called the perineum.*❞

Match these terms with the correct statement or definition:

Anal triangle
Urogenital triangle

_____   1. Anterior portion of the perineum, which contains the external genitalia.

_____   2. Posterior portion of the perineum, which contains the anal opening.

Testes

❝*The testes are small, ovoid organs within the scrotum.*❞

A. Match these terms with the correct statement or definition:

Efferent ductules   Seminiferous tubules
Interstitial cells   Septa
  (cells of Leydig)   Tunica albuginea
Rete testis

_____   1. Thick white capsule surrounding the testis.

_____   2. Connective tissue of the tunica albuginea that divides the testis into lobules.

_____   3. Structures in which sperm cells develop.

_____   4. Endocrine cells located in the connective tissue stroma around the seminiferous tubules.

_____   5. Tubules connected to the rete testis that pierce the tunica albuginea and exit the testis.

B. Match these terms with the correct statement or definition:

Gubernaculum   Process vaginalis
Inguinal canal   Tunica vaginalis

_____   1. Structure that connects the testes to the scrotum.

_____   2. Passageway from the abdominal cavity to the scrotum, through which the testes descend.

_____   3. Outpocketing of peritoneum that precedes the testis as it moves to the scrotum during fetal development.

_____   4. Small, closed sac that covers most of the testis.

The superficial opening of the inguinal canal, called the superficial inguinal ring, is medial, whereas the deep opening, called the deep inguinal ring, is lateral.

C. Match these terms with the correct statement or definition:

Acrosome   Sertoli cells
Dihydrotestosterone Spermatids
Estradiol   Spermatogenesis
Germ cells   Spermatogonia
Primary spermatocytes Spermatozoa
Secondary spermatocytes

_____ 1. Production of sperm cells.

_____ 2. Large cells that nourish the germ cells, produce hormones, and form the blood-testis barrier.

_____ 3. Two hormones produced from testosterone by Sertoli cells; may be the active hormones that promote sperm development.

_____

_____ 4. Most peripheral germ cells; they divide by mitosis.

_____ 5. Germ cells produced from spermatogonia that are passing through the first meiotic division.

_____ 6. Germ cells produced from primary spermatocytes by the first meiotic division.

_____ 7. Germ cells produced from secondary spermatocytes by the second meiotic division.

_____ 8. Cap found on the head of the sperm cell, containing enzymes necessary for penetrating the female oocyte.

 Sertoli cells secrete a protein called androgen-binding protein to which testosterone and dihydrotestosterone bind.

Ducts

After their production in the seminiferous tubules, sperm cells leave the testes through the *efferent ductules and pass through a series of ducts to reach the exterior of the body.*

Match these terms with the correct statement or definition:

Ductus (vas) deferens   Prostatic urethra
Ductus epididymis   Spermatic cord
Ejaculatory duct   Spongy (penile) urethra
Epididymis   Urethral glands
Membranous urethra

_____ 1. Comma-shaped structure on the posterior testis in which maturation of spermatozoa occurs.

_____ 2. Single convoluted ductule located primarily in the body of the epididymis.

_____ 3. Duct running from the tail of the epididymis to the ejaculatory duct; the end enlarges to form the ampulla.

_____ 4. The ductus deferens, the blood vessels and nerves that supply the testis, and the cremaster muscle compose this structure.

_____ 5. Duct formed when the duct of the seminal vesicle joins the ductus deferens.

_____ 6. Portion of the urethra that passes through the prostate gland and into which the ejaculatory ducts empty.

_____ 7. Shortest portion of the urethra; extends from the prostate through the urogenital diaphragm.

_____ 8. Mucus-secreting glands that empty into the urethra.

☞ The ducts of the epididymis have pseudostratified columnar epithelium with microvilli called stereocilia, which increase the surface area of the epithelial cells.

Penis

The penis consists of three columns of erectile tissue.

Match these terms with the correct statement or definition:

Bulb of the penis   External urethral orifice
Corpora cavernosa   Glans penis
Corpus spongiosum   Prepuce (foreskin)
Crus of the penis   Root of the penis
Erection

_____ 1. Engorgement of erectile tissue with blood, which causes the penis to enlarge and become firm.

_____ 2. Two erectile columns that form the dorsum and sides of the penis.

_____ 3. Smallest erectile column, which occupies the ventral portion of the penis; the spongy urethra passes through it.

_____ 4. Cap, formed from the corpus spongiosum, over the distal end of the penis.

_____ 5. Expansion of the corpus spongiosum at the base of the penis.

_____ 6. Expansion of the corpora cavernosa at the base of the penis.

_____ 7. Collective name for the crus of the penis and the bulb of the penis; attaches the penis to the coxae.

_____ 8. Loose fold of skin that covers the glans penis.

Glands

❝_Several glands secrete substances into the reproductive system._**❞**

A. Match these terms with the correct statement or definition:

Bulbourethral glands	Prostate gland
Ejaculation	Semen
Emission	Seminal vesicles

_____ 1. Sac-shaped glands located near the ampullae of the ductus deferentia.

_____ 2. Gland the size and shape of a walnut located dorsal to the symphysis pubis at the base of the bladder; surrounds part of the urethra and the ejaculatory ducts.

_____ 3. Pair of small mucous glands near the membranous portion of the urethra.

_____ 4. Discharge of semen into the prostatic urethra; stimulated by sympathetic impulses.

_____ 5. Glands producing a mucous secretion that lubricates the urethra, neutralizes the contents of the spongy urethra, provides lubrication during intercourse, and reduces the acidity of the vagina.

_____ 6. Glands producing a thick, mucoid secretion containing fructose, fibrinogen, and prostaglandins.

_____ 7. Gland that produces a thin, milky, alkaline secretion that neutralizes the urethra and vagina and contains clotting factors that cause fibrinogen to aggregate.

B. Match these terms with the correct parts of the diagram labeled in Figure 28-1:

Bulbourethral gland
Ductus deferens
Ejaculatory duct
Epididymis
External urethral orifice
Membranous urethra
Penis
Prostate
Prostatic urethra
Scrotum
Seminal vesicle
Spongy urethra
Testis

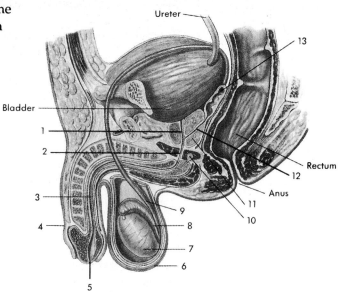

Figure 28-1

1. _____   6. _____   11. _____

2. _____   7. _____   12. _____

3. _____   8. _____   13. _____

4. _____   9. _____

5. _____   10. _____

Regulation of Sex Hormone Secretion

66 *Hormonal mechanisms that influence the male reproductive system involve the hypothalamus,* 99 *the pituitary gland, and the testes.*

Match these terms with the correct statement or definition:

Androgen
Follicle-stimulating hormone (FSH)
Gonadotropins
GnRH

Inhibin
Luteinizing hormone (LH)
Testosterone

_____ 1. Small peptide hormone released from the hypothalamus that stimulates cells in the adenohypophysis.

_____ 2. General term for hormones that affect the gonads; secreted by the adenohypophysis.

_____ 3. Hormone that binds to the cells of Leydig in the testes and stimulates testosterone synthesis and secretion.

_____ 4. Hormone that binds to the Sertoli cells in the seminiferous tubules of the testes and promotes spermatogenesis.

_____ 5. Major male hormone secreted by the testes.

_____ 6. General name for male hormones that stimulate the development of male secondary sexual characteristics.

_____ 7. Released from the testes, this hormone inhibits FSH secretion from the adenohypophysis.

Puberty

"_The reproductive system becomes functional at puberty, which normally begins when a boy is 12_**"** _to 14 years old._

Using the terms provided, complete these statements:

HCG	Spermatogenesis
GnRH	Testosterone
LH and FSH	

 (1) , which is secreted by the placenta, stimulates the synthesis and secretion of testosterone before birth. After birth only small amounts of testosterone are secreted until puberty, when the hypothalamus becomes less sensitive to the inhibitory effect of (2) and the rate of (3) secretion increases, leading to increased (4) release from the adenohypophysis. Elevated FSH levels promote (5) , and elevated LH levels cause the cells of Leydig to secrete (6) .

1. _____

2. _____

3. _____

4. _____

5. _____

6. _____

Effects of Testosterone

"_Testosterone is by far the major androgen in males._**"**

Using the terms provided, complete these statements:

Body fluids	Protein synthesis
Coarser	Rapid bone growth
Hypertrophy	Sebaceous glands
Melanin	Testes

Testosterone causes the enlargement and differentiation of the male reproductive system, is necessary for spermatogenesis, and is required for the descent of the (1) during fetal development. Testosterone stimulates hair growth in several regions and causes the texture of the hair and skin to become (2) . Testosterone increases the quantity of (3) in the skin, causing it to become darker, and increases the rate of secretion by (4) , frequently resulting in acne. Testosterone also causes (5) of the larynx, deepening the voice. Testosterone stimulates red blood cell production, stimulates metabolism, and causes retention of sodium in the body, resulting in an increase in (6) . Testosterone promotes (7) , resulting in an increased skeletal muscle mass, and (8) , resulting in an increase in height.

1. _____

2. _____

3. _____

4. _____

5. _____

6. _____

7. _____

8. _____

Male Sexual Behavior and the Male Sex Act

❝*Testosterone is required to initiate and maintain normal male sexual behavior.***❞**

Match these terms with the
correct statement or definition:

Ejaculation	Impotence
Emission	Orgasm
Erection	Resolution

_____ 1. Pleasurable climax sensation associated with ejaculation.

_____ 2. Period after ejaculation when the penis becomes flaccid and the male is unable to achieve erection and a second ejaculation.

_____ 3. Inability to accomplish the male sex act because of psychic or physical factors.

_____ 4. Occurs when parasympathetic or sympathetic impulses cause the dilation of the arteries that supply blood to the erectile tissues in the penis.

_____ 5. Stimulated by sympathetic centers in the spinal cord as the level of sexual tension increases; secretions of the prostate and seminal vesicles are released.

_____ 6. Rhythmic contractions that force semen out of the urethra, triggered by efferent somatic impulses.

☞ Psychic stimuli (e.g., sight, sound, odor, or thoughts) have a major effect on sexual reflexes.

Ovaries

❝*The ovaries are small organs located on each side of the uterus.***❞**

A. Match these terms with the
correct statement or definition:

Cortex	Ovarian follicles
Medulla	Ovarian ligament
Mesovarium	Suspensory ligament
Oocyte	Tunica albuginea
Ovarian (germinal) epithelium	

_____ 1. Peritoneal fold that attaches ovaries to the broad ligament.

_____ 2. Ligament that extends from the mesovarium to the body wall.

_____ 3. Peritoneum covering the surface of the ovary.

_____ 4. Layer of dense fibrous connective tissue; surrounds the ovary.

_____ 5. Looser inner portion of the ovary.

_____ 6. Small vesicles, each of which contains an oocyte, distributed throughout the cortex of the ovary.

_____ 7. Egg cell.

Match these terms with the correct statement or definition:

Oogenesis   Primary oocyte
Oogonium   Secondary oocyte
Polar body

_____ 1. Production of a secondary oocyte within the ovaries.

_____ 2. A cell from which an oocyte develops.

_____ 3. Oogonium that has started meiosis but stopped at prophase I.

_____ 4. Two structures produced from a primary oocyte by the first meiotic division.

_____

Match these terms with the correct statement or definition:

Antrum   Primordial follicle
Corona radiata   Secondary follicle
Cumulus mass   Theca
Polar body   Vesicular (graafian) follicle
Primary follicle   Zona pellucida

_____ 1. Primary oocyte with the surrounding layer of flat granulosa cells.

_____ 2. Layer of clear, viscous fluid that is deposited around a primary oocyte.

_____ 3. Follicle that is not yet mature, but contains an antrum, cumulus mass, and corona radiata.

_____ 4. Two layers of cells (interna and externa) molded around the secondary follicle or graafian follicle to form a capsule.

_____ 5. Fully mature follicle.

_____ 6. Mass of follicular cells that surrounds the oocyte in a secondary follicle.

_____ 7. Innermost cells of the cumulus mass.

_____ 8. Structure with little cytoplasm produced when first meiotic division is completed.

573

D. Using the terms provided, complete these statements:

Corpus albicans
Corpus luteum
Corpus luteum of pregnancy
Fertilization
Ovulation

Polar body
Progesterone and
  estrogen
Secondary oocyte
Zygote

1. _____

2. _____

3. _____

4. _____

5. _____

6. _____

7. _____

8. _____

9. _____

The release of the secondary oocyte from the follicle is called _(1)_. During ovulation the development of the _(2)_ has stopped at metaphase II. Continuation of the second meiotic division is triggered by _(3)_, the entry of the sperm cell into the secondary oocyte. If fertilization does not occur, meiosis is not completed; if fertilization occurs, meiosis II is completed, and a second _(4)_ is formed. The fertilized oocyte is now called a _(5)_. After ovulation the follicle is transformed into a glandular structure called the _(6)_. The granulosa cells and the theca interna enlarge and begin to secrete _(7)_. If pregnancy occurs, the corpus luteum enlarges and remains throughout pregnancy as the _(8)_. If pregnancy does not occur, the connective tissue cells in the corpus luteum become enlarged and clear and give the whole structure a whitish color; therefore it is called the _(9)_.

Uterine Tubes

66*The two uterine tubes are located along the superior margin of the broad ligament.*99

Match these terms with the correct statement or definition:

Ampulla
Fimbriae
Infundibulum
Isthmus
Mesosalpinx

Mucosa
Muscular layer
Ostium
Serosa
Uterine (intramural) part

_____ 1. Portion of the broad ligament most directly associated with the uterine tube.

_____ 2. Funnel-shaped end of the uterine tube.

_____ 3. Opening in the funnel-shaped end of the uterine tube.

_____ 4. Long thin processes that surround the opening into the uterine tube.

_____ 5. Portion of the uterine tube closest to the infundibulum; the widest and longest part of the tube.

_____ 6. Narrow, thin-walled portion of the uterine tube.

_____ 7. Middle layer of the uterine tube wall.

_____ 8. Inner layer of the uterine tube wall.

Uterus

"The uterus is the size and shape of a medium-sized pear."

A. Match these terms with the correct statement or definition:

Body	Isthmus
Broad ligament	Ostium
Cervical canal	Round ligaments
Cervix	Uterine cavity
Fundus	Uterosacral ligaments

_____ 1. Larger, rounded portion of the uterus, directed superiorly.

_____ 2. Narrower portion of the uterus, directed inferiorly.

_____ 3. Slight constriction that marks the junction between the body of the uterus and the cervix.

_____ 4. Cavity inside the cervix of the uterus.

_____ 5. Opening of the cervix into the vagina.

_____ 6. Major ligaments that extend from the uterus through the inguinal canals to the external genitalia.

_____ 7. Major ligament that spreads on both sides of the uterus and to which the ovaries and uterine tubes are attached.

B. Match these terms with the correct statement or definition:

Basal layer	Myometrium (muscular coat)
Endometrium (mucous membrane)	Perimetrium (serous coat)
Functional layer	

_____ 1. The outer coat of the uterus; peritoneum.

_____ 2. The innermost layer of the uterus.

_____ 3. The thin, deep layer of the endometrium that is continuous with the myometrium.

_____ 4. The thicker, superficial layer of the endometrium; undergoes the greatest change during the menstrual cycle.

C. Match these terms with the
   correct parts of the diagram
   labeled in Figure 28-2:

Body of uterus
Cervix
Endometrium
Fundus of uterus
Myometrium
Ovarian ligament
Ovary
Perimetrium
Round ligament
Uterine tube
Vagina

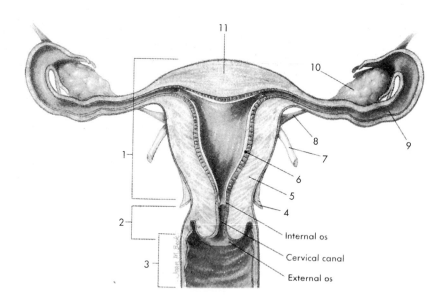

Internal os

Cervical canal

External os

Figure 28-2

1. _____   5. _____   9. _____

2. _____   6. _____   10. _____

3. _____   7. _____   11. _____

4. _____   8. _____

Vagina

❝*The vagina is the female organ of copulation, functions to receive the penis during intercourse,*❞
and allows menstrual flow and childbirth.

Match these terms with the   Columns   Hymen
correct statement or definition: Fornix   Rugae

_____   1. Longitudinal ridges that extend the length of the vaginal
                              walls.

_____   2. Transverse ridges in the vagina.

_____   3. Superior, domed portion of the vagina.

_____   4. Thin mucous membrane that may cover the external vaginal
                              opening.

External Genitalia

" *The external genitalia consist of the vestibule and its surrounding structures.* **"**

A. Match these terms with the correct statement or definition:

Bulb of the vestibule   Lesser vestibular glands
Clitoris   Mons pubis
Greater vestibular glands   Prepuce
Labia majora   Pudendal cleft
Labia minora   Vestibule

_____ 1. Space into which the vagina and urethra open.

_____ 2. Thin, longitudinal skin folds bordering the vestibule.

_____ 3. Erectile structure that corresponds to the corpora cavernosa in the male; located in the anterior margin of the vestibule.

_____ 4. Fold of skin over the clitoris formed by the two labia minora.

_____ 5. Erectile tissue; corresponds to corpus spongiosum in the male.

_____ 6. Ducts of these glands open on either side of the vestibule between the vaginal opening and the labia minora.

_____ 7. Small mucous glands near the clitoris and urethral opening.

_____ 8. Mound located over the symphysis pubis.

B. Match these terms with the correct parts of the diagram labeled in Figure 28-3:

Clitoris
Labia majora
Labia minora
Mons pubis
Prepuce
Urethra
Vagina
Vestibule

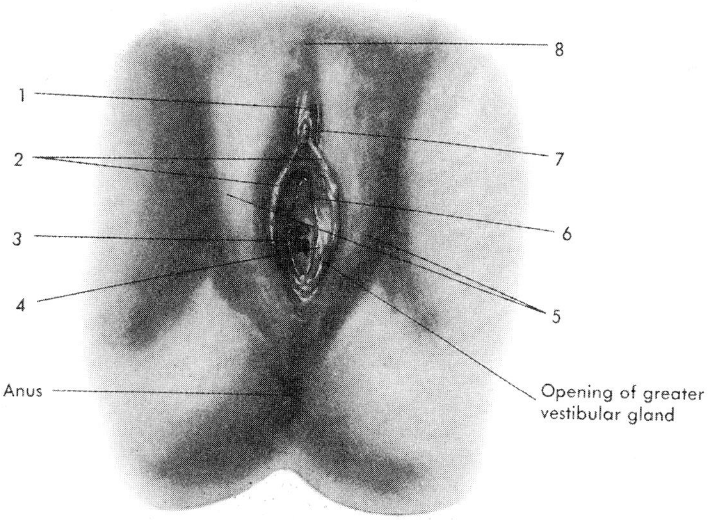

Figure 28-3

1. _____   4. _____   7. _____

2. _____   5. _____   8. _____

3. _____   6. _____

Perineum

"*The area between the thighs, which is bounded by the pubis anteriorly, the coccyx posteriorly,***"**
and the ischial tuberosities laterally, is called the perineum.

Match these terms with the correct statement or definition:

Anal triangle	Episiotomy
Clinical perineum	Urogenital triangle

_____ 1. Anterior portion of the perineum that contains the external genitalia.

_____ 2. Region between the vagina and the anus.

_____ 3. Incision in the clinical perineum to prevent tearing during childbirth.

Mammary Glands

"*The mammary glands are the organs of milk production and are located within the mammae,***"**
or breasts.

A. Match these terms with the correct statement or definition:

Alveoli	Lobes
Areola	Lobules
Areolar glands	Mammary (Cooper's) ligaments
Gynecomastia	Nipple
Lactiferous duct	

_____ 1. Circular, pigmented area surrounding the nipple.

_____ 2. Rudimentary mammary glands located just below the surface of the areola.

_____ 3. Condition in which male breasts become enlarged.

_____ 4. Glandular compartments of the breast, each of which possesses a single lactiferous duct.

_____ 5. Smaller subcompartments of the breast that contain the milk-producing glands.

_____ 6. Duct opening to the nipple that carries milk from one lobe; contains a sinus that accumulates milk.

_____ 7. Secretory sacs in the milk-producing breast.

_____ 8. Ligaments that support and hold the breast in place.

B. Match these terms with the correct parts of the diagram labeled in Figure 28-4:

Alveoli   Lobe
Areola   Lobule
Lactiferous duct   Nipple
Lactiferous sinus

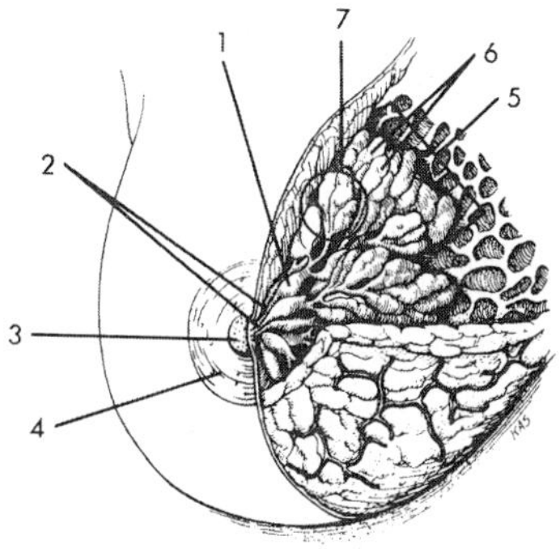

Figure 28-4

1. _____
2. _____
3. _____
4. _____
5. _____
6. _____
7. _____

Puberty

Puberty in the female is marked by the first episode of menstrual bleeding, called menarche.

Using the terms provided, complete these statements:

Estrogen and progesterone   GnRH
FSH and LH

Changes associated with puberty result primarily from elevated levels of __(1)__ secreted by the ovaries. These hormones are secreted in response to an increasing and cyclic pattern of __(2)__ secretion by the adenohypophysis. These hormones in turn are secreted in response to an increasing __(3)__ secretion by the hypothalamus. The cyclic surge of __(4)__ results in ovulation, and the monthly changes in secretion of __(5)__ produce changes in the uterus that characterize the menstrual cycle.

1. _____
2. _____
3. _____
4. _____
5. _____

The Menstrual Cycle

The term menstrual cycle refers to the series of changes that occur in sexually mature, nonpregnant females and culminate in menses.

A. Match these terms with the correct statement or definition:

Follicular (proliferative) phase   Menses
Luteal (secretory) phase

_____ 1. Period of mild hemorrhage during which the uterine epithelium is sloughed and expelled from the uterus.

_____ 2. Time between the ending of menses and ovulation.

_____ 3. Time between ovulation and the beginning of menses.

Typically the menstrual cycle is 28 days long, with the first day of menses considered day 1, and with ovulation occurring about day 14.

B. Using the terms provided, complete these statements:

Adenohypophysis   Luteal
FSH   Ovarian cycle
HCG   Ovulation
LH   Progesterone

The __(1)__ specifically refers to the series of events that occur in a regular fashion in the ovaries of sexually mature, nonpregnant women. A number of follicles begin to mature during each menstrual cycle, under the influence of __(2)__, but normally only one is ovulated. Gonadotropins are released from the __(3)__ in large amounts just before ovulation. __(4)__ causes the primary oocyte to complete the first meiotic division and triggers several events that result in __(5)__. After ovulation, the granulosa cells enlarge and increase in number to become __(6)__ cells, which secrete large amounts of __(7)__ and some estrogen. If fertilization does not occur, the cells of the corpus luteum atrophy. If fertilization of the ovulated oocyte takes place, __(8)__, an LH-like hormone, is secreted by the developing embryo and keeps the corpus luteum from degenerating.

1. _____
2. _____
3. _____
4. _____
5. _____
6. _____
7. _____
8. _____

C. Using the terms provided, complete these statements:

Decrease(s)   Negative feedback
Increase(s)   Positive feedback
FSH and LH surges

Just before ovulation, the theca interna cells of the developing follicle secrete estrogen, which __(1)__ GnRH secretion from the hypothalamus, which in turn __(2)__ FSH and LH secretion from the adenohypophysis. This leads to a __(3)__ system, in which FSH and LH __(4)__ estrogen secretion from the ovary, which __(5)__ GnRH secretion and FSH and LH production. This leads to a large increase in secretion of FSH and LH, called the __(6)__. Shortly after ovulation, estrogen secretion __(7)__, and progesterone secretion __(8)__. Both estrogen and progesterone secretion __(9)__ after the corpus luteum is formed, and this has a __(10)__ effect on GnRH release from the hypothalamus. As a result, LH and FSH secretion by the adenohypophysis __(11)__, the corpus luteum degenerates, and the levels of estrogen and progesterone __(12)__, resulting in cyclic changes in the uterus.

1. _____
2. _____
3. _____
4. _____
5. _____
6. _____
7. _____
8. _____
9. _____
10. _____
11. _____
12. _____

D. Match these terms with the correct statement or definition:

Spiral arteries   Uterine cycle
Spiral glands

_____ 1. Changes that occur primarily in the endometrium of the uterus during the menstrual cycle.

_____ 2. Glands formed when columnar epithelia in the endometrium are thrown into folds; secrete a glycogen-rich fluid.

_____ 3. Arteries that project between spiral glands to supply nutrients to the endometrial cells.

E. Match these terms with the correct statement:

Decrease(s)
Increase(s)

_____ 1. Effect of estrogen on the thickness of the endometrium; affects cell division.

_____ 2. Effect of estrogen on the sensitivity of the endometrial cells to progesterone.

_____ 3. Effect of progesterone on the thickness of the endometrium and the myometrium; affects cell size.

_____ 4. Effect of progesterone on the secretory ability of endometrial cells.

_____ 5. Effect of progesterone on smooth muscle contraction.

_____ 6. Effect of declining progesterone levels on blood supply to the endometrium.

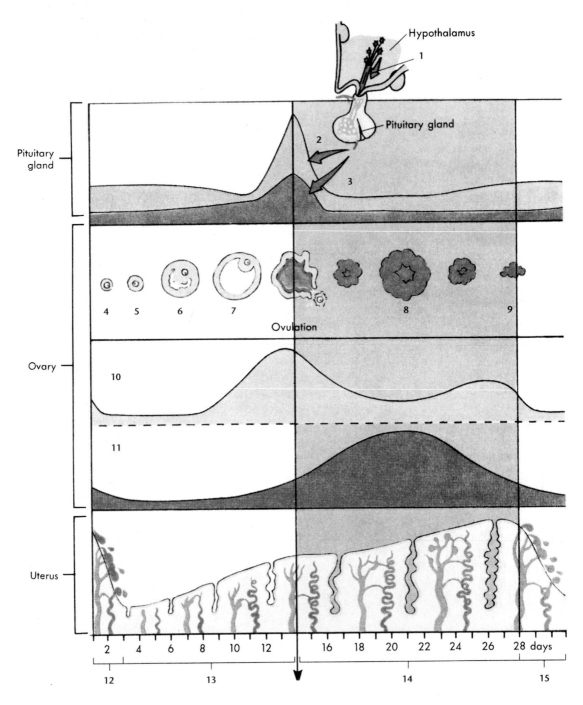

Hypothalamus

1

Pituitary gland

Pituitary gland

2

3

Ovulation

4   5   6   7   8   9

Pituitary
gland

Ovary

10

11

Uterus

2   4   6   8   10   12   16   18   20   22   24   26   28 days

12   13   14   15

Figure 28-5

F. Match these terms with the correct parts of the diagram labeled in Figure 28-5:

Corpus albicans
Corpus luteum
Estrogen level
FSH
GnRH
Mature
   (graafian) follicle
LH

Menses
Primary follicle
Primordial follicle
Progesterone level
Proliferative phase
Secondary follicle
Secretory phase

1. _____
2. _____
3. _____
4. _____
5. _____

6. _____
7. _____
8. _____
9. _____
10. _____

11. _____
12. _____
13. _____
14. _____
15. _____

Female Sexual Behavior and the Female Sex Act

66 *Sexual drive in females, like sexual drive in males, is dependent upon hormones.* 99

Using the terms provided, complete these statements:

Clitoris
Fertilization
Orgasm
Psychic factors

Resolution
Sacral
Vagina and
   vestibular glands

1. _____
2. _____
3. _____
4. _____
5. _____
6. _____
7. _____

Androgens and possibly estrogens affect cells in the brain and influence sexual behavior; however, (1) also play a role in sexual behavior. The (2) region of the spinal cord is the area that integrates sexual reflexes, which are modulated by cerebral influences. During sexual excitement, parasympathetic stimulation causes erectile tissue in the (3) and around the vaginal opening to become engorged with blood. Secretions from the (4) provide lubrication for the movement of the penis. Tactile stimulation during intercourse, as well as psychological stimuli, normally trigger a(n) (5) , the female climax. After the sexual act, there is a period of (6) , characterized by an overall sense of satisfaction and relaxation. Although orgasm is a pleasurable component of sexual intercourse, it is not required for (7) to occur.

Female Fertility and Pregnancy

"_Sperm cells must travel from the vagina to the uterine tube ampulla for fertilization to occur._**"**

Match these terms with the correct statement or definition:

Capacitation   Placenta
HCG   Swimming
Oxytocin and prostaglandins Trophoblast

_____   1. One of the forces that propels the sperm cells.

_____   2. Hormones that stimulate smooth muscle contraction, moving the sperm cells toward the ampulla.

_____   3. Enables the sperm cells to release acrosomal enzymes that allow penetration of the oocyte.

_____   4. Outer layer of the developing embryonic mass responsible for implantation.

_____   5. Hormone secreted by the trophoblast that causes the corpus luteum to remain functional.

_____   6. Organ responsible for secreting estrogen and progesterone during most of pregnancy.

Menopause

"_The cessation of menstrual cycles is called menopause._**"**

Using the terms provided, complete these statements:

Estrogen and progesterone   LH and FSH
Female climacteric   Menopause
Hot flashes   Ovary

The whole time period from the onset of irregular menstrual cycles to their complete cessation is called the _(1)_. The major cause of menopause is age-related changes in the _(2)_. Follicles become less sensitive to stimulation by _(3)_, and fewer mature follicles and corpora lutea are produced. Gradual morphological changes occur in the female in response to reduced amounts of _(4)_ produced by the ovaries. Symptoms include _(5)_, irritability, fatigue, anxiety, and occasionally severe emotional disturbances.

1. _____

2. _____

3. _____

4. _____

5. _____

1. List the stages of development of sperm cells.

2. List these structures in the correct order from the site of sperm cell production to the exterior of the body: ductus deferens, efferent ductules, ejaculatory duct, epididymis, membranous urethra, prostatic urethra, rete testis, seminiferous tubules, spongy urethra.

3. List three glands involved in reproduction in the male, and describe the secretions they produce.

4. For the male, list the hormones involved with the reproductive system that are secreted by the hypothalamus, the adenohypophysis, and the testes.

5. List six effects that testosterone has in the male.

6. Arrange these stages in the development of follicle cells in the correct order: corpus albicans, corpus luteum, primary follicle, primordial follicle, secondary follicle, vesicular (graafian) follicle.

7. List the stages of development of the oocyte, and state at what stage fertilization normally takes place.

8. Starting with the site of milk production, name the structures a drop of milk would pass through on the way to the outside of the woman's body.

9. Name the three phases of the menstrual cycle.

10. List the effect these hormones have on the ovarian cycle: GnRH, FSH, and LH.

11. List the effects of estrogen and progesterone on the uterine cycle.

MASTERY LEARNING ACTIVITY

Place the letter corresponding to the correct answer in the space provided.

_____ 1. If an adult male jumped into a swimming pool of cold water, which of these muscles would be expected to contract?
   a. cremaster muscle
   b. dartos tunic
   c. gubernaculum
   d. a and b
   e. all of the above

_____ 2. Early in development the testes
   a. are found inside the peritoneal cavity.
   b. move through the inguinal canal.
   c. produce a membrane that becomes the scrotum.
   d. all of the above

_____ 3. Testosterone is produced in the
   a. cells of Leydig.
   b. seminiferous tubules of the testes.
   c. anterior lobe of the pituitary.
   d. sperm cells.

_____ 4. Given these structures:
   1. ductus deferens
   2. efferent ductule
   3. epididymis
   4. ejaculatory duct
   5. rete testis

Choose the arrangement that lists the structures in the order a sperm cell would pass through them from the seminiferous tubules to the urethra.
   a. 2, 3, 5, 4, 1
   b. 2, 5, 3, 4, 1
   c. 3, 2, 4, 1, 5
   d. 3, 4, 2, 1, 5
   e. 5, 2, 3, 1, 4

_____ 5. The site of spermatogenesis in the male is the
   a. ductus deferens.
   b. seminiferous tubules.
   c. epididymis.
   d. rete testis.

_____ 6. The site of final maturation and storage of sperm cells before their ejaculation is the
   a. seminal vesicles.
   b. seminiferous tubules.
   c. glans penis.
   d. epididymis.
   e. sperm bank.

_____ 7. Concerning the penis,
   a. the membranous urethra passes through the corpora cavernosa.
   b. the glans penis is formed by the corpus spongiosum.
   c. the penis contains four columns of erectile tissue.
   d. the prepuce is attached to the crus of the penis.

_____ 8. Given these glands:
   1. prostate gland
   2. bulbourethral gland
   3. seminal vesicle

   Choose the arrangement that is in the order the glands would contribute their secretions in the formation of semen.
   a. 1, 2, 3
   b. 2, 1, 3
   c. 2, 3, 1
   d. 3, 1, 2
   e. 3, 2, 1

_____ 9. Which of these glands is correctly matched with the function of gland secretion?
   a. bulbourethral gland - neutralizes acidic contents of the urethra
   b. seminal vesicles - contains large amounts of fructose that nourishes the sperm cells
   c. prostate gland - contains clotting factors that causes coagulation of the semen
   d. all of the above

_____ 10. LH in the male stimulates
   a. development of the seminiferous tubules.
   b. spermatogenesis.
   c. testosterone production.
   d. a and b
   e. all of the above

_____ 11. In the male, before puberty
   a. FSH levels are higher than after puberty.
   b. LH levels are higher than after puberty.
   c. GnRH release is inhibited by testosterone.
   d. all of the above

_____ 12. Which of these cause a decrease in GnRH release?
   a. decreased inhibin
   b. increased testosterone
   c. decreased FSH
   d. decreased LH

_____ 13. Testosterone
   a. stimulates hair growth.
   b. decreases red blood cell count.
   c. prevents closure of the epiphyseal plate.
   d. all of the above

_____ 14. Which of these is consistent with erection?
   a. parasympathetic stimulation
   b. dilation of arterioles
   c. engorgement of sinusoids with blood
   d. occlusion of veins
   e. all of the above

_____ 15. A polar body
   a. is normally formed before fertilization.
   b. is normally formed after fertilization.
   c. is a sunbathing Eskimo.
   d. normally receives most of the cytoplasm.
   e. a and b

_____ 16. After ovulation the graafian follicle collapses, taking on a yellowish appearance to become the
   a. atretic follicle.
   b. corpus luteum.
   c. corpus albicans.
   d. tunica albuginea.

_____17. The infundibulum of the uterine tube
  a. is the opening of the uterine tube
     into the uterus.
  b. has long, thin projections called
     the ostium.
  c. is connected to the isthmus of the
     uterine tube.
  d. is lined with a ciliated mucous
     membrane.

_____18. The layer of the uterus that
     undergoes the greatest change during
     the menstrual cycle is the
  a. serosa.
  b. hymen.
  c. endometrium.
  d. myometrium.

_____19. The vagina
  a. consist of skeletal muscle.
  b. has ridges called rugae.
  c. is lined with simple squamous
     epithelium.
  d. all of the above

_____20. During intercourse, which of these
     structures fill with blood and cause
     the vaginal opening to narrow?
  a. bulb of the vestibule
  b. clitoris
  c. mons pubis
  d. labia majora

_____21. Given these vestibular-perineal
     structures:
  1. vaginal opening
  2. clitoris
  3. urethral opening
  4. anus

     Choose the arrangement that lists
     the structures in their proper order
     from the anterior to the posterior
     aspect.
  a. 2, 3, 1, 4
  b. 2, 4, 3, 1
  c. 3, 1, 2, 4
  d. 3, 1, 4, 2
  e. 4, 2, 3, 1

_____22. Concerning the breasts,
  a. lactiferous ducts open on the
     areola.
  b. each lactiferous duct supplies an
     alveolus.
  c. they are attached to the
     pectoralis major muscles by
     mammary ligaments.
  d. even before puberty, the female
     breast is quite different from the
     male breast.

_____23. The major secretory product of the
     graafian follicle is
  a. estrogen.
  b. progesterone.
  c. LH.
  d. FSH.
  e. relaxin.

_____24. An increase in LH initiates the
     process of ovulation approximately
     at day 14 in the average adult
     female. Shortly after the increase
     in LH, the rate at which it is
     secreted declines to very low levels.
     Which of these events may play a
     role in decreasing the plasma LH
     levels following its rapid increase in
     the blood?
  a. corpus luteum function
  b. increasing progesterone
     concentration in the plasma
  c. very low estrogen and
     progesterone secretion
  d. elevated FSH secretion
  e. a and b

_____25. Which of these processes or phases
     in the monthly reproductive cycle of
     the human female occur at the same
     time?
  a. maximal LH secretion and
     menstruation
  b. early follicular development and
     the secretory phase in the uterus
  c. regression of the corpus luteum
     and an increase in ovarian
     progesterone secretion
  d. ovulation and menstruation
  e. proliferation stage of uterus and
     increased estrogen production

26. During the secretory phase of the menstrual cycle, one would normally expect
    a. the highest levels of estrogen that occur during the menstrual cycle.
    b. a mature graafian follicle to be present in the ovary.
    c. an increase in the thickness of the endometrium.
    d. a and b
    e. all of the above

27. The cause of menses in the menstrual cycle appears to be
    a. increased progesterone secretion from the ovary, which produces blood clotting.
    b. increased estrogen secretion from the ovary, which stimulates the muscles of the uterus to contract.
    c. decreased progesterone and estrogen secretion by the ovary.
    d. decreased production of oxytocin, causing the muscles of the uterus to relax.

28. A woman with a typical 28-day menstrual cycle is most likely to become pregnant as a result of coitus on days
    a. 1 - 3
    b. 5 - 8
    c. 12 - 15
    d. 22 - 24
    e. 24 - 28

29. After fertilization the successful development of a mature, full-term fetus depends on
    a. the release of human chorionic gonadotropin (HCG) by the developing placenta.
    b. production of estrogen and progesterone by the placental tissues.
    c. the maintenance of the corpus luteum for 9 months.
    d. a and b
    e. all of the above

30. Menopause
    a. happens whenever a woman pauses to think about a man.
    b. occurs when a woman stops a man from making a pass.
    c. develops when follicles become less responsive to FSH.
    d. results from high estrogen levels in 40- to 50-year old women.
    e. c and d

FINAL CHALLENGES

Use a separate sheet of paper to complete this section.

1. Two teenage girls wanted to make a douche (a solution used to remove sperm cells from the vagina following intercourse). One girl suggested using a vinegar solution, and the other girl wanted to use a solution of baking soda. Although neither solution promises to be very effective, which solution would be more likely to succeed in preventing a pregnancy?

2. In response to an injection of a large amount of testosterone in an adult male, what would happen to testosterone production in the testes? Explain.

3. Suppose a 9-year-old boy had an interstitial cell tumor that resulted in very high levels of testosterone production. How would this affect his development?

4. Suppose that a man had a complete transection of his spinal cord at level L3. Would it be possible for him to achieve an erection through psychic stimulation or through stimulation of the genitals? Explain.

5. Birth control pills for women usually contain both estrogen and progesterone. Explain how these hormones can prevent pregnancy.

6. Most birth control pills for women are taken for only 21 days. The woman stops taking the birth control pill or takes a placebo pill for 7 days. Then she resumes taking the birth control pill. Why does she do this?

7. Sexually transmitted diseases such as gonorrhea can sometimes cause peritonitis in females. However, in males peritonitis from this cause does not develop. Explain.

8. Predict the consequences of removing the corpus luteum during the first trimester (first 3 months) of pregnancy in humans. What happens if the corpus luteum is removed during the second trimester (months 4 to 6)?

Scrotum
1. Raphe; 2. Dartos muscle; 3. Cremaster muscles

Perineum
1. Urogenital triangle; 2. Anal triangle

Testes
A. 1. Tunica albuginea; 2. Septa; 3. Seminiferous tubules; 4. Interstitial cells (cells of Leydig); 5. Efferent ductules

B. 1. Gubernaculum; 2. Inguinal canal; 3. Process vaginalis; 4. Tunica vaginalis

C. 1. Spermatogenesis; 2. Sertoli cells; 3. Dihydrotestosterone, Estradiol; 4. Spermatogonia; 5. Primary spermatocytes; 6. Secondary spermatocytes; 7. Spermatids; 8. Acrosome

Ducts
1. Epididymis; 2. Ductus epididymidis; 3. Ductus (vas) deferens; 4. Spermatic cord; 5. Ejaculatory duct; 6. Prostatic urethra; 7. Membranous urethra; 8. Urethral glands

Penis
1. Erection; 2. Corpora cavernosa; 3. Corpus spongiosum; 4. Glans penis; 5. Bulb of the penis; 6. Crus of the penis; 7. Root of the penis; 8. Prepuce (foreskin)

Glands
A. 1. Seminal vesicles; 2. Prostate gland; 3. Bulbourethral glands; 4. Emission; 5. Bulbourethral glands; 6. Seminal vesicles; 7. Prostate gland

B. 1. Prostatic urethra; 2. Membranous urethra; 3. Spongy urethra; 4. Penis; 5. External urethral orifice; 6. Scrotum; 7. Testis; 8. Epididymis; 9. Ductus deferens; 10. Bulbourethral gland; 11. Prostate; 12. Ejaculatory duct; 13. Seminal vesicle

Regulation of Sex Hormone Secretion
1. GnRH; 2. Gonadotropins; 3. Luteinizing hormone; 4. Follicle stimulating hormone; 5. Testosterone; 6. Androgen; 7. Inhibin

Puberty
1. HCG; 2. Testosterone; 3. GnRH; 4. LH and FSH; 5. Spermatogenesis; 6. Testosterone

Effects of Testosterone
1. Testes; 2. Coarser; 3. Melanin; 4. Sebaceous glands; 5. Hypertrophy; 6. Body fluids; 7. Protein synthesis; 8. Rapid bone growth

Male Sexual Behavior and the Male Sex Act
1. Orgasm; 2. Resolution; 3. Impotence; 4. Erection; 5. Emission; 6. Ejaculation

Ovaries
A. 1. Mesovarium; 2. Suspensory ligament; 3. Ovarian (germinal) epithelium; 4. Tunica albuginea; 5. Medulla; 6. Ovarian follicles; 7. Oocyte

B. 1. Oogenesis; 2. Oogonium; 3. Primary oocyte; 4. Secondary oocyte, polar body

C. 1. Primordial follicle; 2. Zona pellucida; 3. Secondary follicle; 4. Theca; 5. Vesicular (graafian) follicle; 6. Cumulus mass; 7. Corona radiata; 8. Polar body

D. 1. Ovulation; 2. Secondary oocyte; 3. Fertilization; 4. Polar body; 5. Zygote; 6. Corpus luteum; 7. Progesterone and estrogen; 8. Corpus luteum of pregnancy; 9. Corpus albicans

Uterine Tubes
1. Mesosalpinx; 2. Infundibulum; 3. Ostium; 4. Fimbriae; 5. Ampulla; 6. Isthmus; 7. Muscular layer; 8. Mucosa

Uterus
A. 1. Fundus; 2. Cervix; 3. Isthmus; 4. Cervical canal; 5. Ostium; 6. Round ligaments; 7. Broad ligament

B. 1. Perimetrium (serous coat); 2. Endometrium (mucous membrane); 3. Basal layer; 4. Functional layer

C. 1. Body of uterus; 2. Cervix; 3. Vagina; 4. Perimetrium; 5. Myometrium; 6. Endometrium; 7. Round ligament; 8. Ovarian ligament; 9. Uterine tube; 10. Ovary ; 11. Fundus of uterus

Vagina
1. Columns; 2. Rugae; 3. Fornix; 4. Hymen

External Genitalia
A. 1. Vestibule; 2. Labia minora; 3. Clitoris; 4. Prepuce; 5. Bulb of the vestibule; 6. Greater vestibular glands; 7. Lesser vestibular glands; 8. Mons pubis

B. 1. Prepuce; 2. Labia minora; 3. Vagina; 4. Vestibule; 5. Labia majora; 6. Urethra; 7. Clitoris; 8. Mons pubis

Perineum
1. Urogenital triangle; 2. Clinical perineum; 3. Episiotomy

Mammary Glands
A. 1. Areola; 2. Areolar glands; 3. Gynecomastia; 4. Lobes; 5. Lobules; 6. Lactiferous duct; 7. Alveoli; 8. Mammary (Cooper's) ligaments

B. 1. Lactiferous sinus; 2. Lactiferous duct; 3. Nipple; 4. Areola; 5. Lobe; 6. Alveoli; 7. Lobule

Puberty
1. Estrogen and progesterone; 2. FSH and LH;
3. GnRH; 4. FSH and LH; 5. Estrogen and
progesterone

The Menstrual Cycle
A. 1. Menses; 2. Follicular (proliferative) stage;
3. Luteal (secretory) stage
B. 1. Ovarian cycle; 2. FSH; 3. Adenohypophysis;
4. LH; 5. Ovulation; 6. Luteal; 7. Progesterone;
8. HCG
C. 1. Increases; 2. Increases; 3. Positive
feedback; 4. Increase; 5. Increases; 6. FSH
and LH surges; 7. Decreases; 8. Increases;
9. Increase; 10. Negative feedback;
11. Decrease; 12. Decrease
D. 1. Uterine cycle; 2. Spiral glands; 3. Spiral
arteries
E. 1. Increases; 2. Increases; 3. Increases;
4. Increases; 5. Decreases; 6. Decreases

F. 1. GnRH; 2. LH 3. FSH 4. Primordial follicle;
5. Primary follicle; 6. Secondary follicle;
7. Mature (graafian) follicle; 8. Corpus
luteum; 9. Corpus albicans; 10. Estrogen
level; 11. Progesterone level; 12. Menses;
13. Proliferative phase; 14. Secretory phase;
15. Menses

Female Sexual Behavior and the Female Sex Act
1. Psychic factors; 2. Sacral; 3. Clitoris; 4. Vagina
and vestibular glands; 5. Orgasm; 6. Resolution;
7. Fertilization

Female Fertility and Pregnancy
1. Swimming; 2. Oxytocin and prostaglandins;
3. Capacitation; 4. Trophoblast; 5. HCG;
6. Placenta

Menopause
1. Female climacteric; 2. Ovary; 3. LH and FSH;
4. Estrogen and progesterone; 5. Hot flashes

QUICK RECALL

1. Spermatogonia, primary spermatocyte,
secondary spermatocyte, spermatid, sperm cell
(spermatozoon)
2. Seminiferous tubules, rete testis, efferent
ductules, epididymis, ductus deferens,
ejaculatory duct, prostatic urethra, membranous
urethra, spongy urethra
3. Seminal vesicles: thick, mucoid secretions
containing nutrients, fibrinogen, and
prostaglandins; prostate gland: thin, milky,
alkaline secretions containing clotting factors;
bulbourethral glands: alkaline mucous
secretions
4. Hypothalamus: GnRH; adenohypophysis: FSH
and LH; testes: testosterone and inhibin
5. Enlargement and differentiation of the
reproductive system, descent of testes,
spermatogenesis, hair growth, increased skin
pigmentation, increased sebaceous secretions,
increased muscle mass, increased body fluids,
increased skeletal growth, laryngeal hypertrophy,
increased metabolism, and increased blood cell
count
6. Primordial follicle, primary follicle, secondary
follicle, vesicular (graafian) follicle, corpus
luteum, corpus albicans
7. Oogonia, primary oocyte, secondary oocyte;
fertilization occurs in the secondary oocyte.
8. Alveoli, lobule, lobe, lactiferous sinus, lactiferous
duct, nipple
9. Menses, proliferative (follicular) phase, secretory
(luteal) phase
10. GnRH: stimulates release of FSH and LH from
the adenohypophysis; FSH: stimulates follicle
development; LH: triggers ovulation,
development of corpus luteum
11. Estrogen: thickening of endometrium;
progesterone: thickening, glandularity, and
vascularity of endometrium stimulated;
decreased smooth muscle contraction

MASTERY LEARNING ACTIVITY

1. D. The cremaster muscle, which is a
continuation of the internal abdominal oblique
muscle, extends through the spermatic cord to
the testes. The dartos muscle layer is found
beneath the skin of the scrotal sac. When
temperatures are low, these muscles contract,
bringing the testes closer to the body and
warming them. If temperatures are too high,
they relax, and the testes descend to a cooler
climate. The gubernaculum is a fibromuscular
cord that is involved in the descent of the testes
into the scrotum.

2. B. The testes develop as retroperitoneal organs
in the abdominopelvic cavity and pass through
the inguinal canal into the scrotum. The scrotum
is formed before the testes descend from the
abdominopelvic cavity.

3. A. The cells of Leydig produce testosterone.
The seminiferous tubules produce sperm cells.
Sperm cells do not produce testosterone. The
anterior pituitary produces FSH and LH.

4. E. The sperm cells would pass through the rete testis, efferent ductule, epididymis, ductus deferens, and ejaculatory duct.

5. B. Sperm cells are produced in the seminiferous tubules.

6. D. Sperm cells undergo final maturation and are stored in the epididymis. During ejaculation smooth muscles within the epididymis contract and move sperm cells into the ductus deferens. A small number of sperm cells are also stored in the ampulla of the ductus deferens.

7. B. There are three columns of erectile tissue in the penis: the corpus spongiosum and two corpora cavernosa. The spongy urethra passes through the corpus spongiosum. The distal end of the corpus spongiosum is the glans penis, which is covered by the prepuce. The proximal ends of the corpora cavernosa form the crura of the penis, which attach the penis to the coxae.

8. D. The seminal vesicles empty into the ejaculatory duct, prostate gland into the prostatic urethra, and bulbourethral glands into the membranous urethra.

9. D. The bulbourethral glands neutralize the urethra before ejaculation. The seminal vesicles provide fructose and fibrinogen. The prostatic secretions neutralize semen and contain clotting factors that act on fibrinogen to cause coagulation of the semen.

10. C. LH stimulates the production of testosterone. FSH stimulates the development of the seminiferous tubules and spermatogenesis.

11. C. Before puberty, GnRH release is inhibited by small amounts of testosterone. Consequently, FSH and LH levels are low.

12. B. Testosterone has a negative-feedback effect that decreases GnRH release from the hypothalamus.

13. A. Testosterone stimulates such factors as hair growth, red blood cell production, and increased metabolism. It also stimulates bone growth and closes off the epiphyseal plate.

14. E. When the penis is flaccid, the arterioles that supply it are partially contracted. During sexual excitation parasympathetic impulses from the sacral region of the spinal cord travel to the arterioles, causing relaxation and dilation. This allows more blood to enter the sinusoids in the penis, causing them to enlarge which compresses the veins that drain the penis. The combination of increased blood flow into the penis and reduced blood flow out produces the erection.

15. E. A polar body is formed by the first meiotic division before fertilization. After fertilization the second meiotic division produces a second polar body. Polar bodies contain little cytoplasm.

16. B. A follicle goes through several stages: primordial follicle, primary follicle, secondary follicle, and graafian follicle; after ovulation, the corpus luteum, and later the corpus albicans are formed.

17. D. The infundibulum is the expanded end of the uterine tube next to the ovary. It has long projections, the fimbriae, which are covered by a ciliated mucous membrane. When an oocyte is released from the ovary, it is swept through the ostium, which is the opening of the infundibulum, into the uterine tube. The infundibulum is attached to the ampulla of the uterine tube.

18. C. The functional layer of the endometrium undergoes the greatest change.

19. B. The vaginal walls have longitudinal ridges (columns) and transverse ridges (rugae) that allow the vagina to expand. The vagina consists of smooth muscle, which can stretch to allow expansion of the vagina and is lined with moist, stratified squamous epithelium.

20. A. The bulb of the vestibule, which corresponds to the corpus spongiosum of the male penis, causes the vaginal opening to narrow. The clitoris, which corresponds to the corpora cavernosa of the male penis, also fills with blood during intercourse. Although the clitoris expands in diameter, it does not affect the size of the vaginal opening.

21. A. The order is clitoris, urethral opening, vaginal opening, and anus.

22. C. The mammary ligaments support the breasts. The lactiferous ducts supply the lobes of the breast and open onto the nipple. Before puberty male and female breasts are similar in structure.

23. A. The graafian follicle secretes primarily estrogen. The corpus luteum produces primarily progesterone. Relaxin is produced by the corpus luteum during pregnancy. LH and FSH are both produced in the anterior pituitary.

24. E. The sudden increase in plasma LH initiates ovulation. Following ovulation the ovulated follicle is rapidly converted to a functional corpus luteum and begins to produce large amounts of progesterone. The development of the corpus luteum and the progesterone that it secretes appears to play a role in decreasing the LH levels following the LH surge.

25. E. Increased estrogen levels produced by the developing follicle stimulate the endometrium to proliferate.

26. C. During the secretory phase of the menstrual cycle, the corpus luteum produces mainly progesterone and some estrogen (but less than the graafian follicle produces during the proliferation stage). The estrogen increases cell division, and the progesterone stimulates secretory activity, resulting in increased thickness of the endometrium.

27. C. During the last few days of the menstrual cycle, progesterone and estrogen production by the ovary drops. This results in vasospasms of vessels in the uterus and the cutting off of blood supply to the endometrium. The outer portion of the endometrium dies, and uterine contractions expel the menses.

28. C. Ovulation occurs about day 14 of the typical menstrual cycle, and the ovum remains viable for about 24 hours. Sperm cells remain viable in the female reproductive tract for up to 72 hours. Therefore deposition of sperm cells on days 12 - 15 could result in fertilization.

29. D. HCG stimulates the corpus luteum to produce estrogen and progesterone for the first 3 months of pregnancy. The estrogen and progesterone are necessary for the maintenance of the endometrium. After 3 months, HCG levels decline, and the corpus luteum becomes a corpus albicans. Meanwhile, the placenta produces estrogen and progesterone.

30. C. A decrease in the number of follicles and a decrease in the sensitivity of the remaining follicles to FSH and LH are responsible for menopause.

 FINAL CHALLENGES

1. Sperm cells are destroyed by the acidic environment of the vagina; secretions in semen help neutralize acids. Consequently, the acidic vinegar would be better than an alkaline baking soda solution.

2. Increased testosterone, by a negative feedback mechanism, inhibits the production of gonadotropin-releasing hormone (GnRH) from the hypothalamus. Decreased GnRH results in decreased LH release from the adenohypophysis, which causes a reduction in the production of testosterone by the testes.

3. The testosterone would cause early and pronounced development of the boy's sexual organs. He would also have rapid growth of muscle and bone, but early closure of the epiphyseal plate would result in a shorter height than he would normally achieve.

4. Erection is mediated through parasympathetic reflex centers in the sacral region of the spinal cord. Injury at level L3 would not interrupt the reflex arc from the genitals, so stimulation of the genitals should cause an erection. Erection can also be achieved by sympathetic stimulation through centers at levels T12 to L1. Psychic stimulation can activate the sympathetic centers, which are superior to the spinal cord injury, but not the parasympathetic centers, which are inferior to the injury. It is possible that psychic stimulation could also cause an erection.

5. The synthetic estrogen and progesterone in birth control pills presumably have a negative-feedback effect on the hypothalamus, inhibiting both GnRH and LH secretion. This reduces the LH surge and prevents ovulation. In women who take birth control pills with very low doses of estrogen and progesterone, ovulation occurs but pregnancy is prevented, perhaps because of an unusually rapid rate of oocyte transport through the uterine tube as a result of abnormal contractions.

6. Not taking the birth control pill causes estrogen and/or progesterone levels to fall, resulting in menstruation.

7. In females the microorganisms can travel from the vagina to the uterus, to the uterine tubes, and to the pelvic cavity. Infection of the peritoneum lining the pelvic cavity results in peritonitis. In males the microorganisms can move up the urethra to the bladder or into the ejaculatory duct to the ductus deferens, but there is no direct connection into the pelvic cavity, so peritonitis does not develop.

8. Progesterone and estrogen prepare the uterine endometrium for implantation and maintain the endometrium throughout pregnancy. Progesterone also inhibits uterine contractions. In the first trimester of pregnancy the corpus luteum is the primary source of progesterone. After the first trimester the placenta produces enough progesterone to maintain pregnancy. Therefore, during the first trimester if the corpus luteum is removed, the blood levels of progesterone decrease, the endometrium degenerates, and an abortion results. Removal of the corpus luteum in the second or third trimester does not interrupt pregnancy.

29 Development, Growth, Aging, and Genetics

FOCUS: The lifespan of an individual can be divided into life stages. During the germinal period the fertilized oocyte develops into an embryo with three primary germ layers, which give rise to the organ systems. In the embryonic stage, the major organ systems develop, and in the fetal stage, growth and maturation of the organ systems take place. After birth, during the neonate stage, the fetal circulation pattern changes, and the digestive system becomes functional. In the infant, muscular-nervous coordination increases, enabling the infant to control body movements. During childhood there is further growth, and in adolescence the child becomes a sexually functional adult. Aging results in a decrease in nervous and muscular cells, loss of elasticity in tissues, decreased ability to mount an immune system response, and a general loss of the ability to maintain homeostasis. Eventually some challenge to homeostasis cannot be overcome, and death results.

CONTENT LEARNING ACTIVITY

Prenatal Development

❝*The nine months before birth constitute a critical part of an individual's existence.*❞

Match these terms with the correct statement or definition:

Clinical age
Embryonic period
Fetal period

Germinal period
Postovulatory age
Prenatal period

_____

_____

_____

_____

_____

1. Period from conception to birth.

2. Period of development during which primitive germ layers are formed; the first 2 weeks of development.

3. Period of development during which the major organ systems are formed; second to end of the eighth week of development.

4. Period of development in which the organ systems grow and become more mature; the last 30 weeks of development.

5. The calculation of developmental age that uses the last menstrual period (LMP) as the starting point.

Fertilization

"_Normally one sperm cell penetrates the oocyte cell membrane and enters the cytoplasm in the_**"** _process of fertilization._

Match these terms with the
correct statement or definition:

Female pronucleus	Second meiotic division
Male pronucleus	Zygote

_____ 1. Division triggered by entrance of the sperm cell into the oocyte; a second polar body is formed.

_____ 2. Haploid nucleus of the oocyte, after the second meiotic division.

_____ 3. Enlarged head of the sperm cell, containing the haploid number of chromosomes.

_____ 4. Result of the process of fertilization.

☞ The fusion of the male and female pronuclei restores the diploid number of chromosomes.

Early Cell Division: Morula and Blastocyst

"_Through the process of cell division, the zygote produces an embryonic cell mass._**"**

Match these terms with the
correct statement or definition:

Blastocele	Morula
Blastocyst	Pluripotent
Inner cell mass	Trophoblast

_____ 1. Ability of the embryonic cell mass, up to the 8-cell stage, to develop into any cell in the body.

_____ 2. Dividing embryonic cell mass, once it has 12 or more cells.

_____ 3. Embryonic cell mass, once a cavity begins to form inside.

_____ 4. Single layer of cells that surrounds most of the blastocele and develops into the placenta.

_____ 5. Thickened area, several cell layers thick, at one end of the blastocyst, that will develop into the embryo.

Implantation of the Blastocyst and Development of the Placenta

66*About 7 days after fertilization, the blastocyst attaches itself to the uterine wall.*99

Match these terms with the
correct statement or definition:

Cytotrophoblast   Placenta
Implantation   Syncytiotrophoblast
Lacunae

_____ 1. The organ of metabolic exchange between the fetus and the mother.

_____ 2. Nondividing, multinucleate cell that invades maternal tissues; does not trigger an immune reaction.

_____ 3. Cavities produced when the syncytiotrophoblast surrounds and digests away the walls of maternal blood vessels.

_____ 4. Dividing population of trophoblast cells that protrudes into the lacunae (chorionic villi) and is followed by embryonic blood vessels; later disappears, leaving only a thin layer separating the maternal and fetal blood supplies.

Formation of Germ Layers

66*All the tissues of the adult can be traced to three germ layers.*99

Match these terms with the
correct statement or definition:

Amniotic cavity   Mesoderm
Ectoderm   Notochord
Embryo   Primitive streak
Embryonic disk   Yolk sac
Endoderm

_____ 1. New cavity formed inside the inner cell mass after implantation occurs; lined by the amniotic sac.

_____ 2. Flat disk of tissue composed of two cell layers.

_____ 3. Cell layer of the embryonic disk that is adjacent to the amniotic cavity.

_____ 4. Third cavity produced by the endoderm that forms inside the blastocele.

_____ 5. Thickened line, formed by cells of the ectoderm migrating to the center and caudal end of the embryonic disk.

_____ 6. Third germ layer, formed from ectodermal cells that migrate through the primitive streak and emerge between the ectoderm and endoderm.

_____ 7. Formed when ectoderm, mesoderm, and endoderm are present.

Neural Tube and Neural Crest Formation

❝*The neural tube and neural crest are formed from ectoderm.* **❞**

Match these terms with the
correct statement or definition:

Neural crest cells   Neural tube
Neural crests   Neuroectoderm
Neural plate

_____ 1. Thickened layer of cells produced when the notochord
stimulates overlying ectoderm cells.

_____ 2. Lateral edges of the neural plate that rise and move toward
each other.

_____ 3. Structure formed when the neural crests meet at the midline
and fuse; becomes the brain and spinal cord.

_____ 4. Cells that become part of the peripheral nervous system,
melanocytes, or varied structures in the head.

Somite, Gut, and Body Cavity Formation

❝*During the early embryonic period, many structures are being formed.* **❞**

Match these terms with the
correct statement or definition:

Branchial arches   Oropharyngeal membrane
Celom   Pharyngeal pouches
Cloacal membrane   Somites
Evaginations   Somitomeres

_____ 1. Distinct segments formed from mesoderm adjacent to the neural
tube.

_____ 2. First few somites in the head region that never become clearly
divided.

_____ 3. Membrane formed in the hindgut, where it is in close contact
with ectoderm; this later becomes the urethra and the anus.

_____ 4. Outpocketings from the GI tract; these develop along the
head.

_____ 5. Solid bars of tissue that develop along the head.

_____ 6. Pockets of the pharynx that extend between the branchial
arches; give rise to the tonsils, thymus, and parathyroid
glands.

_____ 7. Collective name for the body cavities, subdivided to form the
pericardial, pleural, and peritoneal cavities.

Limb Bud and Face Development

The arms and legs first appear as limb buds, but the face develops by fusion of five embryonic structures.

Match these terms with the correct statement or definition:

Apical ectodermal ridge
Frontonasal process
Mandibular processes
Maxillary processes

Nasal placodes
Primary palate
Secondary palate

_____ 1. Specialized thickening of the ectoderm that develops on the lateral margin of each limb bud.

_____ 2. Process that forms the forehead, nose, and mid portion of the upper jaw and lip.

_____ 3. Processes that form the lateral portion of the upper jaw and lip.

_____ 4. Develops at the lateral margin of the frontonasal process and become the nose and center of the upper jaw and lip.

_____ 5. Upper jaw and lip, formed by fusion of the nasal placodes and the maxillary processes.

_____ 6. Roof of the mouth; failure to fuse is a cleft palate.

Development of the Organ Systems

The major organ systems appear and begin to develop during the embryonic period.

A. Match these terms with the correct structure that develops from them:

Mesoderm
Neural crest cells

Neural tube
Neural tube cavity

_____ 1. Dermis of the skin.

_____ 2. Melanocytes and sensory receptors in the skin.

_____ 3. Appendicular skeleton.

_____ 4. Ventricles of the brain and the central canal of the spinal cord.

B. Match these terms with the correct statement or definition:

Adenohypophysis
Myoblasts
Neurohypophysis

Olfactory bulb and nerve
Optic stalk and optic vesicle

_____ 1. Multinucleated cells that develop into skeletal muscle fibers.

_____ 2. Develops as an evagination from the telencephalon.

_____ 3. Develops as an evagination from the diencephalon.

_____ 4. Develops as an evagination of the floor of the diencephalon.

599

C. Match these terms with the correct statement or definition:

Adrenal cortex   Parathyroid glands
Adrenal medulla   Thyroid gland
Pancreas

_____ 1. Originates as an evagination from the floor of the pharynx in the region of the developing tongue.

_____ 2. Gland part that arises from neural crest cells and consists of specialized postganglionic neurons of the sympathetic division.

_____ 3. Gland part that originates from mesoderm.

_____ 4. Originates as two evaginations from the duodenum; they later fuse.

D. Match these terms with the correct statement or definition:

Blood islands   Foramen ovale
Bulbus cordis   Sinus venosus

_____ 1. The blood vessels are formed from these structures, which are located on the surface of the yolk sac and inside the embryo.

_____ 2. Site of blood entering the embryonic heart; part of this structure later becomes the SA node.

_____ 3. Site of blood exiting the embryonic heart; this structure is later incorporated into the ventricles.

_____ 4. Opening in the interatrial septum that allows blood to flow from the right to the left atrium.

☞ The lungs develop as a single, midline evagination from the foregut in the region of the future esophagus.

E. Match these terms with the correct statement or definition:

Allantois   Metanephros
Cloaca   Pronephros
Mesonephros   Ureter

_____ 1. Consists of a duct and simple tubules connected to the celomic cavity; it is probably never functional in the embryo.

_____ 2. Consists of tubules that open into the mesonephric duct; the other end forms a glomerulus.

_____ 3. Common junction of the digestive, urinary, and genital systems.

_____ 4. Blind tube that extends into the umbilical cord; develops into the urinary bladder.

_____ 5. Distal end of the ureter enlarges and branches to form the duct system of this structure; becomes the adult kidney.

F. Match these terms with the correct statement or definition:

Gonads   Paramesonephric ducts
Mesonephric ducts Primordial germ cells

_____ 1. Cells that form on the surface of the yolk sac, migrate into the embryo, and enter the gonadal ridge.

_____ 2. Ducts that form just lateral to the mesonephric ducts; if no testosterone is present, these ducts become the internal female reproductive structures.

_____ 3. Under the influence of testosterone, will develop into the epididymis, ductus deferens, seminal vesicles, and prostate gland.

G. Match these terms with the structures that they later become:

Genital tubercle   Urogenital folds
Labioscrotal swellings

_____ 1. Penis

_____

_____ 2. Scrotum or labia majora

_____ 3. Clitoris

_____ 4. Labia minora

☞ If the closure of the urogenital folds is not complete when the penis is formed, a defect known as hypospadias results.

Growth of the Fetus

“*The embryo becomes a fetus about 56 days after fertilization.*”

Using the terms provided, complete these statements:

Embryonic   Subcutaneous fat
Growing   Vernix caseosa
Lanugo

Most morphological changes occur in the _(1)_ phase of development, whereas the fetal period is primarily a _(2)_ phase. Fine, soft hair called _(3)_ covers the fetus, and a waxy coat of sloughed epithelial cells called _(4)_ protects the fetus from the somewhat toxic nature of the amniotic fluid. _(5)_ accumulates in the late fetus; this provides a nutrient reserve, helps insulate the infant, and assists the infant in sucking by supporting the cheeks.

1. _____

2. _____

3. _____

4. _____

5. _____

Parturition

"Parturition refers to the process by which the baby is born."

A. Match these terms with the correct statement or definition:

First stage of labor   Third stage of labor
Second stage of labor

_____   1. Delivery of the placenta.

_____   2. Stage of labor from the onset of regular uterine contractions until maximum cervical dilation occurs.

_____   3. Stage of labor from maximal cervical dilation until delivery is complete.

B. Match these terms with the correct statement:

Decrease(s)
Increases(s)

_____   1. Effect of stress on fetal ACTH production.

_____   2. Effect of ACTH on glucocorticoid production in the fetus.

_____   3. Effect of fetal glucocorticoids on maternal progesterone production.

_____   4. Effect of fetal glucocorticoids on maternal estrogen production.

_____   5. Effect of declining progesterone and increasing estrogen levels on smooth muscle contraction.

_____   6. Effect of prostaglandins on uterine contractions.

_____   7. Effect of oxytocin on uterine contractions.

_____   8. Effect of stretching the uterus on oxytocin production.

_____   9. Effect of decreased progesterone on oxytocin release.

☞ Uterine contractions and oxytocin form a positive-feedback mechanism that ends when delivery occurs and the uterus is no longer stretched.

The Newborn

"The newborn infant, or neonate, experiences several dramatic changes at the time of birth."

A. Match these terms with the correct statement or definition:

Ductus arteriosus   Fossa ovalis
Ductus venosus   Umbilical vein

_____ 1. Foramen ovale closes at birth to become this structure.

_____ 2. Opening that is between the pulmonary trunk and the aorta; it closes 1 or 2 days after birth and becomes the ligamentum arteriosum.

_____ 3. Vessel that flows through the umbilical cord; it becomes nonfunctional at birth, and is called the round ligament of the liver.

_____ 4. Vessel that receives blood from the umbilical vein and passes through the liver; at birth it degenerates into the ligamentum venosum.

B. Using the terms provided, complete these statements:

Amylase   Meconium
Bilirubin   Stomach pH
Lactose

1. _____

2. _____

3. _____

4. _____

5. _____

Swallowed amniotic fluid, cells sloughed from the mucosal lining, mucus, and bile from the liver pass from the GI tract as a greenish discharge called (1) . The pH of the stomach is nearly neutral at birth, but gastric acid secretion occurs, causing (2) to decrease. The newborn digestive system is capable of digesting (3) from the time of birth, but (4) secretion by the salivary glands remains low until after the first year. The neonatal liver also lacks adequate amounts of the enzyme required to excrete (5) , which may lead to jaundice.

 The newborn infant may be evaluated for its physiological condition by the APGAR score; this acronym stands for appearance, pulse, grimace, activity, and respiratory effect.

Lactation

❝*Lactation is the production of milk by the breasts (mammary glands).***❞**

Match these terms with the
correct statement or definition:

Colostrum	Prolactin
Estrogen	Prolactin inhibiting factor (PIF)
Oxytocin	Prolactin releasing factor (PRF)
Progesterone	

_____ 1. Hormone primarily responsible for breast growth during
pregnancy.

_____ 2. Hormone responsible for development of the breast's secretory
alveoli.

_____ 3. Hormone responsible for milk production.

_____ 4. Substance secreted by the hypothalamus in response to
mechanical stimulation of the breast.

_____ 5. Substance produced by the breasts for the first few days after
parturition that contains little fat and less lactose than milk.

_____ 6. Hormone that causes cells around the alveoli to contract;
responsible for milk letdown.

_____ 7. Substance secreted by the neurohypophysis in response to
mechanical stimulation of the breast.

Life Stages

❝*There are considered to be 8 stages of life from fertilization to death.***❞**

Match these life stages with
the correct description:

Adolescent	Fetus
Adult	Germinal
Child	Infant
Embryo	Neonate

_____ 1. Period from fertilization to 14 days.

_____ 2. Period from 14 to 56 days after fertilization.

_____ 3. Period from 56 days after fertilization to birth.

_____ 4. Period from birth to 1 month after birth.

_____ 5. Period from 1 or 2 years after birth to puberty.

_____ 6. Period from puberty to 20 years.

Aging

The process of aging begins at fertilization.

Using the terms provided, complete these statements:

Atherosclerosis	Embolus
Arteriosclerosis	Filtration
Autoimmunity	Genetic
Collagen	Heart
Cytologic aging	Increase
Decrease	Thrombus

1. _____

2. _____

3. _____

4. _____

5. _____

6. _____

7. _____

8. _____

9. _____

10. _____

11. _____

With age, more cross-links are formed between __(1)__ molecules, rendering the tissues more rigid. Death of nondividing cell produces irreversible damage; as a result, the number of muscle cells and neurons __(2)__ with age. The __(3)__ loses elastic recoil ability and muscular contractility, causing a decline in cardiac output, which may also cause a decreased blood flow to the kidneys, resulting in a decrease in __(4)__. __(5)__ is the deposit of lipid in the tunica intima of arteries. These deposits become fibrotic and calcified, causing __(6)__, which interferes with normal blood flow and may lead to a __(7)__ (a clot or plaque formed inside a vessel). A piece of plaque, called an __(8)__, can break loose and lodge in smaller arteries, causing myocardial infarction or stroke. __(9)__, or cellular wear and tear, also contributes to aging. __(10)__ (responding to one's own antigens) or losing the ability to respond to a foreign antigen may be part of the aging process. Progeria may indicate there is a __(11)__ component to aging.

☞ Brain death is irreparable brain damage clinically manifested by the absence of response to stimulation, the absence of spontaneous respiration and heart rate, and an isoelectric electroencephalogram for at least 30 minutes.

Genetics

Genetics is the study of heredity, i.e., those characteristics inherited by children *from their parents.*

Match these terms with the correct statement or definition:

Autosomes	Gene
Congenital	Regulatory genes
DNA	Structural genes

_____ 1. Functional unit of heredity.

_____ 2. DNA sequences that code for specific amino acid sequences in proteins such as enzymes, hormones, and structural proteins.

_____ 3. Disorder that is present at birth.

_____ 4. All chromosomes except those that determine sex.

605

Each normal human somatic cell contains 23 pairs of chromosomes, or 46 total chromosomes.

Chromosomes
❝The DNA of each cell is packed into chromosomes.❞

Match these terms with the correct statement or definition:

Alleles   Homozygous
Dominant   Locus
Genotype   Phenotype
Heterozygous Recessive

_____ 1. Specific location on a chromosome.

_____ 2. Genes occupying the same locus on homologous chromosomes.

_____ 3. Condition in which a person has two copies of the same gene; identical alleles for a trait specified at one gene locus.

_____ 4. Trait that covers up the opposite trait (its allele).

_____ 5. Trait that can be seen in an individual.

_____ 6. Actual genetic condition of an individual.

Consideration of inheritance patterns is based on probability predictions.

QUICK RECALL

1. List the following structures in the order in which they form during development: blastocyst, embryonic disk, mesoderm, morula, primitive streak, zygote.

2. Name the structures derived from the inner cell mass and the trophoblast.

3. Name an organ system that develops entirely from ectoderm.

4. Name an organ system that develops entirely from mesoderm.

5. List the three kidneys that form during the development of the urinary system.

6. List the positive-feedback mechanism that occurs during parturition.

7. List the effects of estrogen, progesterone, prolactin, and oxytocin on the female breast.

8. List three circulatory changes that occur at birth.

9. List the eight life stages.

10. List four factors that may influence aging.

11. Name the two kinds of chromosomes, and give the number of each in humans.

Place the letter corresponding to the correct answer in the space provided.

_____ 1. The major development of organs takes place in the
a. organ period.
b. fetal period.
c. germinal period.
d. embryonic period.

_____ 2. Given the following structures:
1. blastocyst
2. morula
3. zygote

Choose the arrangement that lists the structures in the order in which they are formed during development.
a. 1, 2, 3
b. 1, 3, 2
c. 2, 3, 1
d. 3, 1, 2
e. 3, 2, 1

_____ 3. The embryo develops from the
a. inner cell mass.
b. trophoblast.
c. blastocele.
d. yolk sac.

_____ 4. The placenta
a. develops from the trophoblast.
b. allows maternal blood to mix with embryonic blood.
c. invades the lacunae of the embryo.
d. all of the above

_____ 5. The embryonic disk
a. forms between the amniotic cavity and the yolk sac.
b. contains the primitive streak.
c. becomes a three-layered structure.
d. all of the above

_____ 6. The brain develops from
a. endoderm.
b. ectoderm.
c. mesoderm.

_____ 7. Most of the skeletal system develops from
a. endoderm.
b. ectoderm.
c. mesoderm.

_____ 8. Given the following structures:
1. neural crest
2. neural plate
3. neural tube

Choose the arrangement that lists the structures in the order in which they form during development.
a. 1, 2, 3
b. 1, 3, 2
c. 2, 1, 3
d. 2, 3, 1
e. 3, 2, 1

_____ 9. The somites give rise to the
a. circulatory system.
b. skeletal muscle.
c. skin.
d. kidneys.

_____ 10. The pericardial cavity forms from
a. evagination of the early GI tract.
b. the neural tube.
c. the celom.
d. the branchial arches.

_____ 11. The parts of the limbs develop
a. in a proximal-to-distal sequence.
b. in a distal-to-proximal sequence.
c. at approximately the same time.

_____ 12. Concerning development of the face,
a. the face develops by the fusion of five embryonic structures.
b. the maxillary processes normally meet at the midline to form the lip.
c. the primary palate forms the roof of the mouth.
d. cleft palates normally occur to one side of the midline.

13. Concerning the development of the heart,
    a. the heart develops from a single tube.
    b. the sinus venosum becomes the SA node.
    c. the foramen ovale lets blood flow from the right to the left atrium.
    d. all of the above

14. Given the following structures:
    1. mesonephros
    2. metanephros
    3. pronephros

    Choose the arrangement that lists the structures in the order in which they form during development.
    a. 1, 2, 3
    b. 1, 3, 2
    c. 2, 3, 1
    d. 3, 1, 2
    e. 3, 2, 1

15. A study of the early embryo indicates that the penis of the male develops from the same embryonic structure as which of the following female structures?
    a. labia majora
    b. uterus
    c. clitoris
    d. vagina

16. Which hormone causes differentiation of sex organs in the developing male fetus?
    a. FSH
    b. LH
    c. testosterone
    d. estrogen

17. Onset of labor may be a result of which of the following?
    a. increased estrogen secretion by the placenta
    b. increased glucocorticoid secretion by the fetus
    c. increased secretion of oxytocin
    d. all of the above

18. Following birth,
    a. the ductus arteriosus closes.
    b. meconium is discharged from the GI tract.
    c. stomach pH decreases.
    d. all of the above

19. The hormone involved in milk production is
    a. oxytocin.
    b. prolactin.
    c. estrogen.
    d. progesterone.
    e. ACTH.

20. Which of the following most appropriately predicts the consequence of removing the sensory neurons from the areola of a lactating rat (or human)?
    a. Blood levels of oxytocin decrease.
    b. Blood levels of prolactin decrease.
    c. Milk production and secretion decrease.
    d. all of the above

21. Which of the following life stages is correctly matched with the time that the stage occurs?
    a. neonate - birth to 1 month after birth
    b. infant - 1 month to 6 months
    c. child - 6 months to 5 years
    d. puberty - 10 to 12 years

22. Which of the following occurs as one gets older?
    a. Neurons replicate to replace lost neurons.
    b. Skeletal muscle cells replicate to replace lost neurons.
    c. Cross-links between collagen molecules increase.
    d. The immune system becomes less sensitive to the body's own antigens.
    e. all of the above

23. A gene is
    a. the functional unit of heredity.
    b. a certain portion of a DNA molecule.
    c. a part of a chromosome.
    d. all of the above

_____24. Which of the following is correctly matched with its definition?
a. autosome - an X or Y chromosome
b. phenotype - the genetic makeup of an individual
c. allele - genes occupying the same locus on homologous chromosomes
d. heterozygous - having two identical genes for a trait
e. recessive - a trait expressed when the genes are heterozygous

_____25. Which of the following genotype would be heterozygous?
a. DD
b. Dd
c. dd
d. both a and c

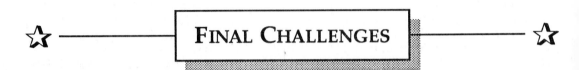

FINAL CHALLENGES

Use a separate sheet of paper to complete this section.

1. If a woman contracts rubella (German measles) while pregnant, the baby may be born with a congenital disorder such as cataracts, deafness, neurologic disturbances, or cardiac malformation. If the mother is infected in the first month of pregnancy versus the third month, what effect would this time difference have on the likelihood that the baby will have a congenital disorder? Explain.

2. A surgical procedure can remove the embryonic yolk sac from early mammalian embryos. How would this affect the development of the reproductive system?

3. Predict the consequences of removing the gonads in an early embryo (8 weeks old) is a genetic female.

4. The ability to taste PTC (phenyl thiocarbamide) is an inherited trait. If t people, each of whom can taste PTC, hav child who cannot, is the trait for PTC dominant or recessive?

5. A woman who does not have hemophilia marries a man who has the disorder. Determine the genotype of both parents half of their daughters and half of their sons have hemophilia.

ANSWERS TO CHAPTER 29

Prenatal Development
1. Prenatal period; 2. Germinal period;
3. Embryonic period; 4. Fetal period; 5. Clinical age

Fertilization
1. Second meiotic division; 2. Female pronucleus;
3. Male pronucleus; 4. Zygote

Early Cell Division: Morula and Blastocyst
1. Pluripotent; 2. Morula; 3. Blastocyst;
4. Trophoblast; 5. Inner cell mass

Implantation of the Blastocyst and Development of the Placenta
1. Placenta; 2. Syncytiotrophoblast; 3. Lacunae;
4. Cytotrophoblast

Formation of Germ Layers
1. Amniotic cavity; 2. Embryonic disk;
3. Ectoderm; 4. Yolk sac; 5. Primitive streak;
6. Mesoderm; 7. Embryo

Neural Tube and Neural Crest Formation
1. Neural plate; 2. Neural crests; 3. Neural tube;
4. Neural crest cells

Somite, Gut, and Body Cavity Formation
1. Somites; 2. Somitomeres; 3. Cloacal membrane; 4. Evaginations; 5. Branchial arches;
6. Pharyngeal pouches; 7. Celom

Limb Bud and Face Development
1. Apical ectodermal ridge; 2. Frontonasal process; 3. Maxillary processes; 4. Nasal placodes; 5. Primary palate; 6. Secondary palate

Development of the Organ Systems
A. 1. Mesoderm; 2. Neural crest cells;
3. Mesoderm; 4. Neural tube cavity
B. 1. Myoblasts; 2. Olfactory bulb and nerve;
3. Optic stalk and optic vesicle;
4. Neurohypophysis
C. 1. Thyroid gland; 2. Adrenal medulla;
3. Adrenal cortex; 4. Pancreas
D. 1. Blood islands; 2. Sinus venosus; 3. Bulbus cordis; 4. Foramen ovale
E. 1. Pronephros; 2. Mesonephros; 3. Cloaca;
4. Allantois; 5. Metanephros
F. 1. Primordial germ cells; 2. Paramesonephric ducts; 3. Mesonephric ducts
G. 1. Genital tubercle and urogenital folds;
2. Labioscrotal swellings; 3. Genital tubercle;
4. Urogenital folds

Growth of the Fetus
1. Embryonic; 2. Growing; 3. Lanugo; 4. Vernix caseosa; 5. Subcutaneous fat

Parturition
A. 1. Third stage of labor; 2. First stage of labor;
3. Second stage of labor
B. 1. Increases; 2. Increases; 3. Decrease;
4. Increase; 5. Increase; 6. Increase;
7. Increases; 8. Increases; 9. Increases

The Newborn
A. 1. Fossa ovalis; 2. Ductus arteriosus;
3. Umbilical vein; 4. Ductus venosus
B. 1. Meconium; 2. Stomach pH; 3. Lactose;
4. Amylase; 5. Bilirubin

Lactation
1. Estrogen; 2. Progesterone; 3. Prolactin;
4. Prolactin releasing factor (PRF); 5. Colostrum;
6. Oxytocin; 7. Oxytocin

Life Stages
1. Germinal; 2. Embryo; 3. Fetus; 4. Neonate;
5. Child; 6. Adolescent

Aging
1. Collagen; 2. Decrease; 3. Heart; 4. Filtration;
5. Atherosclerosis; 6. Arteriosclerosis;
7. Thrombus; 8. Embolus; 9. Cytologic aging;
10. Autoimmunity; 11. Genetic

Genetics
1. Gene; 2. Structural genes; 3 Congenital;
4. Autosomes

Chromosomes
1. Locus; 2. Alleles; 3. Homozygous; 4. Dominant;
5. Phenotype; 6. Genotype

1. Zygote, morula, blastocyst, embryonic disk, primitive streak, mesoderm
2. Inner cell mass: embryo; trophoblast: placenta
3. Nervous system
4. Circulatory system; muscular and skeletal systems are almost all mesodermal in origin, except for some muscles and bones of the head, which are derived from ectoderm
5. Pronephros, mesonephros, and metanephros

6. Stretch of the uterus stimulates oxytocin secretion, which stimulates uterine contraction, which increases uterine stretch.
7. Estrogen: development of duct system and fat deposition; progesterone: development of secretory alveoli; prolactin: milk production; oxytocin: milk letdown.
8. Ductus arteriosus closes, foramen ovale closes, and umbilical arteries and veins and ductus venosus close.

9. Germinal, embryo, fetus, neonate, infant, child, adolescent, adult

10. Cross-linking of collagen, loss of functional cells, atherosclerosis and arteriosclerosis, cytological

aging, immune changes, and genetic components

11. Autosomes: 22 pairs (44); sex chromosomes: one pair (2).

1. D. Most organ development takes place in the embryonic period (second to eighth week). The germ layers develop in the germinal period (first 2 weeks), and growth and maturation occur in the fetal period (the last 30 weeks).

2. E. Fertilization of the secondary oocyte by a sperm cell produces the single-celled zygote. After several divisions the morula is formed; this continues to develop and becomes the blastocyst.

3. A. The embryo develops from the inner cell mass, which is several layers of cells at one end of the blastocyst.

4. A. The placenta develops from the trophoblast. Trophoblast cells protrude into cavities (lacunae) formed within maternal blood vessels. However, there is no mixing of maternal and embryonic blood.

5. D. The embryonic disk is located in the inner cell mass between the amniotic cavity and the yolk sac. It consists of ectoderm and endoderm. The primitive streak forms in the embryonic disk, followed by development of the mesoderm. Thus the embryonic disk has three germ layers: ectoderm, mesoderm, and endoderm.

6. B. The brain, spinal cord, and nerves develop from ectoderm.

7. C. Most of the skeleton is derived from mesoderm. Some of the facial bones are derived from ectoderm.

8. C. The neural plate rises to form the neural crests, which come together to form the neural tube.

9. B. The somites develop into skeletal muscle, the vertebral column, and a portion of the skull.

10. C. The celom, an isolated cavity within the embryo, gives rise to the pericardial, pleural, and peritoneal cavities. Evaginations of the GI tract become, for example, the lungs, liver, anterior pituitary, and the urinary bladder.

11. A. The limb parts develop in a proximal-to-distal sequence.

12. A. The face develops from five embryonic structures. The maxillary processes fuse with the frontonasal process to form the lip and upper jaw. The secondary palate forms the roof of the mouth. A cleft palate normally occurs along the midline.

13. D. The heart develops from a single tube divided into the sinus venosus, an atrium, a ventricle, and the bulbus cordis. The sinus venosus initiates contraction of the tube heart and later becomes the SA node (pacemaker). The foramen ovale allows blood to bypass the lungs.

14. D. Three kidneys form during development in the following order: pronephros, mesonephros, and metanephros. The metanephros becomes the functional kidney, and the others degenerate.

15. C. Many of the sexual organs in males and females have the same embryonic origin but differentiate during development. The penis and clitoris form from the genital tubercle.

16. C. Testosterone stimulates the development of male sexual characteristics. Without testosterone, female characteristics are formed.

17. D. The initiation of labor involves several mechanisms. Production of fetal glucocorticoids causes the placenta to increase estrogen secretion and decrease progesterone secretion. This results in increased uterine excitability. Stretch of the cervix leads to the production of oxytocin, which in turn stimulates uterine contraction, leading to more stretch, and so on.

18. D. The ductus arteriosus closes to prevent blood flow from the pulmonary trunk to the aorta. It later becomes the ligamentum arteriosum. Meconium is a mixture of amniotic fluid, mucus, and bile. Once the alkaline amniotic fluid is removed and gastric secretions increase, stomach pH decreases.

19. B. Prolactin stimulates milk production. Oxytocin is responsible for ejection of milk into the ducts of the breast. Estrogen stimulates duct development, and progesterone causes development of the secretory alveoli.

20. D. Sensory stimuli initiated by suckling is critical in the female rat (also in the human). It is important in initiating oxytocin secretion and maintaining prolactin secretion. The loss of sensory information because of a lack of suckling or because of severed neurons results in reduced oxytocin and prolactin secretion from the anterior pituitary and therefore eventually stops the secretion and production of milk.

21. A. The correct times for each stage are neonate (birth to 1 month), infant (1 month to 1 or 2 years), child (1 or 2 years to puberty), adolescent (puberty to 20 years), and adult (20 years to death). Puberty is from about 11 to 14 years.

22. C. The cross-links between collagen molecules increase, making tissues less plastic. Neurons and skeletal muscle do not replicate after birth, and the immune system becomes more sensitive to self-antigens, resulting in autoimmune disease.

23. D. A gene is a certain portion of a DNA molecule that we consider to be the fundamental unit of heredity. Many genes are located on each chromosome.

24. C. Chromosomes come in pairs called homologues, and the location of the gene for a given trait on the chromosomes is called a locus. Alleles are genes for a given trait found at the same locus.

25. B. Dd is the heterozygous geneotype. DD or dd are homozygous genotypes.

 ☆ FINAL CHALLENGES ☆

1. During the time of formation of the organ systems (first 2 months of development), the individual is most susceptible to damage. About 50% develop congenital malformations if the infection occurs in the first month, 25% in the second month, and 10% if the infection occurs later in the pregnancy.

2. The yolk sac gives rise to the cells that eventually produce gametes (oocytes or sperm cells). These cells migrate from the yolk sac to the gonadal ridge during development. If the yolk sac is destroyed before the cells have migrated, the animal will be sterile.

3. In the normal male the embryonic gonads secrete testosterone and Mullerian-inhibiting hormone. These hormones cause masculinization of the mesonephric duct system and the external genitalia. If the gonads are removed before 8 weeks of development, the embryo develops morphologically as a female regardless of its genotype. Therefore, removing the gonads of a genetic female has little or no effect on her eventual phenotype at birth.

4. The trait is dominant. The parents must be heterozygous (Tt), and the child homozygous-recessive (tt).

5. Hemophilia is an X-linked trait. Because the father has hemophilia he must be X^hY. If the mother were X^HX^H, all their offspring would be normal. Because half the children have hemophilia, she must be X^HX^h.

613

THE AUTHOR

He'd said that he'd do it
And do it he would.
It had to be long,
and preferably good.

He wrote in the morning
And into the night,
His shoulders were aching,
He feared for his sight.

Then one day it happened,
He'd finished it all!
He cracked all his knuckles,
He stared at the wall.

He looked out the window,
No need now to stay.
He climbed on his pony,....
And just wrote away.

Minyon Bond